Nucleosides and Nucleotides as Antitumor and Antiviral Agents

Nucleosides and Nucleotides as Antitumor and Antiviral Agents

Edited by

Chung K. Chu

The University of Georgia
Athens, Georgia

and

David C. Baker

The University of Tennessee
Knoxville, Tennessee

Springer Science+Business Media, LLC

Library of Congress Cataloging-in-Publication Data

Nucleosides and nucleotides as antitumor and antiviral agents / edited
by Chung K. Chu and David C. Baker.
     p.   cm.
   "Proceedings of an American Chemical Society Carbohydrate Division
Symposium on Nucleosides as Antitumor and Antiviral Agents: Tohru
Ueda Memorial Symposium, held April 7, 1992, in San Francisco,
California". -- t.p. verso.
   Includes bibliographical references and index.
   ISBN 978-1-4613-6221-0   ISBN 978-1-4615-2824-1 (eBook)
   DOI 10.1007/978-1-4615-2824-1
   1. Antiviral agents--Testing--Congresses. 2. Antineoplastic
agents--Testing--Congresses. 3. Nucleosides--Derivatives-
-Therapeutic use--Testing--Congresses. 4. Nucleotides--Derivatives-
-Therapeutic use--Testing--Congresses.   I. Chu, Chung K.
II. Baker, David C., 1946- . III. Symposium on Nucleosides as
Antitumor and Antiviral Agents (1992 : San Francisco)
RM411.N79 1993
616.9'25061--dc20   93-13432
                                                             CIP

Proceedings of an American Chemical Society Carbohydrate Division Symposium on Nucleosides as
Antitumor and Antiviral Agents: Tohru Ueda Memorial Symposium, held April 7, 1992, in San Francisco,
California

ISBN 978-1-4613-6221-0

©1993 Springer Science+Business Media New York
Originally published by Plenum Press, New York in 1993
Softcover reprint of the hardcover 1st edition 1993

This book is dedicated to the memory of Professor Tohru Ueda (1931–1990)

PREFACE

Due to the worldwide epidemic of acquired immunodeficiency syndrome (AIDS), the past ten years have witnessed a flurry of activity in the chemotherapy of viral diseases. Unprecedented scientific efforts have been made by scientists and clinicians to combat infections of human immunodeficiency virus (HIV), the causative agent. Looking back over the past ten years, we have made remarkable progress toward the treatment of the viral disease: isolation of HIV only two years after the identification of the disease, plus major strides in the areas of the molecular biology and virology of the retrovirus, etc. More remarkably, the discovery of the chemotherapeutic agent AZT (Retrovir) was made within two years after the isolation and identification of the virus, followed by unprecedented drug development efforts to culminate in the FDA approval of AZT in twenty-three months, which was a record-breaking time for approval of any drug for a major disease.

The last six to seven years have particularly been an exciting and productive period for nucleoside chemists. Since the activity of AZT was established in 1985, nucleoside chemists have had golden opportunities to discover additional anti-HIV nucleosides, which are hoped to be less toxic and more effective than AZT, and the opportunity continues. As we all are aware, AZT possesses extremely potent anti-HIV activity, and no other nucleoside or non-nucleoside has surpassed the potency of AZT in vitro. Due to its bone-marrow toxicity, however, its usefulness for AIDS patients as well as HIV-infected individuals is unfortunately limited. Furthermore, the isolation of resistant strains from patients who have been undergoing AZT chemotherapy for more than six months is an increasingly critical problem in AIDS chemotherapy. More recently, AZT- resistant strains of HIV have been isolated from patients who have never had AZT therapy. Due to the toxicity and viral resistance against AZT, a number of drug firms have been making vigorous efforts to come up with second-generation drugs for HIV infections. Consequently, two other nucleosides, ddI (didanosine) and ddC (zalcitavine), have recently been approved by the FDA. Currently, other nucleosides are also undergoing preclinical and clinical studies to determine their usefulness as anti-HIV drugs, and we are confident that additional clinically useful nucleosides will emerge from these efforts in the near future.

Of course, other classes of compounds, synthetic as well as natural products, have been considered as anti-HIV agents. Although some of these classes of compounds, in particular non-nucleoside reverse transcriptase (RT) inhibitors, exhibit potent anti-HIV activity in vitro, clinical trials suggest that the virus prematurely develops a resistance to these compounds. Therefore, non-nucleoside RT inhibitors may not play an important role in AIDS chemotherapy as single agents. As a result, nucleoside chemists can confidently say that nucleosides are currently the only class of compounds playing a major role in AIDS chemotherapy. Some diehard nucleoside chemists would further extend their statements that nucleosides have always been the frontline armaments for human infectious diseases of viral origin today and will be for years to come. Thus, it is critical to continue exploring the chemistry and biology of nucleosides to prepare for future viral diseases yet to emerge. Obviously, nucleosides are not only effective against viruses, but are also effective against several other disease states such as cancer. This volume also deals with that subject matter, as well as the related chemistry.

This book is particularly significant in that this volume is dedicated to Professor Tohru Ueda of Hokkaido University Faculty of Pharmaceutical Sciences. After dedicating over forty years of his life for nucleoside chemistry, he passed away prematurely at the age of 58 due to lung cancer on September 19, 1990. His outstanding contributions in nucleoside chemistry will remain forever for us, as well as for the future generations that follow in his footsteps. Finally, we would like to extend our appreciation to the Carbohydrate Division of the American Chemical Society for providing us the opportunity to organize the Ueda Memorial Symposium held in San Francisco, California on April 7, 1991, and for allowing us to edit this volume of manuscripts which have been graciously supplied by friends and former students of Professor Ueda. We truly appreciate the effort of these authors for making this book possible. We would like to recognize Yoshitomi, Sankyo, Berlex Biosciences, ISIS Pharmaceuticals, Gilead Sciences, and Syntex for their financial contributions, without which the symposium would not have been possible. Special thanks are also extended to our research groups who thoroughly reviewed each manuscript and to our secretary, Sandra Fields, who assisted in its compilation.

C.K. Chu
Department of Medicinal Chemistry
College of Pharmacy
The University of Georgia
Athens, Georgia

D. C. Baker
Department of Chemistry
The University of Tennessee
Knoxville, Tennessee

CONTENTS

DESIGN OF NEW TYPES OF ANTITUMOR NUCLEOSIDES: THE SYNTHESIS AND ANTITUMOR ACTIVITY OF 2'-DEOXY-(2'-C-SUBSTITUTED)CYTIDINES[1,‡]

Akira Matsuda,*,[a] Atsushi Azuma,[a] Yuki Nakajima,[a] Kenji Takenuki,[a] Akihito Dan,[a] Tomoharu Iino,[a] Yuichi Yoshimura,[a] Noriaki Minakawa,[a] Motohiro Tanaka,[b] and Takuma Sasaki[b]

[a]Faculty of Pharmaceutical Sciences, Hokkaido University, Sapporo 060, Japan

[b]Cancer Research Institute, Kanazawa University, Kanazawa 920, Japan

INTRODUCTION

Interest in certain 2'-deoxyribofuranosylnucleosides and arabinofuranosylnucleosides as potential anticancer chemotherapeutic agents was stimulated by the observation that 1-β-D-arabinofuranosylcytosine (*ara*-C, **1a**) had antileukemic activity. *Ara-C* is one of the most effective drugs for the treatment of adult acute myeloblastic leukemia.[2,3] However, its usefulness is limited by several drawbacks, including its short half-life in plasma, which is due in part to the rapid deamination to chemotherapeutically inactive 1-β-D-arabinofuranosyl-uracil by the action of cytidine deaminase, development of drug resistance, and ineffectiveness on solid tumors.[4-7] Consequently, with the objective of overcoming these problems, efforts have been made to develop a large number of prodrugs[8] of *ara*-C or to substitute certain functional groups other than the hydroxyl group at the 2'-position in place of the hydrogen atom of 2'-deoxycytidine. By a latter approach, 2'-azido- and 2'-amino-2'-deoxy-β-D-arabinofuranosylcytosines (**1b, c**) have been synthesized from D-glucose[9, 10] or uridine[11] and were found to be potent inhibitors of mouse L1210 leukemia cells *in vitro* as well as *in vivo*. These nucleosides are resistant to deamination by human cytidine deaminase.[9,12] Among a series of 2'-deoxy-2'-halogeno-β-D-arabinofuranosylcytosines (**1d-g**),[13] 2'-fluoro derivative **1d** showed a potent cell-growth inhibition of mouse L5178Y

---

‡This paper is dedicated to the memory of Professor Tohru Ueda, deceased September 19, 1990.

Nucleosides and Nucleotides as Antitumor and Antiviral Agents,
Edited by C.K. Chu and D.C. Baker, Plenum Press, New York, 1993

1

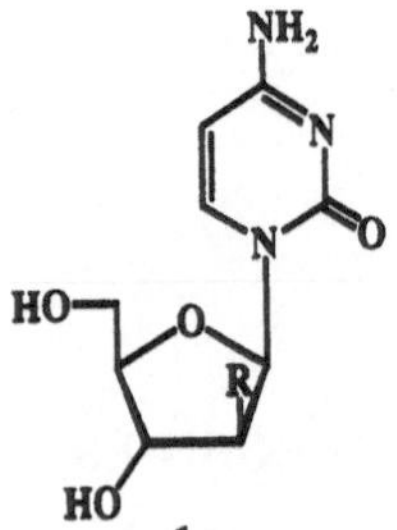

R	
OH	1a (ara-C)
N_3	1b (cytarazid)
NH_2	1c (cytaramin)
F	1d
Cl	1e
Br	1f
I	1g
SMe	1h
Me	1i (SMDC)
CN	1j (CNDAC)
C≡CH	1k (EDAC)
NC	1l (NCDAC)

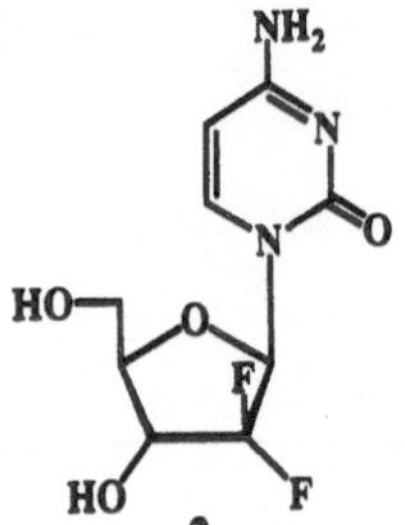

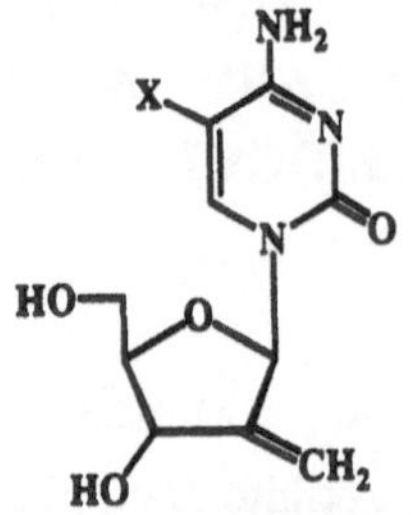

3a: X = H (DMDC)
3b: X = F (FDMDC)

Scheme 1

leukemia cells *in vitro*, although it was found to be susceptible to the deaminase.[14] Similarly, 2'-deoxy-2',2'-difluorocytidine (**2**) was reported to have inhibitory activity toward tumor cells, but it is a substrate of the deaminase.[15-17] On the other hand, 2'-deoxy-2'-(methylthio)-β-D-arabinofuranosylcytosine (**1h**) did not show cytotoxicity against L5178Y cells *in vitro*.[18] When such a nucleoside has antitumor activity, it must be phosphorylated at the 5'-position by deoxycytidine kinase, followed by corresponding nucleotide kinases to produce a 5'-diphosphate or a 5'-triphosphate, which would inhibit certain target enzymes to cause tumor cell death. Therefore, the nature of the substituents being introduced at the 2'-β position of 2'-deoxycytidine, such as bulkiness, electronegativity, and/or hydrogen-bond-forming ability, seems to be critical for enzyme recognition. In addition, these substituents would affect the overall shape of the nucleoside, including sugar conformation, which should influence the spatial position of the 5'-hydroxyl group, as well as that of the 3'-hydroxyl group. Although the significance of the susceptibility of a substrate to cytidine deaminase and deoxycytidine kinase is not fully understood in terms of antitumor activity, this would cause a change in their half-life times in plasma and subsequently affect their efficacy.

In addition to the above considerations, one approach to design nucleoside analogues that exert a broad spectrum of activity toward both leukemic and solid tumors has been the construction of analogues that (i) are not substrates for cytidine deaminase, (ii) are converted into the corresponding 5'-polyphosphates by certain kinases, which inhibit ribonucleotide reductase and/or DNA polymerases, and can also be incorporated into a DNA molecule, (iii) have a chemically reactive functionality at the 2'-β position of 2'-deoxycytidine, which is chemically stable at the nucleoside level but would be expected to be reactive after incorporation into a DNA molecule to cleave phosphodiester linkages and/or to produce

2

abasic sites. As examples of such compounds that fulfill, or partly fulfill, these requirements, we designed several 2'-deoxy-(2'-C-substituted)cytidine analogues, such as (2'S)-2'-
deoxy-2'-C-methylcytidine (SMDC, 1i), 2'-deoxy-2'-methylidenecytidine (DMDC, 3a), 2'-
C-cyano-2'-deoxy-1-β-D-arabinofuranosylcytosine (CNDAC, 1j), 2'-deoxy-2'-C-ethynyl-1-
β-D-arabinofuranosylcytosine (EDAC, 1k), and 2'-deoxy-2'-C-isocyano-1-β-D-arabino-
furanosylcytosine (NCDAC, 1l). In this article, we describe the syntheses of these
nucleosides and their activity *in vitro* as well as *in vivo*.

SYNTHESIS AND ANTITUMOR ACTIVITY OF (2'S)-2'-DEOXY-2'-C-METHYLCYTIDINE (SMDC)

Methods so far reported for the synthesis of branched-chain sugar nucleosides[19] were
rather lengthy, involving formation of a requisite branched-chain sugar then condensation
with a nucleobase, with a few exceptions.[20] Stereoselective alkyl addition of a carbonyl
group is one of the most useful techniques in synthetic organic chemistry. Application of this
methodology to nucleoside chemistry seems to be a straightforward route for the synthesis of
branched-chain sugar nucleosides. Although Cook and Moffatt originally synthesized 2'-
keto-3',5'-di-O-trityluridine, they reported that the alkyl addition to the carbonyl group was
unsuccessful due to its base lability.[21] However, we have recently reported the alkyl
addition reaction of various organometallics to compound 4 producing 2'-branched-chain
sugar nucleosides 5 and 6 in good yields.[22-24]

When 4 was treated with MeMgBr in Et_2O, 5 and 6 were obtained in a ratio of 1:0.8
in 93% combined yield, while 5 was exclusively obtained in the reaction with MeLi or
Me_3Al. These stereochemical differences could be mainly explained by a chelation-controlled
process, in which the Grignard reagent would chelate between the 2'-carbonyl oxygen and
the 2-carbonyl oxygen in 4, resulting in β-attack to give the 2'-C-methylribofuranoside 6,
competing with a non-chelation process which would prefer the less-hindered α-attack on the
2'-C-methylarabinofuranoside 5. Additionally, we found that β-attack of the methyl-
carbanionoid may be correlated with the Lewis basicity of the 2-carbonyl oxygen of the
pyrimidine moiety.[25]

Conversion of 5 and 6, together with their Et congeners, into the corresponding
cytosine derivatives 7-10 was performed. Cytotoxicity of these nucleosides was tested
against murine leukemia L1210 cells *in vitro*. Among them, the most potent inhibitor of cell
growth was (2'R)-2'-C-methylcytidine (8, IC_{50} = 15 μg/mL), which was much less active
than *ara*-C. The ethyl derivative 10 was much less potent (45% inhibition at 400 μg/mL).
Although (2'S)-C-methylcytidine (7) is an analogue of *ara*-C, the inhibitory activity was only
30% even at 400 μg/mL concentration.[23] To improve the activity, deletion of the hydroxyl
group of 7 and 8 seems to be desirable.

Although deoxygenation of the sugar hydroxyl group of nucleosides is generally done
by the Barton method, i.e., the radical deoxygenation of imidazoylthiocarbonyl esters or
phenoxythiocarbonyl esters using tributyltin hydride (Bu_3SnH) in the presence of radical

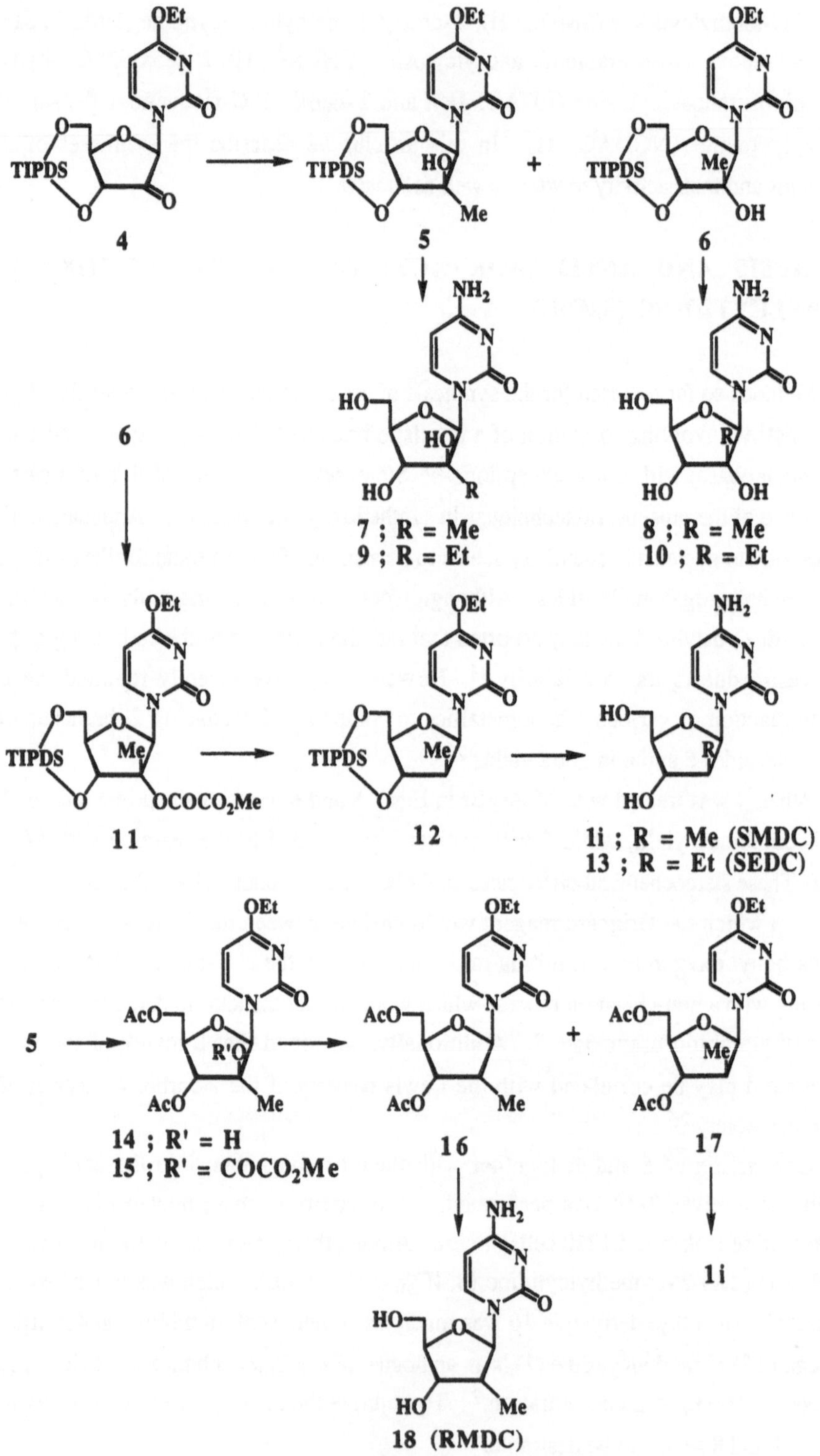

Scheme 2

initiators,[26-28] we used a new method for deoxygenation of *tert*-alcohols via a methoxalyl ester.[29] When **6** was treated with methoxalyl chloride in the presence of 4-(dimethyl-amino)pyridine (DMAP) in CH_3CN, methoxalyl ester **11** was obtained in quantitative yield. Compound **11** was next subjected to radical deoxygenation with Bu_3SnH in the presence of 2,2'-azobis(isobutyronitrile) (AIBN) in toluene to give the desired (2'*S*)-2'-deoxy-2'-*C*-methyl derivative **12**.[22,30] Compound **12** was converted to the corresponding cytosine nucleoside, (2'*S*)-2'-deoxy-2'-*C*-methylcytidine (SMDC, **1i**). (2'*S*)-2'-Deoxy-2'-*C*-ethyl-cytidine (SEDC, **13**) was also prepared similarly.

In contrast, esterification of the 2'-*tert*-hydroxyl group of **5** with methoxalyl chloride under conditions similar to those or even under forced conditions proceeded in low yield, perhaps due to steric hindrance around the hydroxyl group. By changing the protecting group on the sugar moiety from TIPDS to acetyl, the desired methoxalyl ester **15** was easily obtained from **14** under similar conditions described above. However, radical deoxygenation of **15** with Bu_3SnH in the presence of AIBN in toluene gave a 1:3 mixture of **16** and **17**. Compound **16** was likewise converted to (2'*R*)-2'-deoxy-2'-*C*-methylcytidine (RMDC, **18**). Thus, stereoselectivity of the deoxygenation was controlled by the steric hindrance around the 2'-*tert*-radical formed.

In vitro cytotoxicities of the nucleosides **7-10**, SMDC (**1i**), SEDC (**13**), and RMDC (**18**) toward L1210 cells are summarized in Table 1. SMDC had the most potent cytotoxicity to L1210 cells in the series. Its potency was 46 times greater than that of the corresponding ribonucleoside **8**. It is obvious that deletion of the 2'-*tert*-hydroxyl group is important for the cytotoxicity. However, RMDC was more than 150 times less cytotoxic than SMDC. The 2'-ethyl derivative, SEDC did not show any activity at up to 100 µg/mL. These results are consistent with the accumulated results that almost all 2'-modified cytidines having a substituent at the 2'-α position are less active than the corresponding nucleosides bearing a substituent at the 2'-β position. Additionally, the size of the substituent seems to be a critical factor governing the cytotoxicity, the smaller being the better.

Table 1. IC_{50} values of 2'-branched-chain sugar pyrimidine nucleosides on L1210 cell growth *in vitro*.[1]

compd	IC_{50},[2] µg/mL	compd	IC_{50},[2] µg/mL
1i	0.26	**8**	12
13	> 100	**9**	> 100
18	40	**10**	> 100
7	> 100		

[1]Cytotoxicity *in vitro* was done by the method of Carmichael *et al.*[31] L1210 cells (2×10^3/well) were incubated in the presence or absence of test compound for 72 h. Then, 3-(4,5-dimethylthiazol-2-yl)-2,5-diphenyltetrazolium bromide was added and OD (570 nm) was measured. Percent inhibition was calculated as follows: % inhibition = [1 - OD (570 nm) of sample well / OD (570 nm) of control well] x 100. [2]IC_{50} was given as the concentration in µg/mL required for 50% inhibition of cell growth. The data were taken from ref. 30.

We next compared the inhibitory activity spectrum of SMDC with *ara*-C and 5-fluorouracil (5-FU) on the growth of various human tumor cells *in vitro* (Table 2). *Ara*-C showed inhibitory activity against mouse leukemic, human T-cell acute leukemic, and chronic leukemic cell lines, but not against human carcinoma or adenocarcinoma cells except for T-24 human bladder transitional cell line, while 5-FU had a broad spectrum of cytotoxicity to this range of cells. SMDC showed a quite similar inhibitory activity spectrum to that of *ara*-C.

We next examined *in vivo* antileukemic activity of SMDC compared with *ara*-C against implanted mouse leukemia P388 (10^6) in CDF1 mice, with doses indicated in Table 3 given ip once each day on days 1-5. As described in Table 3, *ara*-C was much more potent than SMDC at every dose. We found that SMDC was not a substrate for cytidine deaminase from

Table 2. Inhibitory effects of SMDC (1i), *ara*-C, and 5-FU on the growth of various human tumor cell lines *in vitro*.[1]

cell line		IC_{50},[2] µg/mL		
		SMDC	*ara*-C	5-FU
CCRF CEM	T-cell acute lymphoblastoid leukemia	0.15	0.065	40
MOLT-4	T-cell acute lymphoblastic leukemia	0.032	0.056	3.8
HL-60	promyelocytic leukemia	0.65	> 0.1	ND[3]
K-562	chronic myelogenous leukemia	2.2	3.2	38
U-937	histiocytic lymphoma	0.38	0.31	3.5
PC-10	lung squamous cell carcinoma	81.8	> 100	> 100
PC-14	lung adenocarcinoma	> 100	> 100	10
KATO III	gastric carcinoma	> 100	> 100	3.7
SW-480	colon adenocarcinoma	> 100	> 100	3.3
TE-2	esophagus adenocarcinoma	8.8	> 100	3.9
T-24	bladder transitional cell carcinoma	1.1	0.5	6.1

[1,2]See Table 1. The data were taken from ref. 30. [3]Not determined.

Table 3. Effects of SMDC and *ara*-C on the life span of mice bearing P388 leukemia.[1]

Dose	T/C (%)	
(mg/kg/day)	SMDC	*ara*-C
1	100	ND
3	100	152
10	116	184
30	111	189
100	126	189

[1]Female BDF$_1$ mice were implanted ip with 10^6 P388 leukemia cells on Day 0. Compounds were given ip on Day 1-5. Median survival time of the untreated control group was 9.5 days.

mouse kidney, and its 5'-triphosphate, which was chemically synthesized, was a potent inhibitor of DNA polymerases α, β, and γ (K_i values were 0.05, 0.10, and 0.07 μM, respectively) from calf thymus in a competitive manner to dCTP, those K_m of which for each enzyme was 1.25, 3.90, and 0.77 μM, respectively.[32] Thus, the 5'-triphosphate of SMDC is one of the most potent inhibitors of DNA polymerases known. Therefore, the reason for the ineffectiveness of SMDC *in vivo* would be related to its insusceptibility to deoxycytidine kinase.

SYNTHESIS AND ANTITUMOR ACTIVITY OF 2'-DEOXY-2'-METHYLIDENECYTIDINE (DMDC) AND 2'-DEOXY-2'-METHYLIDENE-5-FLUOROCYTIDINE (FDMDC)

As discussed above, a substituent introduced at the 2'-position of 2'-deoxycytidine should be placed at the β position, and the smaller the better for showing cytotoxicity to tumor cells. Although *in vitro* cytotoxicity of SMDC was comparable to that of *ara*-C, *in vivo* antileukemic activity did not parallel to that of *ara*-C. Therefore, we designed a new type of 2'-deoxycytidine analogue, 2'-deoxy-2'-methylidenecytidine (DMDC, **3a**), which has an allylic alcohol system together with the 3'-secondary alcohol in the sugar moiety.[24,33,34] This allylic alcohol in DMDC could be stable at a nucleoside level, but if it is phosphorylated to the 5'-polyphosphates by cellular kinases, the following chemical reactivities are expected. (i) At the 5'-diphosphate level (DMDCDP), it could be a mechanism-based inhibitor of ribonucleotide reductase if a 3'-radical is formed by its enzyme action.[35] (ii) When its 5'-triphosphate (DMDCTP), although this could be an inhibitor of DNA polymerases, is incorporated into DNA molecules, the allylic alcohol becomes a more reactive allyl phosphate ester. As shown in Fig. 1, such a diester is expected to rearrange intramolecularly and/or strand-break by intermolecular nucleophilic attack. Furthermore, the allylic alcohol system is found in nucleoside antibiotics such as angustmycin A[36,37] and neplanocin A[38,39], in which the structural feature may be important in the biological activity due to enhanced chemical reactivity and/or fixation of the sugar conformation.[40,41]

The synthesis of DMDC is rather straightforward, as outlined in Scheme 3.

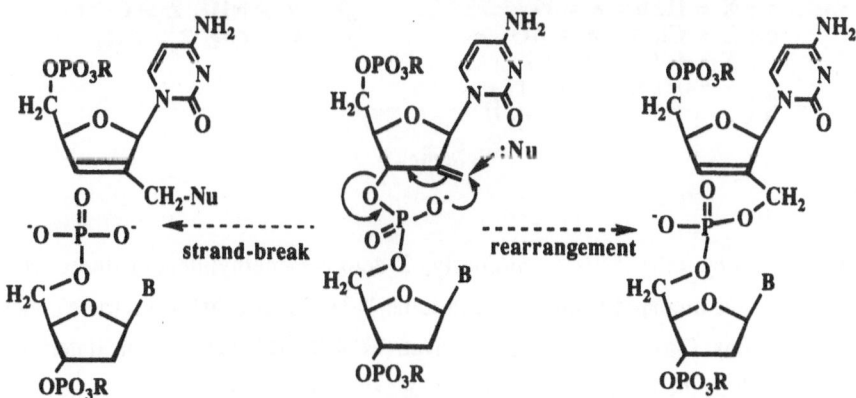

Fig. 1. Possible pathways of degradation of DNA-strand containing DMDC.

Compound **19** was converted to the corresponding 2'-keto derivative **20** by oxidation with a CrO₃/pyridine/Ac₂O complex, which was then treated with methylenetriphenylphosphorane. After complete consumption of the starting material, the reaction mixture was acidified by the addition of aqueous ammonium bromide to give the desired 2'-methylidene nucleoside **21**, along with a large amount of phosphonium salt **22**. To convert **22** into **21**, the crude mixture was further treated with NaH in THF to afford **21** in 80% total yield from **20**. Deblocking of **21** furnished the desired DMDC (**3a**) as a hydrochloride in 36% overall yield

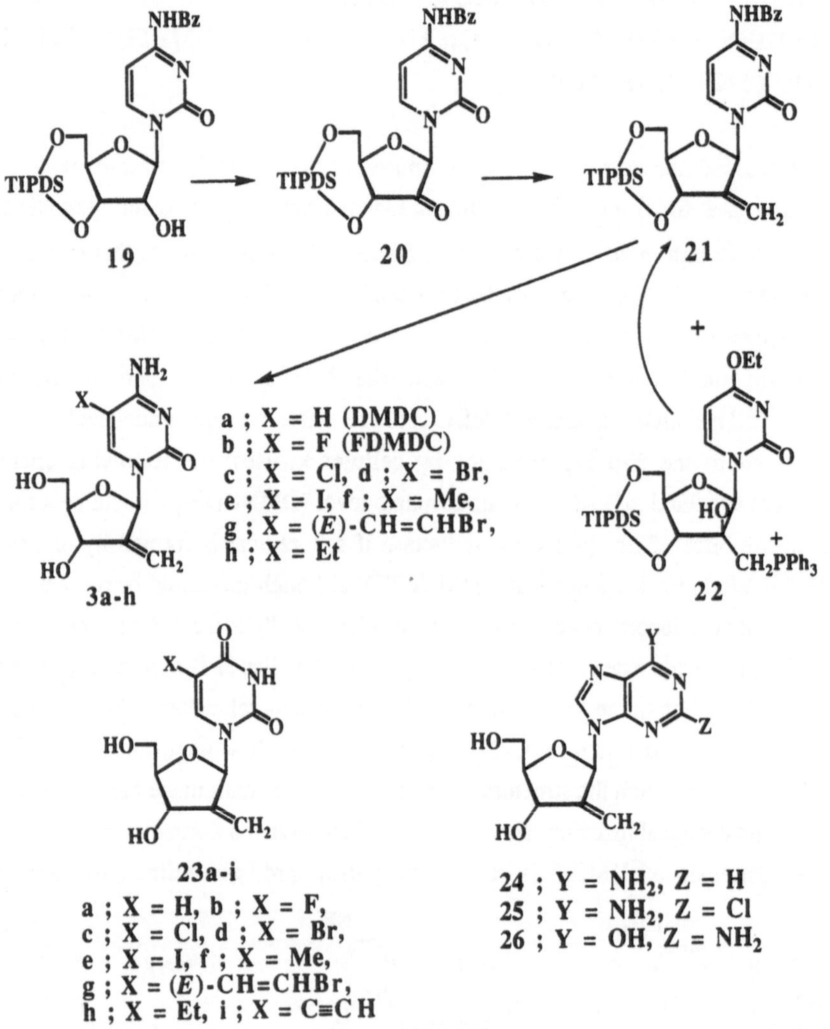

Scheme 3

in six steps from cytidine.[34,41,42] Similarly, 2'-deoxy-2'-methylidene-5-fluorocytidine (FDMDC, **3b**) was prepared from 5-fluorouridine.[34,41] Other 5-halogeno (**3c-e**), methyl (**3f**), (*E*)-bromovinyl (**3g**)[43] and ethyl (**3h**) derivatives of DMDC and their uracil derivatives (**23a-i**),[34] as well as the purine nucleosides, 2'-deoxy-2'-methylideneadenosine (**24**),[44,45] - 2-chloroadenosine (**25**)[32], and -guanosine (**26**)[32] were also prepared.

As shown in Table 4, among the series of 2'-deoxy-2'-methylideneuridines **23**, only the 5-methyl derivative **23f** showed marginal activity against L1210 cells. In contrast with 2'-deoxy-5-fluorouridine, 5-fluoro derivative **23b** was completely inactive at up to 100 μg/mL, probably due to inability of the latter to act as a substrate of nucleoside kinases. As 5-FU is an effective inhibitor of the growth of various tumor cells, this result may imply that **23b** could not be a substrate for nucleoside phosphorylase. The ineffectiveness of the 5-ethynyl derivative **23i** would also be related to insusceptibility to nucleoside kinase, because 2'-deoxy-5-ethynyluridine 5'-monophosphate is known to be a potent inhibitor of thymidylate synthetase.[46]

Table 4. IC$_{50}$ values of various DMDU and DMDC analogues on L1210 cell growth *in vitro*.[1]

compd	IC$_{50}$,[2] μg/mL	compd	IC$_{50}$,[2] μg/mL
3a	0.37	**23a**	> 100
3b	0.33	**23b**	> 100
3c	> 100	**23c**	> 100
3d	> 100	**23d**	> 100
3e	> 100	**23e**	> 100
3f	58	**23f**	50
3h	> 100	**23h**	> 100
		23i	> 100

[1,2]See Table 1. The data were taken from ref. 34.

On the other hand, the most potent among these in the cytosine series are DMDC (**3a**) and FDMDC (**3b**), the cytotoxicities of which are comparable to that of SMDC (**1i**). It is important to note that FDMDC would act as a 5-fluorocytosine derivative but not as a 5-fluorouracil derivative via the deamination by cytidine deaminase since DMDC was reported not to be substrate of cytidine deaminase from mouse kidney.[33] The 5-methylcytosine derivative **3f** was about 170 times less active than DMDC and FDMDC. The other 5-halogeno- and 5-ethyl derivatives, **3c-e** and **h**, were found to be devoid of the activity. Therefore, it is conceivable that the order of cytotoxicity might be related to the substrate specificity of the first activation enzymes, such as thymidine kinase or deoxycytidine kinase.

We next compared the cytotoxicity of DMDC, FDMDC, *ara*-C, and 5-FU against various mouse and human lymphoma, leukemia, melanoma, and carcinoma cells *in vitro*.[34] The results are shown in Table 5. It can be seen that *ara*-C is quite cytotoxic to various leukemia and lymphoma cells, but not to human carcinoma cells, except P-36 and T-24 cells. By contrast, 5-FU had a broad spectrum of activity on this range of tumor cells. Although DMDC is an analogue of 2'-deoxycytidine, its spectrum of activity against tumor cells is quite different from that of either *ara*-C or SMDC, but similar to that of 5-FU. FDMDC showed almost equipotent inhibition of the tumor cells to DMDC.

Table 5. Inhibitory effects of DMDC (**3a**), FDMDC (**3b**), *ara*-C, and 5-FU on the growth of various human tumor cell lines *in vitro*.[1]

cell line		IC$_{50}$,[2] μg/mL			
		DMDC	FDMDC	*ara*-C	5-FU
B16	mouse melanoma	3.0	9.2	0.24	2.6
3LL	mouse Lewis lung carcinoma	16.0	22.0	ND[3]	ND[3]
CCRF CEM	T-cell acute lymphoblastoid leukemia	0.067	0.1	0.065	40
MOLT-4	T-cell acute lymphoblastic leukemia	0.041	0.030	0.056	3.8
HL-60	promyelocytic leukemia	0.040	0.053	> 0.1	1.3
U-937	histiocytic lymphoma	0.44	0.40	0.46	3.5
PC-6	lung small-cell carcinoma	39.0	16.0	50.0	ND[3]
PC-10	lung squamous cell carcinoma	> 100	> 100	> 100	> 100
PC-13	lung large-cell carcinoma	1.0	1.4	> 100	ND[3]
PC-14	lung adenocarcinoma	> 100	> 100	> 100	10
P-36	melanoma	0.36	0.25	11.1	10.7
KATO III	gastric carcinoma	8.3	8.3	> 100	3.7
SW-480	colon adenocarcinoma	3.8	4.4	> 100	3.3
TE-2	esophagus adenocarcinoma	2.9	1.8	> 100	3.9
T-24	bladder carcinoma	1.1	0.94	0.5	6.1

[1,2]See Table 1. The data were taken from ref. 34. [3]Not determined.

In vivo antileukemic activity of DMDC was also examined by using female CD2F$_1$ mice bearing ip inoculated L1210 cells (Table 6). DMDC administered ip once a day for 5 days at 250 mg/kg had a T/C (%) of 235. The activity of DMDC was schedule-dependent with much more therapeutic effect obtained by daily treatment than by a single treatment. DMDC was also effective against colon 26 murine carcinoma, M5076 murine reticulum cell sarcoma, LX-1 human lung cancer xenograft, and SK-Mel-28 human melanoma xenograft, which are less sensitive or refractory to *ara*-C.[47] *In vivo* antileukemic activity of FDMDC was also reported.[41] Thus, these nucleosides are interesting and promising agents for the treatment of cancer.

The *in vitro* cytotoxicity of DMDC against L1210 cells was prevented dose dependently by 2'-deoxycytidine, suggesting that DMDC is phosphorylated by deoxycytidine kinase.[47] Unlike *ara*-C, DMDC was not a substrate of cytidine deaminase from mouse kidney.[33] Together with these characteristics, chemically synthesized DMDCTP was a potent inhibitor of DNA polymerase α, β, and γ from calf thymus with K_i values of 0.42, 2.52, and 1.00 μM, respectively, in a manner competitive with dCTP, while *ara*-CTP inhibited only DNA polymerase α with a K_i of 1.10 μM.[32] It was also found that DMDCTP was incorporated into DNA molecules at the site complementary to guanine by the action of DNA polymerase α using a synthetic template-primer system. In this experiment, DMDCTP apparently acted as a chain-terminator, but whether it was a real chain-terminator or a result of strand-break after a further elongation of the chain, which we postulated, has not been proved. Moreover, DMDCDP was reported to be a time-dependent inhibitor of *E. coli* ribonucleotide reductase, and a possible mechanism is shown in Fig. 2.[48] Thus, DMDC has

Table 6. Effects of DMDC and *ara*-C on the life span of mice bearing L1210 leukemia.[1]

Dose	T/C (%)	
(mg/kg/day)	DMDC	*ara*-C
0.5	103	130
1	106	130
2.5	113	145
5	120	130
10	131	137
25	152	152
50	179	165
100	201	193
250	235	148

[1]Female CD2F$_1$ mice were implanted ip with 10^5 L1210 leukemia cells on Day 0. Compounds were given ip on Day 1-5. Median survival time of the untreated control group was 7.1 days. The data were taken from ref. 47.

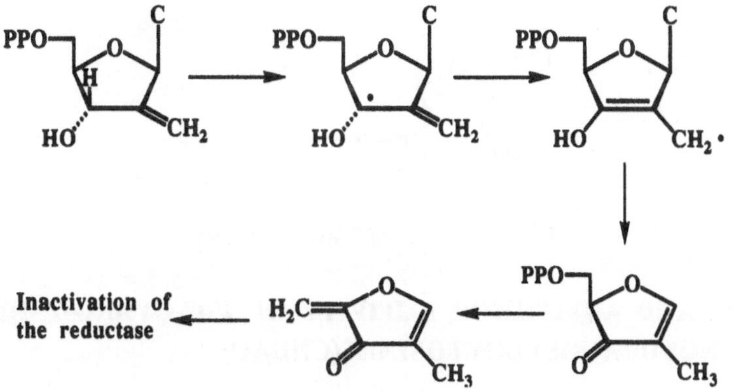

Fig. 2. Ribonucleotide reductase inactivation (from ref. 48).

a dual mechanism at least to inhibit tumor cell proliferation and an overall scheme of its action is shown in Fig. 3.

To elucidate the structure-activity relationships of the sugar structure of DMDC, several analogues have been synthesized, some of which are shown in Fig. 4. 2',3'-Dideoxy-2'-methylidene derivative **27**,[49-51] 3'-amino-2',3'-dideoxy-2'-methylidene derivative **28**,[52] xylofuranosyl derivative **29**,[53] 3'-deoxy-3'-methylidene derivative **30**,[41,42,51] and 2',3'-dideoxy-3'-methylidene derivative **31**[51] have been shown to be inactive on L1210 cells. However, fluoromethylene derivative **32** was reported to be cytotoxic to tumor cells.[54]

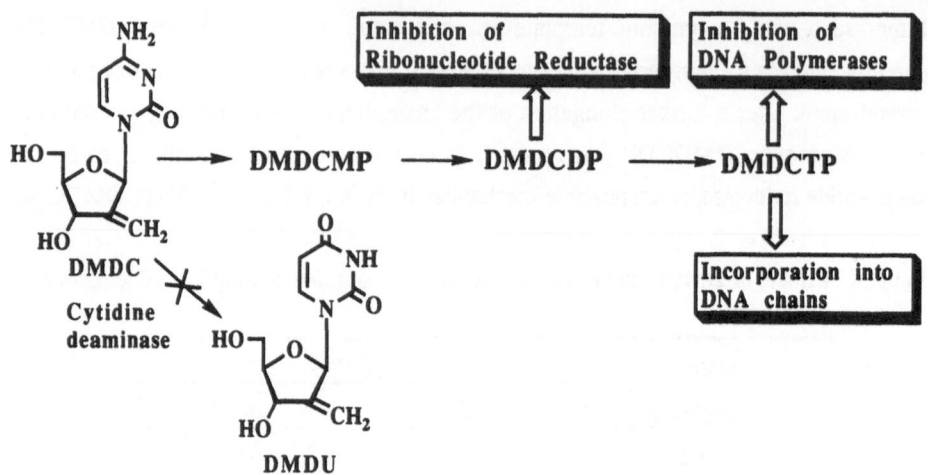

Fig. 3. Possible metabolism and mechanism of DMDC.

Prodrugs of DMDC were also synthesized and evaluated to be active against P388 mouse leukemia *in vivo*.[55] Antiviral activity of DMDC and its derivatives were reported to be effective against HSV-1, VZV, and HCMV.[41,43]

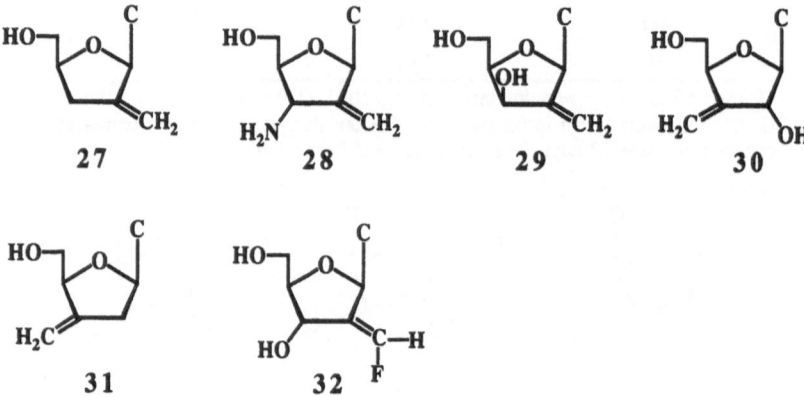

Fig. 4. Structures of DMDC analogues.

SYNTHESIS AND ANTITUMOR ACTIVITY OF 2'-*C*-CYANO-2'-DEOXY-1-β-D-ARABINOFURANOSYLCYTOSINE (CNDAC)

As described before, a smaller substituent at the 2'-β position of 2'-deoxycytidine is tolerant only for cytotoxicity. Introduction of an electron-withdrawing group into the 2'-β position would increase the acidity of the 2'-α proton. If such a nucleoside is enzymatically phosphorylated and incorporated into DNA molecules, the electron-withdrawing group becomes β to the phosphate diester in the DNA molecule. In this case, β-elimination would produce (i) DNA strand-breaks or (ii) abasic site formation as illustrated in Fig. 5. Together with this hypothesis, the nucleoside 5'-polyphosphates would also inhibit ribonucleotide reductase and/or DNA polymerases. Since strand-breaks in DNA by radiation therapy have been hypothesized to produce tumor cell death, it is worth examining whether the nucleoside having such chemical reactivities inhibits tumor cell growth or not. As an example of the nucleoside that has an electron-withdrawing group at the 2'-β position of 2'-deoxycytidine, we synthesized 2'-C-cyano-2'-deoxy-1-β-D-arabinofuranosylcytosine (CNDAC, **1j**) and its thymine analogue **38**.

In scheme 4, the synthetic route of CNDAC is outlined.[56] Since CNDAC is expected to be base-labile, we started with the N^4-acetyl derivative **33**, which can be deprotected by acidic hydrolysis. Compound **33** was treated with NaCN in the presence of $NaHCO_3$ in aqueous ether[57,58] to afford an epimeric mixture of 2'-cyanohydrin **34**. Compound **34**, without purification, was further treated with phenoxythiocarbonyl chloride in the presence

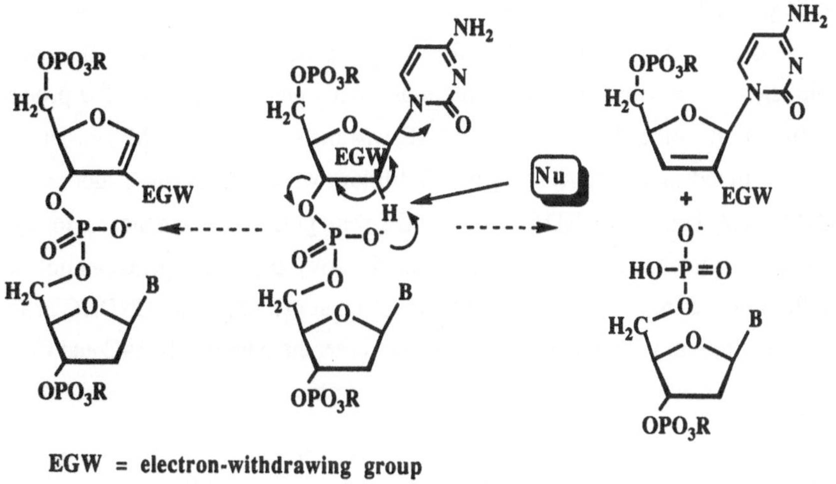

EGW = electron-withdrawing group

Fig. 5. Possible pathways of degradation of DNA-strand containing a nucleoside with an electron-withdrawing group at the 2'-β position of deoxycytidine.

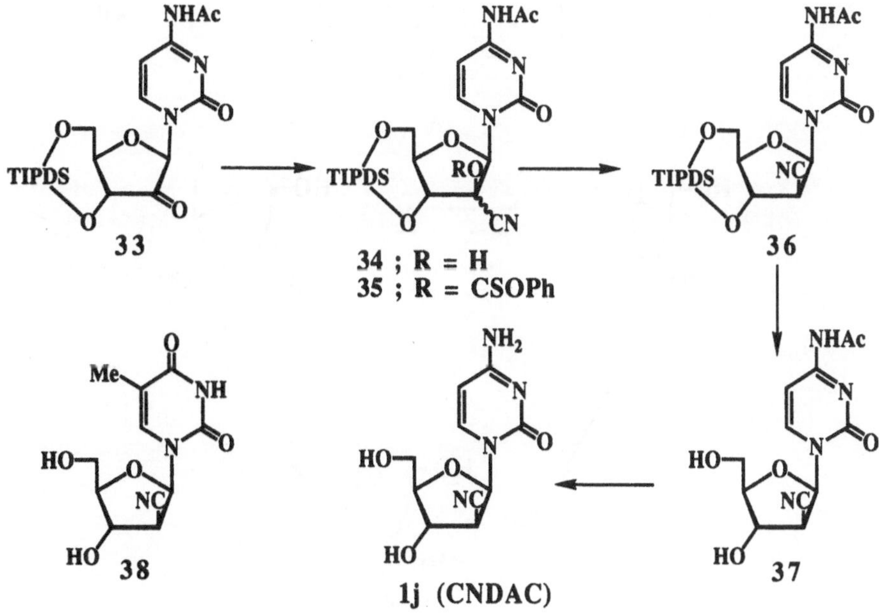

Scheme 4

of one equivalent of DMAP in CH_3CN to afford thiocarbonate **35**, followed by deoxygenation with Bu_3SnH and AIBN in toluene at 100 °C to give the protected desired nucleoside **36** in 57% yield from **33**. Deprotection of the 3',5'-TIPDS group of **36** by TBAF in the presence of AcOH gave **37**, which on treatment with HCl/MeOH furnished CNDAC (**1j**) as a hydrochloride. Thymine derivative **38** was also prepared in a similar manner.

As expected, treatment of **37** with NH_3/MeOH to remove the N^4-acetyl group was unsuccessful, and the only isolable product from the reaction mixture was cytosine. This can be explained as a β-elimination reaction from the 2'-α proton to the cytosine base. Reaction of the 5'-dimethoxytrityl derivative **39** with 1,1'-carbonyldiimidazole in *N,N*-dimethyl-formamide gave a β-elimination product, 2'-*C*-cyano-2',3'-didehydro-2',3'-dideoxy deriva-tive **40**. It is clear that when the 3'-hydroxyl group was converted into the carbonyl-imidazole ester, the carbonyl group may intramolecularly pick up the acidic 2'-α proton in **A** to form **40**. On the other hand, when CNDAC was incubated in 0.01 M Tris-HCl buffer (pH 9.0), we found that CNDAC was in equilibrium with 2'-*C*-cyano-2'-deoxycytidine (CNDC, **41**), and then both CNDAC and CNDC were gradually decomposed to cytosine and the sugar fragment, 1,4-anhydro-2-*C*-cyano-2-deoxy-D-*erythro*-pent-1-enitol (**42**).[59] This equilibrium was reached within about 2 h. Even at pH 7.5, either CNDAC or CNDC reached an equilibrium within about 25 h. In every case, the ratio of CNDAC and CNDC in

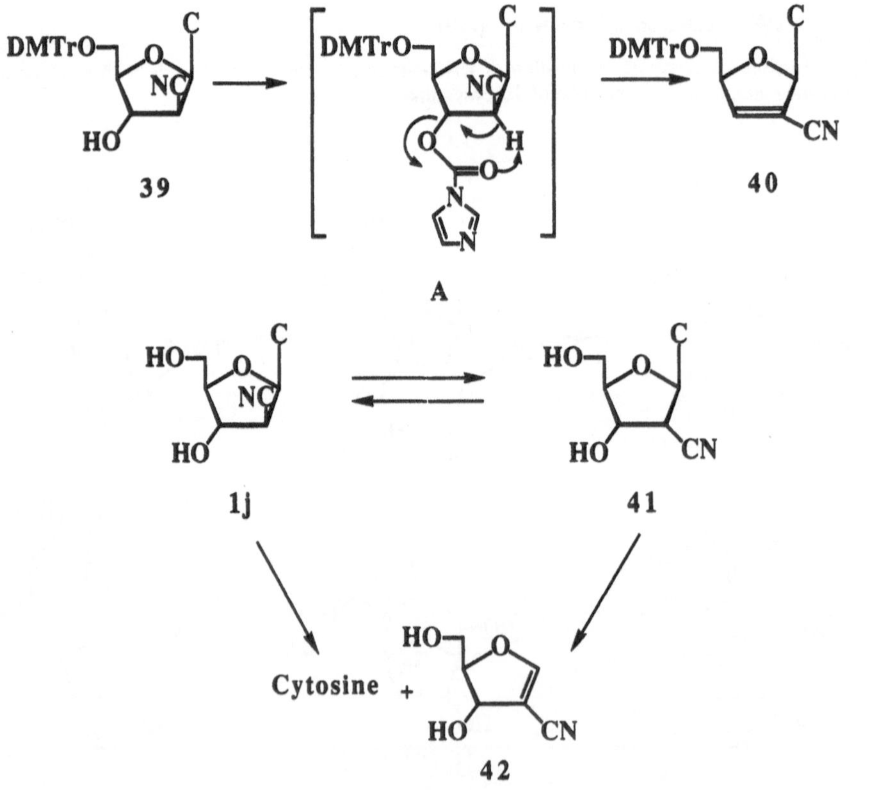

Scheme 5

Table 7. Inhibitory effects of *ara*-C, cytarazid, SDMDC, DMDC, FDMDC, and CNDAC on the growth of various human tumor cell lines *in vitro*.[1]

cell line	origin	IC₅₀,[2] μg/mL					
		ara-C	Cytarazid	SMDC	DMDC	FDMDC	CNDAC
PC-3	lung adenocarcinoma	> 100	18	30	1.1	2.4	5.5
PC-8	lung adenocarcinoma	0.28	1.2	2.5	1.0	4.0	4.6
PC-9	lung adenocarcinoma	1.6	1.0	< 0.3	< 0.3	< 0.3	4.4
PC-6	lung small-cell carcinoma	50	18	ND[3]	39	16	> 100
PC-13	lung large-cell carcinoma	> 100	18	10	1.0	1.4	> 100
QG-56	lung squamous cell carcinoma	> 100	5.4	23	3.4	> 100	40
QG-95	lung squamous cell carcinoma	50	27	58	> 100	> 100	24
Lu-65	lung giant cell carcinoma	ND[3]	6.2	5.0	1.3	2.7	> 100
KKLS	stomach adenocarcinoma	> 100	3.8	8.0	0.9	1.4	> 100
NUGC-4	stomach adenocarcinoma	> 100	1.4	3.7	0.85	0.4	5.2
KATO-III	stomach adenocarcinoma	> 100	6.6	8.8	1.6	2.8	15
NAKAJIMA	stomach adenocarcinoma	> 100	> 100	> 100	51	> 100	2.8
ST-KM	stomach adenocarcinoma	> 100	21	> 100	1.8	> 100	2.8
MKN-45	stomach adenocarcinoma	14	1.7	0.85	0.46	1.3	4.6
STSA-1	stomach adenocarcinoma	0.34	1.2	1.9	1.0	4.2	1.1
NKPS	stomach adenocarcinoma	0.11	30	> 100	0.95	> 100	0.24
SW-48	colon adenocarcinoma	> 100	21	> 100	5.0	> 100	> 100
SW-480	colon adenocarcinoma	> 100	> 100	> 100	3.8	4.4	> 100
MG-63	osteosarcoma	> 100	0.84	1.0	0.45	1.3	28
OST	osteosarcoma	> 100	6.4		4.0	> 100	6.4
G-292	osteosarcoma	> 100	> 100	> 100	> 100	> 100	> 100
HT-1080	fibrosarcoma	0.13	1.6	1.3	0.56	0.6	0.15
RD	rhabdomyosarcoma	> 100	34	11	2.9	5.3	42
T-24	bladder transitional cell carcinoma	0.5	0.56	1.1	1.1	0.94	9.4
KK-47	bladder transitional cell carcinoma	21	37	> 100	4.0	24	> 100

[1,2]See Table 1. [3]Not determined.

15

the equilibrium was about 3:5, suggesting CNDC, the cyano group being *trans* to the glycosylic linkage, is more thermodynamically stable than CNDAC. Thus, the 2'-α proton of CNDAC would be acidic enough to be eliminated by weak bases. From these experiments, therefore, if the compound is incorporated into DNA molecules, it could be hypothesized that the phosphate group or intermolecular bases could pick up the 2'-α proton to result in strand scission and abasic site formation as shown in Fig. 5.

In vitro cytotoxicity against L1210 cells of CNDAC and its thymine analogue **38** was next examined. The IC_{50} value for CNDAC was 0.21 μg/mL,[56] which is comparable to those of SMDC, DMDC, and FDMDC, while **38** showed only 6.7% inhibition at 100 μg/mL. This indicates that CNDAC itself is an active precursor, but degradation side products are not. Again, there are different substrate specificities between nucleoside kinases such as deoxycytidine kinase and thymidine kinase. Quite interestingly, CNDC (**41**), which was isolated from the degradation experiments, was cytotoxic to L1210 cells with an IC_{50} of 0.80 μg/mL after 72 h incubation, while the IC_{50} of CNDAC was 0.36 μg/mL in the same experiment.[59] As has been discussed, 2'-deoxycytidine analogues having a substituent at the 2'-α position were less active than those having a substituent at the 2'-β position. However, since CNDC can be converted to CNDAC in the cell culture medium, the active form would be CNDAC.

We then compared the *in vitro* cytotoxic spectrum of CNDAC with *ara*-C, SMDC, DMDC, FDMDC, and cytarazid in the various human tumor cell lines. The results are listed in Table 7. It is obvious that CNDAC is more active than *ara*-C in several cell lines, especially stomach adenocarcinoma cells. However, in comparison with others such as DMDC, FDMDC, and cytarazid, the data on the cytotoxicity of CNDAC seems not to be impressive. This is because CNDAC would be converted to CNDC either in the medium or in the cells, and only CNDAC would be phosphorylated to its 5'-polyphosphates.

In vivo antileukemic activity of CNDAC, cytarazid, and *ara*-C was also examined against ip-implanted P388 in CDF_1 mice (Table 8). CNDAC administered ip once a day on days 1 and 5 at 100 mg/kg had a T/C of 183%, while cytarazid and *ara*-C on the same

Table 8. Effects of CNDAC, cytarazid, and *ara*-C on the life span of mice bearing P388 leukemia.[1]

Schedule	compd	Dose (mg/kg/day)	T/C (%)
Day 1 and 5	CNDAC	100 x 2	183
	cytarazid	100 x 2	150
	ara-C	100 x 2	163
Day 1-10	CNDAC	20 x 10	> 600 (CR 5/6)[2]
	ara-C	20 x 10	225

[1]Female BDF1 mice were ip transplanted with 10^6 P388 leukemia cells on Day 0. Compounds were given on an appropriate schedule described above. Median survival time of the untreated control group was 10.0 days. [2]Five out of six mice survived for 60 days. The data were taken from refs. 56 and 60.

schedule showed T/C's of 150% and 163%, respectively. With treatment for 10 consecutive days at a dose of 20 mg/kg, the activity of CNDAC was greater than that of *ara*-C. The T/C value of CNDAC was more than 600%, and 5 out of 6 mice survived for 60 days. The T/C of *ara*-C was only 225%. Additionally, CNDAC showed excellent activities against several tumors refractory to *ara*-C, such as HT-1080 human fibrosarcoma implanted in chick embryos or athymic mice, although its cytotoxicity against HT-1080 was almost equal to that of *ara*-C.[60] Although the optimal therapeutic schedule of CNDAC is not yet known, such excellent activity suggests that CNDAC is a promising agent for further evaluation for clinical chemotherapy. The detailed mechanism of action of CNDAC including the DNA strand breakage hypothesis is under investigation in our group.

SYNTHESES OF 2'-DEOXY-2'-*C*-ETHYNYL-1-β-D-ARABINO-FURANOSYLCYTOSINE AND 2'-DEOXY-2'-*C*-ISOCYANO-1-β-D-ARA-BINOFURANOSYLCYTOSINE.

We have described a new type of antineoplastic nucleoside, CNDAC, in the previous chapter. This result prompted us to synthesize 2'-deoxy-2'-*C*-ethynyl-1-β-D-arabinofurano-sylcytosine (EDAC, **1k**) and 2'-deoxy-2'-*C*-isocyano-1-β-D-arabinofuranosylcytosine (NCDAC, **1l**) as potential antimetabolites.

Synthesis of EDAC (**1k**) was achieved by radical deoxygenation of the *tert*-propargyl alcohol in the key step (Scheme 6).[61] Treatment of the 2'-ketouridine derivative **43** with

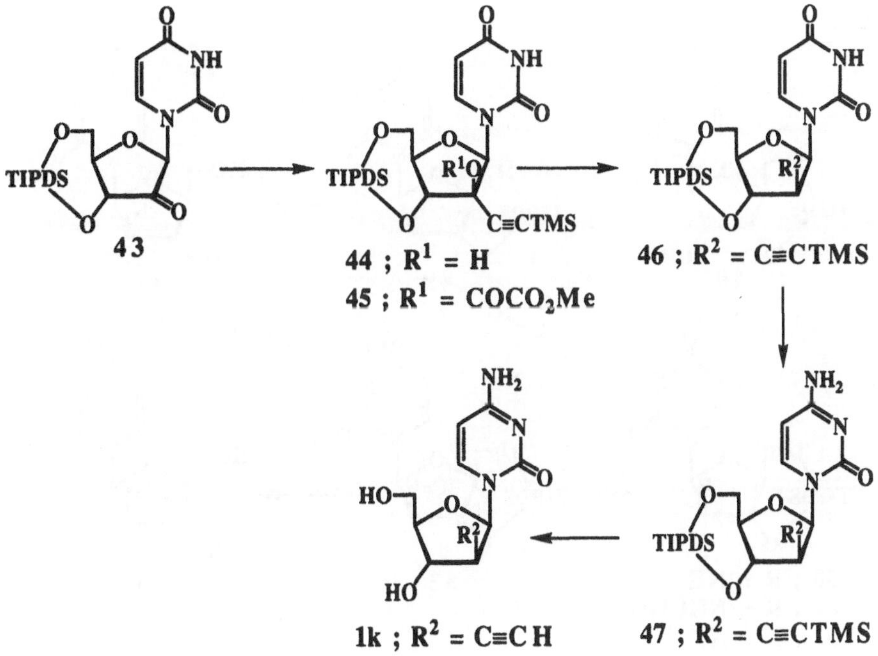

43 → 44 ; R^1 = H
45 ; R^1 = COCO$_2$Me → 46 ; R^2 = C≡CTMS

1k ; R^2 = C≡CH ← 47 ; R^2 = C≡CTMS

Scheme 6

lithium salts of (trimethylsilyl)acetylene in THF at -78 °C afforded stereoselectively 2'-β-C-(trimethylsilyl)ethynyl nucleoside **44** in 86% yield. Esterification of **44** with methoxalyl chloride in the presence of DMAP to give the corresponding methoxalyl ester **45**, which, without purification, was radically deoxygenated with Bu_3SnH in the presence of AIBN in hot toluene.[22,24,30] As expected, the reaction proceeded smoothly to give the desired 2'-deoxynucleoside **46** in 68% yield. On the other hand, the radical deoxygenation of the 2'-β-C-ethynyl derivative, which does not have the trimethylsilyl group, gave a complex mixture (about 9 spots on TLC). Several of these products when isolated were found to have tributylstannyl groups incorporated into the molecules. This may result from hydro-stannylation of the acetylenic bond. Although radical hydrostannylation of alkynes is frequently observed, bulky substituents attached at the ethynyl group can prevent such side-reactions to achieve the desired radical deoxygenation of the *tert*-propargyl alcohols in nucleosides.[62] Finally, **46** was converted into the cytosine derivative **47** via NH_4OH treatment of the corresponding O^4-triisopropylbenzenesulfonate. Compound **47** was further deblocked to furnish EDAC (**1k**).

In Scheme 7, the synthetic route of NCDAC (**1l**) is depicted. It has been generally recognized that the intramolecular nucleophilic attack of the 2-carbonyl group of the pyrimidine base on the 2'-position having a leaving group in such a pyrimidine nucleoside is predominant, rather than the intermolecular nucleophilic substitution. If the nucleophilicity of the 2-carbonyl oxygen could be reduced, the direct S_N2 reaction at the 2'-position would be realized. We have found a benzoyl group to be an effective choice for N^3-protection.[63]

When **48** was treated with diphenyl phosphorylazidate, triphenylphosphine, and diethyl azodicarboxylate in THF, the 2'-β-azido derivative **49** was formed in 70% yield,

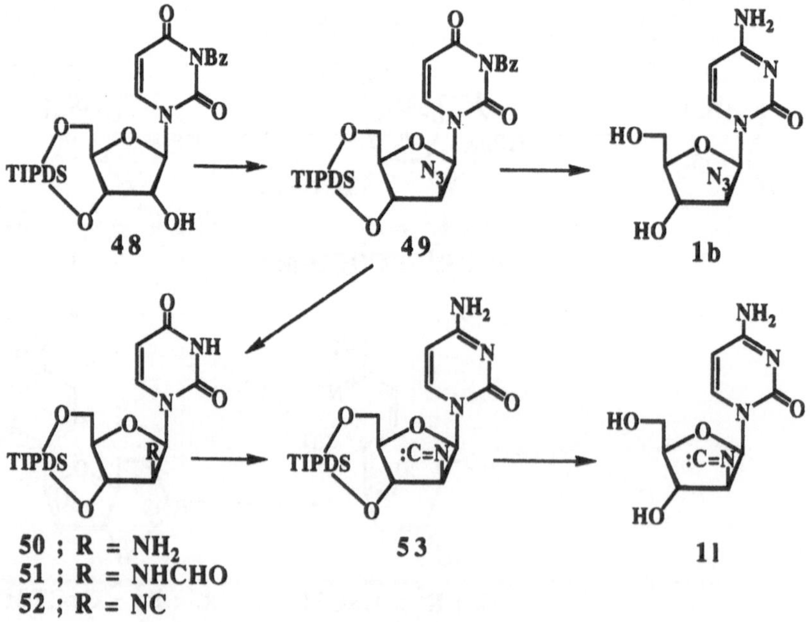

Scheme 7

18

from which cytarazid (**1b**) was prepared.[11,63] After debenzoylation from **49**, the azido group was reduced to afford the amino derivative **50**, followed by treatment with acetic formic anhydride to afford **51**. Dehydration of **51** with *p*-toluenesulfonyl chloride in pyridine at 70 °C gave the desired isonitrile **52**. Compound **52** was likewise converted to the cytosine derivative **53**, from which was deblocked to afford NCDAC (**1l**).[64]

In vitro cytotoxicity of **1k** and **1l** was examined using L1210 and KB (human oral epidermoid carcinoma) cells. Compound **1k** was inactive up to 100 μg/mL, while **1l** was a potent inhibitor of growth of both cell lines with IC_{50}'s of 3.2 and 2.0 μg/mL, respectively. These differences might be a reflection of susceptibility to deoxycytidine kinase. Although the cytotoxicity of **1l** was about 10 times less than that of CNDAC, we plan to compare its activity with other active nucleosides and examine its *in vivo* activity.

In conclusion, we have developed two kinds of new methods for introduction of certain functional groups to the 2'-β position of pyrimidine nucleosides. (i) Addition of the carbon nucleophile to 2'-ketonucleosides, followed by deoxygenation of the methoxalyl or thiocarbonyl ester of the subsequent *tert*-alcohol giving stereoselectively the 2'-β substituted 2'-deoxypyrimidine nucleosides, from which the desired 2'-deoxy-(2'-*C*-substituted)cytidine derivatives, such as SMDC, CNDAC, and EDAC were synthesized. In these radical reactions, the 2'-*tert*-radicals formed from the radical deoxygenations can react with Bu3SnH from the less hindered α-face of the sugar moiety to furnish stereoselectively the *arabino*nucleosides. (ii) A direct S_N2 reaction was possible when the N^3-position of the uridine derivative was benzoylated, in which nucleophilicity of the 2-carbonyl was reduced to prevent the intramolecular substitution reaction. By using this reaction, cytarazid can be synthesized, and the azido group was further transformed to the isonitrile furnishing NCDAC. Addition to these methods, the simple Wittig reaction with methylene-triphenylphosphorane to the 2'-ketonucleosides afforded DMDC and its derivatives.

We have found that DMDC, FDMDC, and CNDAC showed potent antitumor activities *in vitro* as well as *in vivo* against, not only leukemia and lymphomas but also carcinoma, melanoma, and fibrosarcoma cells, unlike *ara*-C. It is quite interesting to study the mechanisms of action of these nucleosides. Whether the mechanism responsible for their antitumor properties is related to the DNA-strand-breakage hypothesis is being studied in our group. If these indeed act as strand-breakers, it would open a new approach for anticancer chemotherapy.

References

1. This paper is part 113 of Nucleosides and Nucleotides; part 112: H. Homma, Y. Watanabe, T. Abiru, T. Murayama, Y. Nomura, and A. Matsuda, 2-(1-Hexyn-1-yl)adenosine-5'-uronamides: A New Entry of Selective A2 Adenosine Receptor Agonists with Potent Antihypertensive Activity, *J. Med. Chem.* 35: 2881 (1992).
2. R.R. Ellison, J.F. Holland, M. Weil, C. Jacquillat, M. Boiron, J. Bernard, A. Sawitsky, F. Rosner, B. Gussoff, R.T. Silver, A. Karanas, J. Cuttner, C.L. Spurr, D.M. Hayes, J. Blom, L.A. Leone, F. Haurani, R. Kyle, J.L. Hutchison, R.J. Forcier, and J.H. Moon, Arabinosyl cytosine: a useful agent in the treatment of acute leukemia in adults, *Blood*, 32: 507 (1968).
3. G.P. Bodey, E.J. Freireich, R.W. Monto, and J.S. Hewlett, Cytosine arabinoside (NSC-63878) therapy for acute leukemia in adults, *Cancer Chemother. Rep.* 53: 59 (1969).
4. G.W. Camiener and C.G. Smith, Studies of the enzymatic deamination of cytosine arabinoside-I.

Enzyme distribution and species specificity, *Biochem. Pharmacol.* 14: 1405 (1969).

5. D.H.W. Ho, Distribution of kinase and deaminase of 1-β-D-arabinofuranosylcytosine in tissues of man and mouse, *Cancer Res.* 33: 2816 (1973).

6. B.A. Chabner, K.R. Hande, and J.C. Drake, Ara-C metabolism: implications for drug resistance and drug interactions, *Bull. Cancer*, 66: 89 (1979).

7. H.N. Prince, E. Grumberg, M. Buck, and R. Cleeland, A comparative study of antitumor and antiviral activity of 1-β-D-arabinofuranosyl-5-fluorocytosine (FCA) and 1-β-D-arabinofuranosylcytosine (CA), *Proc. Soc. Exp. Biol. Med.* 130: 1080 (1969).

8. A.F. Hadfield and A.C. Sartorelli, The pharmacology of prodrugs of 5-fluorouracil and 1-β-D-arabinofuranosylcytosine, *Adv. Pharmacol. Chemother.* 20, 21 (1984).

9. M. Bobek, Y.C. Cheng, and A. Block, Novel arabinofuranosyl derivatives of cytosine: Resistant to enzymatic deamination and possessing potent antitumor activity, *J. Med. Chem.* 21: 597 (1978).

10. M. Bobek, Y.C. Cheng, E. Mihich, and A. Block, Synthesis, biologic effects, and biochemical properties of some 2'-azido- and 2'-amino-2'-deoxyarabinofuranosyl pyrimidines and purines, *Recent Results Cancer Res.* 74: 78 (1980).

11. A. Matsuda, J. Yasuoka, T. Sasaki, and T. Ueda, Improved synthesis of 1-(2-azido-2-deoxy-β-D-arabinofuranosyl)cytosine (Cytarazid) and -thymine. Inhibitory spectrum of cytarazid on the growth of various human tumor cells in vitro, *J. Med. Chem.* 34: 999 (1991).

12. Y.C. Cheng, D. Derse, R.S. Tan, G. Dutschman, M. Bobek, A. Schroeder, and A. Bloch, Biological and biochemical effects of 2'-azido-2'-deoxyarabinofuranosylcytosine on human tumor cells in vitro, *Cancer Res.* 41: 3144 (1981).

13. K.A. Watanabe, U. Reichiman, K. Hirota, C. Lopez, and J.J. Fox, Synthesis and antiherpes virus activity of some 2'-fluoro-2'-deoxyarabinofuranosylpyrimidine nucleosides, *J. Med. Chem.* 22: 21 (1979).

14. W. Kreis, K.A. Watanabe, and J.J. Fox, Structural requirements for the enzymatic deamination of cytosine nucleosides, *Helv. Chim. Acta*, 61: 1011 (1978).

15. V. Heinemann, L.W. Hertel, G.B. Grindey, W. Plunkett, Comparison of the cellular pharmacokinetics and toxicity of 2',2'-difluorodeoxycytidine and 1-β-D-arabinofuranosylcytosine, *Cancer Res.* 48: 4024 (1988).

16. W. Plunkett, V. Gandhi, S. Chubb, B. Nowak, V. Heinemann, S. Mineishi, A. Sen, L.W. Hertel, and G.B. Grindey, 2',2'-Difluorodeoxycytidine metabolism and mechanism of action in human leukemia cells, *Nucleosides Nucleotides*, 8: 775 (1989).

17. V. Heinemann, Y-Z. Xu, S. Chubb, A. Sen, L.W. Hertel, G.B. Grindey, and W. Plunkett, Inhibition of ribonucleotide reduction in CCRF-CEM cells by 2',2'-difluorodeoxycytidine, *Mol. Pharmacol.* 38: 567 (1990).

18. S. Shibuya and T. Ueda, Synthesis of 1-(2-deoxy-2-methylthio-β-D-arabinofuranosyl)-uracil and -cytosine, *J. Carbohydr. Nucleosides Nucleotides*, 7: 49 (1980).

19. J. Yoshimura, Synthesis of branched-chain sugars, *Adv. Carbohydr. Chem. Biochem.* 42: 69 (1984) and references cited therein.

20. A. Rosenthal, M. Sprinzl, and D.A. Baker, Nucleosides of branched-chain nitromethyl, cyanomethyl, and aminoethyl sugars, *Tetrahedron Lett.* 4233 (1970).

21. A.F. Cook and J.G. Moffatt, Carbodiimide-sulfoxide reactions. VI. Synthesis of 2'- and 3-ketouridines, *J. Am. Chem. Soc.* 89: 2697 (1967).

22. A. Matsuda, K. Takenuki, H. Itoh, T. Sasaki, and T. Ueda, Radical deoxygenation of *tert*-alcohols in 2'-branched-chain sugar pyrimidine nucleosides: Synthesis and antileukemic activity of 2'-deoxy-2'(*S*)-methylcytidine, *Chem. Pharm. Bull.* 35: 3967 (1987).

23. A. Matsuda, H. Itoh, K. Takenuki, T. Sasaki, and T. Ueda, Alkyladdition reaction of pyrimidine 2'-ketonucleosides: Synthesis of 2'-branched-chain sugar pyrimidine nucleosides, *Chem. Pharm. Bull.* 36: 945 (1988).

24. T. Ueda, A. Matsuda, Y. Yoshimura, and K. Takenuki, Synthesis and biological activity of branched chain-sugar nucleosides, *Nucleosides Nucleotides*, 8: 743 (1989).

25. K. Takenuki, H. Itoh, A. Matsuda, and T. Ueda, On the stereoselectivity of alkyl addition reaction of pyrimidine 2'-ketonucleosides, *Chem. Pharm. Bull.* 38: 2947 (1990).

26. D. F. R. Barton and S. W. McCombie, A new method for the deoxygenation of secondary alcohols, *J. Chem. Soc., Perkin Trans. 1.* 1574 (1975).

27. M.J. Robins and J.S. Wilson, Smooth and efficient deoxygenation of secondary alcohols. A general procedure for the conversion of ribonucleosides to 2'-deoxynucleosides, *J. Am. Chem. Soc.* 103: 932 (1981).

28. K. Pankiewicz, A. Matsuda, and K.A. Watanabe, Improved and general synthesis of 2'-deoxy *C*-nucleosides. Synthesis of 5-(2-deoxy-β-D-*erythro*-pentofuranosyl)uracil, -1-methyluracil, -1,3-dimethyluracil, and -isocytosine, *J. Org. Chem.* 47: 485 (1982).

29. S.C. Dolan and J. MacMillan, A new method for the deoxygenation of tertiary and secondary alcohols, *J. Chem. Soc., Chem. Commun.* 1588 (1985).

30. A. Matsuda, K. Takenuki, T. Sasaki, and T. Ueda, Radical deoxygenation of tert-alcohols in 1-(2-*C*-alkylpentofuranosyl)pyrimidines: Synthesis of (2'*S*)-2'-deoxy-2'-*C*-methylcytidine, an antileukemic nucleoside, *J. Med. Chem.* 34: 234 (1991).

31. J. Carmichael, W.G. DeGraff, A.F. Gazdar, J.D. Minna, and J.B.Mitchell, Evaluation of a tetrazolium-

based semiautomated colorimetric assay; Assessment of chemosensitivity testing, Cancer Res. 47: 943 (1987).

32. K. Takenuki and A. Matsuda, unpublished results.

33. K. Takenuki, A. Matsuda, T. Ueda, T. Sasaki, A. Fujii, and K. Yamagami, Design, synthesis, and antineoplastic activity of 2'-deoxy-2'-methylidenecytidine, *J. Med. Chem.* 31: 1063 (1988).

34. A. Matsuda, K. Takenuki, M. Tanaka, T. Sasaki, and T. Ueda, Synthesis of new broad spectrum antineoplastic nucleosides, 2'-deoxy-2'-methylidenecytidine (DMDC) and its derivatives, *J. Med. Chem.* 34: 812 (1991).

35. G.W. Ashley and J. Stubbe, Inhibitors of ribonucleoside diphosphate reductase activity, *in:* "International Encyclopedia of Pharmacology and Therapeutics, Section 128," J. G. Cory and A. H. Cory, eds., Pergamon Press, New York, 55 (1989).

36. For a review of nucleoside antibiotics, see: J.G. Buchanan and R.H. Wightman, The chemistry of nucleoside antibiotics, *in:* "Topics in Antibiotic Chemistry," P.G. Sammes, ed., Ellis Horwood Ltd. West Sussex, 6: 229 (1982).

37. R.J. Suhadolnik, "Nucleosides as Biological Probes," John Wiley & Sons, New York, (1979).

38. S. Yaginuma, N. Muto, M. Tsujino, Y. Sudate, M. Hayashi, and M. Otani, Studies on neplanocin A, a new antitumor antibiotic. I. Producing organism, isolation and characterization, *J. Antibiot.* 34: 359 (1981).

39. M. Hayashi, S. Yaginuma, H. Yoshida, and K. Nakatsu, Studies on neplanocin A, new antitumor antibiotic. II. Structure determination, *J. Antibiot.* 34: 675 (1981).

40. Y. Yamagata, K. Tomita, N. Marubayashi, I. Ueda, S. Sakata, A. Matsuda, K. Takenuki, and T. Ueda, Molecular conformation of 2'-deoxy-2'-methylidenecytidine: a potentially antineoplastic nucleoside, *Nucleosides Nucleotides*, 11: 835 (1992).

41. T.S. Lin, M.Z. Luo, M.C. Liu, R.H. Clarkekatzenburg, Y.C. Cheng, W.H. Prusoff, W.R. Mancini, G.I. Birnbaum, E.J. Gabe, J. Giziewicz, Synthesis and anticancer and antiviral activities of various 2'-methylidene-substituted and 3'-methylidene-substituted nucleoside analogues and crystal structure of 2'-deoxy-2'-methylidenecytidine hydrochloride, *J. Med. Chem.* 34: 2607 (1991).

42. V. Samano and M.J. Robins, Stereoselective addition of a Wittig reagent to give a single nucleoside oxaphosphetane diastereoisomer. Synthesis of 2'(and 3')-deoxy-2'(and 3')-methyleneuridine (and cytidine) derivatives from uridine ketonucleosides, *Synthesis*, 283 (1991).

43. H. Machida, S. Sakata, N. Ashida, K. Takenuki, and A. Matsuda, In vitro antiherpesvirus activities of 5-substituted 2'-deoxy-2'-methylidene pyrimidine nucleosides, *Antiviral Chem. Chemother.* in press.

44. H. Usui and T. Ueda, Synthesis of 8,2'-methano- and 8,2'-ethanoadenosines, *Chem. Pharm. Bull.* 34: 1518 (1986).

45. V. Samano and M.J. Robins, Synthesis of 2'(and 3')-deoxy-2'(and 3')-methyleneadenosines and bis(methylene)furan 4',5'-didehydro-5'-deoxy-2'(and 3')-methyleneadenosines. Inhibitors of S-adenosyl-L-homocysteine hydrolase and ribonucleotide reductase, *J. Org. Chem.* 56: 7108 (1991).

46. P.J. Barr, M.J. Robins, D.V. Santi, Reaction of 5-ethynyl-2'-deoxyuridylate with thiols and thymidylate synthetase, *Biochemistry*, 22: 1696 (1983).

47. K. Yamagami, A. Fujii, M. Arita, T. Okumoto, S. Sakata, A. Matsuda, T. Ueda, and T. Sasaki, Antitumor activity of 2'-deoxy-2'-methylidenecytidine, a new 2'-deoxycytidine derivative, *Cancer Res.* 51: 2319 (1991).

48. C.H. Baker, J. Banzon, J.M. Bollinger, J. Stubbe, V. Samano, M.J. Robins, B. Lippert, E. Jarvi, and R. Resvik, 2'-Deoxy-2'-methylenecytidine and 2'-deoxy-2',2'-difluorocytidine 5'-diphosphates: Potent mechanism-based inhibitors of ribonucleotide reductase, *J. Med. Chem.* 34: 1879 (1991).

49. A. Matsuda, H. Okajima, and T. Ueda, Synthesis of 2',3'-dieoxy-3'-methylidenethymidine and 2',3'-didehydro-2',3'-dideoxy-3'-methylthymidine: Deoxygenation of the allylic alcohol system in 3'-deoxy-3'-methyidene-5-methyluridine, *Heterocycles*, 29: 25 (1989).

50. M. Sharma and M. Bobek, Synthesis of 2',3'-dideoxy-3'-methylene pyrimidine nucleosides as potential anti-AIDS agents, *Tetrahedron Lett.* 31: 5839 (1990).

51. A. Matsuda, H. Okajima, A. Masuda, A. Kakefuda, Y. Yoshimura, and T. Ueda, Radical and palladium-catalyzed deoxygenation of the allylic alcohol systems in the sugar moiety of pyrimidine nucleosides, *Nucleosides Nucleotides*, 11: 197 (1992).

52. A.A.A. Hassan and A. Matsuda, [2,3]-Sigmatropic rearrangement of the allylic selenides to allylic amines in sugar moiety of pyrimidine nucleosides: Synthesis of 3'-amino-2',3'-dideoxy-2'-methylidenecytidine, *Heterocycles*, 34: 657 (1992).

53. A.A.A. Hassan and A. Matsuda, unpublished results.

54. J.R. McCarthy, D.P. Matthews, D.M. Stemerick, E.W. Huber, P.Bey, B.J. Lippert, R.D. Snyder, and P.S. Sunkara, Stereospecific method to *E* and *Z* terminal fluoro olefins and its application to the synthesis of 2'-deoxy-2'-fluoromethylene nucleosides as potential inhibitors of ribonucleoside diphosphate reductase, *J. Am. Chem. Soc.* 113: 7439 (1991).

55. T. Miyashita, N. Ashida, K. Kondoh, S. Sakata, H. Machida, A. Fujii, T. Ueda, and A. Matsuda, Synthesis and antitumor activity of acyl and benzyl types of prodrugs of 2'-deoxy-2'-methylidenecytidine, *Nucleosides Nucleotides*, 11: 495 (1992).

56. A. Matsuda, Y. Nakajima, A. Azuma, M. Tanaka, and T. Sasaki, 2'-*C*-Cyano-2'-deoxy-1-β-D-arabinofuranosylcytosine (CNDAC): Design of a potential mechanism-based DNA-strand-breaking antineoplastic nucleoside, *J. Med. Chem.* 34: 2917 (1991).

21

57. A. Calvo-Mateo, M-J. Camarasa, A. Diaz-Oriz, F.G. De las Heras, 3'-*C*-Cyano-3'-deoxythymidine, *Tetrahedron Lett.* 29: 941 (1988).
58. M-J. Camarasa, A. Diaz-Ortiz, A. Calvo-Mateo, F.G. De las Heras, J. Balzarini, E. De Clercq, Synthesis and antiviral activity of 3'-*C*-cyano-3'-deoxynucleosides *J. Med. Chem.* 32: 1732 (1989).
59. A. Azuma and A. Matsuda, unpublished results.
60. M. Tanaka, A. Matsuda, T. Terao, and T. Sasaki, Antitumor activity of a novel nucleoside, 2'-*C*-cyano-2'-deoxy-1-β-D-arabinofuranosylcytosine (CNDAC) against murine and human tumors, *Cancer Lett.*, 64: 67 (1992).
61. Y. Yoshimura, T. Iino, and A. Matsuda, Stereoselective radical deoxygenation of *tert*-propargyl alcohols in sugar moiety of pyrimidine nucleosides: Synthesis of 2'-*C*-alkynyl-2'-deoxy-1-β-D-arabinofuranosylpyrimidines, *Tetrahedron Lett.* 32: 6003 (1991).
62. T. Iino and A. Matsuda, unpublished results.
63. A. Matsuda, J. Yasuoka, and T. Ueda, A new method for synthesizing the antineoplastic nucleosides 1-(2-azido-2-deoxy-β-D-arabinofuranosyl)cytosine (cytarazid) and 1-(2-amino-2-deoxy-β-D-arabino-furanosyl)cytosine (cytaramin) from uridine, *Chem. Pharm. Bull.* 37:1659 (1989).
64. N. Minakawa, A. Dan, and A. Matsuda, unpublished results.

LITHIATION CHEMISTRY OF URIDINE DERIVATIVES: ACCESS TO A NEW ANTI-HIV-1 LEAD

Hiromichi Tanaka, Hiroyuki Hayakawa, and Tadashi Miyasaka

School of Pharmaceutical Sciences, Showa University
1-5-8 Hatanodai, Shinagawa-ku, Tokyo 142, Japan

INTRODUCTION

Direct lithiation of aromatic compounds started with the discovery of *ortho*-lithiation of anisole independently reported by two groups in 1939-1940.[1,2] Numerous experimental results accumulated thereafter, together with the availability of different types of lithiating agents have led to a general understanding of the reaction mechanism.[3] Since a number of functional groups have been recognized to have *ortho*-directing ability,[4] and since lithiated species usually react with a wide range of electrophiles under mild conditions, lithiation strategy has become an efficient alternative to classical electrophilic aromatic substitution.

In the nucleoside field, chemical modification at the heteroaromatic base moiety has mostly been achieved by nucleophilic substitution reactions.[5] The first attempt to lithiate nucleosides reported by Ulbricht in 1959 was a halogen–lithium exchange of 5-bromo-2'-deoxyuridine (1) with butyllithium (BuLi). Successive reaction of the tetra-lithio

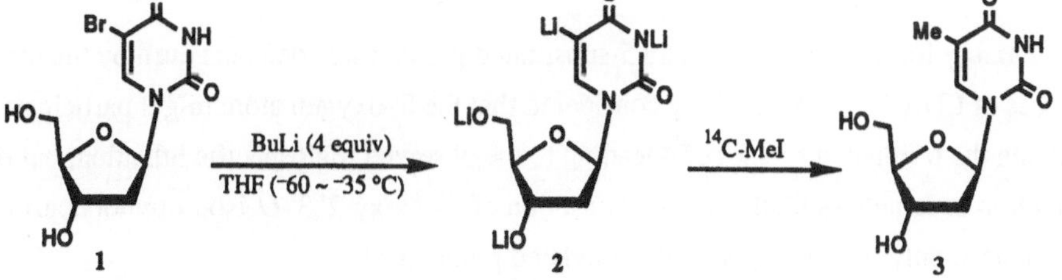

Scheme 1

Nucleosides and Nucleotides as Antitumor and Antiviral Agents,
Edited by C.K. Chu and D.C. Baker, Plenum Press, New York, 1993

23

intermediate (2) with ^{14}C-labelled MeI, furnished [^{14}C-Me]thymidine (3) in 10% yield (Scheme 1).[6,7] It was not until 1973 that the first report on the direct lithiation of nucleosides appeared. In this instance, tris-O-trimethylsilylated uridine was used as a substrate. However, its deprotonation with BuLi resulted in a lithiation level of only 42% with rather poor regioselectivity (C-5, 28 % vs. C-6, 14%).[8] Presumably because of these facts, lithiation strategy was used mainly for the preparation of labelled nucleosides, and the procedure thus was not regarded as a practical synthetic method in the field of nucleoside chemistry.

In this review article, we will focus on the direct lithiation of uridine derivatives, showing principally our own results, to demonstrate that this approach provides a highly general entry to the modification of the base moiety.[9,10] As an application of these studies, the lithiation of acyclouridines was also undertaken. Through this study, a unique anti-HIV-1 (Human Immunodeficiency Virus type 1) acyclonucleoside was found. The synthesis and activity of its analogues are also described in detail in the latter part of this review.

REGIOSPECIFIC C-6 LITHIATION OF URIDINE: A SIMPLE AND HIGHLY GENERAL METHOD FOR 6-SUBSTITUTED URIDINES

In contrast to the poor regioselectivity observed when 2',3',5'-O-(trimethylsilyl)uridine was lithiated with BuLi,[8] we found that treatment of 2',3'-O-isopropylideneuridine (4) with LDA (lithium diisopropylamide) followed by MeI gave the 6-methyl- (5) and 6-ethyl- (6) derivatives (Scheme 2). Although 6, which resulted from further deprotonation of 5,

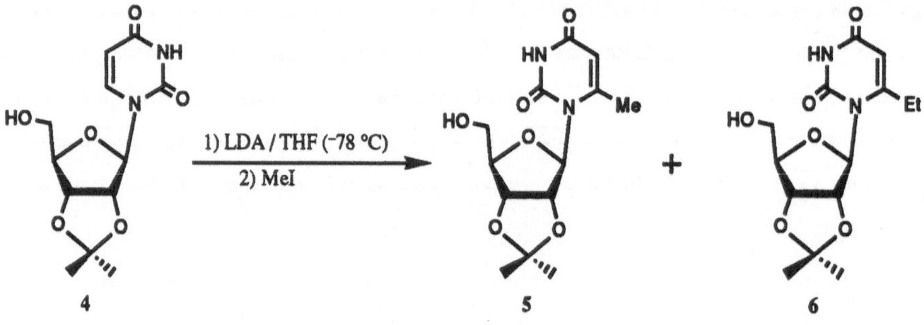

Scheme 2

was invariably formed, no trace of any 5-substituted product was detected even by the use of an excess of LDA.[11-13] We initially considered that the 5'-oxygen atom might participate in stabilizing the 6-lithiated species of 4 leading to the observed regiospecific lithiation, but this explanation became less likely since the reaction of 5'-deoxy-2',3'-O-isopropylideneuridine also produced only the corresponding 6-alkylated products.[14]

It is well known that condensation of 6-substituted pyrimidines with appropriately protected sugar derivatives almost always results in the predominant formation of N^3-ribosylated products due to the steric hindrance of the C-6 substituent.[15,16] Therefore, synthesis of their 6-substituted derivatives would be best carried out starting from naturally occurring nucleosides. As the C-6 position of the pyrimidine ring can be considered to be the β-position of an enone system, there were methods available for this purpose which involved nucleophilic addition–elimination, a general approach of which is illustrated in Scheme 3. There is, however, a severe limitation to this approach. To regenerate the 5,6-double bond

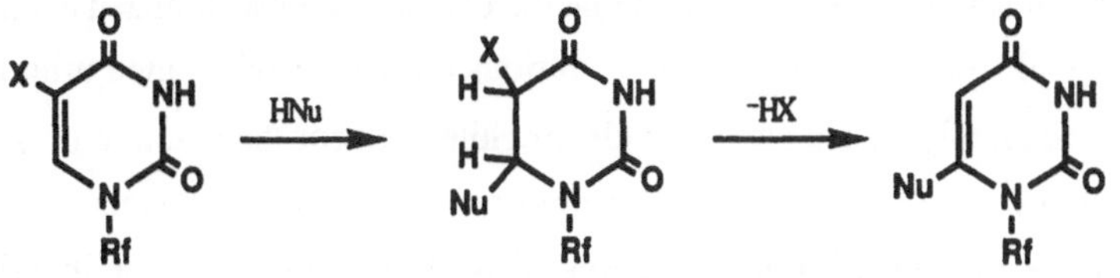

Rf = protected β-D-ribofuranosyl moiety

Scheme 3

from the intermediate having a 5,6-dihydro structure, "*Nu*" must be highly electron-withdrawing to make H-6 sufficiently acidic. Thus, only a few substituents such as cyano- and formyl- groups can be introduced to the C-6 position.[17-19]

As organolithium intermediates are known to react with a wide variety of electrophiles, we anticipated that the regiospecific C-6 lithiation with LDA could provide a general entry to

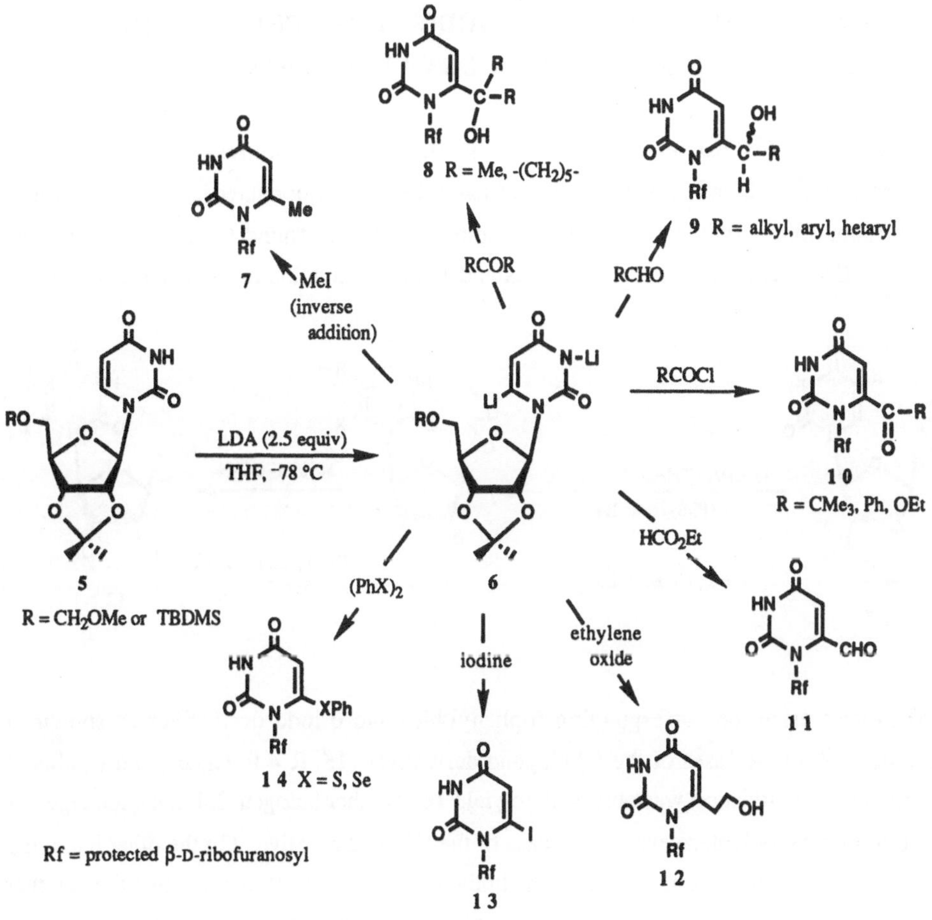

Scheme 4

6-substituted uridines. To give enough solubility to the 6-lithiated species, the 5'-hydroxyl group in 4 was protected with the methoxymethyl group through an acetal exchange reaction using dimethoxymethane to afford 5 (R = CH$_2$OMe).[20] It was found later that the TBDMS, [*tert*-(butyldimethylsilyl)], group is a better protecting group for the preparation of 5.[21] After the LDA (2.5 equiv) lithiation of 5 and subsequent treatment of 6 with MeOD, the ^{1}H NMR spectrum of the deuterated product showed that the C-6 position was lithiated exclusively (to the extent of 87%). As shown in Scheme 4, various types of 6-substituted uridines (7–14) were synthesized simply by changing the electrophile used for the reaction of 6.[20,22] The 2',3'-*O*-isopropylidene and 5'-*O*-methoxymethyl groups in these products were removed simultaneously by treating with 50% aqueous CF$_3$CO$_2$H. As 6 is soluble in THF even at low temperature, the preparation of 7 can be carried out by an inverse addition technique without forming further alkylated side products.[23,24] Some of these 6-substituted uridines serve as versatile starting materials for further transformations.[25,26] When the C-6 substituent has a leaving ability as in the cases of 13 and 14, nucleophilic addition–elimination at the C-6 position proceeds under very mild conditions, and hence it compensates for some of the limitations associated with the LDA lithiation.[27,28]

SYNTHESIS OF ANTILEUKEMIC 6-IODO- AND 6-PHENYLTHIO-URIDINE DERIVATIVES BASED ON LDA LITHIATION

Although 6-substituted pyrimidine nucleosides are known to have a *syn*-conformation,[29] which is reverse to that of naturally occurring nucleosides, 6-phenylthio- and 6-iodouridines synthesized in our lithiation study were found to show antileukemic activity. Encouraged by this result, we extended LDA lithiation to 5-substituted uridines

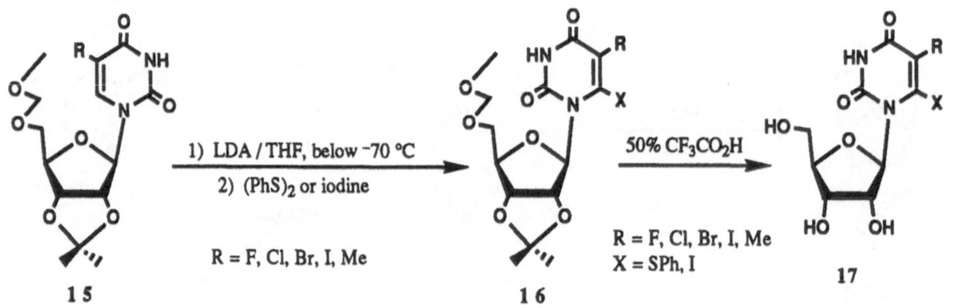

1) LDA / THF, below −70 °C
2) (PhS)$_2$ or iodine

R = F, Cl, Br, I, Me

50% CF$_3$CO$_2$H

R = F, Cl, Br, I, Me
X = SPh, I

15   16   17

Scheme 5

(15) to synthesize the corresponding 6-phenylthio- and 6-iodo derivatives as shown in Scheme 5.[30] In the cases of the 5-halogeno derivatives (15, R = F, Cl, Br, and I), almost quantitative C-6 lithiation was observed, to yield 16. Neither halogen–lithium exchange nor aryne formation took place even in the case of the 5-iodo derivative. On the other hand, the presence of a 5-methyl group decreased the lithiation level to *ca*. 30 %, reflecting the fact that the electron-donating inductive effect of the substituent makes the H-6 less acidic.

Compound **17**, obtained after deprotection, uniformly inhibited the growth of mouse leukemia L5178Y cells in culture. These data are summarized in Table 1.[30] Among the compounds, 5-fluoro-6-iodouridine (**17**, R = F and X = I) was found to be almost five times as potent as 5-fluorouracil.[31]

Table 1. Activity of **17** against mouse leukemia L5178Y cells in culture.

R	X	EC_{50} (μg/mL)[a]
H	SPh	70
Me	SPh	60
F	SPh	30
Cl	SPh	8.0
Br	SPh	20
I	SPh	40
H	I	8.0
Me	I	4.0
F	I	0.02
Cl	I	55
Br	I	52
I	I	55
5-Fluorouracil		0.1

a) Effective concentration of compound required to achieve 50% suppression of the cell proliferation.

C-5 LITHIATION OF 6-PHENYLTHIOURIDINE AND A NEW TYPE OF DESULFURIZATIVE STANNYLATION

Having established a simple synthetic method for the preparation of 6-substituted uridines, we reasoned that, given a practical method for the preparation of C-5 lithiated derivatives and a method for the successful removal of a C-6 substituent, this combination could be used for the preparation of 5-substituted uridines. Although the iodine atom and

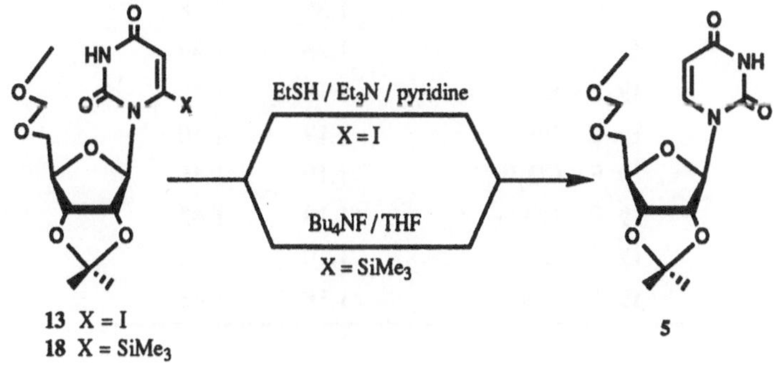

13 X = I
18 X = SiMe$_3$

5

Scheme 6

the trimethylsilyl group can be removed quantitatively from the C-6 position (Scheme 6), the C-5 lithiation of both **13** and **18**[32] failed. Compound **13** was highly susceptible to halogen–lithium exchange, forming approximately 50% of the 6-lithiated species **6** even with the use of LDA. On the other hand, no lithiation took place following treatment of **18** with several lithiating agents such as LDA, BuLi, and LTMP (lithium 2,2,6,6-tetramethyl-piperidide).

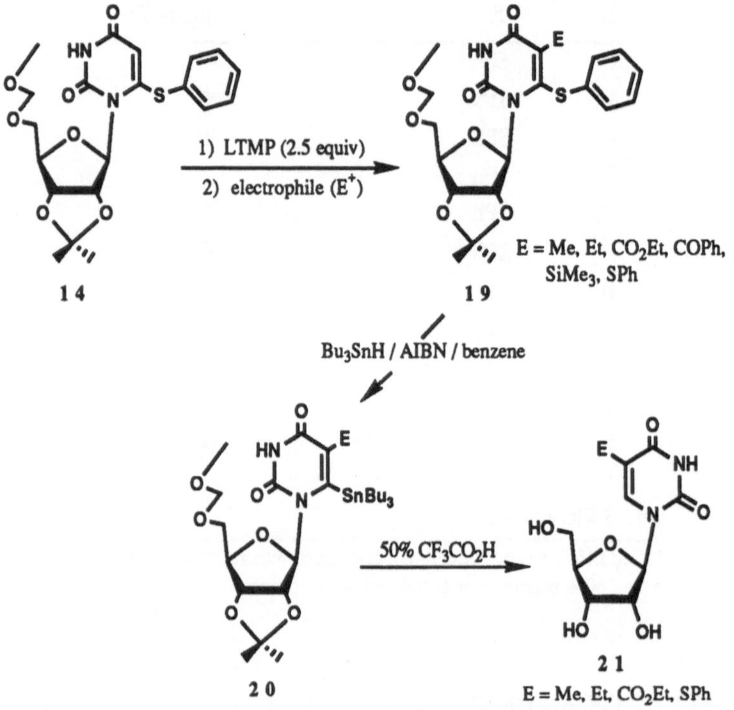

Scheme 7

Table 2. Chemical shifts δ (ppm) of isopropylidene methyl groups of **5**, **14**, and **19** in CDCl$_3$ (100 MHz).

Compd.	exo	endo
5	1.36	1.59
14	1.35	1.58
19 E = Me	1.23	1.46
19 E = Et	1.18	1.40
19 E = Bu	1.17	1.40
19 E = CO$_2$Et	1.19	1.45
19 E = COPh	1.19	1.43
19 E = SiMe$_3$	1.10	1.34
19 E = SPh	1.18	1.43

We found that almost quantitative C-5 lithiation can be achieved following treatment of the 6-(phenylthio)uridine derivative **14** with LTMP.[33,34] Successive reactions with electrophiles yielded 5-substituted products **19** as shown in Scheme 7. Attempted removal of the 6-phenylthio group in **19** with Raney-Ni or Al-Hg met with reduction of the 5,6-double bond. When **19** was subjected to a radical reaction in the presence of Bu3SnH and a catalytic amount of 2,2'-azobisisobutyronitrile (AIBN), the phenylthio group was replaced by the tributylstannyl group to give **20**. Interestingly, the conditions required for deprotection of the sugar moiety effected concomitant protonolysis of the stannyl group to furnish **21**.   During this study, we noticed from the ^{1}H NMR spectra of **19** that the presence of a substituent at the C-5 position caused a conformational change of the 6-phenylthio group. In the ^{1}H NMR spectrum of 2',3'-O-isopropylidene-5'-O-(methoxy-methyl)uridine (**5**), the resonance of H-5 appeared at 5.70 ppm. However compound **14** having a phenylthio group at the C-6 position had its H-5 resonance at 4.93 ppm, this is considerably shifted upfield due to the magnetic anisotropy of the substituent. However, no significant difference can be seen between **5** and **14** in terms of chemical shifts of the isopropylidene methyl groups. This suggests that the 6-phenylthio substituent in **14** lies perpendicular to the uracil ring. As shown in Table 2, when the C-5 position was substituted, both *exo* and *endo* methyl signals of the isopropylidene group were shielded by *ca*. 0.1–0.2 ppm. This observation is clearly attributable to the conformational change of the 6-phenylthio group and the bulkier group is likely to cause the greater change.

Since there had been no precedent regarding both direct desulfurizative stannylation[35] and protonolysis of the stannyl group in an enone or a vinyl system, we studied these

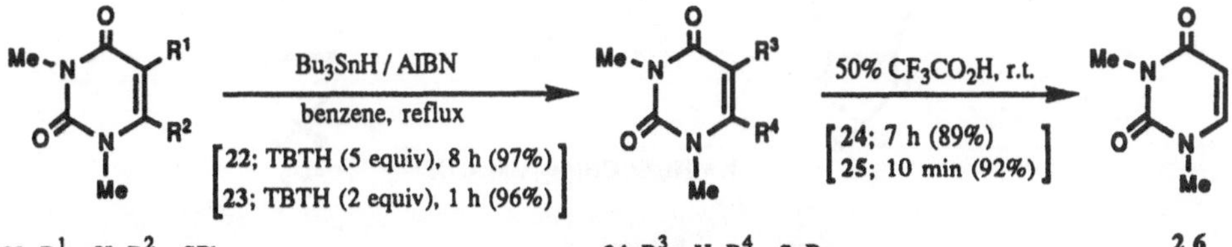

22 R^1 = H, R^2 = SPh
23 R^1 = SPh, R^2 = H

24 R^3 = H, R^4 = SnBu$_3$
25 R^3 = SnBu$_3$, R^4 = H

26

Scheme 8

reactions further by using 1,3-dimethyluracil derivatives (**22** and **23**) as model compounds. The results are summarized in Scheme 8 with detailed reaction conditions. Despite an apparent difference in electron density between the two positions, both the 6- and 5-phenylthio groups can be substituted by a tributylstannyl group to give **24** and **25**, respectively. It is also interesting to see that both desulfurizative stannylation and protonolysis of the stannyl group proceed much faster at the C-5 position.

SYNTHESIS OF 5-SUBSTITUTED URIDINES *via* THE AMIDE α-ANION OF 5,6-DIHYDROURIDINE

So far we have been discussing the direct lithiation of uridine derivatives which necessitates deprotonation at an sp^2 carbon. In this section, the lithiating agent is being used to deprotonate at an sp^3 carbon to generate a carbanion.

It is well known that, in contrast to lithium dialkylamides such as LDA, alkyllithiums have some degree of Lewis acid character and thus react through a coordination mechanism.[3]

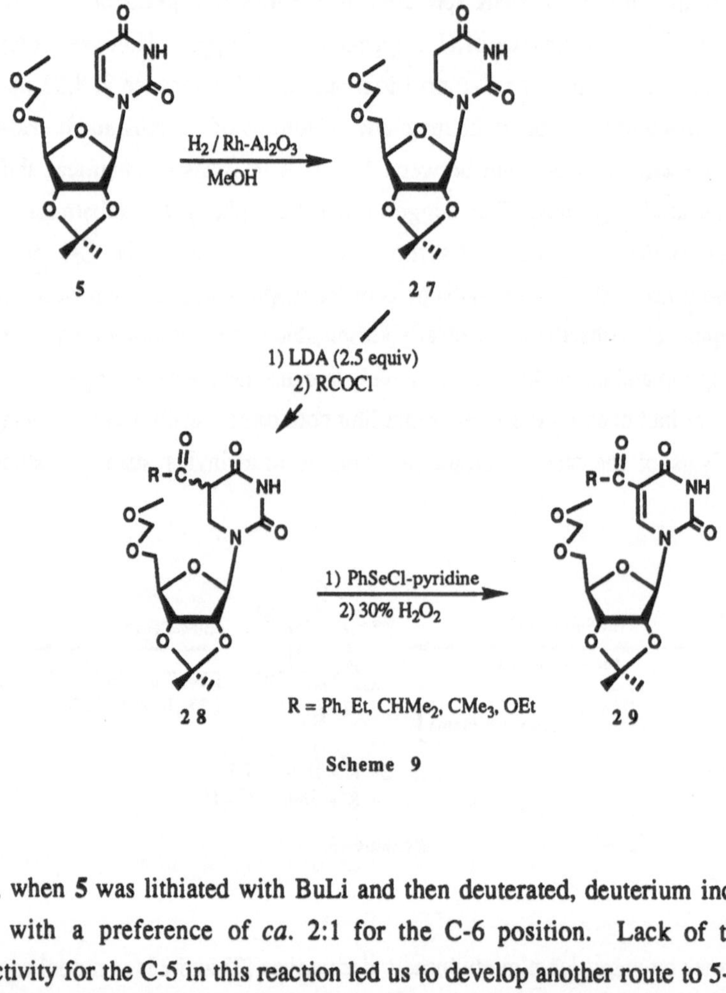

Scheme 9

However, when **5** was lithiated with BuLi and then deuterated, deuterium incorporation occurred with a preference of *ca.* 2:1 for the C-6 position. Lack of the desired regioselectivity for the C-5 in this reaction led us to develop another route to 5-substituted uridines. As the C-5 position is alpha to the carbonyl group, we expected that saturation of the 5,6-double bond would provide an "amide α-anion" despite the presence of the acidic N^3-H.

Catalytic hydrogenation effected virtually quantitative conversion of **5** to the 5,6-dihydrouridine derivative **27** (Scheme 9). The amide α-anion generated from **27** using LDA gives high yields of **28** upon reaction with RCOCl. Due to the presence of a β-dicarbonyl system in **28**, regeneration of the 5,6-double bond leading to **29** can be readily accomplished by treating with the PhSeCl–pyridine complex and subsequent oxidative elimination. The whole sequence constitutes a general method for the synthesis of 5-acyluridines.[36] This method has been successfully employed for the total synthesis of octosyl acid A.[37]

A different route has to be used to synthesize 5-alkyluridines (Scheme 10) because introduction of an electron-donating alkyl group at the C-5 position of **27** decreases the acidity of H-5. Thus, while **30** can be prepared in high yield by methylation of the amide α-anion, its reaction with the PhSeCl–pyridine complex only resulted in complete recovery

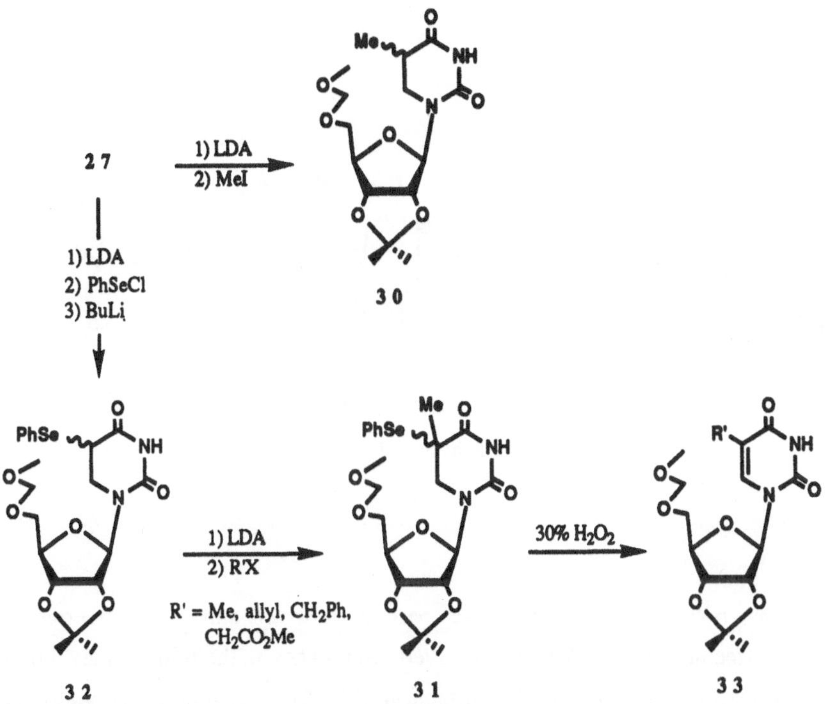

Scheme 10

of the starting material. Compound **30** underwent phenylselenation only when treated with LDA followed by PhSeCl, but the yield of **31** was poor. A solution to this problem is to introduce a phenylseleno group, which is known to stabilize an α-anion, before alkylation. When the anion derived from **27** was reacted with PhSeCl, a mixture of the requisite product (**32**) and the bis(phenylseleno) derivative was formed. On the basis of the reported cleavage of selenoketals with BuLi,[38] the reaction mixture containing the two products was further treated with BuLi to provide **32** in high yield. Various types of 5-alkyl-5-phenylseleno derivatives (**31**) can be prepared through alkylation of **32**. Finally, 5-alkyluridine derivatives (**33**) were obtained following oxidative elimination of the phenylseleno group from **31**.[39]

EFFECT OF SUGAR CONFORMATION ON THE LITHIATION: REGIOSELECTIVE SYNTHESIS OF 5-SUBSTITUTED URIDINES *via* DIRECT LITHIATION

As mentioned in an earlier part of this review, lithiation of 2',3'-*O*-isopropylidene-5'-*O*-(methoxymethyl)uridine (**5**, R = CH$_2$OMe) with LDA (2.5 equiv) occurs at the C-6 position to the extent of 87%. When the same amount of LDA was used for the reaction of a 2'-deoxyuridine derivative **34** (Scheme 11), the C-6 lithiation level was found to be only 20%. The highest lithiation level of 53% was attained by the use of 5 equivalents of LDA.

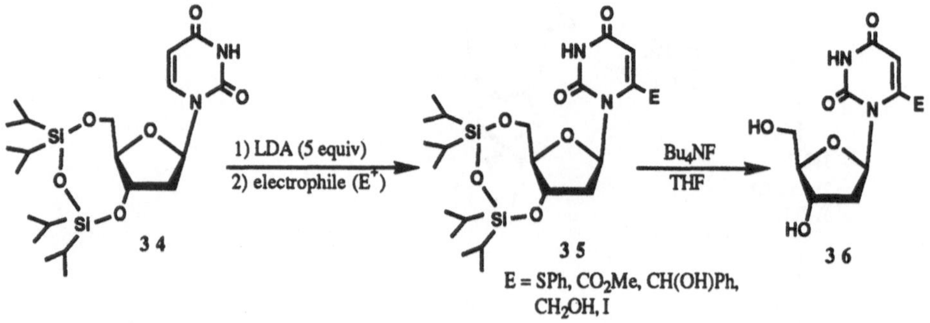

E = SPh, CO$_2$Me, CH(OH)Ph, CH$_2$OH, I

Scheme 11

Although some 6-substituted products **35** were synthesized and then converted to the corresponding free nucleoside **36**,[40] the observed discrepancy in the lithiation level of **5** and **34** prompted us to reinvestigate the C-6 lithiation of **5**. Since uridine is known to exist in *anti*-conformation even in solution,[29b] we initially thought the efficiency of the lithiation of 2',3'-*O*-isopropylidene-5'-*O*-(methoxymethyl)uridine (**5**, R = CH$_2$OMe) would be affected by changing its 5'-*O*-protecting group to a bulky TBDMS group. This turned out not to be the case, exactly the same extent of C-6 lithiation was observed.[41]   Interestingly, when

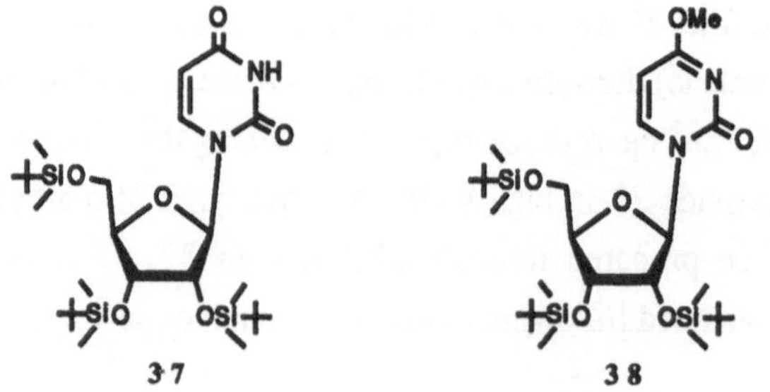

2',3',5'-tris-*O*-TBDMS uridine (**37**) was used as a substrate and reacted with LDA, neither the C-6 nor the C-5 position was lithiated. In contrast to this observation, application of the same protection to the 4-methoxy-2-pyrimidinone derivative **38** allowed preferential lithiation at the C-6 position (C-6, 71% *vs.* C-5, 4%) by the use of LDA.[41]

Although there has been no clear experimental evidence on the mechanism of these LDA lithiations, one possible explanation would be the intervention of an intramolecularly coordinated mono-lithio species that is fixed in the *syn*-conformation as depicted in Scheme 12. If this species is present, the subsequent deprotonation of the thermodynamically more acidic H-6 would be sterically restricted. Another assumption which can be drawn from this

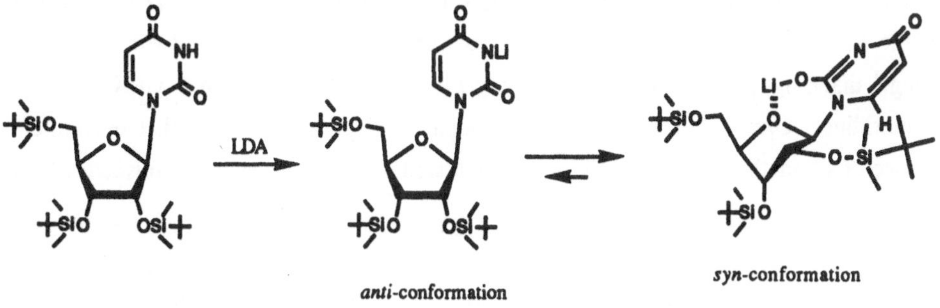

anti-conformation

syn-conformation

Scheme 12

hypothesis is that LDA would be not basic enough to deprotonate the less acidic H-5 of the uracil moiety.[42]

Based on these assumptions, we reasoned that a more basic lithiating agent could effect kinetic deprotonation of **37** at the C-5 position. This turned out to be the case as shown in Scheme 13. A practical lithiation level with a high regioselectivity (C-5, 87% *vs.* C-6, 5%) was attained when *sec*-BuLi was used in combination with *N,N,N',N'*-tetramethylethylene-diamine (TMEDA).[43] Since reaction of the C-5 lithiated species of **37** with electrophiles readily produces various types of 5-substituted products **39**, this method provides a short and general route to 5-substituted uridines.[44,45]

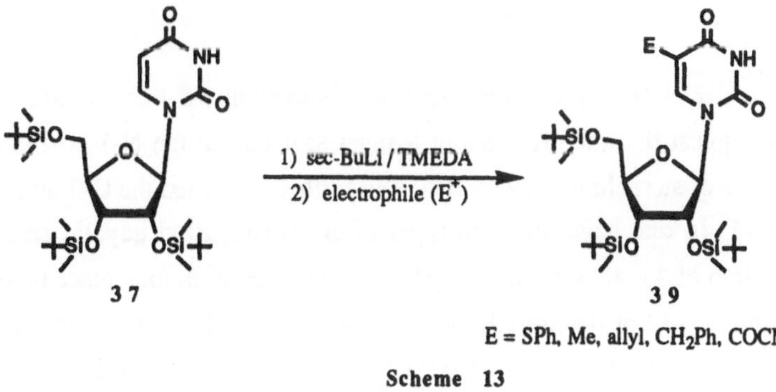

3 7   3 9

E = SPh, Me, allyl, CH_2Ph, $COCMe_3$, $SiMe_3$

Scheme 13

LITHIATION OF ACYCLOURIDINES:
FINDING OF A NEW ANTI-HIV-1 LEAD

Emergence of acyclovir, 9-[(2-hydroxyethoxy)methyl]guanine, as an excellent antiviral agent has stimulated extensive syntheses of a wide variety of acyclic nucleosides modified in either the base moiety or the acyclic structure.[46] However, base-modified pyrimidine acyclonucleosides had always been substituted at the C-5 position, presumably due to the difficulty of substitution at the C-6 position.

As an extension of our LDA lithiation strategy for synthesizing 6-substituted uridine derivatives, 1-[(2-hydroxyethoxy)methyl]uracils 40 were converted to the 6-phenylthio and 6-iodo derivatives 41 and then to the requisite 6-substituted acyclouridines 42 as shown in Scheme 14.[47] The choice of these substituents was, of course, motivated by our previous finding[30] that a significant antileukemic activity was observed in a series of 6-phenylthio- and 6-iodouridines. One salient point revealed through the preparation of 41 is that, even in the presence of a 5-methyl group, a high-yield conversion of 40 to 41 was accomplished, which

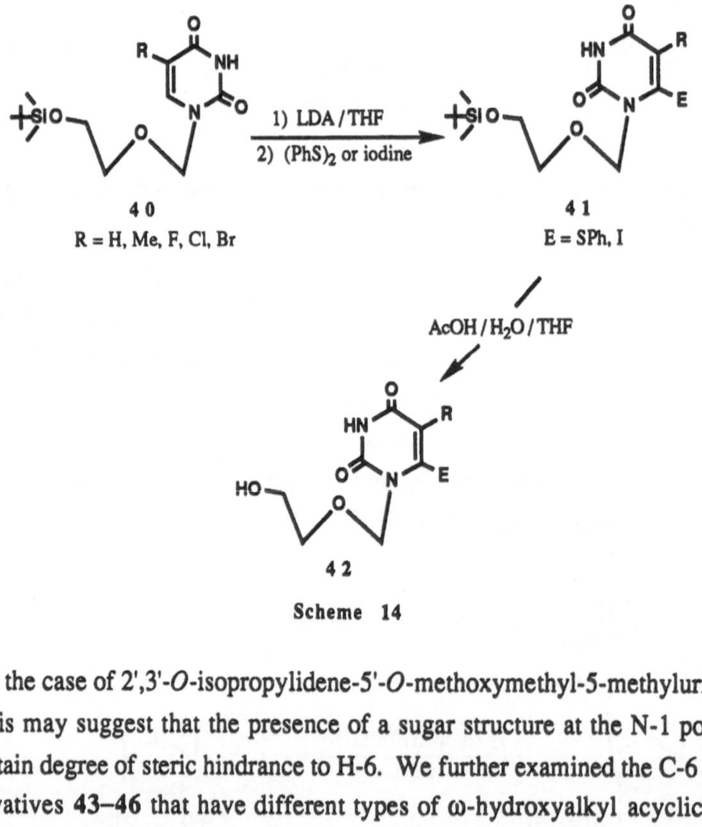

Scheme 14

contrasts to the case of 2',3'-O-isopropylidene-5'-O-methoxymethyl-5-methyluridine (15, R = Me). This may suggest that the presence of a sugar structure at the N-1 position itself exerts a certain degree of steric hindrance to H-6. We further examined the C-6 lithiation of uracil derivatives 43–46 that have different types of ω-hydroxyalkyl acyclic structures. While the C-6 lithiation of 43–45 with LDA is possible,[48] no lithiation took place in the case of 46 even following the use of the more basic lithiating agent LTMP. These observations may also be explicable in terms of the steric environment around the C-6 position.

Table 3. Inhibition of HIV-1 replication in MT-4 cells by **42**.

R	E	EC$_{50}$ (μM)	CC$_{50}$ (μM)	SI
H	SPh	>500	1850	<3.7
Me	SPh	7.0	740	106
F	SPh	>100	288	<2.9
Cl	SPh	>30	100	<3.4
Br	SPh	>20	44	<2.2
H	I	>30	75	<2.5
Me	I	>80	400	<5
F	I	>15	32	<2.1
Cl	I	>6	13	<2.2
Br	I	>5	11.5	<2.3

EC$_{50}$: effective concentration required to achieve 50% protection of
    MT-4 cells against the cytopathic effect of HIV-1.

CC$_{50}$: cytotoxic concentration required to reduce viability of
    mock-infected MT-4 cells by 50%.

SI: selectivity index (ratio of CC$_{50}$/EC$_{50}$).

Compounds **42**, synthesized as shown in Scheme 14, were evaluated for their activity against HIV-1 (Human Immunodeficiency Virus type-1), the results of which are summarized in Table 3. Among these compounds, 1-[(2-hydroxyethoxy)methyl]-6-(phenylthio)thymine (**42**: R = Me, E = SPh) abbreviated as HEPT was found to show an inhibitory effect on the cytopathogenicity of HIV-1 (HTLV-III$_B$ strain) in MT-4 cells.

35

Table 4. Anti-HIV-1 activity of HEPT
and nucleoside analogues in MT-4 cells.

Compd	EC_{50} (µM)	CC_{50} (µM)	SI
HEPT	7.0	740	106
AZT	0.016	20	1250
ddCyd	0.3	40	133
ddAdo	6.3	890	141

Table 5. Inhibitory effect of HEPT on replication
of HIV and other retroviruses.

Virus	Cell	EC_{50} (µM)	CC_{50} (µM)
HIV-1 (HTLV-III$_B$)	MT-4	7.0	740
	HUT-78	5.2	>250
	CEM	1.3	720
	PBL	1.6	>250
HIV-1 (HTLV-III$_{Ba-L}$)	MP	1.0	>100
HIV-2 (ROD)	MT-4	>250	–
	CEM	>250	–
HIV-2 (EHO)	MT-4	>250	–
SIV$_{MAC}$	MT-4	>500	–
SRV	Raji	>500	>500
MSV	C3H	>300	>300

When the activity of HEPT was compared to that of known anti-HIV nucleosides (Table 4), AZT and ddCyd are more active but more toxic than HEPT.[47]   In terms of activity (EC_{50}) and cytotoxicity (CC_{50}), HEPT is almost comparable to ddAdo. The anti-HIV-1 activity of HEPT was further confirmed by changing both the virus strain and cell line used (Table 5).[49]

Unlike other anti-HIV nucleosides such as AZT and ddAdo, HEPT was found to be only active against HIV-1. DNA viruses and other retroviruses, including HIV-2, are totally unaffected by this compound (Table 5). Thus, HEPT can be considered a highly specific anti-HIV-1 agent. Other unique properties of this compound deserve some mention.
1) HEPT does not compete with [^{3}H-*Me*]thymidine for phosphorylation by thymidine kinase derived from MT-4 cells. 2) The synthetic triphosphate of HEPT does not inhibit HIV-1 RT (reverse transcriptase) at concentrations much higher than that of its EC_{50}, irrespective of the template-primer used [either poly(rA)·oligo(dT) or poly(c)·oligo(dG)]. Thus, HEPT can be considered a highly unique and specific lead for anti-HIV-1 agents. This prompted us to carry out synthesis of its analogues with a directed aim to improve its activity.

Since our finding of HEPT as a specific anti-HIV-1 agent, several heterocyclic compounds (**47–49**) have been reported to show a similar specificity against HIV-1.[50-52]

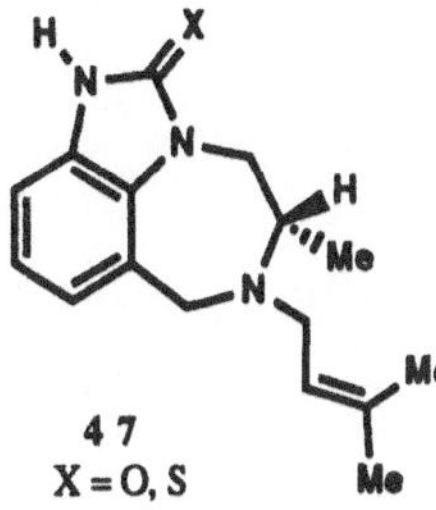

47
X = O, S

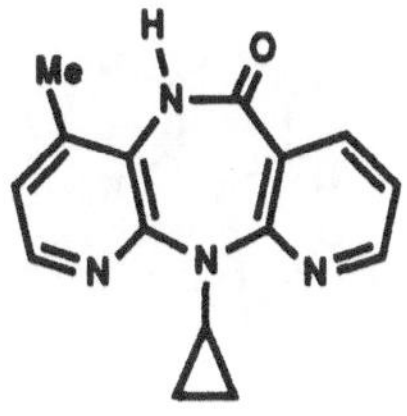

48

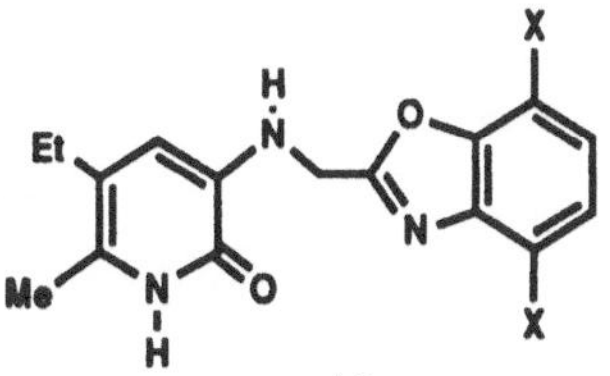

49
X = Me, Cl

SYNTHESIS OF 5- OR 6-SUBSTITUTED ANALOGUES OF 1-[(2-HYDROXYETHOXY)METHYL]-6-(PHENYLTHIO)THYMINE

Basically, three methods were employed for synthesizing HEPT analogues. One is the C-6 lithiation method with LDA. The other two are, as shown in Scheme 15, 1) lithiation of the C-5 position with LTMP and 2) a nucleophilic addition-elimination reaction of 6-phenylsulfinyl derivatives prepared by MCPBA oxidation.[53] The latter reaction proceeds under very mild conditions and is suitable for the introduction of oxygen- and nitrogen-containing substituents to the C-6 position, which is difficult to accomplish by the lithiation strategy.

In Table 6, the results of an anti-HIV-1 assay of the C-6 modified HEPT analogues (**50–66**) are summarized. Introduction of a simple alkylthio group to the C-6 position uniformly results in complete loss of the activity. It is interesting, however, to see that a saturated analogue of HEPT, the 6-cyclohexylthio derivative **54**, retains some activity. Although replacement of the sulfur atom in HEPT by oxygen or nitrogen gave rather discouraging results, the methylene counterpart **65** showed considerable activity.

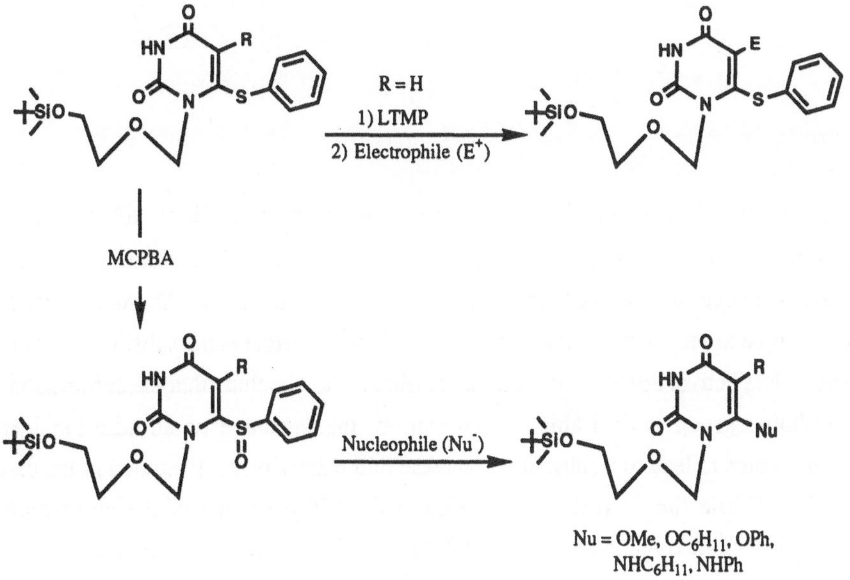

$Nu = OMe, OC_6H_{11}, OPh,$
$NHC_6H_{11}, NHPh$

Scheme 15

37

50-66

Table 6. Anti-HIV-1 activity of HEPT analogues modified at the C-6 position.

Compd.	R	EC$_{50}$ (μM)	CC$_{50}$ (μM)	SI
HEPT	SPh	7.0	740	106
50	SMe	>250	>250	~1
51	SEt	>250	>250	~1
52	SBu	130	>250	>1.9
53	SBu-t	>366	366	<1
54	SC$_6$H$_{11}$	8.2	664	81
55	SCH$_2$Ph	>46	46	<1
56	SOPh	>5.4	5.4	<1
57	OC$_6$H$_{11}$	>408	>408	~1
58	OPh	85	345	4
59	NHPh	>327	>327	~1
60	CCPh	>14	14	<1
61	CCH	>5.5	5.5	<1
62	CH=CH$_2$	>250	>250	~1
63	CH=CHMe	>250	>250	~1
64	COPh	>366	366	<1
65	CH$_2$Ph	23	352	15.3
66	CH$_2$CH$_2$Ph	>444	444	<1

To see the effect of substitution in the phenylthio ring of HEPT, 67–84 were next synthesized.[54] Compounds substituted at the *ortho*- or *para*-position did not show significant activity, except for 70. As can be seen, typically in the case of 72, *meta*-substitution turned out to be appropriate for this type of modification. Most *meta*-substituted analogues are more or less active; however, it might be reasonable to conclude that the substitution with a hydrophobic group is desirable. Furthermore, the presence of an additional *meta*-substituent seems to lead to a substantial increase in the activity, as illustrated in the cases of 83 and 84. While the SI (selectivity index: CC$_{50}$/EC$_{50}$) value of the *meta*-dichloro derivative 84 remained much the same as HEPT due to the increased cytotoxicity, the *meta*-dimethyl derivative 83 showed a SI value which is 9-fold greater than that of HEPT.

67–84

Table 7. Anti-HIV-1 activity of HEPT analogues modified in the 6-phenylthio ring.

Compd.	R	EC_{50} (μM)	CC_{50} (μM)	SI
HEPT	H	7.0	740	106
67	o-Me	71	>250	>3.5
68	o-Cl	>130	130	<1
69	o-NO$_2$	140	>250	>1.8
70	o-OMe	19	>250	>13
71	p-Me	220	>250	>1.1
72	m-Me	2.6	420	162
73	m-Et	2.7	181	67
74	m-Cl	13	210	16
75	m-Br	5.7	141	25
76	m-F	3.3	282	85
77	m-NO$_2$	34	170	5
78	m-OMe	22	>250	>11
79	m-CO$_2$Me	7.9	221	28
80	m-CO$_2$H	>352	352	<1
81	m-CH$_2$OH	>292	292	<1
82	m-OH	82	446	5.4
83	m-Me$_2$	0.26	243	935
84	m-Cl$_2$	1.2	133	111

In the preceding part dealing with the anti-HIV-1 activity of acyclouridines (**42**), it was shown in Table 3 that replacement of the 5-methyl group in HEPT by F, Cl, or Br gave no active compounds. In contrast to this, the 5-iodo derivative **85** listed in Table 8 was slightly active. This may be related to a change in conformation of the 6-phenylthio group. As evidenced from ^{1}H NMR spectral data (Table 2) of 5-substituted 6-(phenylthio)uridines (**19**), steric hindrance of the C-5 substituent causes a conformational change in the 6-phenylthio group. Based on this assumption, HEPT analogues variously substituted at the C-5 position (**86–100**) were synthesized. As expected, the analogues bearing a bulkier C-5 substituent than that in HEPT, such as the 5-ethyl- **98** and 5-isopropyl- **100** derivatives, proved to be potent inhibitors of HIV-1 replication (Table 8).[53,54] Although their activities are not so high when compared with that of AZT, these compounds are much less cytotoxic than AZT. As a result, their SI values are more than 3000.

85-100

Table 8. Anti-HIV-1 activity of HEPT analogues modified at the C-5 position.

Compd.	R	EC$_{50}$ (µM)	CC$_{50}$ (µM)	SI
HEPT	Me	7.0	740	106
85	I	3.6	20	5.6
86	SPh	>21	21	<1
87	CCPh	>3.4	3.4	<1
88	CCH	>18	18	<1
89	CH=CH$_2$	>250	>250	–
90	(Z)-CH=CHPh	6.0	95	16
91	CH=CPh$_2$	0.84	21	25
92	COCHMe$_2$	>12	12	<1
93	COPh	>13	13	<1
94	CONHPh	>18	18	<1
95	CO$_2$Me	>6.6	6.6	<1
96	CH$_2$Ph	>23	23	<1
97	CH$_2$CH=CH$_2$	2.5	183	73
98	Et	0.12	400	3300
99	Pr	3.4	244	72
100	*i*-Pr	0.063	231	3670
AZT		0.006	7.8	1300

MISCELLANEOUS MODIFICATIONS IN THE BASE MOIETY OF HEPT

Anti-HIV-1 activity of the 4-amino, 4-thio, and 2-thio counterparts of HEPT and that of the corresponding analogues lacking the 5-methyl group were evaluated.[55] Because we had encountered difficulty in lithiating a cytosine nucleoside,[56] synthesis of the 4-amino analogues (**101** and **102**) was carried out starting from the respective *O*-TBDMS derivative by way of their 4-(3-nitro-1,2,4-triazol-1-yl) intermediates according to the published method.[57] The 4-thio analogues (**103** and **104**) were prepared by thiation of the *O*-benzoyl derivatives with Lawesson's reagent. For the preparation of the 2-thio analogues **105** and

106, C-6 lithiation with LDA was successfully applied. This indicates that the lithiation also works with the 2-thiouracil system.[44]

As shown in Table 9, only the 2-thio counterpart of HEPT (**105**) showed significant activity. The data for **106** provides additional evidence that the presence of a C-5 substituent is essential for activity.

101–106

Table 9. Anti-HIV-1 activity of the 4-amino, 4-thio, and 2-thio analogues of HEPT.

Compd.	R	X	Y	EC$_{50}$ (μM)	CC$_{50}$ (μM)	SI
HEPT	Me	O	OH	7.0	740	106
101	Me	O	NH$_2$	>91	91	<1
102	H	O	NH$_2$	>250	>250	–
103	Me	O	SH	>27	27	<1
104	H	O	SH	>4.1	4.1	<1
105	Me	S	OH	0.98	123	126
106	H	S	OH	>198	198	<1

As another example of modification at the base moiety of HEPT, methylation and benzylation of the N-3 position were carried out.[55] From the fact that both *N*-alkylated products were totally inactive in the replication assay of HIV-1 in MT-4 cells (data not shown), it can be concluded that the presence of an NH is indispensable for activity. Also, it is quite interesting to see a lactam structure in common in the reported HIV-1-specific heterocyclic compounds (**47–49**) and in the HEPT analogues, all of which have no substituent at the nitrogen.

MODIFICATION OF HEPT IN THE ACYCLIC STRUCTURE: SYNTHESIS OF ITS DEOXY ANALOGUES

In the past few years, a large number of acyclonucleosides have been synthesized in an attempt to develop new antiviral agents.[46,58] Active compounds are generally considered to exhibit antiviral activities after being converted to their triphosphates. For example, an anti-HSV agent acylovir, 9-[(2-hydroxyethoxy)methyl]guanine, is monophosphorylated by HSV thymidine kinase[59] and then further converted to the corresponding triphosphate by cellular enzymes.

Although the acyclic side chain of HEPT is exactly the same as the one found in acyclovir, this compound does not compete with [^{3}H-*Me*]thymidine for phosphorylation by thymidine kinase derived from MT-4 cells, as mentioned earlier.[47,49] Moreover, the triphosphate of HEPT, which was chemically synthesized, does not inhibit HIV-1 RT. These experimental results suggest that the presence of an hydroxyl group in HEPT is not

107-120

Table 10. Anti-HIV-1 activity of deoxy HEPT analogues.

Compd.	R	EC$_{50}$ (µM)	CC$_{50}$ (µM)	SI
HEPT	CH$_2$OH	6.5	>500	>77
107	CH$_2$OAc	5.7	301	53
108	CH$_2$OCOPh	7.6	53	7
109	CH$_2$OMe	8.7	299	34
110	CH$_2$OCH$_2$Ph	>20	45	<2.3
111	CH$_2$OC$_6$H$_{11}$	>55	55	<1
112	Me	0.33	231	700
113	CH$_2$F	1.1	209	190
114	CH$_2$Cl	1.5	196	131
115	CH$_2$N$_3$	5.8	186	32
116	CH$_2$SiMe$_3$	>32	32	<1
117	Et	3.6	147	41
118	Pr	4.7	83	18
119	Ph	0.088	95	1080
120	H	2.1	224	107
AZT		0.0030	7.8	2600

neccessary for its anti-HIV-1 activity. Based on this assumption, a variety of HEPT analogues (**107–120**) listed in Table 10 were synthesized, the acyclic portions of which were altered so they could not be phosphorylated.[60]

Compounds **107** and **108** were prepared by acylation of HEPT. Preparation of the *O*-alkyl derivatives (**109-111**) was carried out either by a selective alkylation of the N^3,O-dianion of HEPT or based on the C-6 lithiation of a preformed acyclonucleoside with LDA. The latter method was successfully used for synthesizing other compounds, including deoxy HEPT analogues such as **113–115**. This indicates that the LDA lithiation works even when the starting material included a halogeno or azido functionality.

The *O*-acylated analogues **107** and **108** showed almost comparable activity to HEPT, while an increase in cytotoxicity results in the considerably decreased SI values. Among the *O*-alkylated analogues **109–111**, only the *O*-methyl derivative **109** appeared to retain activity. As can be seen from the data of **112-116**, the deoxy HEPT analogues are more or less active, except for **116**. The observation that the genuine deoxy analogue **112** is 20-fold more active than HEPT itself, clearly shows that the hydroxyl function of the lead compound HEPT does not contribute to its anti-HIV-1 activity. Another possible conclusion drawn

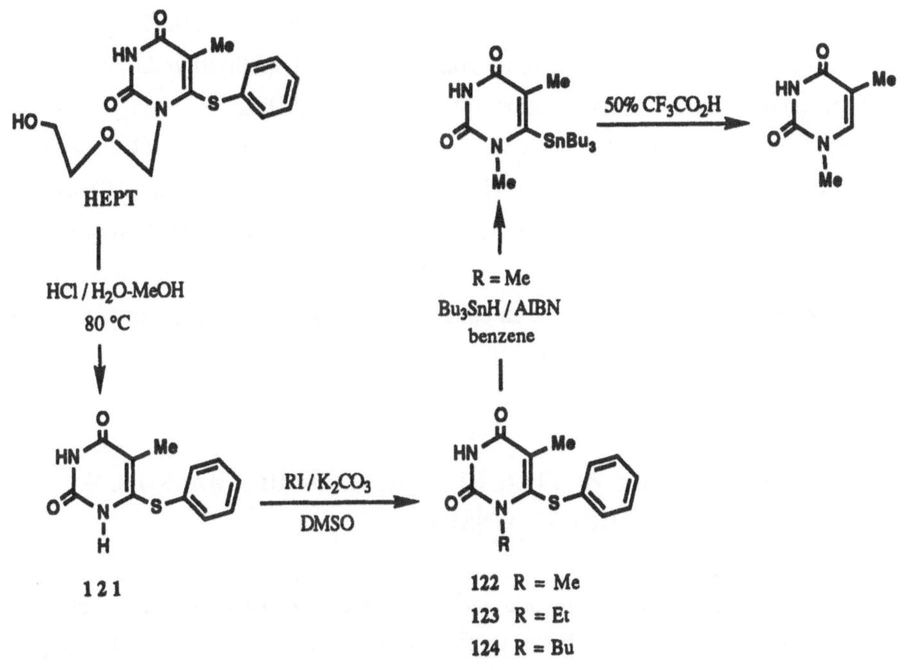

Scheme 16

from these data is that the value of the EC_{50} may correlate with the size of the acyclic side chain. The effect of the carbon-chain length in the alkoxymethyl structure on the anti-HIV-1 activity can be seen in the data of **117–120**. Thus, with the exception of the benzyloxy-methyl derivative **119**, compounds bearing a longer chain turned out to be not only less active but more cytotoxic than the ethoxymethyl derivative **112**. It should be mentioned that **120**, which bears the simplest alkoxymethyl group actually has a higher activity than HEPT itself.

The above positive results from the deoxy HEPT analogues prompted us to further simplify the acyclic structure as shown in Scheme 16.[61]   6-(Phenylthio)thymine (**121**)

Table 11. Anti-HIV-1 activity of HEPT analogues having a simple alkyl group at the N-1 position.

Compd.	EC_{50} (μM)	CC_{50} (μM)	SI
HEPT	7.0	740	106
121	>250	250	<1
122	>150	150	<1
123	2.2	94	43
124	1.2	89	74

obtained by acidic hydrolysis of HEPT, was alkylated with primary alkyl iodides to give the N-1 alkyl derivatives (**122–124**) along with the N^1,N^3-dialkylated products. The alkylation site was determined, in the case of **122**, by converting it to 1-methylthymine by way of radical-mediated desulfurizative stannylation, followed by protonolysis.[34]

In Table 11, the anti-HIV-1 activity of these *N*-alkyl derivatives (**122–124**) is summarized together with that of the base **121** derived from HEPT. Although **121** and the *N*-methyl derivative **122** did not show any activity, introduction of an alkyl group longer than ethyl group to the N-1 position seems to contribute to the activity to some extent. These results indicate that even the presence of an ether oxygen in the alkoxyalkyl structure is not essential for activity.

THE DESIGN OF HIGHLY ACTIVE HEPT ANALOGUES BASED ON STRUCTURE–ACTIVITY RELATIONSHIPS

Several factors which enhance the original activity found in HEPT had become apparent through extensive synthetic studies. Based on the available structure–activity relationships, a series of HEPT analogues expected to have a highly promising anti-HIV-1 activity were designed and synthesized. The factors taken into consideration were the following. 1) Substitution at the *meta*-position of the 6-phenylthio ring with two methyl groups. 2) Replacement of the 6-phenylthio group with a benzyl group. 3) Replacement of the 2-oxo function by a 2-thione function. 4) Replacement of the 5-methyl group with a 5-ethyl group. 5) Modification of the acyclic structure.

The 5-ethyl HEPT analogues synthesized along these lines were evaluated for their anti-HIV-1 activity and the results are summarized in Table 12.[62,63] The activity of these compounds is uniformly much higher than that of the parent compound HEPT. Most of these compounds, such as **127**, **129**, **130–133**, and **135**, show a comparable EC_{50} to that of AZT. The lowest EC_{50} value of 0.0022 μM was noted with **135** which was abbreviated as E-EBU-dM. In addition, these compounds are much less toxic than the reference compounds AZT and D4T, while **131–133** having a benzyloxymethyl group as an acyclic structure are rather cytotoxic.

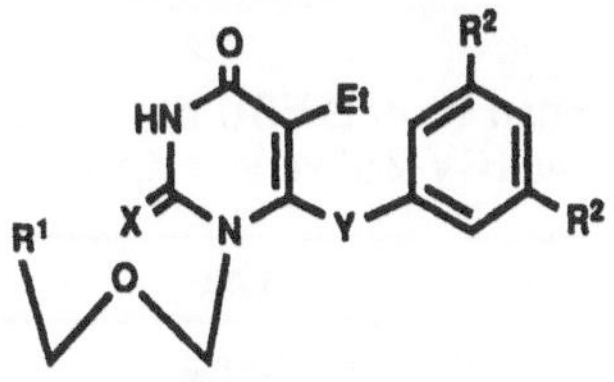

125-135

Table 12. The inhibitory effect of highly active 5-ethyl HEPT analogues on HIV-1 (HTLV-III$_B$) replication in MT-4 cells.

Compd.	X	Y	R^1	R^2	EC$_{50}$ (µM)	CC$_{50}$ (µM)	SI
125	O	S	CH$_2$OH	H	0.12	400	3300
126	O	S	CH$_2$OH	Me	0.016	155	9700
127	S	S	CH$_2$OH	Me	0.0075	172	23000
128	O	S	Me	H	0.022	146	6600
129	O	S	Me	Me	0.0062	>100	>16100
130	S	S	Me	Me	0.0056	>100	>17800
131	O	S	Ph	H	0.0049	30	6100
132	O	S	Ph	Me	0.0024	>20	>8300
133	S	S	Ph	Me	0.0076	>20	>2600
134	O	CH$_2$	CH$_2$OH	Me	0.015	256	17100
135	O	CH$_2$	Me	Me	0.0022	249	113000
HEPT					6.5	>500	>77
AZT					0.0030	7.8	2600
D4T					0.034	15	440

The complete data given in Table 12 indicates that each factor is at work cooperatively in improving the activity in terms of EC$_{50}$. It may be reasonable, therefore, to assume that HEPT analogues exert their anti-HIV-1 activity without being subjected to any metabolic activation.

MECHANISM OF ACTION OF HEPT ANALOGUES

Unlike the known anti-HIV 2',3'-dideoxynucleosides such as AZT and D4T, the HEPT analogues discriminate between HIV-1 and HIV-2, as illustrated in Table 13 for E-EBU-dM **135** which has the highest activity.[63] Since the analogues that lack an hydroxyl group, like E-EBU-dM, show an even higher activity , it is apparent that these compounds function as anti-HIV-1 agents in a totally different manner from the nucleoside derivatives.

In Table 14, the inhibitory effects of E-EBU-dM and AZT 5'-triphosphate (AZT-TP) on the reverse transcriptase (RT) of HIV-1 and HIV-2, either recombinant (rRT) or native (nRT), are shown. Quite unexpectedly, E-EBU-dM appeared to be inhibitory against the RT

45

Table 13. Comparison of E-EBU-dM **135** and AZT against HIV-1 and HIV-2 of various strains.

Compd.	Virus	Strain	Cell	EC_{50} (μM)	CC_{50} (μM)
E-EBU-dM (135)	HIV-1	HTLV-III$_B$	MT-4	0.0022	249
			MOLT-4	0.0012	168
			PBL	0.00045	73
		HTLV-III$_{RF}$	MT-4	0.0013	
		HIV-1$_{HE}$	MT-4	0.0076	
		A012D	MT-4	0.0012	
		HIV-1$_{JR-FL}$	MP	0.00074	>20
	HIV-2	LAV-2$_{ROD}$	MT-4	>249	
		LAV-2$_{EHO}$	MT-4	>249	
AZT	HIV-1	HTLV-III$_B$	MT-4	0.0030	7.8
			MOLT-4	0.0023	90
			PBL	0.0014	26
		HTLV-III$_{RF}$	MT-4	0.0020	
		HIV-1$_{HE}$	MT-4	0.0035	
		A012D	MT-4	0.33	
	HIV-2	LAV-2$_{ROD}$	MT-4	0.0028	
		LAV-2$_{EHO}$	MT-4	0.0020	

Table 14. Inhibitory effects of E-EBU-dM **135** and AZT-TP on HIV RT activity.

Compd.	Enzyme	Template-primer	Substrate	IC_{50} (μM)[a]
E-EBU-dM (EC$_{50}$ 0.0022)	HIV-1 rRT-A	Poly(A):oligo(dT)	dTTP	0.11
		Poly(C):oligo(dG)	dGTP	0.044
	HIV-1 rRT-B	Poly(A):oligo(dT)	dTTP	0.16
		Poly(C):oligo(dG)	dGTP	0.036
	HIV-1 nRT	Poly(A):oligo(dT)	dTTP	0.36
	HIV-2 nRT	Poly(A):oligo(dT)	dTTP	>500
		Poly(C):oligo(dG)	dGTP	>500
AZT-TP (EC$_{50}$ 0.0030)	HIV-1 rRT-A	Poly(A):oligo(dT)	dTTP	0.0082
		Poly(C):oligo(dG)	dGTP	>100
	HIV-1 rRT-B	Poly(A):oligo(dT)	dTTP	0.0032
		Poly(C):oligo(dG)	dGTP	>100
	HIV-1 nRT	Poly(A):oligo(dT)	dTTP	0.0048
	HIV-2 nRT	Poly(A):oligo(dT)	dTTP	0.0081
		Poly(C):oligo(dG)	dGTP	>100

[a] 50% Inhibitory concentration.

of HIV-1, irrespective of the source of the enzyme and the template-primer used. However, reflecting its specificity, this compound did not affect the activity of HIV-2 RT at concentrations up to 500 μM. In contrast to this, AZT-TP showed inhibitory effects on both the RTs when Poly(A):oligo(dT) was used as a primer. Another distinct difference between the two compounds can be seen in a comparison between their EC_{50} and IC_{50}: while the values of AZT-TP are comparable, there is a significant discrepancy between the two values of E-EBU-dM. Although we have no clear explanation at the present time, this observation may suggest the intervention of additional mechanisms of action for the HEPT analogues.

When the inhibition of HIV-1 RT by E-EBU-dM was further analyzed with varying concentrations of substrate and compound, double-reciprocal plots showed that this compound inhibits competitively with respect to dTTP and noncompetitively with respect to dGTP (data not shown).[63] The K_m of HIV-1 RT for dTTP and dGTP was 27 and 7.7 μM, respectively. The K_i of the enzyme for E-EBU-dM with dTTP as substrate was 0.42 μM. Some benzodiazepine (TIBO) derivatives (47)[50] reported recently behave quite similarly to HEPT derivatives. Considering their common specificity for HIV-1, the HEPT and TIBO derivatives might work through a similar mechanism of action. However, the TIBO derivatives apparently differ from the HEPT analogues in that they do not act as competitive inhibitors of HIV-1 RT with respect to dTTP.

CONCLUSIONS

The study of lithiation chemistry of uridine derivatives in this review has shown that, by selecting a suitable combination of protecting groups and a lithiating agent, a highly general and regioselective modification of the base moiety can be accomplished. Above all, the LDA lithiation strategy has disclosed an entry to the synthesis of a variety of 6-substituted uridines, which had heretofore been difficult to synthesize. Although 6-substituted pyrimidine nucleosides are generally considered not to be biologically active due to their opposite glycosylic conformation to that of naturally occurring nucleosides, 6-phenylthio- and 6-iodouridines turned out to have antileukemic activity in cell culture. This led us to synthesize a series of 6-phenylthio- and 6-iodo derivatives of 1-[(2-hydroxyethoxy)methyl]-uracils, among which a highly HIV-1 specific lead HEPT was discovered. We believe that several unique properties of HEPT may be attributable to its "glycosyl" conformation.

An extensive synthetic study to improve the original anti-HIV-1 activity of HEPT has furnished a series of highly potent analogues. Some of these analogues are as active as AZT and, at the same time, much less cytotoxic. Our recent investigation indicates that these compounds have a synergistic inhibitory activity against HIV-1 replication when used with alpha interferon or AZT.[64,65] These compounds are revealed to be less toxic to bone marrow cells than AZT,[66] but still effective against AZT-resistant strains of HIV-1.[62] Although several experiments including toxicology and pharmacokinetics[67] need to be carried out before their clinical studies are undertaken, we believe these compounds constitute highly promising candidates for the chemotherapy of AIDS (Acquired Immunodeficiency Syndrome).

ACKNOWLEDGEMENTS

Much of this review have been published in Japanese in the *J. Syn. Chem., Jpn.* 49: 1142-1155 (1991). The authors are grateful to The Society of Synthetic Organic Chemistry, Japan, for generous permission to reproduce this material here. The lithiation study of uridine derivatives has been done with the aid of many undergraduate and postgraduate students. The authors deeply appreciate their intense and persistent efforts. The study of HEPT and its analogues has been carried out as an academic collaboration with Dr. E. De Clercq, Rega Institute for Medical Research (Belgium), Dr. M. Baba, Fukushima Medical College (Japan), and Dr. R. T. Walker, Birmingham University (United Kingdom). The synthesis of a vast number of HEPT analogues has been accomplished by a team of skilled chemists at the Mitsubishi Kasei Corporation Research Center.

REFERENCES

1. H. Gilman and R.L. Bebb, Relative reactivities of organometallic compounds. XX. Metalation, *J. Am. Chem. Soc.* 61: 109 (1939).

2. G. Wittig and G. Fuhrmann, Über das Verhalten der halogenierten Anisole gegen Phenyl-lithium, Chem. Ber. 73: 1197 (1940).

3. H.G. Gschwend and H.R. Rodriguez, Heteroatom-facilitated lithiation, *in*: "Organic Reactions," W.G. Dauben, ed., John Wiley and Sons, Inc., New York (1979).

4. P. Beak and V. Snieckus, Directed lithiation of aromatic tertiary amides: an evolving synthetic methodology for polysubstituted aromatics. *Acc. Chem. Res.* 15: 306 (1982).

5. T. Ueda, Synthesis and reaction of pyrimidine nucleosides, *in*: "Chemistry of Nucleosides and Nucleotides," L.B. Townsend ed., Plenum Press, New York and London (1988)

6. T.L.V. Ulbricht, Syntheses with pyrimidine-lithium compounds, *Tetrahedron* 6: 225 (1959).

7. Later, this reaction was reexamined by using 5-bromo-2'-deoxy-3',5'-bis-*O*-(trimethylsilyl)uridine: L. Pichat, J. Godbillon, and M. Herbert, Lithiation par le *n*-butyllithium de bromo-5 uracile nucléosides silylés. Préparation de méthyl (^{14}C-5) et méthyl (^{14}C-6) uridine, de désoxy-2'-éthyl (^{14}C-5) uridine et désoxy-2'-éthyl (^{14}C-6) uridine, *Bull. Soc. Chim. Fr.* 2712 (1973).

8. L. Pichat, J. Godbillon, and M. Herbert, Lithiations directes par le *n*-butyllithium d'uracile et thymine nucléosides silylés. Méthylation de ces lithiens. Préparation de thymidine-(D-6), *Bull. Soc. Chim. Fr.* 2715 (1973).

9. For some reports dealing with halogen-lithium exchange reactions of pyrimidine nucleosides: a) R.F. Schinazi and W.H. Prusoff, Synthesis and properties of boron and silicon substituted uracil or 2'-deoxyuridine,*Tetrahedron Lett.* 4981 (1978). b) *Idem.* Synthesis of 5-(dihydroxyboryl)-2'-deoxyuridine and related boron-containing pyrimidines, *J. Org. Chem.* 50: 841 (1985). c) P.L. Coe, M.R. Harnden, A.S. Jones, S.A. Nobel, and R.T. Walker, Synthesis and antiviral properties of some 2'-deoxy-5-(fluoroalkenyl)uridines, *J. Med. Chem.* 25: 1329 (1982).

10. For a review on organometallic chemistry of nucleosides: D.E. Bergstrom, Organometallic intermediates in the synthesis of nucleoside analogs, *Nucleosides Nucleotides* 1: 1 (1982).

11. H. Tanaka, I. Nasu, and T. Miyasaka, Regiospecific C-alkylation of uridine: a simple route to 6-alkyluridines, *Tetrahedron Lett.* 4755 (1979).

12. Formation of 6-isopropyl-2',3'-*O*-isopropylideneuridine was also observed especially when this reaction was carried out on a large scale.

13. For an application of the LDA lithiation of 4: B. Wang, J.R. Kagel, T.S. Rao, and M.P. Mertes, A novel cyclization reaction of a C-6 substituted uridine analog: an entry to 5,6-dialkylated uridine derivatives, *Tetrahedron Lett.* 30: 7005 (1989).

14. H. Tanaka, I. Nasu, H. Hayakawa, and T. Miyasaka, A facile and regiospecific preparation of 6-alkyl-uridines, *Nucleic Acids Symp. Ser.* 8: 33 (1980).

15. For examples: a) W.V. Curran and R.B. Angier, The synthesis of orotidine and its isomer, 3-β-D-ribofuranosylorotic acid, and the methylation of orotic acid, *J. Org. Chem.* 31: 201 (1966). b) T. Ueda and H. Tanaka, Synthesis of 2-thiouridine and 6-methyl-3-(β-D-ribofuranosyl)-2-thiouracil, *Chem. Pharm. Bull.* 18: 1491 (1970). c) U. Niedballa and H. Vorbrüggen, A general synthesis of *N*-glycosides. II. Synthesis of 6-methyluridine, *J. Org. Chem.* 39: 3660 (1974).

16. Improved condensation methods for the preparation of 6-methylcytidine and 6-methyluridine have been reported: a) M.W. Winkley and R.K. Robins, The synthesis of 6-methylcytidine, 6-methyluridine, and related 6-methylpyrimidine nucleosides, *J. Org. Chem.* 33: 2822 (1968). b) H. Vorbrüggen, U. Niedballa, K. Krolikiewicz, B. Bennua, and G. Höfle, On the mechanism of nucleoside synthesis, *in*: "Chemistry and Biology of Nucleosides and Nucleotides," R.E. Harmon, R.K. Robins, and L.B. Townsend, eds., Academic Press, New York (1978).

17. H. Inoue and T. Ueda, Synthesis of 6-cyano- and 5-cyano-uridines and their derivatives, *Chem. Pharm. Bull.* 26: 2657 (1978).

18. A. Rosenthal and R.H. Dodd, Synthesis of 6-substituted uridines. Synthesis of (*R* or *S*)-6-(3-amino-2-carboxypropyl)uridine, *Carbohydr. Res.* 85: 15 (1980).

19. Photochemical addition of nucleophilic α-hydroxyalkyl radicals has been used for the synthesis of 6-alkyluridines: J.-L. Fourrey, G. Henry, and P. Jouin, New synthesis of 5,6-cyclothymine- and 6-alkyluracil-1-N-β-D-ribosides, *Tetrahedron Lett.* 951 (1979).

20. a) H. Tanaka, H. Hayakawa, and T. Miyasaka, Synthesis of 6-aroyluridine from uridine *via* regiospecific lithiation, *Chem. Pharm. Bull.* 29: 3565 (1981). b) *Idem*, A general entry to 6-substituted uridines from uridine *via* regiospecific lithiation, *Nucleic Acids Symp. Ser.* 10: 1 (1981). c) *Idem*, "Umpolung" of reactivity at the C-6 position of uridine: a simple and general method for 6-substituted uridines, *Tetrahedron* 38: 2635 (1982).

21. For the preparation and LDA lithiation of 5'-O-TBDMS derivative of **5**: H. Tanaka, H. Hayakawa, S. Shibata, K. Haraguchi, T. Miyasaka, and K. Hirota, Synthesis of 6-methyluridine *via* palladium-catalyzed cross-coupling between a 6-iodouridine derivative and tetramethylstannane, *Nucleosides Nucleotides*, 11: 319 (1982).

22. For an application of the LDA lithiation of **5**: Y. Yamamoto, T. Seko, F.G. Rong, and H. Nemoto, Boron-10 carriers for NCT. A new synthetic method *via* condensation with aldehydes having boronic moiety, *Tetrahedron Lett.* 30: 7191 (1989).

23. H. Tanaka, H. Hayakawa, and T. Miyasaka, unpublished result.

24. For an improved synthesis of 6-methyluridine by the use of **13**: see reference 21.

25. a) H. Ikehira, T. Matsuura, and I. Saito, Photochemistry of 6-halo- and 5,6-dihalouracils. A simple synthesis of fluorescent uracil derivatives, *Tetrahedron Lett.* 26: 1743 (1985). b) K. Satoh, H. Tanaka, A. Andoh, and T. Miyasaka, Photochemical synthesis of 6-aryluridines, *Nucleosides Nucleotides* 5: 461 (1986).

26. H. Tanaka, K. Haraguchi, Y. Koizumi, M. Fukui, and T. Miyasaka, Synthesis of 6-alkynylated uridines, *Can. J. Chem.* 64: 1560 (1986).

27. H. Tanaka, S. Iijima, A. Matsuda, H. Hayakawa, T. Miyasaka, and T. Ueda, The reaction of 6-phenylthiouridine with sulfur nucleophiles: a simple and regiospecific preparation of 6-alkylthiouridines and 6-alkylthiouridylic acids, *Chem. Pharm. Bull.* 31: 1222 (1983).

28. For the synthesis of 6-azidouridine from **13** and its photochemical reaction: a) H. Tanaka, H. Hayakawa, K. Haraguchi, and T. Miyasaka, Introduction of an azido group to the C-6 position of uridine by the use of a 6-iodouridine derivative, *Nucleosides Nucleotides* 4: 607 (1985). b) T. Miyasaka, H. Tanaka, K. Satoh, M. Imahashi, K. Yamaguchi, and Y. Iitaka, Photochemical intramolecular nitrene insertion of 6-azidouridine derivatives, *J. Heterocycl. Chem.* 24: 873 (1987).

29. a) A.M. Kapuler, C. Monny, and A.M. Michelson, The relationship of mono- and polynucleotide conformation to catalysis by polynucleotide phosphorylase, *Biochim. Biophys. Acta* 217: 18 (1970). b) M.P. Schweizer, J.T. Witkowski, and R.K. Robins, Nuclear magnetic resonance determination of syn and anti conformations in pyrimidine nucleosides, *J. Am. Chem. Soc.* 93: 277 (1971).

30. H. Tanaka, A. Matsuda, S. Iijima, H. Hayakawa, and T. Miyasaka, Synthesis and biological activities of 5-substituted 6-phenylthio and 6-iodouridines, a new class of antileukemic nucleosides, *Chem. Pharm. Bull.* 31: 2164 (1983).

31. An application of LDA lithiation for an attempted preparation of 5-fluoro-6-vinyluracil has been reported: G.-J. Koomen, H. Wolschrijn, J.B.M. van Rhijn, M.J. Wanner, and U.K. Pandit, Synthesis of 5-fluoro-3-(2-tetrahydrofuryl)-6-vinyluracil. Design of a 5-FU-derivative with extended conjugation, *Hetero*cycles 24: 939 (1986).

32. Synthesis of **18** based on the LDA lithiation of **5** has been reported: H. Ikehira, T. Matsuura, and I. Saito, Synthesis and reaction of 6-trialkylsilyl substituted uracils and uridines, *Tetrahedron Lett.* 25: 3325 (1984).

33. Practically no lithiation occurred by the use of LDA. The use of BuLi gave a complex mixture of products from which a small amount of 6-butyl derivative was isolated.

34. a) H. Tanaka, H. Hayakawa, K. Obi, and T. Miyasaka, Desulfurizative stannylation of uracil derivatives, *Tetrahedron Lett.* 26: 6229 (1985). b) *Idem*, Synthetic route to 5-substituted uridines *via* a new type of desulfurizative stannylation, *Tetrahedron* 42: 4187 (1986).

35. Just after our preliminary communication (reference 34a) was published, homolytic stannylation of vinylsulfones was reported, and an electron transfer mechanism is proposed for the reaction: Y. Watanabe, Y. Ueno, T. Araki, T. Endo, and M. Okawara, A novel homolytic substitution on vinylic carbon. A new route to vinyl stannane, *Tetrahedron Lett.* 27: 215 (1986).

36. H. Hayakawa, H. Tanaka, and T. Miyasaka, A novel and regiospecific route to 5-acyluridines *via* an amide α-anion derived from 5,6-dihydrouridine, *Chem. Pharm. Bull.* 30: 4589 (1982).

37. S. Hanessian, J. Kloss, and T. Sugawara, Stereocontrolled access to the octosyl acids: total synthesis of octosyl acid A, *J. Am. Chem. Soc.* 108: 2758 (1986).

38. W. Dumont, P. Bayet, and A. Krief, Cleavage of selenium compounds by butyllithium. A new, regiospecific, allyl alcohol synthon, *Angew. Chem. Int. Ed. Engl.* 13: 804 (1974).

39. H. Hayakawa, H. Tanaka, and T. Miyasaka, Lithiation of 5,6-dihydrouridine: a new route to 5-substituted uridines, *Tetrahedron* 41: 1675 (1985).

40. H. Tanaka, H. Hayakawa, S. Iijima, K. Haraguchi, and T. Miyasaka, Lithiation of 3',5'-*O*-(tetraisopropyl-disiloxan-1,3-diyl)-2'-deoxyuridine: synthesis of 6-substituted 2'-deoxyuridines, *Tetrahedron* 41: 861 (1985).

41. H. Hayakawa, H. Tanaka, Y. Maruyama, and T. Miyasaka, Regioselectivity in the lithiation of uridine. Effect of the sugar protecting groups, *Chem. Lett.* 1401 (1985).

42. J.A. Rabi and J.J. Fox, Facile base-catalyzed hydrogen isotope labeling at position 6 of pyrimidine nucleosides, *J. Am. Chem. Soc.* 95: 1628 (1973).

43. H. Hayakawa, H. Tanaka, K. Obi, M. Itoh, and T. Miyasaka, A simple and general entry to 5-substituted uridines based on the regioselective lithiation controlled by a protecting group in the sugar moiety, *Tetrahedron Lett.* 28: 87 (1987).

44. For an application of this method: B. Nawrot and A. Malkiewicz, The t RNA "wobble position" uridines. III. The synthesis of 5-[*S*-methoxycarbonyl(hydroxy)methyl]uridine and its 2-thio analogue, *Nucleosides Nucleotides* 8: 1499 (1989).

45. For an application of this method to the C-5 lithiation of 2'-deoxyuridine: R.W. Armstrong, S. Gupta, and F. Whelihan, Synthesis of 5-substituted nucleosides *via* the regioselective lithiation of 2'-deoxy-uridine, *Tetrahedron Lett.* 30: 2057 (1989).

46. For a review: C.K. Chu and S.J. Cutler, Chemistry and antiviral activities of acyclonucleosides, *J. Heterocycl. Chem.* 23: 289 (1986).

47. T. Miyasaka, H. Tanaka, M. Baba, H. Hayakawa, R.T. Walker, J. Balzarini, and E. De Clercq, A novel lead for specific anti-HIV-1 agents: 1-[(2-hydroxyethoxy)methyl]-6-(phenylthio)thymine, *J. Med. Chem.* 32: 2507 (1989).

48. H. Tanaka, T. Miyasaka, K. Sekiya, H. Takashima, M. Ubasawa, I. Nitta, M. Baba, R.T. Walker, and E. De Clercq, Synthesis of some analogues of 1-[(2-hydroxyethoxy)methyl]-6-(phenylthio)thymine (HEPT) which have different types of acyclic structures, *Nucleosides Nucleotides* 11: 447 (1992) .

49. M. Baba, H. Tanaka, E. De Clercq, R. Pauwels, J. Balzarini, D. Schols, H. Nakashima, C.-F. Perno, R.T. Walker, and T. Miyasaka, Highly specific inhibition of human immunodeficiency virus type 1 by a novel 6-substituted acyclouridine derivative, *Biochem. Biophys. Res. Commun.* 165: 1375 (1989).

50. a) R. Pauwels, K. Andries, J. Desmyter, D. Schols, M.J. Kukla, H.J. Breslin, A. Raeymaekers, J. Van Gelder, R. Woestenborghs, J. Heykants, K. Schellekens, M.A.C. Janssen, E. De Clercq, and P.A.J. Janssen, Potent and selective inhibition of HIV-1 replication *in vitro* by a novel series of TIBO derivatives, *Nature* 343: 470 (1990). b) M.J. Kukla, H.J. Breslin, R. Pauwels, C.L. Fedde, M. Miranda, M.K. Scott, R.G. Sherrill, A. Raeymaekers, J. Van Gelder, K. Andries, M.A.C. Janssen, E. De Clercq, and P.A.J. Janssen, Synthesis and anti-HIV-1 activity of 4,5,6,7-tetrahydro-5-methyl-imidazo[4,5,1-*jk*][1,4]benzodiazepin-2(1*H*)-one (TIBO) derivatives, *J. Med. Chem.* 34: 746 (1991).

51. V.J. Merluzzi, K.D. Hargrave, M. Labadia, K. Grozinger, M. Skoog, J.C. Wu, C.-K. Shih, K. Eckner, S. Hattox, J. Adams, A.S. Rosenthal, R. Faanes, R.J. Eckner, R.A. Koup, and J.L. Sullivan, Inhibition of HIV-1 replication by a nonnucleoside reverse transcriptase inhibitor, *Science* 250: 1411 (1990).

52. M.E. Goldman, J.A. O'Brien, T.L. Ruffing, A.M. Stern, S.L. Gaul, W.S. Saari, J.S. Wai, J. Hoffman, C.S. Rooney, J.C. Quintero, W.A. Schleif, E.A. Emini, and J.H. Nunberg, HIV-1 specific pyridinone RT inhibitors: preclinical biological characterization of two investigational new drugs, Abstract of 7th International Conference on AIDS, TU.A. 67 (1991).

53. H. Tanaka, M. Baba, H. Hayakawa, T. Sakamaki, T. Miyasaka, M. Ubasawa, H. Takashima, K. Sekiya, I. Nitta, S. Shigeta, R.T. Walker, J. Balzarini, and E. De Clercq, A new class of HIV-1 specific 6-substituted acyclouridine derivatives: synthesis and anti-HIV-1 activity of 5- or 6-substituted analogues of 1-[(2-hydroxyethoxy)methyl]-6-(phenylthio)thymine (HEPT), *J. Med. Chem.* 34: 349 (1991).

54. H. Tanaka, H. Takashima, M. Ubasawa, K. Sekiya, I. Nitta, M. Baba, S. Shigeta, R.T. Walker, E. De Clercq, and T. Miyasaka, Structure-activity relationships of 1-[(2-hydroxyethoxy)methyl]-6-(phenyl-thio)thymine analogues: effect of substitutions at the C-6 phenyl ring and at the C-5 position on anti-HIV-1 activity, *J. Med. Chem.* 35: 337 (1992).

55. H. Tanaka, M. Baba, M. Ubasawa, H. Takashima, K. Sekiya, I. Nitta, S. Shigeta, R.T. Walker, E. De Clercq, and T. Miyasaka, Synthesis and anti-HIV activity of 2-, 3-, and 4-substituted analogues of 1-[(2-hydroxyethoxy)methyl]-6-(phenylthio)thymine (HEPT), *J. Med. Chem.* 34: 1394 (1991).

56. M. Shimizu, H. Tanaka, H. Hayakawa, and T. Miyasaka, Dynamic aspects in the LDA lithiation of arabinofuranosyl derivative of 4-ethoxy-2-pyrimidinone: regioselective entry to both C-5 and C-6 substitutions, *Tetrahedron Lett.* 31: 1295 (1990).

57. C.B. Reese and A. Ubasawa, Reaction between 1-arenesulphonyl-3-nitro-1,2,4-triazoles and nucleoside base residues. Elucidation of the nature of side-reactions during oligonucleotide synthesis, *Tetrahedron Lett.* 21: 2265 (1980).

58. For a bibliography: R.J. Remy and J.A. Secrist III, Acyclic nucleosides other than Acyclovir as potential antiviral agents, *Nucleosides Nucleotides* 4: 411 (1985).

59. J.A. Fyfe, P.M. Keller, P.A. Furman, R.L. Miller, and G.B. Elion, Thymidine kinase from herpes simplex virus phosphorylates the new antiviral compound, 9-(2-hydroxyethoxymethyl)guanine, *J. Biol. Chem.* 253: 8721 (1978).

60. H. Tanaka, M. Baba, S. Saito, T. Miyasaka, H. Takashima, K. Sekiya, M. Ubasawa, I. Nitta, R.T. Walker, H. Nakashima, and E. De Clercq, Specific anti-HIV-1 "acyclonucleosides" which cannot be phosphorylated: synthesis of some deoxy analogues of 1-[(2-hydroxyethoxy)methyl]-6-(phenylthio)-thymine, *J. Med. Chem.* 34: 1508 (1991).

61. H. Tanaka, H. Takashima, M. Ubasawa, K. Sekiya, M. Baba, S. Shigeta, R.T. Walker, E. De Clercq, and T. Miyasaka, manuscript in preparation.

62. M. Baba, E. De Clercq, H. Tanaka, M. Ubasawa, H. Takashima, K. Sekiya, I. Nitta, K. Umezu, H. Nakashima, S. Mori, S. Shigeta, R.T. Walker, and T. Miyasaka, Potent and selective inhibition of human immunodeficiency virus type 1 (HIV-1) by 5-ethyl-6-phenylthiouracil derivatives through their interaction with the HIV-1 reverse transcriptase, *Proc. Natl. Acad. Sci. U.S.A.* 88: 2356 (1991).

63. M. Baba, E. De Clercq, H. Tanaka, M. Ubasawa, H. Takashima, K. Sekiya, I. Nitta, K. Umezu, R.T. Walker, S. Mori, M. Ito, S. Shigeta, and T. Miyasaka, Hihgly potent and selective inhibition of human immunodeficiency virus type 1 by a novel series of 6-substituted acyclouridine derivatives, *Mol. Pharmacol.* 39: 805 (1991).

64. M. Ito, M. Baba, S. Shigeta, E. De Clercq, R.T. Walker, H. Tanaka, and T. Miyasaka, Synergistic inhibition of human immunodeficiency virus type 1 (HIV-1) replication in vitro by 1-[(2-hydroxy-ethoxy)methyl]-6-(phenylthio)thymine (HEPT) and recombinant alpha interferon, *Antiviral Res.* 15: 323 (1991).

65. M. Baba, M. Ito, S. Shigeta, H. Tanaka, T. Miyasaka, M. Ubasawa, K. Umezu, R.T. Walker, and E. De Clercq, Synergistic inhibition of human immunodeficiency virus type 1 replication by 5-ethyl-1-ethoxymethyl-6-(phenylthio)uracil (E-EPU) and azidothymidine in vitro, *Antimicrob. Agents Chemother.* 35: 1430 (1991).

66. K. Umezu, M. Tsurufuji, S. Yuasa, N. Tsutsui, S. Yabuuchi, and Y. Ikeda, No toxic effects of a new anti-HIV-1 agent 6-substituted scyclouridine derivative on murine bone marrow cell growth, Abstracts of 7th International Conference on AIDS, W.A. 1010 (1991).

67. For a preliminary pharmacokinetics of certain HEPT analogues: M. Baba, E. De Clercq, S. Iida, H. Tanaka, I. Nitta, M. Ubasawa, H. Takashima, K. Sekiya, K. Umezu, H. Nakashima, S. Shigeta, R.T. Walker, and T. Miyasaka, Anti-human immunodeficiency virus type 1 activities and pharmacokinetics of novel 6-substituted acyclouridine derivatives, *Antimicrob. Agents Chemother.* 34: 2358 (1990).

A SYNTHESIS OF 2'-FLUORO- AND 3'-FLUORO-SUBSTITUTED PURINE

NUCLEOSIDES *via* A DIRECT APPROACH

Krzysztof W. Pankiewicz* and Kyoichi A. Watanabe

Sloan-Kettering Institute for Cancer Research
Memorial Sloan-Kettering Cancer Center
Sloan-Kettering Division
Graduate School of Medical Sciences
Cornell University
New York, NY 10021

INTRODUCTION

There are many examples of purine nucleosides fluorinated in the 2'- and 3'-positions of the sugar moiety (Figure 1) that are making an impact in chemistry, biochemistry, and drug development.

Recently, the synthesis and biological evaluation of acid stable 2'-ß-fluoro purine dideoxynucleosides as active agents against HIV have been reported[1-4]. Since it was established that introduction of fluorine to the 2'-ß-position of nucleosides decreases chemical and phosphorylase (PNP) catalyzed hydrolysis in general[5], and for purine nucleosides in particular[6] (1, Figure 1), these 2'-fluoro analogues were designed to overcome the chemical and metabolic instability of the parent purine dideoxynucleosides.

On the basis of the same rationale, the 2'-ß-fluoro analogues of fludarabine and other 2-halo derivatives of 9-ß-D-arabinofuranosyladenine have been synthesized and evaluated for their biological activity and metabolic stability[7,8].

Nucleosides 1 and 2, when incorporated into oligonucleotides, may improve their stability against nucleases. Such studies have rarely been reported, except for studies of the hammerhead ribozyme sequences containing 2'-deoxy-2'-fluoro-adenosine (2a) and guanosine (2b) have been reported very recently.[9,10] In order to synthesize such modified oligonucleotides, large amounts of 1 and 2 are needed. The 3'-fluorinated purine nucleosides 3 and 4 are also intriguing compounds that can be viewed as even closer analogues of ribonucleosides than cordycepin, although their utilization in biochemistry and medicinal chemistry has been only sporadic, probably due to their very limited synthetic accessibility. To our best knowledge, in the purine series only adenosine analogues 3a[11] and 4a[11-13] have been synthesized in rather low yields. A

Nucleosides and Nucleotides as Antitumor and Antiviral Agents,
Edited by C.K. Chu and D.C. Baker, Plenum Press, New York, 1993

55

communication on incorporation of **3a** into 2',5'-oligoadenylate 5'-triphosphate has been reported recently[14] as a potential interferon inducer, but the activity of the modified oligomer has not yet been published.

Although, all the above studies show that sugar-fluorinated purine nucleosides have attracted attention in recent years, the methods of their synthesis are neither simple nor efficient.

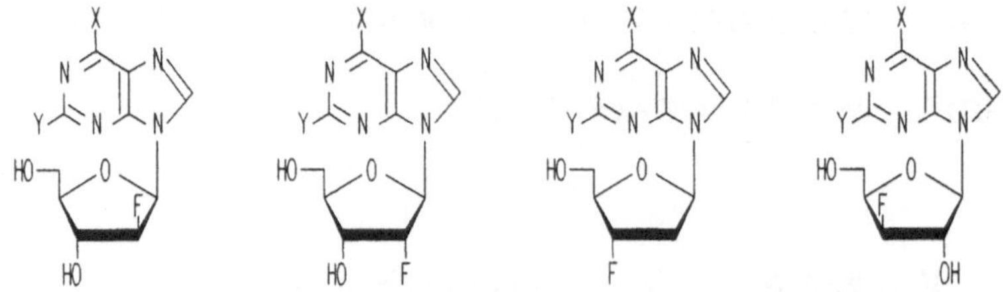

a: Adenine X = NH$_2$, Y = H; b: Guanine X = OH, Y = NH$_2$; c: Hypoxanthine X = OH, Y = H

Figure 1

In this paper we wish to report some new approaches to the synthesis of 2'-fluoro- and 3'-fluoro-substituted purine nucleosides, which may stimulate further progress in this area.

2'-ß-FLUORO-SUBSTITUTED PURINE NUCLEOSIDES via A DIRECT APPROACH.

9-(2-Deoxy-2-fluoro-ß-D-arabinofuranosyl)adenine (F-ara-A, **1a**) was originally synthesized in our laboratory in 1969 by condensation of 1,3-di-O-acetyl-5-O-benzyl-2-deoxy-2-fluoro-D-arabinose with 2,6-dichloropurine and further conversion of the dichloropurine into adenine[15].

The guanine and hypoxanthine analogues of F-ara-A, F-ara-G (**1b**) and F-ara-H (**1c**), were synthesized by us[16], as well as by others[17], by condensation of 3-O-acetyl-5-O-benzoyl-2-deoxy-2-fluoro-D-arabinofuranosyl bromide with N[6]-benzoyladenine, 6-chloropurine, 2,6-dichloropurine, or 2-acetamido-6-chloropurine, followed by appropriate functional group replacement and deprotection. It was found that the guanine nucleoside **1b** is selectively toxic to T-cells[16-18] while the hypoxanthine congener **1c** is a potent inhibitor of the growth of *Leishmania tropica* promastigotes[19].

The synthesis of the 9-(2-deoxy-2-fluoro-ß-D-ribofuranosyl)adenine (**2a**) was first reported some 15 years ago by Ranganathan[20] and by Ikehara *et al.*[21-23] This 2'-α-fluoro substituted purine nucleoside was originally prepared *via* nucleophilic displacement of the 2 -triflate function of the 3',5'-di-O-THP-*ara*-A derivative **5** (Scheme 1) with fluoride by treatment with tetrabutylammonium fluoride (TBAF), followed by deprotection. The fluorination reaction gave the desired product **6** in 50 % yield, as well as elimination product **7**, which was isolated in 30% yield. The fluoride ion is a poor nucleophile but a strong base, which frequently catalyses elimination reactions. Although successful displacement of the triflate **8** with nucleophiles such as LiN$_3$, KSAc, and NaOAc to give **9, 10, 11** has been described,[24] a similar synthesis of F-*ara*-A (**1a**) has not been reported.

So far, all 2'-ß-fluoro purine nucleosides were synthesized by condensation of a purine base and fluorinated sugar. Unfortunately, in contrast to the simple and efficient glycosylation of pyrimidines,[25-27] condensation of purines with 2-deoxy-2-fluoro-D-arabinofuranosyl halide is rather difficult[16]. In fact, some purine bases do not react with the sugar halide. For example, F-*ara*-A (1a) was recently prepared[1,16] by

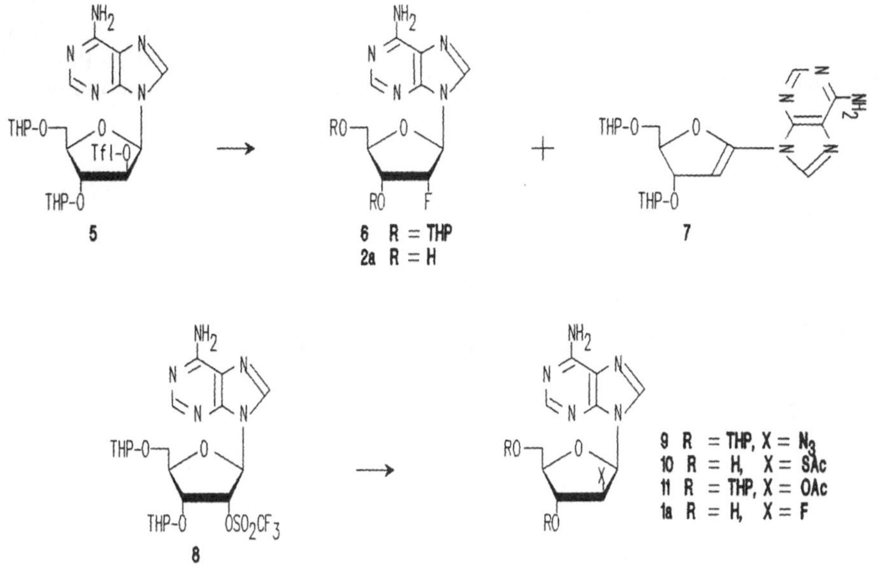

Scheme 1

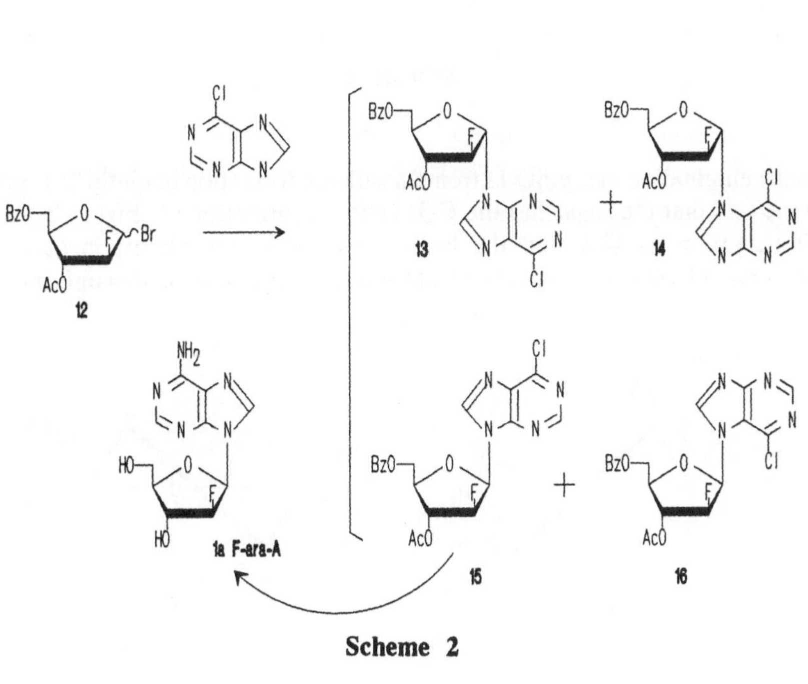

Scheme 2

condensation of 6-chloropurine (Scheme 2) with 3-O-acetyl-5-O-benzoyl-2-deoxy-2-fluoro-D-arabinofuranosyl bromide[28] (12) followed by conversion of the purine 15 into adenine. The glycosylation reaction afforded a mixture of four isomers (13-16) from which the desired isomer 15 was separated in very low yield.

Since TASF and DAST were reported as efficient reagents in fluorination of sugars[29] and nucleosides[11,29-31] we studied if a good leaving group at C-2' of a purine nucleoside such as 8 could be displaced with fluorine by these reagents. Thus, we synthesized 3',5'-di-O-benzyl-2'-O-triflyl-1-benzylinosine for TASF treatment. As depicted in Scheme 3 the 3',5'-O-(1,1,3,3-tetraisopropyl)disiloxanyl-1-benzylinosine (17) was treated with dihydropyran (DHP)/TsOH and then desilylated with Et₃NHF-THF to obtain 19. Benzylation of 19 followed by 2'-O-deprotection and triflylation afforded 20-22. Finally, when 22 was treated with TASF, only elimination products 23 and 24 were isolated. No traces of F-ara-H were detected in the reaction mixture.

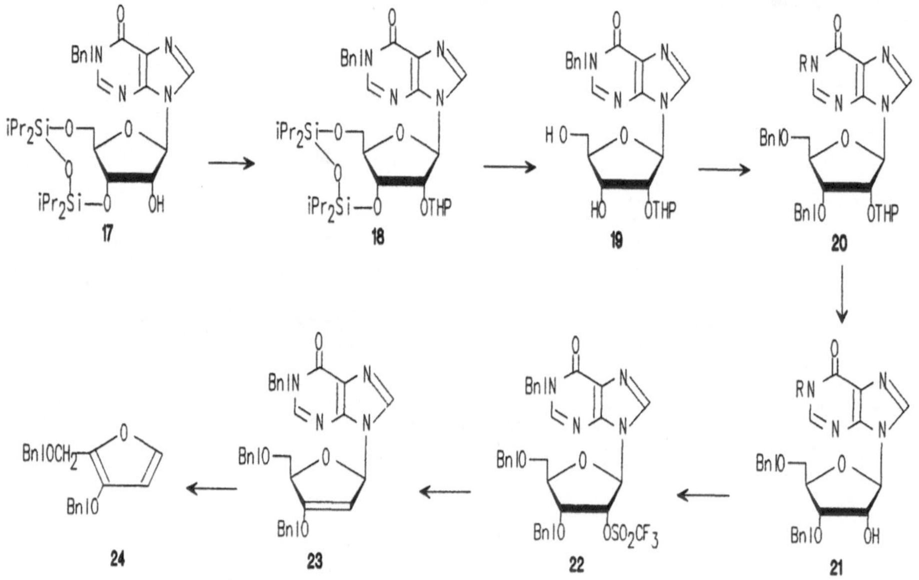

Scheme 3

Facile elimination of CF_3SO_3H from 22 with the formation of olefin 23 is probably due to the fact that the sugar has the C-3' endo conformation (A, Figure 2), in which the triflate group on C-2' and the hydrogen on C-3' are almost in trans diaxial configuration. Uesugi et al. reported[32] that the amount of C-3' endo conformer in 2'-

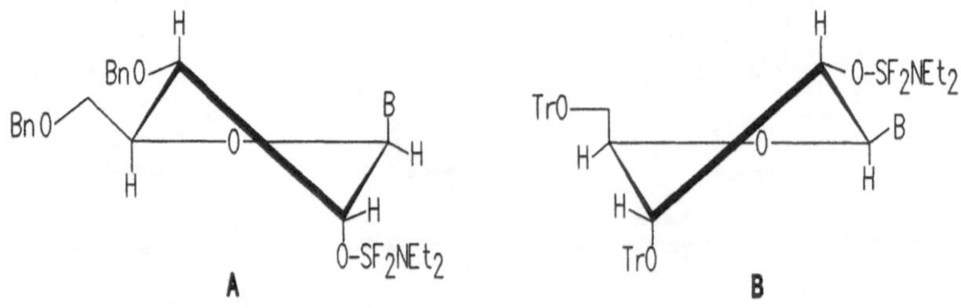

Figure 2

substituted adenosines increases linearly with the electronegativity of 2'-substituent. Thus, the presence of the electronegative 2'-triflate group may force **22** to assume the C-3' *endo* conformation, which favors elimination (**A**, Figure 2).

The furanose ring conformation may be shifted toward C-2' *endo* as in **B** using bulky protecting groups at C 5' and C 3' of the purine nucleoside 2'-triflate. It has been suggested that trityl group may force the furanose ring to assume an unfavorable conformation [33,34] for *trans* elimination. We therefore synthesized 3',5'-di-O-trityl-2'-O-triflyl-1-benzylinosine.

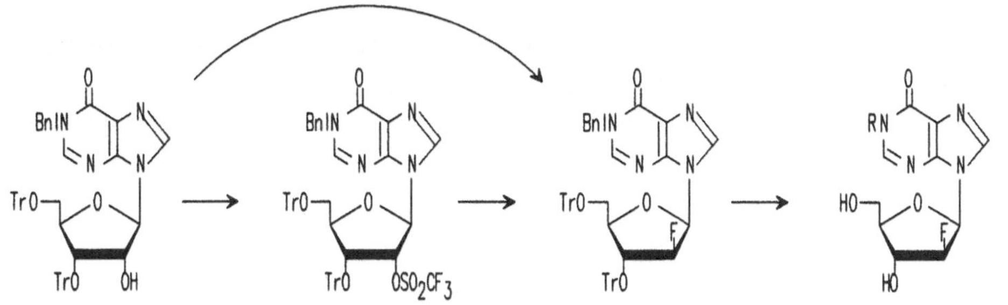

25 26 27 28 R = Bnl
1c F-ara-H, R = H

Scheme 4

Thus, 1-benzylinosine was directly tritylated with TrCl in pyridine, and the 2',5'- and 3',5'-regioisomers were separated on a silica gel column. The relatively large $J_{1,2}$ value of 7.1 Hz for the 3',5'-di-O-trityl derivative **26** versus small 1',2' coupling of 2.5 Hz for **22** indicates that trityl groups did cause the desired conformational change in **26** toward the C-2' *endo* conformer. When **26** was treated with TASF, the desired 2'-ß-fluorinated *arabino* nucleoside **27** was obtained in 30% yield. Treatment of nucleoside **25** with DAST afforded the same F-*ara*-H derivative **27** in 79% yield. Detritylation of **27** with $CF_3CO_2H/CHCl_3$[35], followed by hydrogenolysis with $Pd(OH)_2$, afforded a good yield of F-*ara*-H (**1c**).[36]

These results prompted us to investigate the applicability of the similar direct approach to the synthesis of F-*ara*-A (**1a**) and F-*ara*-G (**1b**).

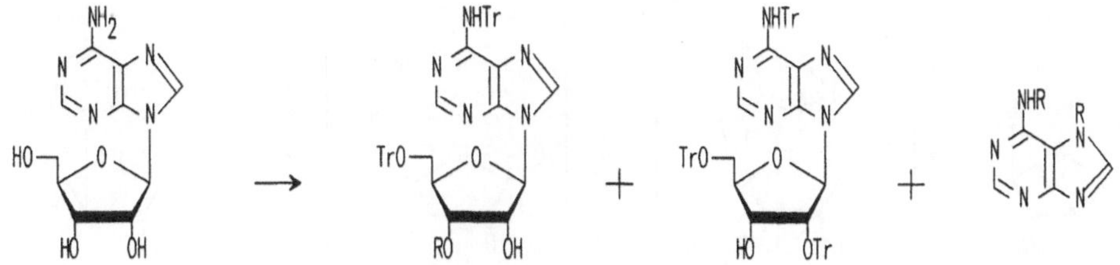

29 30 R = H
31 R = Tr
32 33 R = Tr
34 R = H

Scheme 5

Thus, treatment of adenosine (29, Scheme 5) with TrCl in pyridine containing 4-dimethylaminopyridine[37] (DMAP) at 80° C for 2-3 days afforded a mixture of 5'-O-trityl-N[6]-trityladenosine (30), 3',5'-di-O-trityl-N[6]-trityladenosine (31), 2',5'-di-O-trityl-N[6]-trityl-adenosine (32), and N[6],N[7]-ditrityladenine (33). These compounds were separated on a silica gel column, and the desired nucleoside 31, was obtained in crystalline form in 20% yield. Further tritylation of 30 afforded additional amount of 31 increasing the total yield of 31 up to 34%.

Treatment of the trityl derivative 31 with DAST gave a mixture of two components. A minor product 35 (Scheme 6) was isolated in 30% yield. Detritylation of 35 with CF$_3$CO$_2$H/CHCl$_3$ gave the desired F-ara-A (1a) in good yield. To our surprise, the major product was not the expected elimination product 38 but an isomer of 35 containing a fluorine atom. Its structure was established on the basis of spectral and elemental analyses as 2-deoxy-2-(N[6]-trityladenin-3-yl)-3,5-di-O-trityl-α-D-arabinofuranosyl fluoride (36).[38]

A plausible mechanism of the 31 to 36 conversion may be schematically illustrated

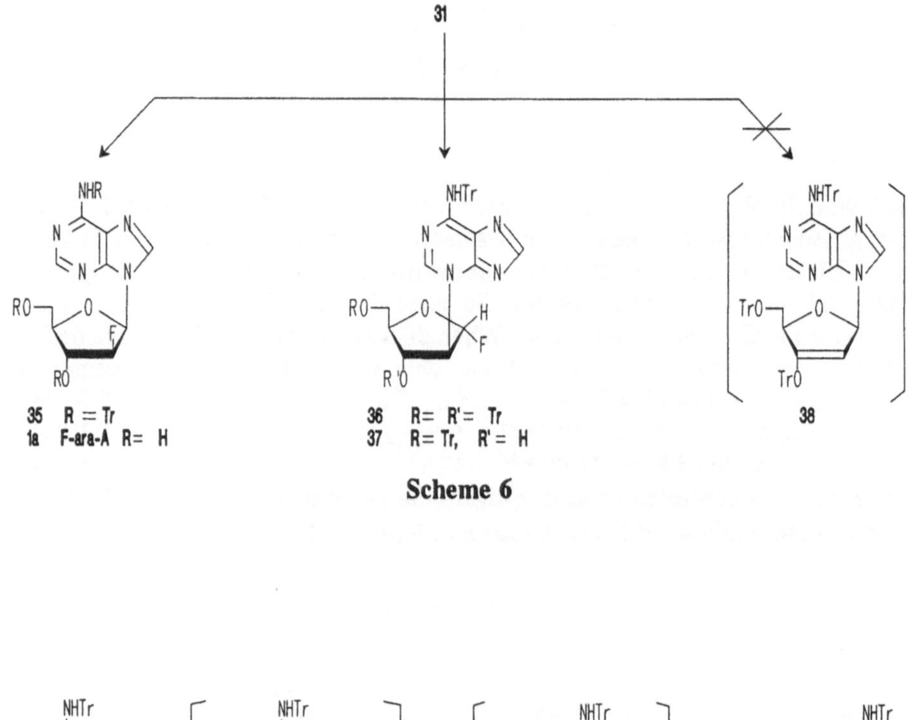

Scheme 6

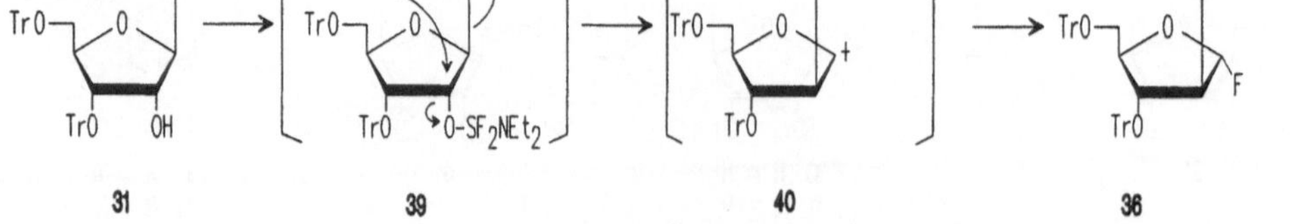

Scheme 7

60

as shown in Scheme 7.   Attack of N-3 of the aglycon on the activated C-2' of alkoxy(dimethyl-amino)sulfur difluoride intermediate [39] **39** may result in formation of carbocation **40** *via* cleavage of the glycosyl bond of **39**. Subsequently, fluoride ion attacks from less hindered α-side to give **36**. Alternatively, the fluoride ion may cleave the glycosyl bond of the intermediate **39** with inversion of configuration at the anomeric position. It is well established that adenine N-3 participation in sugar transformation in adenosine series results in undesired formation of intramolecular cyclization products between N-3 and the sugar moiety.[40-47] Recently. the attack of N-3 on C-3' of the sugar of 2'-deoxy-5'-O-monomethoxytrityladenosine upon DAST treatment was also reported[48].

A similar tritylation of guanosine and its derivatives **41 - 43** did not afford the desired product but led only to decomposition. We found, however, that 2-N-acetyl-6-O-(4-nitrophenylethyl)guanosine[49] (**46**) upon treatment with TrCl gave the desired mixture of 2',5'- and 3',5'-di-O-trityl derivatives (**47** and **48**). Again, fluorination of the 3',5'-di-O-trityl derivative **47** (Scheme 8) with DAST, followed by deprotection, afforded F-*ara*-G (**1b**).[50]

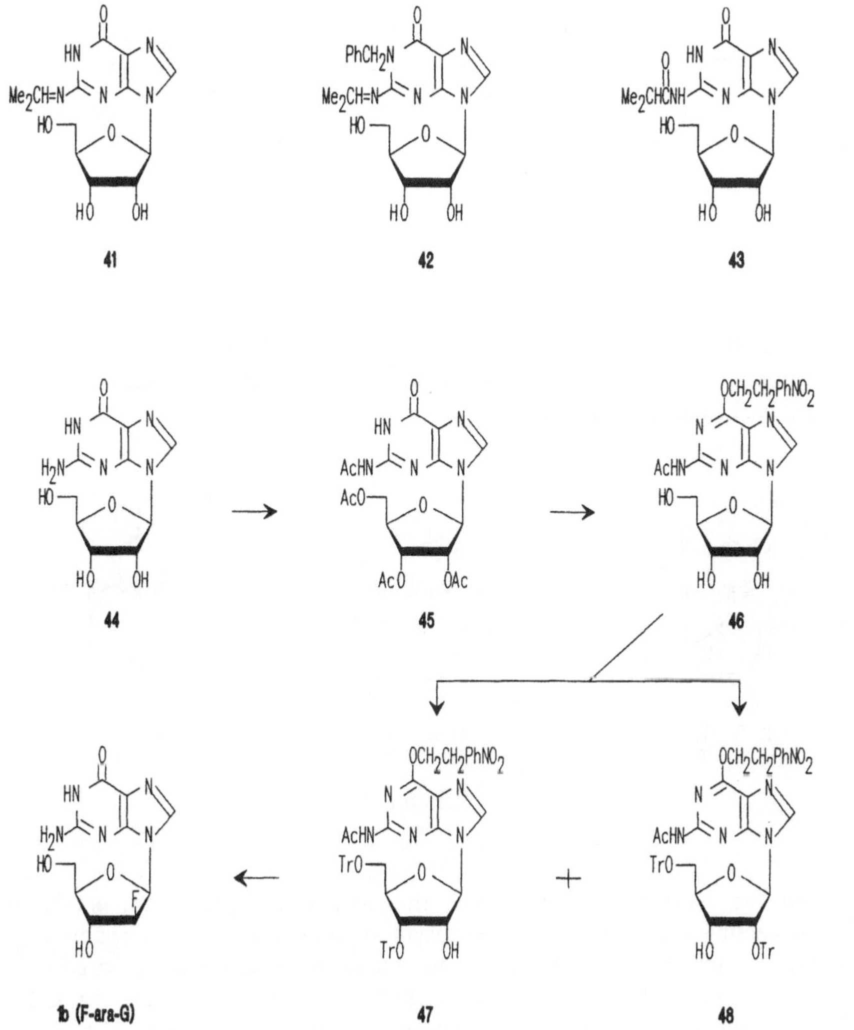

Scheme 8

2'-α-FLUORO-SUBSTITUTED PURINE NUCLEOSIDES.

In comparison to the direct introduction of fluorine at C-2' of pre formed purine nucleosides from the β-side of the sugar ring by nucleophilic reactions, substitution at C-2' from the α-side is much less difficult. It should be noted that conformational factors also play an important role in nucleophilic displacement of a leaving group at C-2' in the ß (*arabino*) configuration to form the 2'-α-fluoro (*ribo*) product. In this case, the C-2' *endo* conformation of the starting nucleoside is not advantageous for fluorination with DAST, since the anomeric proton and the leaving group at C-2' are in the *trans diaxial* disposition (Figure 3, C), making elimination imminent. Indeed, such elimination was reported when 2'-O-triflyl-3',5'-di-O-THP-*ara*-A (**5**, Scheme 1) was treated with Bu₄NF in THF[20]. Recently, Eckstein *et al.*[9] published the synthesis of **2a** *via* DAST treatment of N⁶,O³',O⁵'-tripixyl-9-ß-D-*ara*-A (**49**, Scheme 9), followed by deprotection. The reported yield was 43%. This is in agreement with our observation at the conversion of N⁶,O³',O⁵'-tritrityl-9-ß-D-*ara*-A (**50**) into the 2'-α-fluoro substituted derivative **52** with DAST[38]. The reaction gave a 2:1 mixture

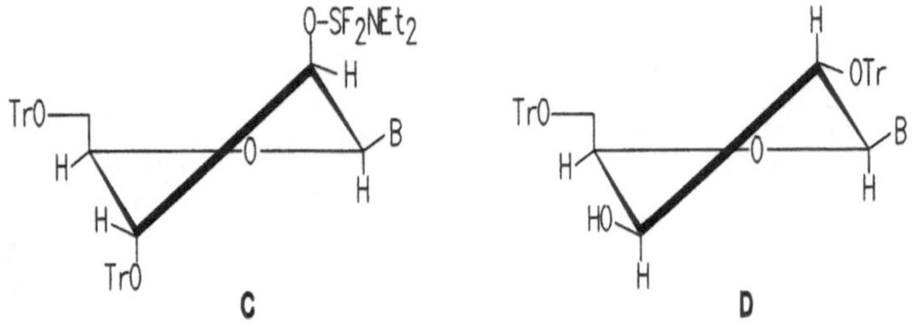

Figure 3

of the desired nucleoside **52** and elimination product **53**. Thus, it appears that conformational shift toward C-3' *endo* is important for efficient displacement of the 2'-ß (*arabino*) function with DAST.

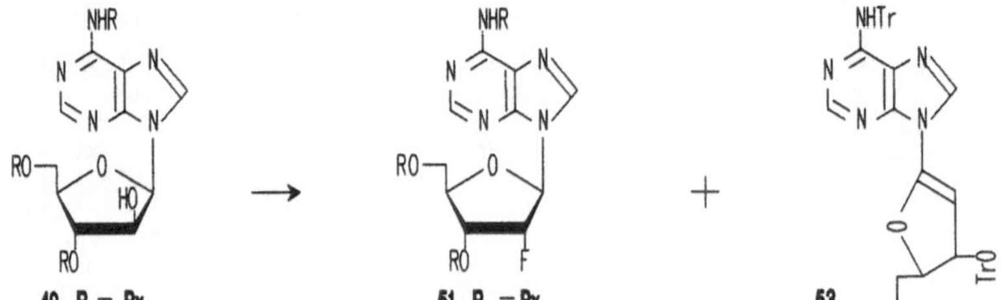

Scheme 9

For the synthesis of 2'-deoxy-2'-fluoroadenosine (**2a**) from *ara*-A, smaller 3',5'- protecting groups (such as methoxymethyl, benzyl, or methoxybenzyl groups) appear to be more favorable for *ara*-A to assume the presumably favored C-3' *endo* conformation than the trityl groups (which promote C-3' *endo* to C-2' *endo* conformational shift). In order to prepare 3',5'-protected *ara*-A, 3',5'-O-(1,1,3,3-tetraisopropyl)disiloxanyl-N⁶-benzoyladenosine (**54**, Scheme 10) was triflylated, and the triflate **55** was then treated with BzONa/HMPA to afford the 2'-O-benzoylated *ara*-A derivative **56**. Desilylation of **56** with Et₃NHF/THF gave O²',N⁶-di-benzoyl-*ara*-A (**57**). Surprisingly, when **57** was treated with MeOCH₂Cl, (MeO)₂CH₂, PhCH₂Cl or MeOPhCH₂Cl under various conditions, only extensive decomposition resulted. None of the desired ethers were obtained.

Nucleoside **57**, however, could be converted to the 3',5'-di-O-tetrahydropyranyl derivative **58**. Debenzoylation of **58** with NH₃/MeOH followed by DAST treatment afforded the desired 2'-α-fluoro substituted nucleoside **60** in 50% yield. Upon treatment with pyridinium Dowex 50, **60** was further deprotected to give a good yield of **2a**.

Although O²',N⁶-dibenzoyladenosine **57** resisted benzylation, O²',N⁶-ditrityl-adenosine (**61**, obtained by selective 5'-O-detritylation of **32**, Scheme 5) was smoothly converted into the desired 3',5'-di-O-benzyl derivative **62** by treatment with BnlCl/KOH (Scheme 11). Such differences in reactivity of **61** versus that of **57** may be explained by the steric effect of the bulky N⁶-trityl group of **61**, which in contrast to the N⁶-benzoyl group of **57**, may prevent alkylation at N⁷ of the adenine moiety and hence glycosyl cleavage. Detritylation of **62** with CF₃CO₂H/CHCl₃ gave 3',5'-di-O-

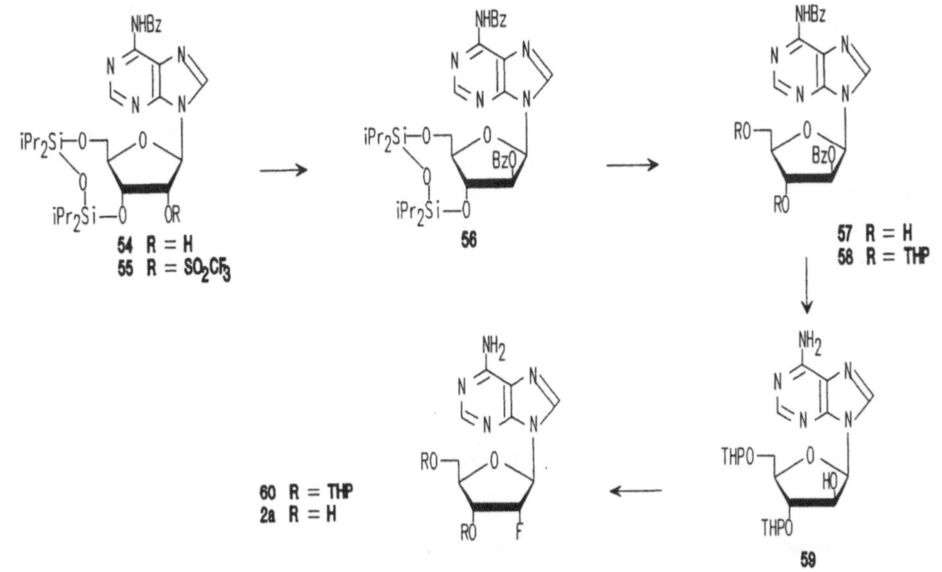

Scheme 10

benzyladenosine (**63**), which under treatment with triflyl chloride, followed by sodium acetate, afforded the *arabino* derivative **65**. After deacetylation, the product **66** was treated with DAST. The desired 2'-fluoroadenosine derivative **67** was isolated in 82% yield. No evidence for the formation of the corresponding elimination product was obtained. Hydrogenolytic debenzylation of **67** afforded the quantitative yield of 2'-deoxy-2'-fluoroadenosine (**2a**).

The importance of the use of non-participating 3',5'-protecting groups in the synthesis of **2a** may be illustrated in Scheme 12. Thus, 3',5'-O-tetraisopropyl-disiloxan-1,3-yl-adenosine[51] (**54**) was oxidized to the 2'-keto derivative by treatment with DMSO/Ac₂O, and the product (without isolation) was directly reduced with NaBH₄ to give the *arabino* nucleoside **68**.[52] Reaction of **68** with 3,4-dihydro-2H-pyran -

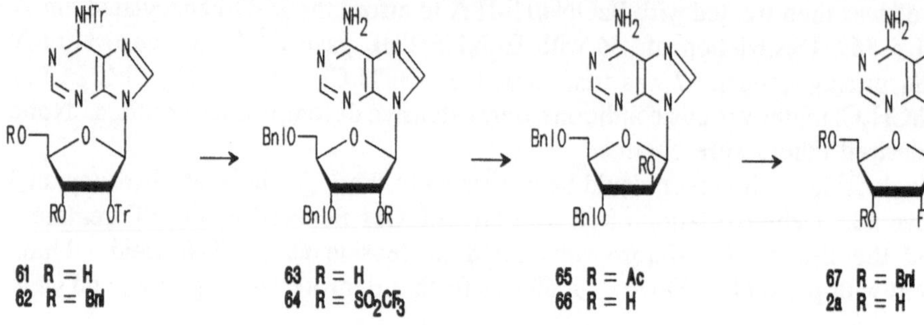

Scheme 11

afforded 9-(2-O-THP-3,5-O-TIPDS-ß-D-arabinofuranosyl)adenine (**69**), which without isolation, was desilylated with Et₃NHF/THF to give the 2'-O-THP derivative **70** in good yield. Compound **70** was benzoylated with benzoyl chloride in pyridine to the fully protected nucleoside **71** from which the THP group was removed by treatment with TsOH/MeOH to give **72**. Upon treatment of **72** with DAST, a mixture of less polar nucleoside **74** (35%) and two more inseparable polar compounds (50%) were obtained. Debenzoylation of **74** with NH₃/MeOH afforded the desired 2'-deoxy-2'-fluoroadenosine (**2a**), whereas similar treatment of the mixture of the more polar compounds gave a good yield of adenosine.

These results may be explained by the assumption that the 3'-O-benzoyl group of the alkoxy(dimethylamino)sulfur difluoride intermediate[39] (**73**) participates in the reaction giving the cation **75**, which was hydrolyzed on a silica gel to the mixture of 3',5'- and 2',5'-di-O-benzoyl-N⁶,N⁶-dibenzoyladenosine. These compounds were further hydrolyzed with NH₃/MeOH to give adenosine.

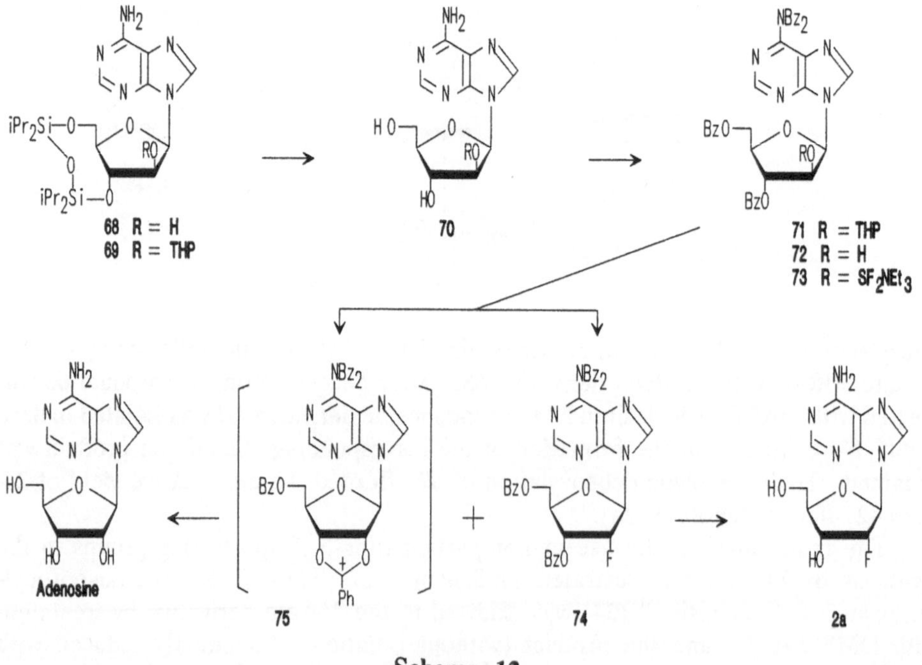

Scheme 12

64

3'-α-FLUORO-SUBSTITUTED PURINE NUCLEOSIDES.

It was reported that DAST treatment of 1,2:5,6-di-O-isopropylidene-α-D-allo-furanose 76 afforded 3-fluoro-1,2:5,6-di-O-isopropylidene-D-glucofuranose (77) in high yield. The fluorinated sugar 77 can also be prepared efficiently by treatment of 3-O-activated allofuranose with a fluorinating agent, such as TASF[54] or TBAF[55]. On the other hand, when 1,2:5,6-di-O-isopropylidene-α-D-glucofuranose (78) or 5-O-protected-1,2-O-isopropylidene-α-D-xylofuranose 80 or 81 was treated with DAST[52], only elimination occurred leading to formation of olefin 79 or 82 (Scheme 13). The difficulty in displacing the hydroxyl group in 76 and 78 has been explained to be due to the difficulty for the nucleophile to approach C 3 from the highly hindered side of the *cis*-fused bicyclo[3.3.0] system. Since there is no fused bicyclic system in the sugar moiety in 2',5'-di-O-substituted *xylo* nucleoside, *e.g.*, 87 (Scheme 14), conversion of 87 to the corresponding 3'-fluoro-*ribo* nucleoside by nucleophilic reaction may be possible.

Actually, the synthesis of 3 -fluoro-xylosyladenine and 3'-fluoroadenosines (3a and 4a, respectively) have recently been reported by Herdewijn *et al.*[11] These authors used 9-ß-D-xylofuranosyladenine as the starting material, which was tritylated in 50% yield to the corresponding 2',5'-di-O-triyl-N⁶-trityl derivative and then treated with DAST. This reaction yielded the expected 3'-α-fluoro compound in 78% yield. Detritylation, however, was rather disappointing (46%).[11]

We also synthesized 3'-fluoroadenosine. In our case, the starting material was again 2',5'-di-O-trityl-N⁶-trityladenosine 32 (Scheme 5), which had already served as the starting material for the synthesis of 2'-deoxy-2'-fluoroadenosine (2a, Scheme 11).

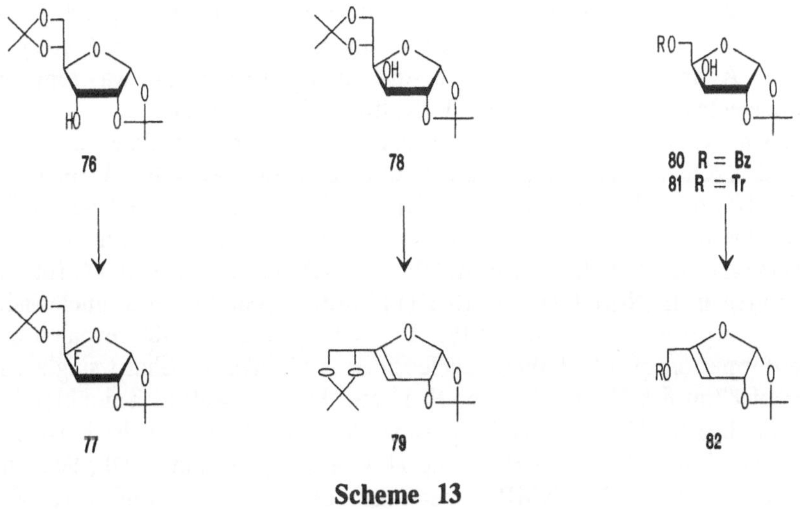

Scheme 13

In contrast to the readily conversion of di-O-isopropylidene-D-allose (76) into the 3-ß-fluoro derivative 77 with DAST, the same treatment of 32 did not afford the corresponding 3'-ß-fluoro nucleoside 84 (Scheme 14) but only decomposition products. Molecular modeling studies revealed[38] that the conformation of 32 is the same as that of 3',5'-di-O-trityl isomer 31, *i.e.*, C2'-*endo*. In this conformation H 2' and HO at C 3' are in the *trans quasi diaxial* disposition, favoring elimination. This conformational feature may account for decomposition of 32 upon treatment with DAST, although the participation of N³ of the adenine aglycon in reaction involving

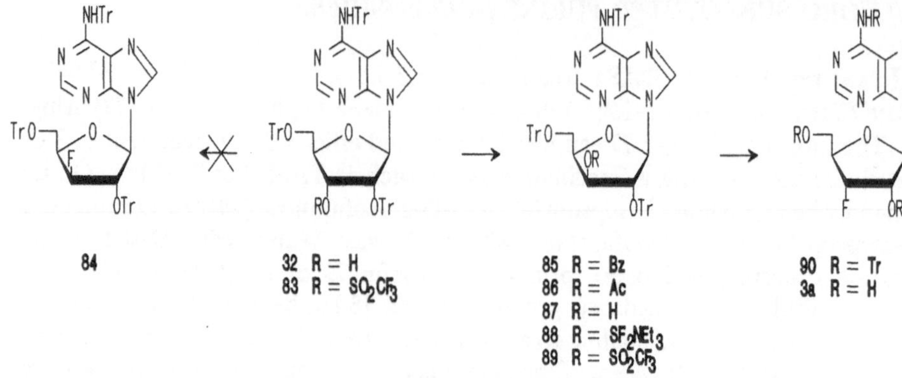

Scheme 14 (structures):

84

32 R = H
83 R = SO₂CF₃ → as SO_2CF_3

NHTr ... TrO ... F ... OTr

32 R = H
83 R = SO_2CF_3

85 R = Bz
86 R = Ac
87 R = H
88 R = SF_2NEt_3
89 R = SO_2CF_3

90 R = Tr
3a R = H

Scheme 14

C 3' cannot be ruled out since treatment of 2'-deoxy- 5'-O-monomethoxytrityl-adenosine with DAST afforded N^3,3'-cyclonucleoside in good yield.[48] 3'-Fluoro-xylosyladenine 4a, however, should be prepared readily by opening the epoxide ring of 2',3'-anhydroadenosine with fluoride.[12]

We expected that inversion of configuration at C 3' of the trityl derivative 32 (*ribo* to *xylo* conversion) will give nucleoside 87 (Scheme 14) with the same C 2'-*endo* conformation in which H 2' and the hydroxyl group at C 3' are in *cis*, and H 4' and 3'-OH are in *quasi diequatorial* disposition as in D (Figure 3), hence less amenable to elimination.

Our attempted conversion of 32 to 87 *via* oxidation-reduction[52] was unsuccessful. Although, the corresponding 3'-keto derivative was formed in good yield, the subsequent reduction with NaBH₄ yielded mainly the *ribo* derivative 32 (*ribo-xylo* ratio was 75:25). A similar reduction of 2',5'-di-O-trityl-3-ketouridine was reported[33] to be poorly selective reaction (*xylo-ribo* derivative ratio was 66:34).

As an alternative route, we synthesized the nucleoside triflate 83, which was treated with BzONa/HMPA to give the 3'-O-benzoyl derivative 85. Debenzoylation of 85 with NH₃/MeOH gave, however, a poor yield of deprotected nucleoside 87, apparently due to formation of elimination products and decomposition. Fortunately, the conversion of the triflate 83 with AcONa to 3'-O-acetyl derivative 86, followed by mild hydrolysis in Et₃N-MeOH-H₂O (3:3:1) afforded a good yield of nucleoside 87.

To our surprise, examination of the ¹H NMR spectra of 85-87 revealed that the sugar conformation of all of these nucleosides is C3'-*endo*. Sharp singlets in the spectrum of 87 at δ 5.45 and 4.57 for H 1' and H 2' as well as a doublet ($J_{3',4'}$ = 3.1 Hz) at δ 3.95 for H 3', respectively, show that the dihedral angles between H 1' and H 2', H 2' and H 3', and H 3' and H 4' are approximately 90°, 90°, and 50°, respectively. The ¹H NMR spectrum of the 2',5'-di-O-trityl-N^6-trityl-xylofuranosyladenine reported by Herdewijn *et al.*[11] shows the same chemical shifts for all resonances but a different couplings, *i.e.*, doublets for H 1' and H 2' and doublet of doublets for H 3' (coupling constants were not reported).

Since nucleoside 87 is in C 3'-*endo* conformation, elimination rather than nucleophilic substitution was expected in the reaction of 87 with DAST. The elimination was observed when 3'-O-benzoyl derivative 85 was deprotected with NH₃/MeOH. Unexpectedly, 81 afforded the desired 3'-α-fluoro-substituted nucleoside 90 in 70% yield upon treatment with DAST. In order to find if the conformation of the DAST activated intermediate 88 remains the same as that of nucleoside 87, the latter was converted into the triflate 89. The triflate 89 was viewed as an analogue of

88, since it contains strongly electronegative leaving group. The ^{1}H NMR spectrum of **89** did not show any conformational changes. Sharp singlets at δ 6.42, 4.38, and doublet at δ 3.61 ($J_{3',4'}$ = 2.0 Hz) confirmed the C3'-*endo* conformation of this triflate derivative.

In conclusion, it appears that conformational factors play the most important role in nucleophilic substitution with fluorine at C 2' of the sugar moiety but are not necessarily crucial in a similar nucleophilic displacement at C 3'. It was reported[48] that nucleophilic displacement of the 3'-hydroxyl group in 9-(2-deoxy-ß-D-xylofuranosyl)adenine with DAST was completed within 1 hour, whereas the same reaction with the 2'-hydroxy group of 9-(3-deoxy-ß-D-arabinofuranosyl)adenine required 14 hours. This reflects an unfavorable electronic factors at C 2' due to the proximity of the anomeric center. Also, steric factors are in favor of the S_N2 conversion of nucleoside **87** with DAST, since nucleophilic attack from the α-side of the nucleoside is much more efficient than that from the ß-side. Therefore, even unfavorable C 3 -*endo* conformation of **87** did not cause elimination and 3'-α-fluoro substituted derivative **90** was obtained in good yield. In contrast, displacement of the 2'-hydroxyl group in the *ribo* configuration, such as conversion of **25** to **27**, **31** to **35**, or **41** to **42**, is much more difficult, which requires the favorable C 2'-*endo* conformation in order to prevent undesired elimination.

Finally, the 3'-deoxy-3'-fluoroguanosine (**3b**) was prepared in a similar manner. As a starting material we used the 2-N-acetyl-6-O-(4-nitrophenylethyl)-2',5'-di-O-tritylguanosine (**48**, Scheme 15), which was produced as a by-product in our synthesis of F-*ara*-G (**1b**, Scheme 8). Thus, triflylation of **48** to **91**, followed by NaOAc/DMF treatment gave the 3'-O-acetyl derivative **92**. Treatment of **91** with sodium acetate in HMPA, however, afforded the 3'-O-acetyl derivative **93**, which did not contain the *p*-nitrophenethyl group. Apparently, the HMPA is basic enough to eliminate the *p*-nitrophenethyl group. Deacetylation of **93** with Et$_3$N-MeOH-H$_2$O afforded nucleoside **94** which, upon treatment with DAST, was smoothly converted into the protected 3'-deoxy-3'-fluoroadenosine **95**. Deprotection of **95** afforded **3b** in high yield.

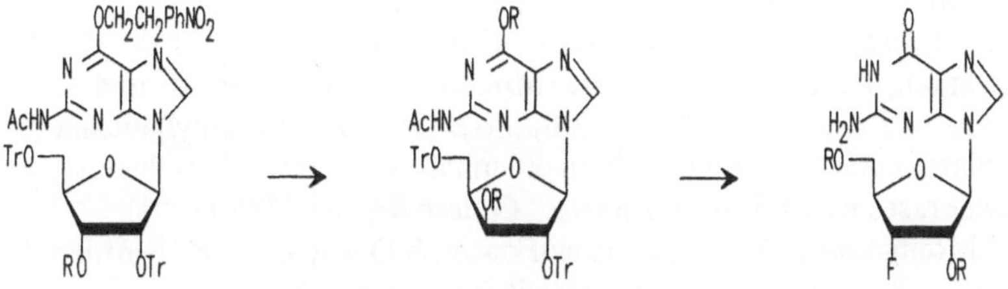

48 R = H	**92** R = Ac, R' = CH$_2$CH$_2$PhNO$_2$	**95** R = Tr
91 R = SO$_2$CF$_3$	**93** R = Ac, R' = H	**3b** R = H
	94 R = H, R' = H	

Scheme 15

ACKNOWLEDGMENTS

The authors acknowledge support of funds from the National Cancer Institute and from the National Institute of General Medical Sciences, National Institutes of Health, U.S. Department of Health and Human Services (Grants Nos. CA-08748, 18601, 33907

The authors acknowledge support of funds from the National Cancer Institute and from the National Institute of General Medical Sciences, National Institutes of Health, U.S. Department of Health and Human Services (Grants Nos. CA-08748, 18601, 33907 and GM 42010). We thank Dr. Barbara Nawrot, now at the Institute of Organic Chemistry, Technical University (Politechnika), Lodz, Poland and Dr. Jacek Krzeminski currently at the American Health Foundation, Valhalla, New York, for their invaluable contribution to the area of research described in this article.

REFERENCES

1.   V.E. Marquez, C.K-H. Tseng, H. Mitsuya, S. Aoki, J.A. Kelly, H. Ford,Jr., J.S. Roth, S. Broder, D.G. Johns, and J.S. Driscoll,   Acid-stable 2'-fluoro purine dideoxynucleosides as active agents against HIV. *J. Med. Chem.* 33:978 (1990).

2.   J.J. Barchi,Jr., V.E. Marquez, J.S. Driscoll, H. Ford,Jr., H. Mitsuya, T. Shirasaka, S. Aoki, and J.A. Kelly,   Potential anti-AIDS drugs. Lipophilic, adenosine deaminase-activated prodrugs. *J. Med. Chem.* 34:1647 (1991).

3.   R. Masood, G.S. Ahluwalia, D.A. Cooney, A. Fridland, V.E. Marquez, J.S. Driscoll, Z. Hao, H. Mitsuya, C-F. Perno, S. Broder, and D.G. Johns,   2'-Fluoro-2',3'-dideoxyarabinosyladenine: a metabolically stable analogue of the antiretroviral agent 2',3'-dideoxyadenosine. *Mol. Pharmacol.* 37:590 (1990).

4.   M.J.M. Hitchcock, K. Woods, H. De Boeck, and H-T. Ho,   Biochemical pharmacology of 2'-fluoro-2',3'-dideoxyarabinosyladenine, an inhibitor of HIV with improved metabolic and chemical stability over 2',3'-dideoxyadenosine. *Antiviral Chemother.* 1:319 (1990).

5.   T-C. Chou, A. Feinberg, A.J. Grant, P. Vidal, U. Reichman, K.A. Watanabe, J.J. Fox, F.S. Philips,   Pharmacological disposition and metabolic fate of 2'-fluoro-5-iodo-β-D-arabinofuranosylcytosine in mice and rats.   *Cancer Res.* 41:3336 (1981).

6.   J.D. Stoeckler, C.A. Bell, R.E. Parks, Jr., C.K. Chu, J.J. Fox, and M. Ikehara, C(2')-Substituted purine nucleoside analogs.   Interactions with adenosine deaminase and purine nucleoside phosphorylase and formation of analog nucleotides. *Biochem. Pharmacol.* 31:1723 (1982).

7.   W..B. Parker, S.C. Shaddix, C-H. Chang, E.L. White, L.M. Rose, R.W. Brockman, A. T. Shortnacy, J.A. Montgomery, J.A. Secrist, and L. Bennett, Effects of 2-chloro- 9-(2-deoxy-2-fluoro-β-D-arabinofuranosyl)adenine on K562 cellular metabolism and inhibition of human ribonucleotide reductase and DNA polymerases by its 5'-triphosphate. *Cancer Res.* 51:2386 (1991).

8.   J.A. Montgomery, A. T. Shortnacy-Fowler, S.D. Clyton, J.M. Riordan, and J.A. Secrist,   Synthesis and biologic activity of 2'-fluoro-2-halo derivatives of 9-β-D-arabinofuranosyladenine. *J. Med. Chem.* 35:397 (1992).

9.   D.B. Olsen, F. Benseler, H. Aurup, W.A. Pieken, F. Eckstein,   Study of a hammerhead ribozyme containing 2'-modified adenosine residues. *Biochemistry* 30:9735 (1991).

10.   D.M. Williams, W.A. Pieken, and F. Eckstein,   Function of specific 2'-hydroxyl groups of guanosines in a hammerhead ribozyme probed by 2'-modifications. *Proc. Nat. Acad. Sci., USA.* 89:918 (1992).

11.   P. Herdewijn, A. Van Aerschot, and L. Kerremans,   Synthesis of nucleosides fluorinated in the sugar moiety.   The application of diethylaminosulfur trifluoride to the synthesis of fluorinated nucleosides. *Nucleosides Nucleotides* 8:65 (1989).

12.   M.J. Robins, Y. Fouron, and R. Mengel,   Nucleic acid related compounds. 11.

13. J.A. Wright and N.F. Taylor, Fluorinated carbohydrates. Part XVIII. 9-(3-Deoxy-3-fluoro-β-D-xylofuranosyl)adenine and 9-(3-Deoxy-3-fluoro-α-D-xylofuranosyl)adenine *Carbohydr. Res.* 6:347 (1968).

14. T. Kovacs, K. Lesiak, A. Pabuccougolu, B. Uznanski, A. Van Aerschot, P. Herdewijn, P.F. Torrence, Can fluorine mimic 3'-hydroxyl groups of 2-5A trimers? Abstract MEDI. 31, 199th ACS National Meeting, Boston, Massachusetts, 1990.

15. J.A. Wright, N.F. Taylor, and J.J. Fox, Nucleosides LX. Fluorocarbohydrates. XXII. Synthesis of 2-deoxy-2-fluoro-D-arabinose and 9-(2-deoxy-2-fluoro-α- and -β-arabinofuranosyl)adenines. *J. Org. Chem.* 34:2632 (1969).

16. C.K. Chu, J. Matulic-Adamic, J-T. Huang, T-C. Chou, J.H. Burchenal, J.J. Fox, and K.A. Watanabe, Nucleosides 135. Synthesis of some 9-(2-deoxy-2-fluoro-β-D-arabinofuranosyl)-9*H*-purines and their biological activities. *Chem. Pharm. Bull.* 37:336 (1989).

17. J.A. Montgomery, A.T. Shortnacy, J.A. Carson, and J.A. Secrist, 9-(2-Deoxy-2-fluoro-β-D-arabinofuranosyl)guanine. A metabolically stable cytotoxic analogue of 2'-deoxyguanosine. *J. Med. Chem.* 29:2389 (1986).

18. T. Priebe, O. Kandil, M. Nakic, B. Fang Pang, and J.A. Nelson, Selective modulation of antibody response and natural killer cell activity by purine nucleoside analogues. *Cancer Res.* 48:4799 (1988).

19. P. Rainey, P.A. Nolan, L.B. Townsend, R.K. Robins, J.J. Fox, and D.V. Santi, Inosine analogs as anti-leishmanial agents. *Pharm. Res., J. Pharm. Soc.* 195 (1985).

20. R. Ranganatan, Modification of the 2'-position of purine nucleosides: Synthesis of 2'-α-substituted-2'-deoxyadenosine analogs. *Tetrahedron Lett.* 1291 (1977).

21. M. Ikehara, H. Miki, Studies of nucleosides and nucleotides. LXXXII. Cyclonucleosides (39). Synthesis and properties of 2'-halogeno-2'-deoxyadenosines. *Chem. Pharm. Bull.* 26:2449 (1978).

22. M. Ikehara, A. Hasegawa, J. Imura, Studies of nucleosides and nucleotides LXXXV. Purine cyclonucleosides 41. A new synthesis of 2'-deoxy-2-fluoro-adenosine by the use of tetrahydropyranyl protecting group. *J. Carbohyd. Nucleosides Nucleotides* 7:131 (1980).

23. S. Uesugi, T. Kaneyasu, J. Matsugi, M. Ikehara, Improved synthesis of 2'-fluoro-2'-deoxyadenosine and synthesis and carbon-13 NMR spectrum of its 3',5'-cyclic phosphate derivative. *Nucleosides Nucleotides* 2:373 (1983).

24. R. Ranganatan and D. Larwood, Facile conversion of adenosine into new 2'-substituted-2'-deoxy-arabinofuranosyladenine derivatives: Stereospecific synthesis of 2'-azido-2'-deoxy-, 2'-amino-2'-deoxy-, and 2'-mercapto-2'-deoxy-β-D-arabinofuranosyladenines. *Tetrahedron Lett.* 4341 (1978).

25. K.A. Watanabe, U. Reichman, K. Hirota, C. Lopez, and J.J. Fox, Nucleosides. 110. Synthesis and antiherpes virus activity of some 2'-fluoro-2'-deoxyarabino-furanosylpyrimidine nucleosides. *J. Med. Chem.* 22:21 (1979).

26. K.A. Watanabe, T-L. Su, R.S. Klein, C.K. Chu, A. Matsuda, M.W. Chun, C. Lopez, and J.J. Fox, Nucleosides. 123. Synthesis of antiviral nucleosides: 5-substituted 1-(2-deoxy-2-halogeno-β-D-arabinofuranosyl)cytosines and -uracils. Some structure-activity relationships. *J. Med. Chem.* 28:152 (1983).

27. T-L. Su, K.A. Watanabe, R.F. Schinazi, and J.J. Fox, Nucleosides. 136. Synthesis and antiviral effects of several 1-(2-deoxy-2-fluoro-β-D-arabinofuranosyl)-5-alkyluracils. Some structure-activity relationships. *J. Med. Chem.* 29:151 (1983).

28. U. Reichman, K.A. Watanabe, J.J. Fox, Nucleosides. 90. A practical synthesis

of 2-deoxy-2-fluoro-D-arabinofuranose derivatives. *Carbohydr. Res.* 42:233 (1975).

29. T. Tsuchiya, Chemistry and development of fluorinated carbohydrates. *Adv. Carbohydr. Chem. Biochem.* 48:91 (1990).

30. D.E. Bergstrom and D.J. Swartling, *in* "Fluorine Substituted Analogs of Nucleic Acid Components" in Fluorine Containing Molecules," J.F. Liebman, A. Greenberg, and W.R. Dolbier, VCH Publishers, New York, 1988, p. 259.

31. H. Hayakawa, F. Takai, H. Tanaka, T. Miyasaka, and K. Yamaguchi, Diethylaminosulfur trifluoride (DAST) as a fluorinating agent of pyrimidine nucleosides having a 2',3'-vicinal diol system. *Chem. Pharm. Bull.* 38:1136 (1990).

32. S. Uesugi, H. Niki, M. Ikehara, H. Iwahashi, and Y. Kyogoku, A linear relationships between electronegativity of 2'-substituents and conformation of adenine nucleosides. *Tetrahedron Lett.* 4073 (1979).

33. A.F. Cook and J.G. Moffatt, Carbodiimide-sulfoxide reactions. VI. Synthesis of 2'- and 3'-ketouridines. *J. Am. Chem. Soc.* 89:2697 (1967).

34. J. Thiem and D. Rash, Synthesis and Perkow reaction of uridine derivatives. *Nucleosides Nucleotides* 4:487 (1985).

35. M. McCoss and D.J. Cameron, Facile detritylation of nucleoside derivatives by using trifluoroacetic acid. *Carbohydr. Res.* 60:206 (1978).

36. J. Krzeminski, B. Nawrot, K.W. Pankiewicz, and K.A. Watanabe, Synthesis of 9-(2-deoxy-2-fluoro-β-D-arabinofuranosyl)hypoxanthine. The first direct introduction of fluorine in the 2'-"up" (arabino) configuration of preformed purine nucleosides. Studies directed toward the synthesis of 2'-deoxy-2'-substituted arabino nucleosides. 7. *Nucleosides Nucleotides* 10:781 (1991).

37. S.K. Chaudhary and O. Hernandez, A simplified procedure for the preparation of triphenylmethylethers. *Tetrahedron Lett.* 95 (1979).

38. K.W. Pankiewicz, J. Krzeminski, L.A. Ciszewski, W-Y. Ren, and K.A. Watanabe, A synthesis of 9-(2-deoxy-2-fluoro-β-D-arabinofuranosyl)-adenine and -hypoxanthine. An effect of C3'-endo and C2'-endo conformational shifts on the reaction course of 2'-hydroxyl group with DAST. *J. Org. Chem.* 57:553 (1992).

39. Detection of the alkoxy(dimethylamino)sulfur difluoride intermediate of 2'-deoxy-5'-O-dimethoxytritylcytidine has been recently reported by K. Agyei-Aye, S. Yan, A.K. Hebbler, and D.C. Baker, Preparation of 2,3'-anhydropyrimidine nucleosides using N,N-dimethylaminosulfur trifluoride (DAST). *Nucleosides Nucleotides* 8:327 (1989).

40. J.P.H. Verheyden and J.G. Moffatt, Halo sugar nucleosides. 1. Iodination of the primary hydroxyl groups of nucleosides with methyltriphenylphosphonium iodide. *J. Org. Chem.* 35:2319 (1970).

41. J.P.H. Verheyden and J.G. Moffatt, Halo sugar nucleosides. 2. Iodination of the secondary hydroxyl groups of nucleosides with methyltriphenylphosphonium iodide. *J. Org. Chem.* 35:2868 (1970).

42. W. Jahn, Synthese 5'-substituierte Adenosinderivate. *Chem. Ber.* 98:1705 (1965).

43. V.M. Clark, A.R. Todd, and J. Zussman, Nucleotides. Part VIII. Cyclonucleoside salts - rearrangement of some *p*-toluenesulphonyl nucleosides. *J. Chem. Soc.* 2952 (1951).

44. M.G. Stout, M.J. Robins, R.K. Olsen, and R.K. Robins, Purine nucleosides. XXV. The synthesis of certain derivatives of 5'-amino-5'-deoxy- and 5'-amino-2',5'-dideoxy-β-D-ribofuranosylpurines as purine nucleotide analogs. *J. Med. Chem.* 12:658 (1969).

45. S.D. Dimitrijevich, J.P.H. Verheyden, and J.G. Moffatt, Halo sugar nucleosides. 6. Synthesis of some 5'-deoxy-5'-iodo and 4',5'-unsaturated purine nucleosides. *J. Org. Chem.* 44:400 (1979);

46. T. Endo and J. Zemlicka, Oxidative transformations of minor components of nucleic acids. An anomalous reaction course of oxidation of N^6,N^6-dialkyladenosine and related compounds with *m*-chloroperbenzoic acid. *J. Org. Chem.* 53:1887 (1988).

47. M. MacCos, E.K. Ryu, R.S. White, and R.L. Last, A new synthetic use of nucleoside N^1-oxides. *J. Org. Chem.* 45:788 (1980).

48. P. Herdewijn, R. Pauwels, M. Baba, J. Balzarini, and E. DeClerq, Synthesis and anti-HIV activity of various 2'- and 3'-substituted 2',3'-dideoxyadenosine: a structure- activity analysis. *J. Med. Chem.*, 30:2131 (1987).

49. M.J. Gait, "*Oligonucleotide Synthesis A Practical Approach* ", p. 169, IRL Press, Oxford (1984).

50. K.W. Pankiewicz, J. Krzeminski, and K.A. Watanabe, A new synthesis of 9-(2-deoxy-2-fluoro-β-D-arabinofuranosyl)guanine. Abstract CARB 45, 203rd ACS National meeting, San Francisco, April, 1992.

51. W.T. Markiewicz, Tetraisopropyldisiloxane-1,3-diyl, a group for simultaneous protection of 3' and 5' hydroxy functions of nucleosides. *J. Chem. Res. Synop.* 24 (1979).

52. F. Hansske, D. Madej, and M.J. Robins, 2' and 3'-Ketonucleosides and their *arabino* and *xylo* reduction products. *Tetrahedron* 40:125 (1984).

53. T.J. Tewson and M.J. Welch, New approaches to the synthesis of 3-deoxy-3-fluoro-D-glucose. *J. Org. Chem.* 43:1090 (1978).

54. W.A. Szarek, G.W. Hay, and B. Doboszewski, Utility of tris(dimethylamino)-sulphonium difluorotrimethylsilicate (TASF) for the rapid synthesis of deoxyfluoro sugars. *J.Chem. Soc., Chem. Commun.*, 663 (1985).

55. A.B. Foster, R. Hems, and J.M. Weber, Fluorinated carbohydrates. Part 1. 3-Deoxy-3-fluoro-D-glucose. *Carbohydr. Res.* 5:292 (1967).

ALLENOLS DERIVED FROM NUCLEIC ACID BASES - A NEW CLASS OF ANTI-HIV AGENTS: CHEMISTRY AND BIOLOGICAL ACTIVITY

Jiri Zemlicka

Department of Chemistry, Michigan Cancer Foundation and Departments of
Internal Medicine and Biochemistry, Wayne State University School of
Medicine, Detroit, Michigan 48201

INTRODUCTION

The last 15 years have witnessed a revival of interest in analogues of nucleosides following the discoveries that several of such compounds exhibit powerful antiviral effects. In this context, two compounds are of particular significance. The first of such analogues is acyclovir (**1a**, Zovirax) which was developed in the late seventies[1] (Chart 1). Acyclovir (**1a**) is an analogue of 2'-deoxyguanosine of clinical importance as a drug for treatment of herpesvirus infections. The second compound is 3'-azido-3'-deoxythymidine (**3b**, AZT, zidovudine or Retrovir), an analogue of thymidine, and, until recently, the only approved drug for treatment of acquired immunodeficiency syndrome (AIDS). AZT (**3b**) was originally prepared in the sixties,[2] its antiretroviral potential was recognized in the seventies,[3] and it was developed as a therapeutic agent against AIDS in the eighties.[4] Drugs **1a** and **3b** are examples of two structurally different classes of nucleoside analogues. Acyclovir (**1a**) belongs to a series of open-chain compounds derived by cleavage of at least one C-C bond or deletion of one or more carbon atoms of the sugar ring, whereas AZT (**3b**) comprises an intact furanose moiety.

Both analogues can serve to illustrate an important problem frequently associated with design of new biologically active nucleoside analogues, namely, base selectivity. Thus, the antiviral effect of acyclovir (**1a**), a guanine derivative, is unique. The corresponding adenine analogue (**1c**), which is historically the first compound of this series,[5] is much less effective.[1] The thymine and cytosine derivatives **1b**, **1d** have virtually no antiviral activity.[6]

Nucleosides and Nucleotides as Antitumor and Antiviral Agents,
Edited by C.K. Chu and D.C. Baker, Plenum Press, New York, 1993

73

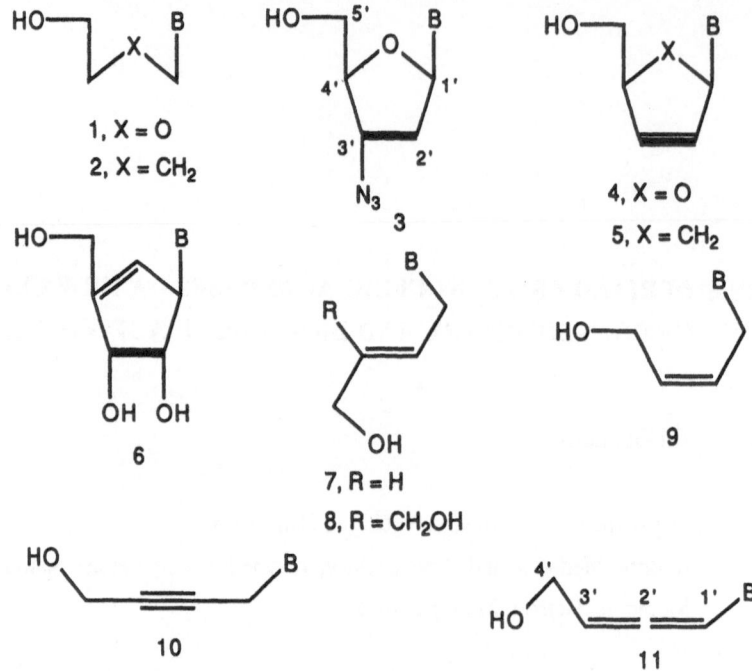

1, X = O
2, X = CH₂

3

4, X = O
5, X = CH₂

6

7, R = H
8, R = CH₂OH

9

10

11

Unless stated otherwise the formulas are as follows: Series a: B = guanin-N^9-yl (Gua), series b: B = thymin-N^1-yl (Thy), series c: B = adenin-N^9-yl (Ade), series d: B = cytosin-N^1-yl (Cyt), series e: B = 2-amino-6-chloropurin-N^9-yl, series f: N^2-dimethylaminomethyleneguanin-N^9-yl, series g: B = 5-methyl-cytosin-N^1-yl, series h: B = 2-aminoadenin-N^9-yl, series i: B = hypoxanthin-N^9-yl (Hyp)

Chart 1

Similarly, the presence of thymine is crucial for a high activity of AZT (**3b**). The adenine and cytosine derivatives **3c**, **3d** are substantially less active.[7] The guanine analogue **3a** has an anti-HIV effect,[8] although of lower magnitude than the parent compound **3b**. Thus, for any new series of analogues, it is wise to prepare all four compounds containing DNA bases unless there are reasons for a different rationale.

Over the years, many analogues belonging to both cyclic and acyclic series were investigated and excellent reviews[9,10] of the subject are available.

CHEMISTRY

1. Design

Unsaturated nucleosides, cyclic and acyclic, have also become a focus of much attention as antiviral and antitumor agents. Thus, analogues of pyrimidine 2'-deoxyribonucleosides **4b** and **4d** are effective against human immunodeficiency virus[11,12] (HIV), an etiologic agent of AIDS. Another anti-HIV agent, carbovir[13] (**5a**), is a carbocyclic mimic of 2'-deoxyguanosine. Antibiotic neplanocin A (**6c**) and the respective cytosine derivative **6d** are active[14] against a broad spectrum of viruses and tumors. By contrast, acyclic unsaturated analogues such as compounds **7a**, **7c**, **8a** and **8c** derived from the skeleton of neplanocin A (**6c**), or related butenols **9a**, **9c** and butynols **10a** and **10c** do not possess any significant biological activity.[15-22]

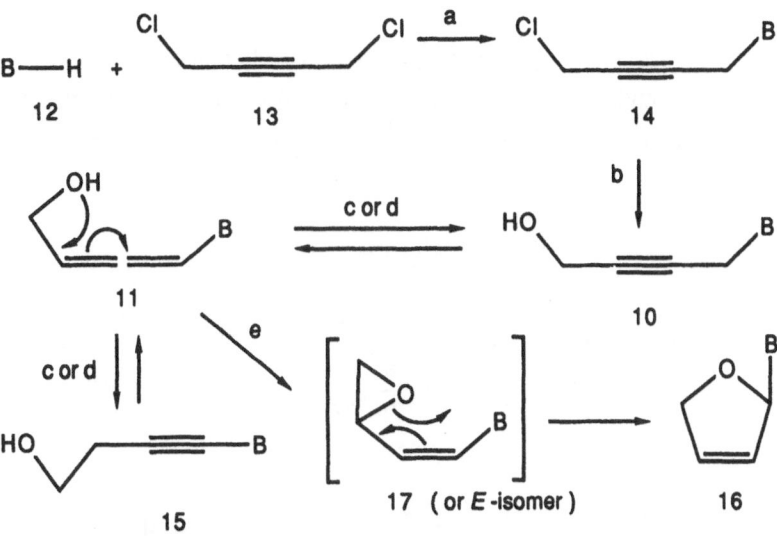

a. K_2CO_3, DMSO.   b. 0.1 M HCl, reflux.   c. 0.1 M NaOH, reflux.
d. t-BuOK, DMF.   e. 1 M NaOH, reflux.

Scheme 1

A common feature of many acyclic nucleoside analogues with antiviral activity, including acyclovir (**1a**), is the presence of a primary alcoholic group. The latter function and a nucleic acid base are essential for phosphorylation catalyzed by nucleoside kinases which is an essential part in mechanism of action of such analogues. It is likely that the presence of ether oxygen or similar function capable of mimicking the arrangement found in furanose ring of nucleosides plays also a role in the effectivity of nucleoside analogues. A distance between the hydroxymethyl group and heterocyclic moiety could also be critical.[23] The latter parameter is almost identical in an extended form of acyclovir (**1a**), the related 4-hydroxybutyl derivative **2a**, which exhibits antiherpetic activity,[24] and the 3'-

endo form of adenosine. Among the unsaturated analogues (compounds of type **7, 9, 10** and **11**), the distance between hydroxymethyl group and heterocycle in acyclovir (**1a**) is matched only in allenols **11**. Moreover, examinations of molecular models[23,25] indicate that the latter analogues resemble to a significant extent 2',3'-dideoxyribonucleosides which are effective anti-HIV agents.[26] Also, the allenic *sp* carbon is isoelectronic with ether oxygen of a ribofuranose ring. The rationale based on all these factors was considered strong enough to warrant synthesis and biological investigation of allenols (**11a** - **11d**) derived from nucleic acid bases.

2. Synthesis

Allenylamines containing a heterocyclic base and additional functionality were unknown before the inception of our study. The literature listed only a few *N*-allenyl heterocycles.[27,28] The first three derivatives of current interest, adenallene (**11c**), cytallene

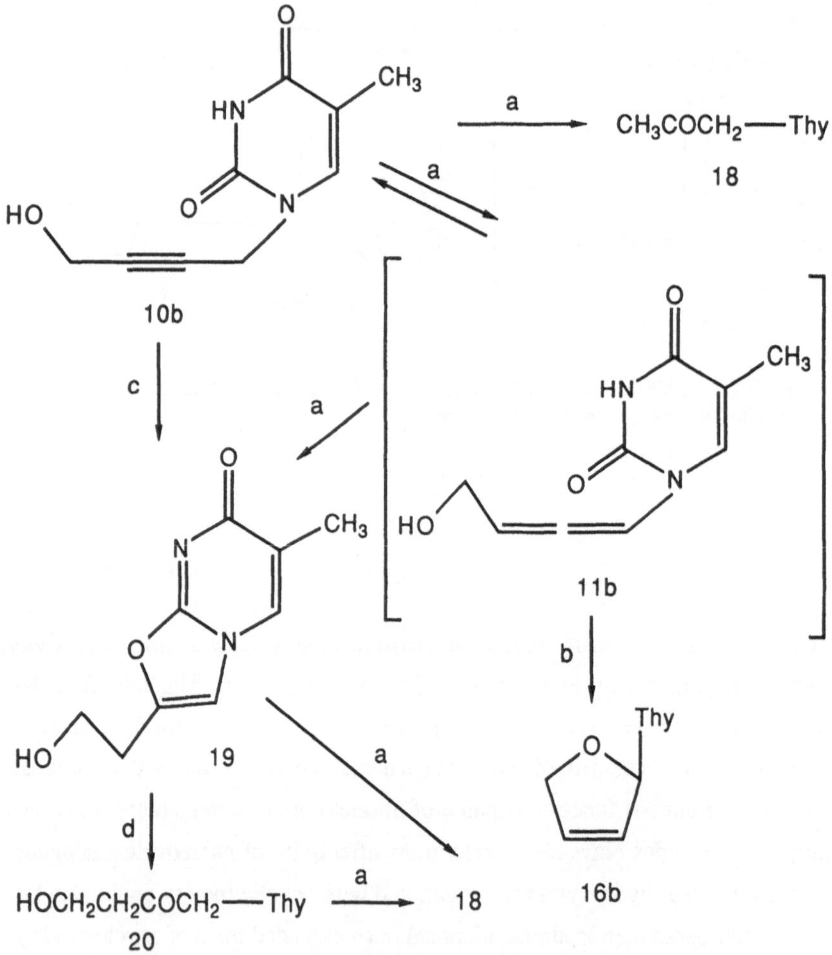

a. 0.1 M NaOH, reflux.   b. 1 M NaOH, reflux.   c. DBN, DMF, reflux.
d. 0.1 M NaOH, room temperature. DBN = 1,5-diazabicyclo[4.3.0]non-5-ene

Scheme 2

(11d) and guanallene (11a), were prepared[23,25] as described in Scheme 1. The nucleic acid bases, adenine (12c), cytosine (12d) or a suitable precursor (2-amino-6-chloropurine, 12e) were alkylated with 1,4-dichloro-2-butyne (13) using K_2CO_3 in DMSO to give the corresponding chlorobutynes 14c - 14e in 50% yield. Hydrolysis in refluxing 0.1 M HCl, accompanied in case of 14e by " restoration " of the guanine moiety, gave butynols 10a, 10c and 10d in 60 - 70% yields. In the last step, acetylene - allene isomerization led to mixtures containing allenols 11a, 11c, 11d as the major products, tentatively identified 1-butynols 15a, 15c and 15d and starting 2-butynols 10a, 10c and 10d. The 0.1 M NaOH (usually 30 min reflux) or, more conveniently, t-BuOK in DMF were used as base catalysts.

Allenols 11c and 11d were obtained in 30 - 50% yield after column chromatography and crystallization. The latter procedure is the only effective method found thus far for purification of cytallene (11d). Isomerization of butynol 10a gave 12% yield of product containing only ca. 35 % of guanallene (11a). It is then preferable to convert 10a into the N^2-dimethylaminomethylene derivative 10f, which was readily isomerized to allenol 11f (80% pure, 50% yield). Deprotection of 11f with ammonia in methanol gave guanallene (11a, 90% pure, 55% yield).

The procedures outlined above were also applied for synthesis of 5-methylcytallene (11g)[25] and 2-aminoadenallene (11h).[22] A simple deamination of 11h with adenosine deaminase[22] may become a method of choice for preparation of guanallene (11a) in high isomeric purity. A similar deamination of adenallene (11c) gave hypoxallene (11i) in 90% yield.[25] Recently reported[29-31] methods for synthesis of adenallene (11c) and cytallene (11d) are modifications of the approach described in Scheme 1. These methods make use of 4-acyloxy-1-chloro-2-butynes instead of 1,4-dichloro-2-butyne (13) as alkylating agents. Their advantages are higher yields and simplification of the synthesis of acetylenic precursors[19,20] such as butynols 10a and 10c.

Acetylene - allene isomerization of butynols 10a, 10c and 10d must be conducted under a strict control of reaction conditions to minimize side reactions. Thus, employing 1 M instead of 0.1 M NaOH led to formation of a different set of products - oxacyclopentenes 16a, 16c and 16d isolated in 17 - 38% yields (Scheme 1). It is likely that allenols 11a, 11c and 11d are intermediates in this transformation. Thus, in a stronger base (1 M NaOH), ionization of the primary alcohol can provide a more effective driving force for cyclization. The reaction is an interesting example of how an open-chain derivative can be transformed in a simple fashion to a nucleoside-like compound. The molecular models indicate that a direct cyclization of 11a, 11c or 11d is strongly disfavored. More probably, oxiranes 17a, 17c and 17d are formed first, and the formation of an oxacyclopentene skeleton of 16a, 16c and 16d is then the result of a [1,3] sigmatropic shift.[32]

The approach outlined in Scheme 1 was effective for butynols 10 containing a more basic heterocycle. The method failed in case of thymallene (11b) comprising an ionizable CONH function. The synthesis of butynol 10b following the procedures described in

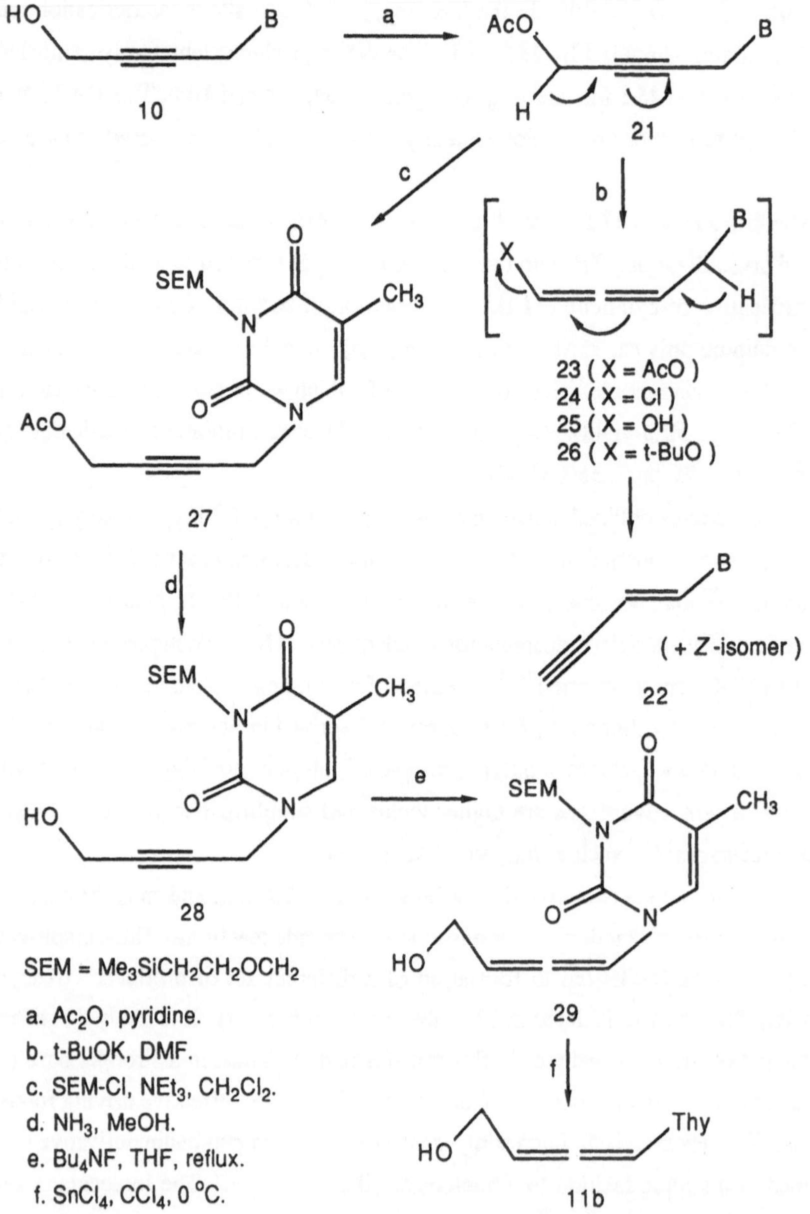

Scheme 3

SEM = Me₃SiCH₂CH₂OCH₂

a. Ac₂O, pyridine.
b. t-BuOK, DMF.
c. SEM-Cl, NEt₃, CH₂Cl₂.
d. NH₃, MeOH.
e. Bu₄NF, THF, reflux.
f. SnCl₄, CCl₄, 0 °C.

Scheme 1 was uneventful,[32] but attempted isomerization in 0.1 M NaOH gave only N^1-acetonylthymine (**18**) in 38% yield (Scheme 2). By contrast, reaction in 1 M NaOH gave the expected oxacyclopentene (**16b**) in 58 % yield, indicating possible involvement of thymallene (**11b**) as an intermediate. Events leading to the formation of **18** became more clear when butynol **10b** was transformed to oxazole **19** (45%) after refluxing with DBN in DMF for 30 min. The latter product was hydrolyzed with 0.1 M NaOH at room temperature to give β-hydroxyketone **20** (74%). Reflux of **19** or **20** in 0.1 M NaOH afforded N^1-acetonylthymine (**18**) in 66 and 55% yield, respectively.

It became obvious that synthesis of the elusive thymallene (**11b**) could not be accomplished by a procedure outlined in Scheme 1 without protection of one or both reactive functions in butynol **10b** (CH_2OH and CONH). Nevertheless, an attempted isomerization of acetate[32] **21b** with t-BuOK in DMF led only to E-butenyne **22b** (Scheme 3) in 50% yield. Apparently, the intermediate allene **23b** underwent an elimination of acetate. In a similar fashion, treatment of chlorobutyne **14c** with 1 M NaOH in 50% dioxane for 30 min. at room temperature afforded a 1:1 mixture of E- and Z-butenynes[33] **22c** in 90% yield. Thus, the presence of a poor leaving group (hydroxy function) in butynols **10a**, **10c** and **10d** may be favorable for a successful course of acetylene - allene isomerization described in Scheme 1. Nevertheless, such a " protective " role of hydroxy groups is limited as indicated by a reaction of butynol **10c** with t-BuOK in DMF at room temperature for 1.5 h to give a 3:1 mixture of E- and Z-butenynes **22c** in 55% yield. As mentioned above, milder conditions (30 min at -15°C) led to adenallene (**11c**, Scheme 1). Similarly, compound **10f** and t-BuOK in DMSO (room temperature, 20 min) furnished 4:1 mixture of E- and Z-butenynes **22f**. Formation of allenes **24c**, **25c** and **25f** can be invoked to account for the reaction course in all these transformations. In case of allenols **25c** and **25f**, which are tautomeric with the respective aldehydes, hemiacetals of t-BuOH and allenes **26c** and **26f** can be additional intermediates in the process.

Protection of acetate **21b** with the β-(trimethylsilyl)ethoxymethyl (SEM) group afforded compound **27**, which was readily deacetylated to give butynol **28**. An attempted isomerization of **27** or **28** with t-BuOK in DMF resulted in an expulsion of the SEM-substituted thymine. The key to a successful preparation of thymallene (**11b**) was the isomerization of **28** using Bu₄NF in THF. Under such conditions, the SEM group is surprisingly stable and the protected allenol **29** was obtained in 60% yield. In the last step, the SEM function was readily removed with $SnCl_4$ in CCl_4 to give thymallene (**11b**, 45%).

3. *R*- and *S*-Enantiomers of Adenallene

The allenols **11** described above belong to a class of 1,3-disubstituted allenes, and they are thus resolvable into *R*- and *S*-enantiomers. Because racemic adenallene (**11c**) and cytallene (**11d**) exhibit a strong anti-HIV activity[34] (*vide supra*) availability of optically pure enantiomers is of utmost importance. Initial experiments[35] indicated a possibility of

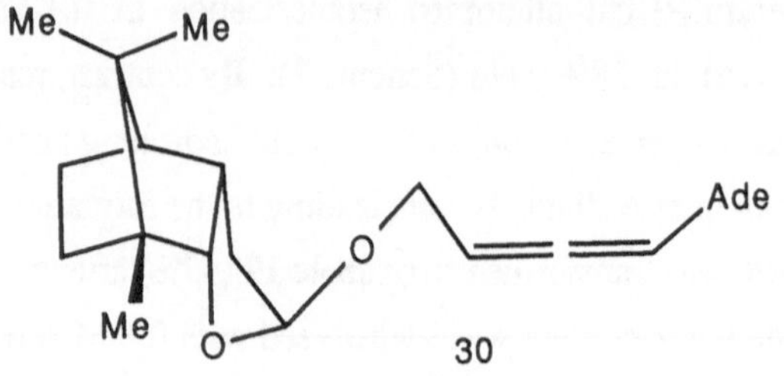

Formula 30

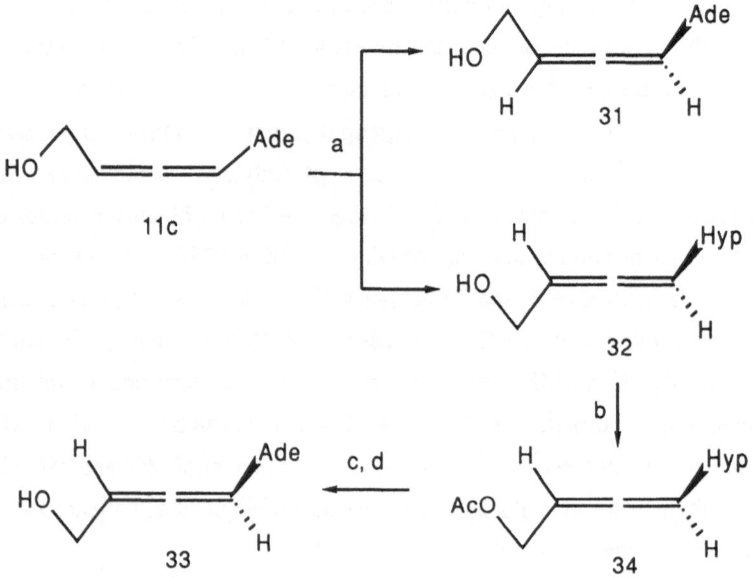

a. Adenosine deaminase, pH 7.5. c. (CF₃SO₂)₂O, pyridine, CH₂Cl₂.
b. Ac₂O, pyridine. d. NH₃, MeOH.

Scheme 4

resolution of racemic **11c** by a low-pressure chromatography on triacetylcellulose or HPLC on the same carrier anchored on silica gel (Chiralcel CA-1). However, the first method gives poor resolution, whereas the suitability of the second for preparation of larger quantities of enantiomers is questionable. Derivatization of adenallene (**11c**) with Noe's lactol[36] was also not fruitful. Although the reaction of the latter reagent with **11c** catalyzed by 4-toluenesulfonic acid in DMF gave the corresponding diastereoisomeric acetal[33] **30** in 35% yield, it was not possible to resolve the diastereoisomers by TLC or column chromatography on silica gel. As mentioned above, racemic adenallene (**11c**) is a substrate for adenosine deaminase, and it can be completely deaminated to give (±)-hypox-allene[25] (11i). Under controlled conditions, deamination of 11c afforded[35] (-)-aden-allene (31, 78%) and (+)-hypoxallene (32, 90%, Scheme 4).* The (-)-enantiomer 31, a direct product of the reaction, is the most active[35] anti-HIV component of the racemate 11c. (+)-Hypoxallene (32) was transformed to (+)-adenallene (33) as follows. Compound 32 was acetylated to give acetate 34 (82%). Activation with trifluoromethanesulfonic anhydride in CH_2Cl_2 - pyridine (4:1) and subsequent ammonolysis furnished an optically pure (+)- enantiomer 33 in 20% yield.

The absolute configuration of (-)-adenallene (31) is R as determined by X-ray crystallography.[35] The product of a controlled adenosine deaminase-catalyzed deamination is then S-(+)-hypoxallene (32) as predicted before.[25] The molecule of 31 in crystal forms a dimer involving an unusual pyrimidine - pyrimidine hydrogen bonding (Fig. 1) which is pseudosymmetric owing to a different rotameric composition of hydroxymethyl groups. The torsional angles of allenic bonds (91 and 97°) depart from an ideal 90° stipulated by sp hybridization. The conformations of adenine moieties in both molecules are anti-like.

Molecular models indicate an essentially free rotation of the heterocyclic base and hydroxymethyl group. Our previous molecular modeling studies[23,25] showed a similarity between S-adenallene (33) and the 3'-endo form of adenosine. Alternate studies suggested that adenallene[38] (presumably the S-enantiomer 33) and the S-cytallene[39] mimic the 3'-exo form and anti-periplanar (ap) rotamer of the $C_{4'}$-$C_{5'}$ bond in nucleosides. The X-ray data of 31 clearly show that differences in energy of preferred rotamers of the hydroxymethyl group are small (ca. 0.5 kcal). It is then concluded that more than one nucleoside conformation can be mimicked by R- or S-adenallene (31 and 33).

4. Physico-Chemical Properties

Adenallene (11c) and cytallene (11d) are unique nucleoside analogues in terms of structure and biological activity. As mentioned in the previous chapter, the crystal structure of compound 31 (Fig. 1) shows departures from the ideal 90° torsional angle associated with a system of cumulated double bonds in allenes. However, there are no significant de-viations from the standard sp - sp^2 carbon bond length found in allenes[40] (1.31 Å). Thus, the former values in both molecules of the dimer (Fig. 1) vary[35] between 1.27 - 1.33 Å. As other allene derivatives[41] R- and S-adenallene (31 and 33) have high values of optical rotation,[33] $[\alpha]_D$ ± 179 - 181° (c 0.2, methanol). Circular dichroism (CD) maxima of 31 and 33 (pH 7) are at 235 nm with molar ellipticities $[\Theta]$ ± 22,400 - 23,100. The R-

---

*Availability of enantiomers 31, 32 and 33 made possibile to re-evaluate the results with products 31 and 32 prepared by deamination of racemic adenallene (11c) at a 50 % conversion.[25,37] Although the CD spectra indicate a high optical purity, the optical rotations are low. The polarimeter quit shortly after these measurements were completed. There is then every reason to believe that optical purity of samples of 31 and 32 obtained under such arbitrary conditions of deamination was high.

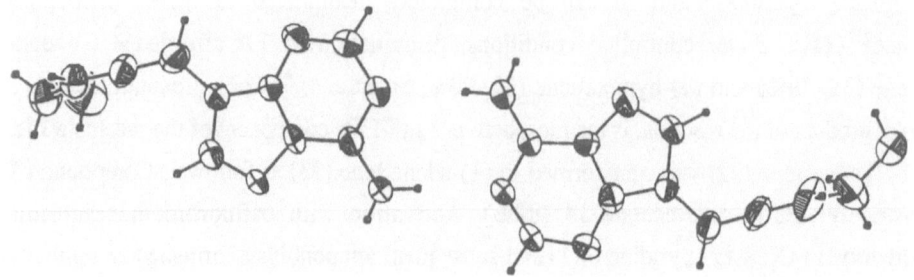

Fig. 1. ORTEP drawing of a crystal structure of *R*-(-)-adenallene (**31**).[35]

Note that two molecules of **31** form a pseudosymmetric dimer associated by pyrimidine - pyrimidine hydrogen bonds. Both molecules are rotamers with a different orientation of the hydroxymethyl groups.

Formulas 35 - 38

enantiomer **31** with a negative rotation exhibits Cotton effect of the same sign and, thus, it formally resembles adenosine.[42] Interestingly, IR spectra of allenols[22,25] **11a - 11i** do not exhibit, with an exception of thymallene[32] (**11b**), a well-defined band at 1960 cm^{-1}, which is typical for the asymmetric vibration of the C=C=C grouping of many allenes, including heterocyclic[27,28] allenylamines. It appears that the presence of a hydroxy group plays some, at present unspecified, role in this effect.

The ^{13}C chemical shifts of an *sp*-hybridized carbon of allenols **11b - 11d** and **11g - 11i** (193.82 - 195.93 ppm)[22,25,32] are located upfield from the usual range of 200 - 220 ppm found in allenes[43]. This may reflect a contribution of canonical form **36** to the resonance hybrid **35** ↔ **36**. Interestingly, allenes substituted with strongly electronegative groups represent another extreme[44] with chemical shifts as high as 227.4 ppm.

Mass spectra of allenols **11a - 11d** are of importance for structure confirmation and in case of the biologically active adenallene (**11c**) and cytallene (**11d**) as tools which may serve for detection and analysis of these potential drugs. The mass spectrum of adenallene (**11c**) indicates a considerable stability of allene grouping toward electron impact. This is especially apparent from a highly abundant fragment of *m/z* 176 (Table 1) resulting from

Table 1. Electron-impact mass spectra (EI-MS) of adenallene[25] (**11c**) and related[22,25,45] unsaturated analogues.[a]

Cmpd.	M	M - OH	M - HCN	M - CH$_2$OH	M - HCN - OH	M - HCN - CH$_2$OH	B[b]
11c	203 (73.4)	186 (100.0)	176 (81.2)	172 (13.2)	159 (28.0)	145 (14.8)	135 (33.9)
10c	203[c] (31.5)	186 (41.6)	176 (5.2)	172 (0.8)	159 (13.6)	145 (3.1)	135 (23.4)
7c	205 (3.4)	188 (1.8)	178 (0.0)	174 (100.0)	161 (0.0)	147 (3.9)	135 (10.9)
9c	205 (6.3)	188 (3.6)	178 (0.0)	174 (100.0)	161 (0.5)	147 (4.6)	135[d] (30.5)
37	205 (34.1)	188 (55.5)	178 (0.8)	174[e] (100.0)	161 (3.3)	147 (26.9)	135 (31.9)
38	205 (47.2)	188 (100.0)	178 (0.7)	174 (42.7)	161 (5.2)	147 (16.3)	135 (23.1)

[a]Ion / (relative intensity), M = molecular ion. [b]B = adenine. [c]202 (100.0, M - H). [d]136 (69.6, B + H). [e]175 (96.7, M - CH$_2$O).

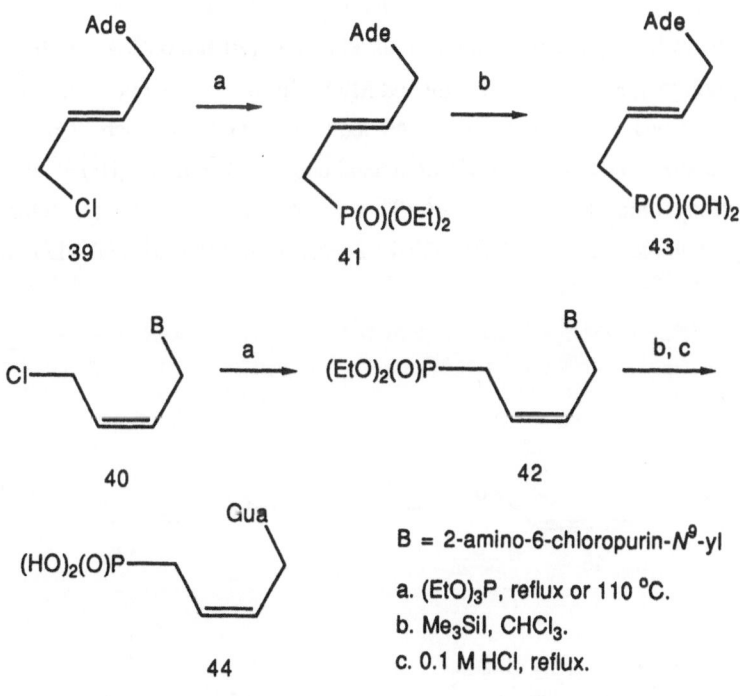

Scheme 5

cleavage of the heterocyclic moiety of molecular ion M. The former species (M - HCN) which was previously assigned to an isobaric M - CH=CH$_2$ fragment,[23] is of low intensity or altogether missing in the spectra of the unsaturated analogues **7c**, **9c** and **10c** including compounds[45] **37** and **38** containing a double bond conjugated with the adenine moiety. A low abundance of ion m/z 176 in butynol **10c** suggests that acetylene - allene isomerization does not occur to any appreciable extent under the conditions of electron impact. A lack of allene - acetylene interconversion was also found in the mass spectra of allenic hydrocarbons.[46]

5. Allenes and Acetylenes Derived from Nucleic Acid Bases as Synthons for Unsaturated Phosphonates

Allenes are reactive species which have found an extensive use in synthetic organic chemistry.[47] Some of the transformations of guanallene (**11a**), thymallene (**11b**), adenallene (**11c**) and cytallene (**11d**) were already discussed (Scheme 1 - 3). The chloro intermediates of allylic type **14a - 14i** could also undergo Michaelis-Arbuzov[48] and Michaelis-Becker[49] reactions. Since several nucleotide analogues, cyclic and acyclic phosphonates, exhibit high antiviral activities,[50] studies of unsaturated nucleotide analogues, especially phosphonates comprising an allenic system, are of interest.

The E- and Z-chloroalkenes **39** and **40** were transformed without difficulty to the corresponding phosphonates **41** and **42** by reaction with triethylphosphite in 61 and 68% yield, respectively[51,52] (Scheme 5). Dealkylation with Me$_3$SiI, accompanied in case of **42** by an acid hydrolysis, gave phosphonic acids **43** and **44** (70 and 65%, respectively).

By contrast, Michaelis-Arbuzov and Michaelis-Becker reactions of chlorobutynes **10c** and **10e**, as well as chloroadenallene **45**, led to anomalous products. Compound **45** is readily accessible by chlorination[25] of adenallene (**11c**) with (C$_6$H$_5$)$_3$P and CCl$_4$ or methanesulfonyl chloride in pyridine[52] (93 and 40% yield, respectively, Scheme 6). Interestingly, attempted reaction of **11c** with 4-toluenesulfonyl chloride and NEt$_3$ in CH$_2$Cl$_2$

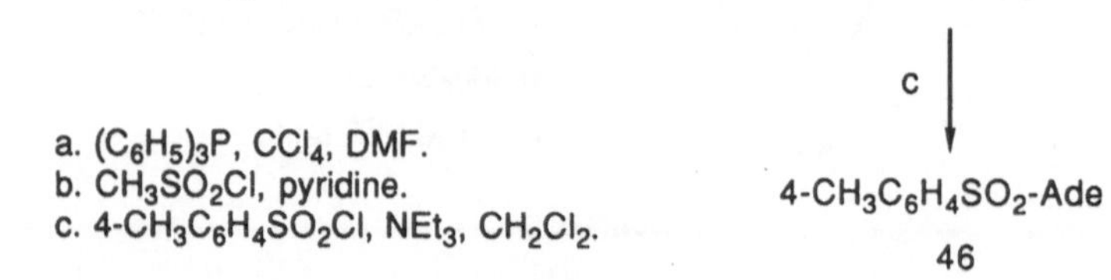

a. (C$_6$H$_5$)$_3$P, CCl$_4$, DMF.
b. CH$_3$SO$_2$Cl, pyridine.
c. 4-CH$_3$C$_6$H$_4$SO$_2$Cl, NEt$_3$, CH$_2$Cl$_2$.

4-CH$_3$C$_6$H$_4$SO$_2$-Ade

46

Scheme 6

led only to elimination of the allene moiety and formation of N^9-(4-toluenesulfonyl)adenine (**46**) in 50% yield.[52]

Chloroadenallene **45** gave, after refluxing with triethylphosphite[52] for 30 min, *E*-2'-phosphonate **47** in 30% yield (Scheme 7). A routine dealkylation afforded phosphonic acid **48** (60%). Thus, the usual reaction at the reactive allylic terminal was suppressed in favor of a nucleophilic attack at the *sp* carbon atom. The mechanism may include an intramolecular dealkylation as indicated in formula **49**, although an intermolecular reaction is also possible.

Even more surprising is the Michaelis-Becker reaction of allene **45** with sodium diethyl phosphite in THF - HMPA mixture which gave *E,E*-4'-phosphonate **50** in 23% yield (Scheme 8). Dealkylation of **50** furnished phosphonic acid **51**. The formation of 2'-phosphonate **47** by a rearrangement of intermediate **52** was expected according to the lite-

a. (EtO)$_3$P, reflux.
b. Me$_3$SiI, CHCl$_3$.

Scheme 7

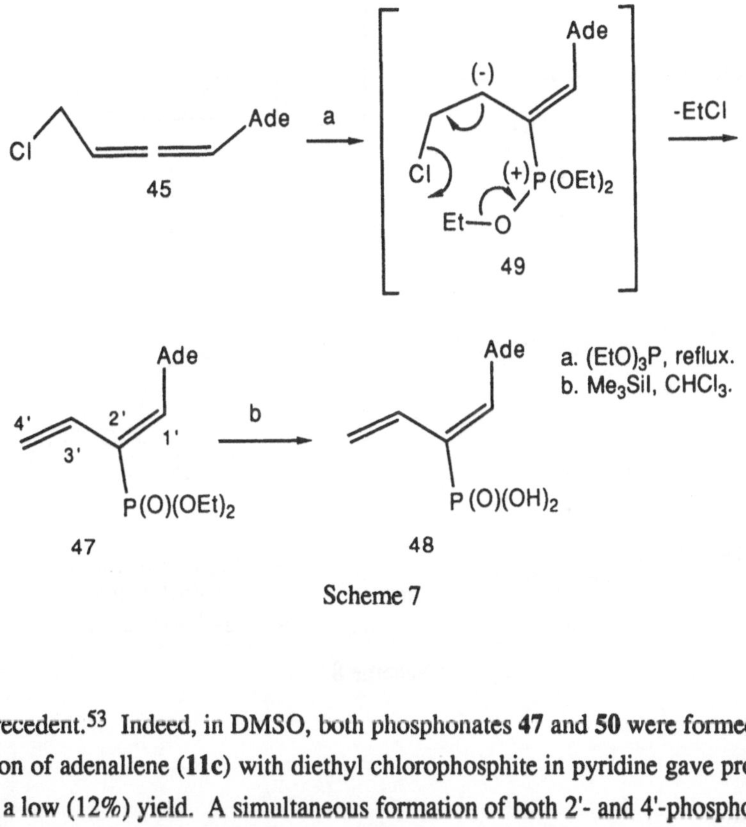

rature precedent.[53] Indeed, in DMSO, both phosphonates **47** and **50** were formed, and an interaction of adenallene (**11c**) with diethyl chlorophosphite in pyridine gave product **47**, albeit in a low (12%) yield. A simultaneous formation of both 2'- and 4'-phosphonates **47** and **50**, separable by chromatography, in the former transformation may have a preparative advantage. Allenic phosphonate **53c** is a likely intermediate in the formation of 4'-phosphonate **50**. Apparently, a strongly activated methylene group of allene **53c** is responsible for a formation of the respective carbanion under basic reaction conditions. Isomerization then leads to a stable conjugated system of double bonds in **50**. Frequently, these reactions and some others that will be described later are accompanied by elimination of nucleic acid bases such as adenine (**12c**) or alkylation of the released heterocycle with a dialkyl phosphonate fragment (e. g., formation of N^9-ethyladenine, **54c**).

The most advantageous method for synthesis of 4'-phosphonate **50** starts from chlorobutyne **14c** and sodium diethyl phosphite in DMF (Scheme 9). Compound **50** was obtained in the same (23%) yield as from chloroadenallene **45** (Scheme 8), but the procedure is shorter. Acetylenic phosphonate **55** and isomerization product **56** are likely intermediates which, under the conditions employed, are of limited stability. By contrast, but-

a. (EtO)$_2$PONa, DMSO.
b. (EtO)$_2$PCl, pyridine.
c. Me$_3$SiI, CHCl$_3$.

Scheme 8

ynol **10c** and diethyl chlorophosphite in pyridine gave only adenine (**12c**).

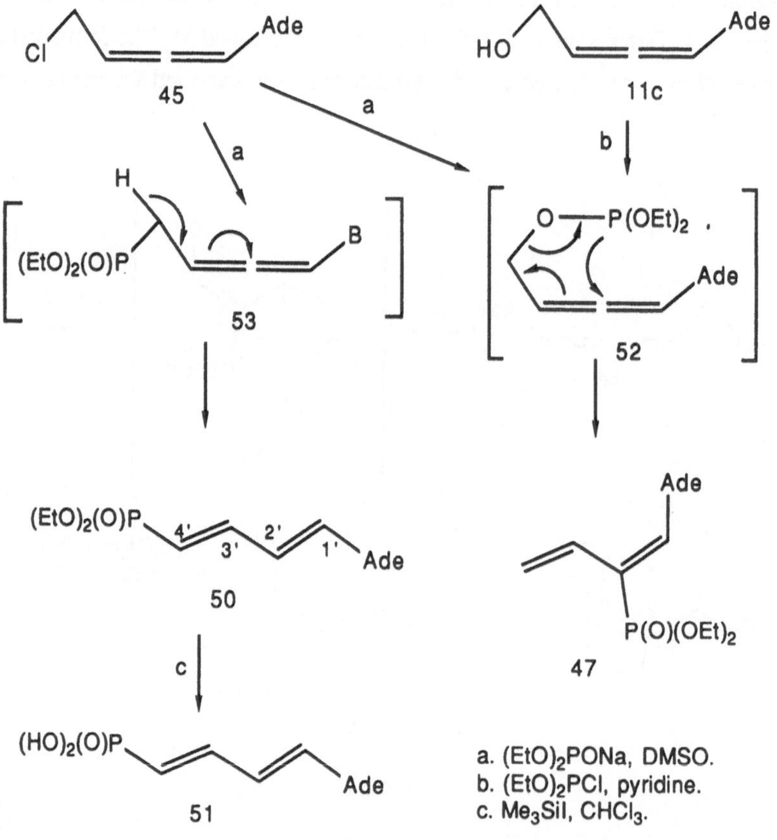

The Michaelis-Arbuzov reaction of chlorobutynes **14c** and **14e** afforded still another line of interesting products.[51,52] The reaction of compound **14e** with triethylphosphite (110 °C, 3 h) gave only 2-amino-6-chloro-N^9-ethylpurine (**54e**, 39%, Scheme 10). In a similar fashion, chlorobutyne **14c** (reflux 2 h) afforded N^9-ethyladenine (**54c**, 27%). The reaction course is best explained by invoking the formation of intermediates **57c** (X = Cl) and **57e** (X = Cl), which then suffer elimination to give cumulene **58** (X = Cl) and the respective heterocycle **12c** or **12e**. Species **58** (X = Cl) then serves as an alkylating agent to give the final product **54c** or **54e** and phosphonate **59**. Alternately, the latter cumulene formed by dealkylation of **58** (X = Cl) can also alkylate **12c** or **12e**.

Introduction of a strong nucleophile (iodide ion) into the reaction mixture in order to suppress elimination of the heterocyclic base from phosphonate intermediates[51,52] **57c** or **57e** entirely changed the reaction course (Scheme 11). Thus, chlorobutyne **14e** and triethylphosphite (110 °C, 2 h) in the presence of Bu_4NI furnished E-3',5'-diphosphonate **60e** in 58% yield. This is an example of a simultaneous introduction of two phosphonyl residues into an acyclic (unsaturated) nucleoside analogue. A similar reaction of compound

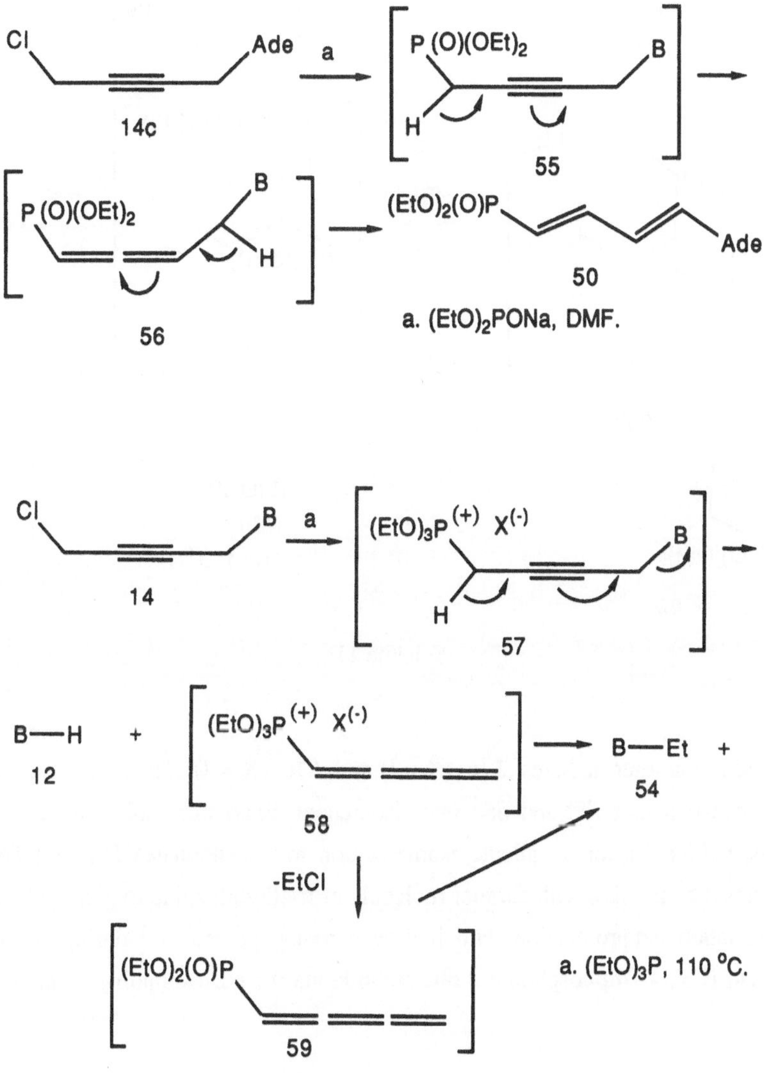

a. (EtO)₂PONa, DMF.

a. (EtO)₃P, 110 °C.

Scheme 9 and 10

14c (80 °C, 16 h) afforded **60c** (20%) and a new type of heterocyclic allene, N^9-(2,3-butadien-1-yl)adenine (**61**, 10%, Scheme 12). The latter compound became a major product (30% yield) under more vigorous conditions (120 °C, 40 min.). It is noteworthy that allenic or acetylenic phosphonates **53c**, **53e**, **55c** and **55e** were isolated in neither case.

The formation of 3',4'-diphosphonates **60c** and **60e** is best explained by an ylid mechanism. Thus, the first phosphonyl residue is introduced in a routine Michaelis-Ar-

Scheme 11

buzov fashion via intermediates **57c** (X = I) and **57e** (X = I). However, the resultant acetylene phosphonates **55c** and **55e** (see also Scheme 9) are not stable under the reaction conditions, and they suffer an allenic isomerization to phosphonates **56c** and **56e**. The latter allenes then interact with another molecule of triethylphoshite to give ylids **62c** or **62e**. Dealkylation and protonation with hydrogen iodide generated[54] during the reaction then leads to *E*-3', 4'- diphosphonates **60c** and **60e** via the corresponding carbanions **63c** and **63e**.

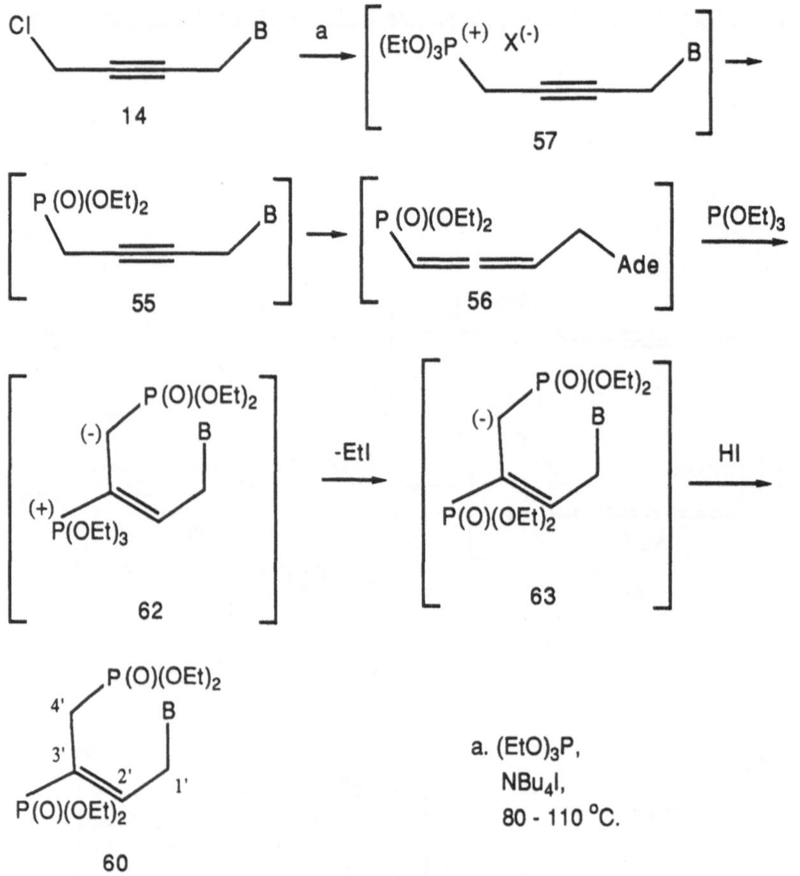

The probable reaction course leading to allene **61** is as follows (Scheme 12). Chlorobutyne **14c** is first isomerized to chloroallene **24c**. Addition of HI affords intermediate **64** and elimination of ICl catalyzed possibly with triethylphosphite gives then allene **61**.

The following conclusions can be drawn from this investigation: (i) Allenic and acetylenic phosphonates **53c**, **53e**, **55c** and **55e** are not stable under the conditions of Michaelis-Arbuzov and Michaelis-Becker reactions; (ii) chloroadenallene **45**, chlorobutyne **14c** and **14e** can serve as synthons for stereo- and regioselective syntheses of phosphonates **47**, **50**, **60c** and **60e**; (iii) the scope and limitations of allene formation (compound **61**) during Michaelis-Arbuzov reaction certainly deserve attention.

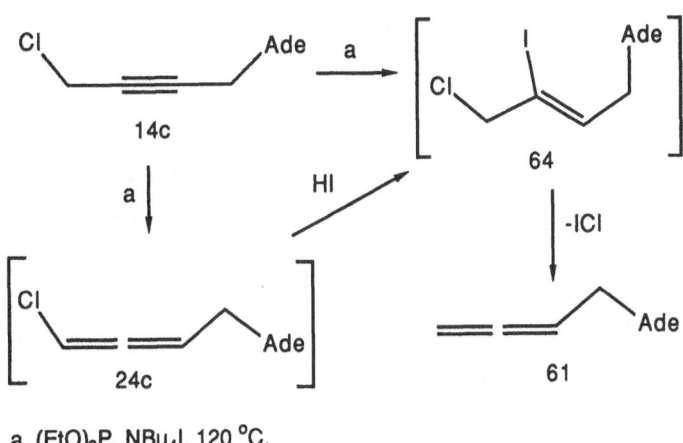

a. (EtO)$_3$P, NBu$_4$I, 120 °C.

Scheme 12

BIOLOGICAL ACTIVITY

1. Antiretroviral Effects of Adenallene and Cytallene

Initial antiviral assays were performed[34] in helper/inducer T-cell line ATH8 infected with HIV-1. (±)-Adenallene (**11c**) and -cytallene (**11d**) exhibit a high inhibitory activity against replication and cytopathic effect of HIV-1. Both compounds thus became the first acyclic **nucleoside** analogues active against the latter virus. Acyclovir (**1a**) has little activity in cell cultures infected with HIV-1, but it is capable of potentiating the antiretroviral activity of AZT (**3b**).[55] Cytallene (**11d**) is the most active allenol giving a virtually complete protection of ATH8 cells against the virus at ≤ 0.5 μM. In quantitative terms, this activity is very similar to that of both 2',3'-dideoxycytidine (ddCyd), which is under active consideration as a drug against AIDS, and AZT (**3b**). Adenallene (**11c**) has a lower activity with a partial protective effect at 10 μM and complete protection at 100 μM. Again, these results closely parallel the effect of the corresponding 2',3'-dideoxyribonucleoside, 2',3'-dideoxyadenosine (ddAdo). However, the latter analogue was somewhat more active at 10 μM and less toxic at concentrations ≥ 500 μM. Also, AZT (**3b**), ddCyd and ddAdo retain their protective ability against cytopathic effect of HIV-1 through the 14-day period of incubation, whereas effectivity of both allenols **11c** and **11d** is often lost by day 14. The protective effect of **11c** and **11d** against HIV-2 was approximately equal to that observed in HIV-1 infected culture.

The antiretroviral activity of adenallene (**11c**) and cytallene (**11d**) was confirmed in a different target-cell system, the normal helper/inducer T-cell clone TM11. Thus, 0.2 - 0.5 μM of **11d** and 50 - 100 μM of **11c** provides a total protection against HIV-1 infection without affecting the cellular growth. Both adenallene (**11c**) and cytallene (**11d**) inhibit also HIV-1 *gag* expression in CD4+ H9 cells. Thus, **11c** gives a partial inhibition of expression of *gag*-encoded protein p24 at 50 μM and total inhibition at 100 μM, but a viral breakthrough occurred by day 9. However, at 200 μM of **11c** no virus replication resumed in this period of time. Cytallene (**11d**) is partially protective at 1 μM, and virus formation is totally inhibited at 5 μM. Allenols **11c** and **11d** also inhibit viral DNA synthesis in ATH8 cells. Thus, 250 μM **11c** and 5 μM **11d** caused 89 - 97% decreases in DNA synthesis after 1 - 3 days in culture.

Adenallene (**11c**) also inhibits expression of HIV-1 p24 *gag* protein in monocytes/macrophages in vitro.[56] The formation of p24 protein was completely suppressed at 100 μM of **11c**. However, after 23-day incubation the inhibition was nullified. Effects of adenallene (**11c**) and cytallene (**11d**) on immune reactions of TM11 and peripheral blood monocyte (PBM) cells from normal individuals were also examined.[34] Allenol **11c** does not inhibit antigen-induced proliferation and pokeweed mitogen (PWM)- or phytohemagglutinin (PHA)-induced T-cell activation in concentrations 50 - 100 μM. Similarly, the latter parameters are little affected with cytallene (**11d**) at 10 μM. Only a moderate suppression of proliferation of PHA-activated PBM cells was noted at this concentration.

The ability of adenallene (**11c**) and cytallene (**11d**) to effectively interfere with HIV infections in AZT-resistant cell lines is of importance as a step in development of both agents as potential drugs against AIDS. The activity of **11c** and **11d** against HIV-1, HIV-2, early-treatment and late-treatment virus isolates was therefore studied by a plaque reduction assay in HT4-6C cells[57] (Table 2). These results confirmed independently the activity of adenallene (**11c**) and cytallene (**11d**) in a different assay. No significant cross-resistance was noted with allenols **11c** and **11d**. It must be noted that in any assay racemic adenallene (**11c**) and cytallene (**11d**) may be at a disadvantage against the respective optically pure 2',3'-dideoxy-D-ribonucleosides.

Availability of *R*- and *S*-enantiomers of adenallene **31** and **33** (Scheme 4) made possible to study the enantioselectivity of anti-HIV effect.[35] It should be noted that **adenallene (11c) and cytallene (11d) are the first nucleoside analogues possessing an axial chirality.** Many other biologically active mimics of nucleic acid components are either achiral (acyclovir, **1a**) or they comprise one or more centers of chirality such as AZT (**3b**) and other cyclic as well as acyclic analogues. **_R_-adenallene (31), a direct product of adenosine deaminase-catalyzed deamination** of the racemic **11c** (Scheme 4), **gives a significantly greater protective effect against HIV-1 infection of ATH8 cells than the _S_-enantiomer 33.** In quantitative terms, the effective concentrations (EC$_{50}$) of

Table 2. Activity of AZT-susceptible and -resistant isolates of HIV-1 and HIV-2 towards antiretroviral agents.[57]

Compound	EC$_{50}$(μM)			
	HIV-1	HIV-2	A[a]	B[b]
AZT (**3b**)	0.05	0.08	0.03	6
ddAdo	3	3.5	-	-
ddIno[c]	2.1	5.6	0.7	0.7
ddCyd	0.2	0.35	0.5	0.5
Adenallene (**11c**)	10	18	20	50
Cytallene (**11d**)	0.2	0.3	0.6	1.2

[a]AZT-susceptible HIV isolate. [b]AZT-resistant HIV isolate. [c] 2',3'-Dideoxyinosine.

racemic adenallene (**11c**), _R_- and _S_-enantiomer **31, 33** are 14, 5.8 and >200 μM, respectively. Thus, _S_-enantiomer (**33**) has a small but distinct anti-HIV activity which is higher than that of _L_-ddAdo.[58] The results[35] seem to indicate that there is a potentiating effect of the _S_-form **33** on the efficacy of _R_-adenallene (**31**) at higher concentrations (20 μM and above).

It is then tempting to predict that the most active form of cytallene (**11d**) is the _R_-enantiomer. However, a definite conclusion must await the availability of both enantiomers and determination of their absolute configuration.

In order to establish the extent of antiretroviral effect of allenols **11c** and **11d**, the activity of both analogues was investigated in SC-1 cells infected with Friend and Rauscher murine leukemia viruses[22] (F-MuLV and R-MuLV). The results are compared with those of other agents (Table 3). It is obvious that AZT is by far the most active analogue. Somewhat surprising is a lack of significant antiretroviral activity of adenallene (**11c**) in both assays. The ddAdo is much more active, and it is also more effective antiretroviral agent against murine leukemia viruses than HIV. By contrast, cytallene (**11d**) is a somewhat more potent antiretroviral agent in both assays than ddCyd. It is also clear that AZT is an effective agent of broad antiretroviral selectivity, whereas activities of other agents listed in Table 2 including allenols **11c** and **11d** are substantially lower.

As in the case of 2',3'-dideoxyribonucleosides, no appreciable activity of allenols

(11a - 11d, 11g and 11h) was noted against several tumor cell lines and DNA or RNA viruses[22] *in vitro*.

2. Structure - Activity Relationships

As already mentioned, the presence of the hydroxy function in adenallene (11c) and cytallene (11d) is necessary for an anti-HIV effect.[34] Thus, chloroadenallene 45 is inactive. Not surprisingly, the antiretroviral activity of allenols 11 is base-selective. Only adenallene (11c) and cytallene (11d) are active, whereas other allenols investigated thus far

Table 3. Antiretroviral effects and cytotoxicity of allenols 11c and 11d.[22]

Compound	F-MuLV		R-MuLV		Cytotoxicity
	EC_{50} (µM)	SI[a]	EC_{50} (µM)	SI[a]	IC_{50} µM
11c	124	1.4	87.3	1.9	168
11d	8.1	60.4	11.5	42.5	489
ddAdo	1	83.8	2.6	32.2	83.8
ddCyd	23	>4.4	13.8	>7.3	>100
AZT (3b)	0.035	17,000	0.046	12,900	594
Virazole	8.3	6.3	32.1	1.6	52.0
PFA[b]	110	9.1	42.8	23.4	1,000

[a] Selectivity index (SI) = IC_{50}/EC_{50}. [b] Phosphonoformic acid.

(compounds 11a, 11b, 11g and 11h) are devoid of any effect. The 5-methylcytallene (11g) and thymallene (11a), which are inactive in ATH8 cell culture,[59] resemble the corresponding nucleoside analogues - 3'-deoxythymidine[26] and 2',3'-dideoxy-5-methylcytidine.[60] By contrast, 2',3'-dideoxyguanosine[26] and 2-amino-2',3'-dideoxyadenosine[61] are effective anti-HIV agents. It is then clear that allenols of type 11 have a narrower range of base selectivity than the respective 2',3'-dideoxyribonucleosides. A comparison of 2',3'-dideoxyinosine (ddIno) with hypoxallene (11i) provides the most striking example of differences between both series of analogues. Whereas ddIno is a recently approved drug for AIDS, compound 11i is totally inactive[34] against HIV. Apparently, a mechanism that provides for conversion of ddIno to ddAdo[62] (enzymatic aminosuccinylation and elimination of fumarate at the 5'-phosphate level) is not available for a similar activation of hypoxallene (11i). It is not clear whether this is a consequence of lack of cellular phosphorylation of 11i or poor binding ability of the resultant phosphate to adenylosuccinate synthetase or lyase.

An intact system of cumulated double bonds is a foremost requirement for antiretroviral activity of adenallene (11c) and cytallene (11d). This is confirmed by the fact that isomeric acetylenes (**10a** - **10d**), analogues containing only one double bond in the side-chain (**7a** - **7d, 9a** - **9d, 37** and **38**) or saturated N^9-(4-hydroxybutyl)adenine are all inactive[22,34,59] as antiretroviral agents. **The grouping -C=C=C- then represents a new isoelectronic bioisoster of the -C-O-C- function** found in nucleosides and a number of acyclic analogues including acyclovir (**1a**). We can view both moieties as spacers separating the hydroxymethyl group and a nucleic acid base which differ substantially in terms of conformational rigidity. Thus, the CH_2OCH_2 group of acyclovir (**1a**) is very flexible,[63] whereas severe orbital constraints are imposed on the allenic function. As already mentioned, molecular models show that there is a substantial freedom of rotation of both nucleic acid base and hydroxymethyl moiety not generally found in nucleosides. This is probably important for antiretroviral effect of analogues **11c** and **11d**. It is interesting that two different rotamers of the $C_{3'}$-$C_{4'}$ bond are observed in the dimeric structure of R-adenallene (**31**), but the conformation of adenine is *anti*-like in both molecules (Fig. 1). The function of the cumulated system of double bonds in receptor binding is not clear. There is no evidence that analogues **11c** and **11d** can act as irreversible inhibitors. The possibility of a hydrogen bonding from receptor to the π-electron system of allene is a subject for speculation.[25] It should be emphasized that several unsaturated nucleoside analogues[11-13] comprising a double bond between 2' and 3' carbon atoms, e.g., **4b, 4d** and **5a**, exhibit a strong anti-HIV effect. Whether this is a consequence of an increased rigidity of the furanose (cyclopentane) moiety or a favorable interaction of receptor with the π-electron system is not clear.

It was concluded on the basis of molecular modeling studies[38,39] that adenallene (**11c**) and cytallene (**11d**) closely resembled an unusual 3'-*exo* furanose conformation and *ap* rotamer deemed important for antiretroviral activity of nucleoside analogues[64]. Nevertheless, nucleoside analogues with a different preferred conformation and active against HIV were recently described.[65,66] In addition, another study[25] indicated that the natural 3'-*endo* form of adenosine could also be mimicked by adenallene (**11c**). An X-ray diffraction (Fig. 1) shows two rotamers of an almost equal energy in the dimer of R-adenallene (**31**). Either of them can be involved in the viral inhibition. It was recently noted[67] that any attempted correlation of nucleoside conformation with anti-HIV effect was limited by a possibility of different conformational requirements of enzymes (nucleoside and nucleotide kinases as well as reverse transcriptase) involved in the mechanism of action of a given analogue.

3. Mechanism of Antiretroviral Action of Adenallene and Cytallene

At the outset, it was mentioned that the presence of the hydroxymethyl group was considered important in the design of allenols of type **11**. It was anticipated that were such analogues active, their mechanism of action would follow the conventional pattern found in other nucleoside analogues such as AZT (**3b**), i. e., phosphorylation to the triphosphate level, incorporation of the triphosphate into the growing DNA chain catalyzed by reverse transcriptase, and resultant chain termination.[10] It is then not surprising that a considerable effort has been devoted to clarify the first step of these events, intracellular phosphorylation of adenallene (11c) and cytallene (11d).

The reversal experiments provide indirect evidence that cytallene (11d) is phosphorylated by a reaction catalyzed with 2'-deoxycytidine (dCyd) kinase.[56] Thus, the protective effect against HIV-1 infection and cellular toxicity of 11d are completely reversed by dCyd. By contrast, neither antiretroviral activity nor toxicity of adenallene (11c) is affected by 2'-deoxyadenosine (dAdo) or dCyd. These results show that a nucleoside kinase different from those using dAdo and dCyd as substrates may be responsible for phosphorylation of 11c in ATH8 cells.

Indications that there may be significant differences between the mechanism of action of adenallene (11c) and cytallene (11d) came also from the studies of inhibition of incorporation of labeled precursors (thymidine, uridine, leucine) into macromolecules of non-infected SC-1 cells[22] with adenallene (11c) and cytallene (11d). The latter cell line was used for tests of antiretroviral activity toward F-MuLV and R-MuLV (Table 3). Thus, adenallene (11c) inhibits macromolecular synthesis in the order: DNA >> protein > RNA. By contrast, the order of inhibitory activity of cytallene (11d) is RNA > DNA > protein.

According to more recent observations, cytallene (11d) is readily phosphorylated with ATP and 6,000-fold purified dCyd kinase from human leukemic spleen cells.[68] In addition, cytallene (11d) but neither adenallene (11c) nor ddAdo inhibit phosphorylation of dCyd and dAdo catalyzed with dCyd kinase.[69] The efficiency of phosphorylation of 11d is lower than that of ddCyd. Results obtained with human T lymphoblastoid CEM cell line deficient in adenosine (Ado) kinase indicate that the latter enzyme may be important for phosphorylation of ddAdo. However, another study[70] shows that ddAdo is not efficiently phosphorylated in a reaction catalyzed by Ado kinase from human thymus. It should also be noted that ddAdo is a substrate for dCyd kinase[70] from the latter source whereas neither racemic adenallene (11c) nor R-enantiomer (31) inhibit phosphorylation of dAdo and dCyd catalyzed by the respective kinases from calf and rabbit thymus.[71]

It is then clear that cytallene (11d) undergoes a cellular phosphorylation catalyzed by dCyd kinase as a part of its activation mechanism. The assumption that adenallene (11c) is also phosphorylated by nucleoside kinase(s) seems reasonable, but the specific enzyme(s) responsible for this event have not been identified.

4. Adenallene and Related Analogues as Substrates for Adenosine Deaminase

Adenosine (ADA) and cytidine (CD) deaminase are important catabolic enzymes which convert many biologically active analogues into inactive metabolites. Therefore, investigation of substrate activity of new analogues of adenosine or cytidine toward ADA or CDA is of utmost importance. We have already mentioned that adenallene (11c) is a substrate for ADA. Depending on the reaction conditions, deamination can either be exhaustive giving racemic hypoxallene[23,25] (11i) or selective affording optically pure[35] R-(-)-adenallene (31) and S-(+)-hypoxallene (32, Scheme 4). The fact that deamination of adenallene (11c) gives **directly** the R-enantiomer, the most potent antiretroviral component of the racemic mixture, is very fortunate. The deamination of enantiomers 31 and 33 catalyzed with ADA follows the Michaelis-Menten kinetics.[35] Quite surprisingly, the K_M values of the R- and S-forms 31 and 33 do not differ substantially (0.52 and 0.41 mM, respectively). This suggests a similar binding affinity of both enantiomers for ADA. The corresponding V_{max} values are 18.5 (31) and 530 (33) nmol/min. Most surprisingly, **the enantioselectivity of anti-HIV effect and substrate activity of ADA are reversed.** This indicates that both enantiomers can serve as nucleoside analogues but with a different enantioselectivity for particular enzymes or receptors. Another favorable factor is the fact that R-enantiomer (31) exhibits anti-HIV effect at concentrations substantially below its K_M value for ADA. Thus, inactivation of R-form (31) by deamination is of little significance except for high concentrations. Cytallene (11d) is resistant to the action of CDA and hypoxallene (11i) is not a substrate for purine nucleoside phosphorylase.[25]

Several other unsaturated adenosine analogues[22,25,45] such as 2-aminoadenallene (11h), 7c, 9c, 10c, 37, 38 and the stucturally simple N^9-(ethynyl)adenine[72] are also substrates for ADA of varying potency. As a rule, E-isomers 7c and 37 are more readily deaminated than the corresponding Z-isomers 9c and 38.

CONCLUSIONS

Adenallene (11c), its R-enantiomer (31) and cytallene (11d) are highly potent anti-HIV agents belonging to a new class of nucleoside analogues comprising an allene function with axial chirality. Their synthesis is simple and it does not require any nucleoside or carbohydrate starting material. Given their high activity in culture, significant acid stability, low cross-resistance with AZT (3b) and high resistance of the most active R-enantiomer 31 to deamination, compounds 11c, 11d and 31 are of interest as potential experimental drugs for AIDS and AIDS-related complex (ARC) either alone or in combination with other antiretroviral agents.

ACKNOWLEDGMENTS

The interdisciplinary effort described in this account would not have been possible without effective cooperation of many distinguished colleagues whose names appear in the appropriate references. A generous financial support by U. S. Public Health Service Research Grant CA32779 from the National Cancer Institute, Bethesda, Maryland and by an institutional grant to the Michigan Cancer Foundation from the United Way of Southeastern Michigan is also acknowledged.

REFERENCES

1. H. J. Schaeffer, L. Beauchamp, P. de Miranda, G. B. Elion, D. J. Bauer, and P. Collins, 9-(2-Hydroxyethoxymethyl)guanine activity against viruses of the herpes group, *Nature* 272:583 (1978).

2. J. P. Horwitz, J. Chua, and M. Noel, Nucleosides. V. The monomesylates of 1-(2'-deoxy-β-D-lyxofuranosyl)thymine, *J. Org. Chem.* 29:2076 (1964).

3. W. Ostertag, G. Roesler, C. J. Krieg, J. Kind, T. Cole, T. Crozier, G. Gaedicke, G. Steinheider, N. Kluge, and S. Dube, Induction of endogenous virus and thymidine kinase by bromodeoxyuridine in cell cultures transformed by Friend virus, *Proc. Natl. Acad. Sci. U.S.A.* 71:4980 (1974).

4. H. Mitsuya, K. J. Weinhold, P. A. Furman, M. H. St. Clair, S. W. Lehrman, R. C. Gallo, D. Bolognesi, D. W. Barry, and S. Broder, 3'-Azido-3'-deoxythymidine (BWA 5090): An antiviral agent that inhibits the infectivity and cytopathic effect of human T-lymphotropic virus type III/lymphoadenopathy - associated virus in vitro, *Proc. Natl. Acad. Sci. U.S.A.* 82:7096 (1985).

5. H. J. Schaeffer, S. Gurwara, R. Vince, and S. Bittner, Novel Substrate of Adenosine Deaminase, *J. Med. Chem.* 14:367 (1971).

6. J. L. Kelly, L. E. Kelsey, W. R. Hall, M. P. Krochmal, and H. J. Schaeffer, Pyrimidine acyclic nucleosides. 1-[(2-Hydroxyethoxy)methyl]pyrimidines as candidate antivirals, *J. Med. Chem.* 24:753 (1981).

7. H. Mitsuya, M. Matsukura, and S. Broder, Rapid in vitro systems for assessing activity of agents against HTLV-III/LAV, *in*: "AIDS, Modern Concepts and Therapeutic Challenges," S. Broder, ed., Marcel Dekker, New York (1987), p. 303.

8. M. Baba, R. Pauwels, J. Balzarini, P. Herdewijn, and E. De Clercq, Selective inhibition of human immunodeficiency virus (HIV) by 3'-azido-2',3'-dideoxyguanosine, *Biochem. Biophys. Res. Commun.* 145:1080 (1987).

9. R. Dolin, Antiviral chemotherapy and chemoprophylaxis, *Science* 227:1296 (1985).

10. H. Mitsuya, R. Yarchoan, and S. Broder, Molecular targets for AIDS therapy, *Science* 249:1533 (1990).

11. M. M. Mansuri, J. E. Starrett, Jr., I. Ghazzouli, M. J. M. Hitchcock, R. Z. Sterzycki, V. Brankovan, T.-S. Lin, E. M. August, W. H. Prusoff, J.-P. Sommadossi, and J. C. Martin, 1-(2,3-Dideoxy-β-D-*glycero*-pent-2-enofuranosyl)thymine. A highly potent and selective anti-HIV agent, *J. Med. Chem.* 32:461 (1989).

12. M. Baba, R. Pauwels, P. Herdewijn, E. De Clercq, J. Desmyter, and J. Vandeputte, Both 2',3'-dideoxythymidine and its 2',3'-unsaturated derivative (2',3'-dideoxythymidinene) are potent and selective inhibitors of human immunodeficiency virus replication in vitro, *Biochem. Biophys. Res. Commun.* 142:128 (1987).

13. R. Vince and M. Hua, Synthesis and anti-HIV activity of carbocyclic 2',3'-didehydro-2',3'-dideoxy 2,6-disubstituted purine nucleosides, *J. Med. Chem.* 33:17 (1990).

14. V. E. Marquez and M.-I. Lim, Carbocyclic nucleosides, *Med. Res. Rev.* (London) 6:1 (1986).

15. A. Larsson, K. Stenberg, A.-C. Ericsson, U. Haglund, W. Yisak, N. G. Johansson, B. Öberg, and R. Datema, Mode of action, toxicity, pharmacokinetics, and efficacy of some antiherpes virus analogs related to buciclovir, *Antimicrob. Agents Chemother.* 30:598 (1986).

16. M. Hua, P. M. Korkowski, and R. Vince, Synthesis and biological evaluation of acyclic neplanocin A analogues, *J. Med. Chem.* 30:198 (1987).

17. S. Phadtare and J. Zemlicka, Synthesis and biological properties of 9-(*trans*-4-hydroxy-2-buten-1-yl)adenine and guanine: Open-chain analogues of neplanocin A, *J. Med. Chem.* 30:437 (1987).

18. D. R. Haines, C. K. H. Tseng, and V. E. Marquez, Synthesis and biological activity of unsaturated carboacyclic purine nucleoside analogues, *J. Med. Chem.* 30:943 (1987).

19. D. R. Borcherding, S. Narayanan, M. Hasobe, J. G. McKee, B. T. Keller, and R. T. Borchardt, Potential inhibitors of S-adenosylmethionine-dependent methyltransferases. 11. Molecular dissections of neplanocin A as potential inhibitors of S-adenosylhomocysteine hydrolase, *J. Med. Chem.* 31:1729 (1988).

20. W. T. Ashton, L. C. Meurer, C. L. Cantone, A. K. Field, J. Hannah, J. D. Karkas, R. Liou, G. F. Patel, H. C. Perry, A. F. Wagner, E. Walton, and R. L. Tolman, Synthesis and antiherpetic activity of (±)-9-[[(Z)-2-(hydroxymethyl)cyclopropyl]-methyl]guanine and related compounds, *J. Med. Chem.* 31:2304 (1988).

21. S. Phadtare, D. Kessel, and J. Zemlicka, Unsaturated nucleoside analogues: Synthesis and antitumor activity, *Nucleosides Nucleotides* 8:907 (1989).

22. S. Phadtare, D. Kessel, T. H. Corbett, H. E. Renis, B. A. Court, and J. Zemlicka, Unsaturated and carbocyclic nucleoside analogues: Synthesis, antitumor, and antiviral activity, *J. Med. Chem.* 34:421 (1991).

23. S. Phadtare and J. Zemlicka, Allenic derivatives of nucleic acid components and their transformation products: a new class of biologically active nucleoside analogues, *Nucleic Acids Res., Symp. Ser.* No. 18:25 (1987).

24. A. Larsson, S. Alenius, N.-G. Johansson, and B. Öberg, Antiherpetic activity and mechanism of action of 9-(4-hydroxybutyl)guanine, *Antiviral Res.* 3:77 (1983).

25. S. Phadtare and J. Zemlicka, Nucleic acid derived allenols: Unusual analogues of nucleosides with antiretroviral activity, *J. Am. Chem. Soc.* 111:5925 (1989).

26. H. Mitsuya and S. Broder, Inhibition of the *in vitro* infectivity and cytopathic effect of human T-lymphotropic virus type III/lymphoadenopathy-associated virus (HTLV-III/LAV) by 2',3'-dideoxynucleosides, *Proc. Natl. Acad. Sci. U.S.A.* 83:1911 (1986).

27. A. J. Hubert and H. Reimlinger, Base-catalysed prototropic isomerisations. Part II. The isomerisation of *N*-prop-2-ynyl heterocycles into *N*-substituted allenes and acetylenes, *J. Chem. Soc.* C:606 (1968).

28. D. Ranganathan, R. Rathi, K. Kesavan, and W. P. Singh, The demonstration of nor-

mal O → N Claisen rearrangement in purines, *Tetrahedron* 42:43873 (1986).

29. A. A. Khorlin, I. P. Smirnov, S. V. Kochetkova, T. L. Tsilevich, I. L. Shchaveleva, B. P. Gottikh, and V. L. Florent'ev, Compounds similar to acyclovir. IV. Convenient method of synthesising adenallene, *Bioorg. Khim.* 15:530 (1989); English translation 15:291 (1990).

30. M. V. Kochetkova, A. V. Tsytovich, and B. I. Mitsner, A convenient approach to the synthesis of nucleic acid allene derivatives, possessing anti-HIV activity, *Nucleic Acids Res., Symp. Ser.* No. 24:233 (1991).

31. A. V. Tsytovich, M. V. Kochetkova, E. V. Kuznetsova, B. I. Mitsner, and V.I. Shvets, Acyclic nucleoside analogues. I. Development of allenic nucleoside derivatives synthesis, *Bioorg. Khim.* 17:1086 (1991) (in Russian).

32. S. Phadtare and J. Zemlicka, Synthesis of N^1-(4-hydroxy-1,2-butadien-1-yl)thymine, an analogue of 3'-deoxythymidine, *J. Org. Chem.* 54:3675 (1989).

33. S. Phadtare and J. Zemlicka, unpublished experiments.

34. S. Hayashi, S. Phadtare, J. Zemlicka, M. Matsukura, H. Mitsuya, and S. Broder, Adenallene and cytallene: Acyclic nucleoside analogues that inhibit replication and cytopathic effect of human immunodeficiency virus in vitro, *Proc. Natl. Acad. Sci. U.S.A.* 85:6127 (1988).

35. S. Megati, Z. Goren, J. V. Silverton, J. Orlina, H. Nishimura, T. Shirasaki, H. Mitsuya, and J. Zemlicka, *R*-(-)- and *S*-(+)-adenallene: Synthesis, absolute configuration, antiretroviral effect, and enzymic deamination, *J. Med. Chem.* 35:4098 (1992).

36. C. R. Noe, Chirale Lactole, II. Racematspaltung und enantioselektive Acetalisierung mit der 2,3,3a,4,5,6,7,7a-Octahydro-7,8,8-trimethyl-4,7-methanobenzofuran-2-yl-Schutzgruppe, *Chem. Ber.* 115:1591 (1982).

37. S. Phadtare and J. Zemlicka, Allenic derivatives of nucleic acid bases - new acyclic nucleoside analogues with antiretroviral activity, *Nucleic Acids Res., Symp. Ser. No.* 20:39 (1988).

38. P. Van Roey, E. W. Taylor, C. K. Chu, and R. F. Schinazi, Correlation of molecular conformation and activity of reverse transcriptase inhibitors, *Ann. N. Y. Acad. Sci.* 616:29 (1990).

39. E. W. Taylor, P. Van Roey, R. F. Schinazi, and C. K. Chu, A stereochemical rationale for the activity of anti-HIV nucleosides, *Antiviral Chem. Chemother.* 1:163 (1990).

40. J. March, "Advanced Organic Chemistry", McGraw-Hill, New York, 1977, p. 24.

41. W. Runge, Stereochemistry of allenes, *in* : "The Chemistry of the Allenes", Vol. 3, S. R. Landor, ed., Academic Press, New York, 1982, p. 579.

42. N. N. H. Teng, M. S. Itzkowitz, and I. Tinoco, Jr., Calculation of the rotational strength of mononucleosides, *J. Am. Chem. Soc.* 93:6257 (1971).

43. W. Runge, Spectroscopic properties of allenes, *in*: "The Chemistry of the Allenes", Vol. 3, S. R. Landor, ed., Academic Press, New York, 1982, p. 777.

44. R. Gompper and U. Wolf, Synthesen und Reaktionen electronarmer Allene, *Liebigs Ann. Chem.* 1388 (1979).

45. S. Phadtare and J. Zemlicka, Synthesis of (Z)- and (E)-N^9-(4-hydroxy-1-buten-1-yl)-adenine - new unsaturated analogues of adenosine, *Tetrahedron Lett.* 31:43 (1990).

46. J. R. Wiersig, A. N. H. Yeo, and C. Djerassi, Mass spectrometry in structural and stereochemical studies. 247. Electron-impact induced fragmentation of allenes, *J. Am. Chem. Soc.* 99:532 (1977).

47. "The Chemistry of the Allenes", Vol. 2, S. R. Landor, ed., 1982.

48. R. Engel, "Synthesis of Carbon-Phosphorus Bond", CRC Press, Boca Raton, Florida, 1988, p. 21.

49. Ref. 48, p. 7.

50. "Nucleotide Analogues as Antiviral Agents", ACS Symposium Series 401, J. C. Martin, ed., American Chemical Society, Washington, D. C., 1989.

51. S. Phadtare and J. Zemlicka, Unsaturated analogues of acyclic nucleoside phosphonates: An unusual Arbuzov reaction with unactivated triple bond, *Nucleosides Nucleotides* 10:275 (1991).

52. S. Megati, S. Phadtare, and J. Zemlicka, Unsaturated phosphonates as acyclic nucleotide analogues. Anomalous Michaelis-Arbuzov and Michaelis-Becker reactions with multiple bond systems, *J. Org. Chem.* 57:2320 (1992).

53. M. Huché and P. Cresson, Réactions sigmatropiques d'ordre (2,3) au niveau d'un atome de phosphore, *Bull. Soc. Chim. Fr.* No. 3-4:800 (1975).

54. A. J. Kirby and S. G. Warren, "The Organic Chemistry of Phosphorus", Elsevier, New York, 1967, p.39.

55. H. Mitsuya and S. Broder, Strategies for antiviral therapy in AIDS, *Nature* 325:773 (1987).

56. S. Hayashi, S. Phadtare, J. Zemlicka, M. Matsukura, H. Mitsuya, and S. Broder, Adenallene and cytallene, two novel acyclic nucleoside derivatives active against human immunodeficiency virus (HIV) in T-cells and monocytes/macrophages in vitro: Further characterization of anti-viral and cytotoxic activity, *in*: "Mechanisms of Action and Therapeutic Applications of Biologicals in Cancer and Immune Deficiency Disorders", Alan R. Liss, Inc., 1989, p. 371.

57. B. A. Larder, B. Chesebro, and D. D. Richman, Susceptibilities of zidovudine-susceptible and -resistant human immunodeficiency virus isolates to antiviral agents determined by using a quantitative plaque reduction assay, *Antimicrob. Agents Chemother.* 34:436 (1990).

58. M. M. Mansuri, V. Farina, J. E. Starrett, Jr., D. A. Benigni, V. Brankovan, and J. C. Martin, Preparation of the geometric isomers of DDC, DDA, D4C and D4T as potential anti-HIV agents, *Bioorg. Med. Chem. Lett.* 1:65 (1991).

59. S. Hayashi, M. Matsukura, H. Mitsuya, and S. Broder, unpublished results.

60. C.-H. Kim, V. E. Marquez, S. Broder, H. Mitsuya, and J. S. Driscoll, Potential AIDS

drugs. 2',3'-Dideoxycytidine analogues, *J. Med. Chem.* 30:862 (1987).

61. R. Pauwels, M. Baba, J. Balzarini, P. Herdewijn, J. Desmyter, M. J. Robins, R. Zhou, D. Madej, and E. De Clercq, Investigations on the anti-HIV activity of 2',3'-dideoxyadenosine analogues with modifications in either the pentose or purine moiety. Potent and selective anti-HIV activity of 2,6-diaminopurine 2',3'-dideoxyriboside, *Biochem. Pharmacol.* 37:1317 (1988).

62. M. A. Johnson, G. Ahluwalia, M. C. Connelly, D. A. Cooney, S. Broder, D. G. Johns, and A. Fridland, Metabolic pathways for the activation of the antiretroviral agent 2',3'-dideoxyadenosine in human lymphoid cells, *J. Biol. Chem.* 263:15354 (1988).

63. G. I. Birnbaum, M. Cygler, and D. Shugar, Conformational features of acyclonucleosides: structure of acyclovir, an antiherpes agent, *Can. J. Chem.* 62:2646 (1984).

64. P. Van Roey, J. M. Salerno, C. K. Chu, and R. F. Schinazi, Correlation between preferred sugar conformation and activity of nucleoside analogues against human immunodeficiency virus, *Proc. Natl. Acad. Sci. U.S.A.* 86:3929 (1989).

65. C. K.-H. Tseng, V. E. Marquez, G. W. A. Milne, R. J. Wysocki, Jr., H. Mitsuya, T. Shirasaki, and J. S. Driscoll, A ring-enlarged oxetanocin A analogue as an inhibitor of HIV infectivity, *J. Med. Chem.* 34:343 (1991).

66. C. K. Chu, S. K. Ahn, H. O. Kim, J. W. Beach, A. J. Alves, L. S. Jeong, Q. Islam, P. Van Roey, and R. F. Schinazi, Asymmetric synthesis of enantiomerically pure (-)-(1'R,4'R)-dioxolane-thymine and its anti-HIV activity, *Tetrahedron Lett.* 32:3791 (1991).

67. Y.-C. Liaw, Y.-G. Gao, V. E. Marquez, and A. H.-J. Wang, Molecular structures of two new anti-HIV nucleoside analogs: 9-(2,3-dideoxy-2-fluoro-β-D-*threo*-pentofuranosyl)adenine and 9-(2,3-dideoxy-2-fluoro-β-D-*threo*-pentofuranosyl)hypoxanthine, *Nucleic Acids Res.* 20:459 (1992).

68. S. Eriksson, B. Kierdaszuk, B. Munch-Petersen, B. Öberg, and N. G. Johansson, Comparison of the substrate specificities of human thymidine kinase 1 and 2 and deoxycytidine kinase toward antiviral and cytostatic nucleoside analogues, *Biochem. Biophys. Res. Commun.* 176:586 (1991).

69. B. Kierdaszuk, C. Bohman, B. Ullman, and S. Eriksson, Substrate specificity of human deoxycytidine kinase toward antiviral 2',3'-dideoxynucleoside analogues, *Biochem. Pharmacol.* 43:197 (1992).

70. D. A. Cooney, G. Ahluwalia, H. Mitsuya, A. Fridland, M. Johnson, Z. Hao, M. Dalal, J. Balzarini, S. Broder, and D. G. Johns, Initial studies on the cellular pharmacology of 2',3'-dideoxyadenosine, an inhibitor of HTLV-III infectivity, *Biochem. Pharmacol.* 36:1765 (1987).

71. D. Kessel, unpublished results.

72. R. V. Joshi, D. Kessel, T. H. Corbett, and J. Zemlicka, Ynamines derived from nucleic acids bases: synthesis, reactivity and biological activity, *J. Chem. Soc., Chem. Commun.* No. 6:513 (1992).

STRUCTURE–ACTIVITY RELATIONSHIPS AMONG HIV INHIBITORY 4'-SUBSTITUTED NUCLEOSIDES

Ernest J. Prisbe, Hans Maag, Julien P.H. Verheyden, and Robert M. Rydzewski

Syntex Research, Institute of Bio-Organic Chemistry
3401 Hillview Ave.
Palo Alto, CA 94303

The success of AZT in ameliorating the condition and prolonging the life of AIDS patients has stimulated intense efforts to discover other nucleosides with equal or better efficacy. As a result of these efforts, literally dozens of nucleoside analogues have been reported to inhibit HIV *in vitro*.[1] Of these, only a few have entered clinical trials and, at this time, only 2',3'-dideoxyinosine has achieved FDA approval. The mechanism of action of these nucleosides involves anabolism to their 5'-triphosphates which can act as inhibitors of the viral reverse transcriptase and/or be incorporated into the viral DNA where they lead to chain termination.

We have recently reported the potent *in vitro* HIV inhibitory activity of 4'-azido-thymidine.[2] This compound is unique among all other nucleoside inhibitors because it retains the 3'-hydroxyl group. However, in spite of the presence of a 3'-hydroxyl group, it too can act as a chain terminator besides being a potent reverse transcriptase inhibitor.[3] In order to delineate the breadth of activity as it extends over other 4'-substituted nucleosides, we have synthesized a series of analogues. In this series, we have varied the base while an azido group remained at the 4'-position, we have made substitutions at the 3'-position of 4'-azidothymidine, and we have examined the effect of other substituents at the 4'-position of thymidine. From these studies some general trends in structure–activity relationships are evident.

SYNTHESIS OF 4'-AZIDO-2'-DEOXYNUCLEOSIDES

The general method for synthesizing 4'-azido-2'-deoxynucleosides has been previously described[2] and is depicted in Scheme I. Selective iodination at the 5'-position of the starting 2'-deoxynucleosides with triphenylphosphine, iodine and pyridine in dioxane, followed by sodium methoxide induced elimination, afforded the 4',5'-unsaturated derivatives. The addition of iodine azide across the double bond was carried out by first preparing a solution of iodine azide in DMF via the reaction of iodine monochloride with excess sodium azide. To this freshly prepared solution was added the unsaturated

Nucleosides and Nucleotides as Antitumor and Antiviral Agents,
Edited by C.K. Chu and D.C. Baker, Plenum Press, New York, 1993

101

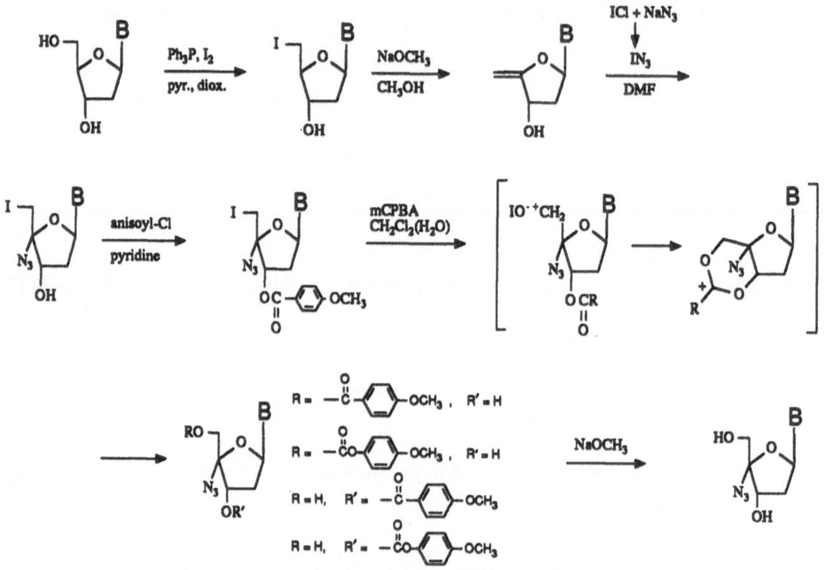

Scheme I

nucleoside. The IN_3 additions went in high yield, achieving both the desired regio- and stereospecificity. This was established unequivocally by ^{13}C nmr.

The most problematic aspect of this synthesis was the conversion of the 5'-iodo group to a hydroxyl group. Presumably, because of the combined inductive effects of the 3'-hydroxyl group, the 4'-azido group, and the 4'-oxygen, nucleophilic displacements at 5' were extremely refractory. This difficulty was overcome by first anisoylating the 3'-hydroxyl using anisoyl chloride in pyridine and then treating the product with 4 equivalents of m-chloroperoxybenzoic acid in dichloromethane saturated with water. The reaction proceeds by first generating a highly reactive hypoiodite, which is attacked intramolecularly by the 3'-ester. The 3',5'-benzoxonium intermediate is hydrolysed to a mixture of 3'- and 5'-anisoate esters of the 4'-azidonucleoside. Interestingly, some of the presumed benzoxonium intermediate escapes hydrolysis and is instead attacked by the excess peracid to furnish a mixture of 3'- and 5'-p-methoxyphenylcarbonates via a Baeyer-Villiger rearrangement. Instead of resolving the mixture of ester and carbonate products at this stage, it proved to be most efficacious to treat the crude mixture with sodium methoxide and separate out the free nucleoside from the now less complex mixture. The yields of the combined iodine displacement-deprotection ranged from 35% to 71%.

The cytidine and inosine analogues in the 4'-azido-2'-deoxy series were made by well known procedures from the uridine and adenosine analogues, respectively (Scheme II). Thus, following the method of Reese,[4] 4'-azido-2'-deoxyuridine was acetylated and then converted to the 4-triazolide, which on treatment with ammonium hydroxide furnished 4'-azido-2'-deoxycytidine. Using a procedure of Herdewijn,[5] 4'-azido-2'-deoxyadenosine was enzymatically deaminated to the inosine analogue.

4'-AZIDO-2'-DEOXY-3'-MODIFIED THYMIDINES

The 3'-deoxy derivative of 4'-azidothymidine was a prime synthetic target based on the anti-HIV activity present in a great many 3'-deoxynucleosides. However, a simple direct synthesis of 4'-azido-3'-deoxythymidine was not feasible. For instance, commonly used deoxygenation methods which rely on hydrogenation or a radical mechanism are not compatible with an azido group and, therefore, could not be used on 4'-azidothymidine.

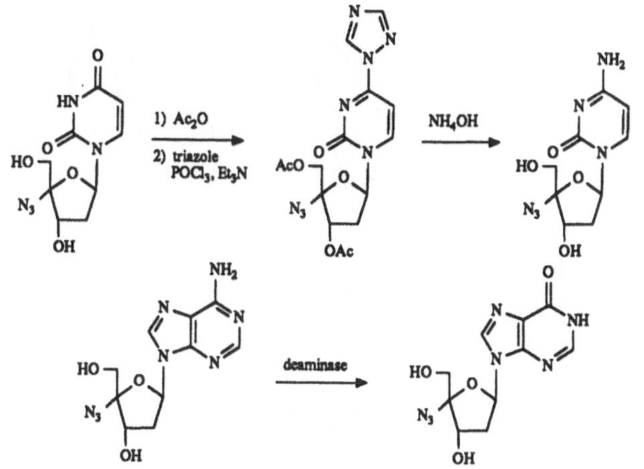

Scheme II

Likewise, the synthesis described above for the 4′-azidonucleosides had, as a key step, participation of a 3′-ester in order to displace a 5′-iodide. When this route was tried starting from 3′-deoxythymidine, the synthesis failed at this step.

A new route to 4′-azido-3′-deoxythymidine was developed which did not use an IN$_3$ addition (Scheme III). Conversion of 3′-deoxythymidine[6] to the 4′,5′-unsaturated derivative was carried out using the iodination/dehydrohalogenation methodology as described above. On treatment of the unsaturated nucleoside with m-chloroperoxybenzoic acid in methanol, a 4′,5′-epoxide was transiently formed, and then opened at the 4′-position with methanol to furnish a 4′-epimeric mixture of methoxy nucleosides. After protection of the 5′-alcohol as the $tert$-butyldimethylsilyl ether, trimethylsilyl azide in the presence of trimethylsilyl triflate was used to displace the 4′-methoxy group with azide and concomitantly cleave the silyl ether. The resulting mixture of 4′-azido epimers was separated by silica gel chromatography and the configuration at C-4′ was determined by NOE NMR experiments.

Since 3′-deoxy-3′-fluorothymidine is one of the most potent inhibitors of HIV known,[7] we were anxious to learn if the presence of a fluorine at the 3′-position of

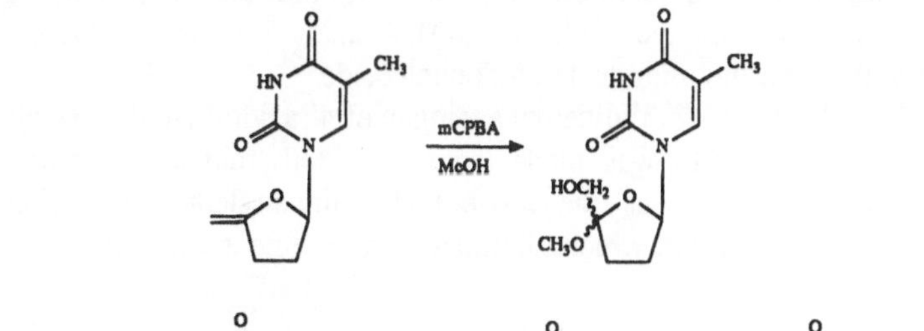

Scheme III

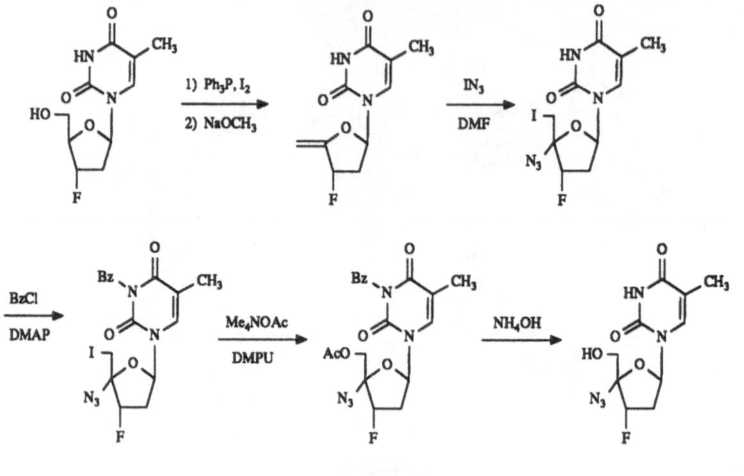

Scheme IV

4′-azidothymidine would lead to greater potency. As in the case of 4′-azido-3′-deoxy-thymidine, however, the lack of a 3′-ester function precluded the use of the orignial methodology to synthesize the 3′-fluoro analogue. For this reason we used the route which proved successful for 4′-azido-3′-deoxythymidine. The synthesis commenced with the 5′-iodination of 3′-deoxy-3′-fluorothymidine followed by methoxide induced elimination. Formation of the 4′-methoxy derivative using *m*-chloroperoxybenzoic acid in methanol proceeded as in the case of the 3′-deoxy analogue, but, to our consternation, attempted displacement of the 4′-methoxy group with trimethylsilylazide and trimethyl-silyltriflate led only to intractable mixtures.

Efforts were redirected back to the original synthetic strategy. The iodine azide addition reaction was applied to the 4′,5′-unsaturated-3′-fluorothymidine resulting in a 57% yield of 4′-azido-3′,5′-dideoxy-3′-fluoro-5′-iodothymidine. The subsequent oxidative displacement of the 5′-iodide failed as expected. However, this synthesis was rescued by a new method,[8] shown in Scheme IV, for carrying out the iodide to hydroxyl conversion. After protecting the N^3-position with a benzoyl group using benzoyl chloride and DMAP in pyridine, the iodonucleoside was heated with 4 equivalents of tetramethylammonium acetate in N^1,N^3-dimethyltetrahydropyrimidone. Displacement of the 5′-iodide by acetate was accompanied by a partial loss of the N^3-benzoyl group. This mixture of 5′-O-acetyl-4′-azido-3′-deoxy-3′-fluorothymidines, with and without the N^3-benzoyl group, was deprotected with ammonium hydroxide to furnish 4′-azido-3′-deoxy-3′-fluorothymidine in 26% overall yield from the unprotected iodonucleoside.

The 2′,3′-didehydro-2′,3′-dideoxy analogue of 4′-azidothymidine (which might be thought of as 4′-azido-D4T) was made from 4′-azidothymidine (Scheme V). After protection of the 5′-alcohol as the *tert*-butyldimethylsilylether, the 3′-alcohol was derivatized as the triflate ester, which eliminated on treatment with mild base. Fluoride

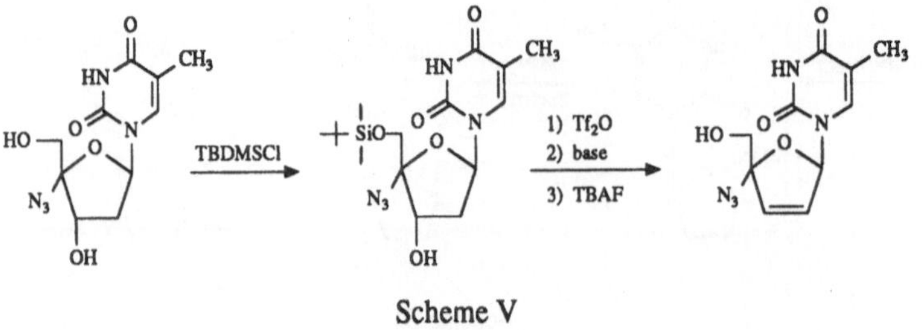

Scheme V

104

ion induced cleavage of the 5'-silylether furnished the final product.

A direct, but low yielding, methylation procedure was used to make 4'-azido-3'-O-methylthymidine. As depicted in Scheme VI, 4'-azidothymidine was tritylated at the 5'-position, methylated using methyl iodide and potassium hydroxide, and then detritylated with acetic acid. Not unexpectedly, the major product of this reaction was 4'-azido-N^3-methylthymidine.

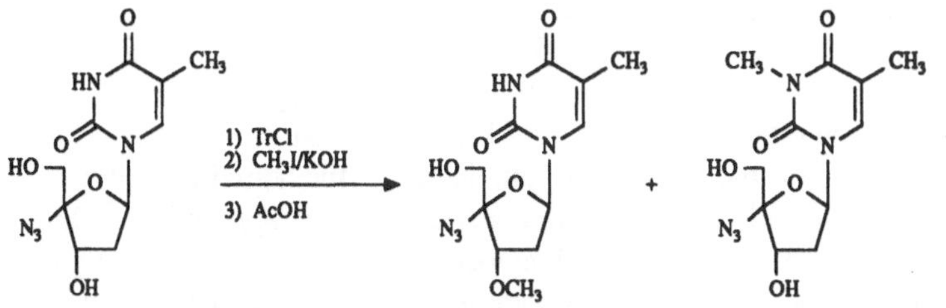

Scheme VI

4'-SUBSTITUTED THYMIDINES

Nucleosides having an azido group at the 4'-position were the primary, but not the only focus of our attention. A few examples with other substituents at the 4'-position of thymidine were also examined.

The synthesis of 4'-azido-3'-deoxythymidine described above made use of a 4'-methoxy intermediate. Essentially the same methodology led to 4'-methoxythymidine (Scheme VII). Thus, 4',5'-unsaturated thymidine was epoxidized at room temperature with m-chloroperoxybenzoic acid in methanol and a 3:2 mixture of 4'-methoxythymidine and its 4'-epimer was directly isolated. It might be noted that the 3'-hydroxyl group has little influence on the stereoselectivity of this epoxidation. When the 3'-hydroxyl group of 4',5'-unsaturated thymidine was protected as a silyl ether prior to epoxidation, the ratio of the resulting 4'-methoxy, β-D and α-L isomers was essentially the same as when the reaction was performed with the hydroxyl group unprotected. The assignment of configuration at 4' was made following the same ^{13}C NMR rules used to establish the 4'-configuration of the 4'-azidonucleosides.

Thymidine analogues having a carbon branch at the 4'-position were synthesized by other workers at Syntex. Extending the methodology first used to synthesize many 4'-hydroxymethyl ribosides,[9a] 4'-hydroxymethylthymidine was made via a formaldehyde condensation and Cannizarro reduction of 3'-O-benzylthymidine 5'-aldehyde, followed by debenzylation.[9b] Recently, this product served as the starting material for the synthesis of 4'-methyl,[10] 4'-azidomethyl,[10] and 4'-cyanothymidine.[11]

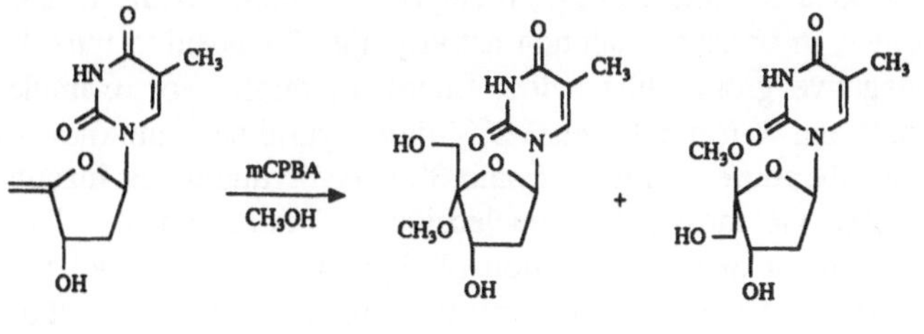

Scheme VII

In order to facilitate an examination of the effect of substitution in this series, the 4'-modified nucleosides are divided by the type of modification and are presented in Tables 1-3 with their anti-HIV activity and cellular toxicity. Testing was carried out using the lab virus strain HIV-1 LAV_β grown in the $CD4^+$ T cell line, A3.01. IC_{50}'s were determined as the concentration of compound which reduced virus levels 50% compared with control cultures. Cytotoxicity was determined by examination of drug treated, mock infected, A3.01 cells, and is shown as a CC_{25} which indicates the concentration with which about 25% cell destruction was observed, and as a CC_{100} which, accordingly, indicates the concentration producing complete destruction of the cell layer. A selectivity index is also shown which is the ratio CC_{25} / IC_{50}. AZT was included in every test as a control.

The effect of the change of the base on the antiviral activity of 4'-azido-nucleosides is illustrated in Table 1. In every case, the IC_{50} remained below 1μM. The most potent analogues were those with cytosine (IC_{50} = 0.004 μM) and guanine (IC_{50} = 0.003 μM) as the base. However, they were the most toxic, both displaying partial (CC_{25}) toxicity at 0.21 μM. Thymine, 5-chlorouracil, and 5-methylcytosine imparted intermediate activity with decreased toxicity. Notable among this triade is the 5-chlorouracil derivative, which displays the least toxicity to this cell line of any of the base modified nucleosides. This correlates with the observation of others[12] that the substitution of chlorine into the 5-position of anti-HIV 2',3'-dideoxypyrimidine nucleosides reduces their toxicity thereby increasing the selectivity. It is generally found that among nucleoside inhibitors of HIV, those having uracil, adenine, or hypoxanthine are less active than the parent compound having thymine, cytosine, or guanosine. This trend continued in the 4'-azido series as these analogues were the weakest inhibitors.

Table 2 underscores the importance of the 3'-hydroxyl group to the activity of 4'-azidothymidine. Replacement of the hydroxyl with hydrogen, or elimination to form the 2',3'-unsaturated analogue completely abolishes antiviral activity although some cellular toxicity, particularly in the case of the olefin, remains. A substantial drop in activity, also occurs when the 3'-hydroxyl is replaced with a 3'-O-methyl or 3'-fluoro group. These changes increase the IC_{50} 2000 fold as compared to the 3'-alcohol. This intolerance to modification at the 3'-position sharply contrasts with the structure-activity relationship of most other anti-HIV nucleosides where a variety of 3'-modifications are present. This may suggest that the 3'-hydroxyl group of 4'-azidothymidine is required for recognition by the kinases that catalyze its conversion to the triphosphate.[13]

Unlike the 3'-position of 4'-azidothymidine, the 4'-position is much less discriminatory. Table 3 lists the variation in activity with changes at the 4'-position of thymidine. Azido and cyano impart the greatest inhibition of HIV having IC_{50}'s in the nanomolar range. The substitution of 4'-methyl or 4'-azidomethyl decreases the activity 2 to 3 orders of magnitude. A further decrease of 50 to 75% occurs when a hydroxy-methyl or methoxy group occupies the 4'-position. The toxicity of this group of nucleosides roughly parallels their antiviral activity.

Another contrast with the 2',3'-dideoxynucleoside family of HIV inhibitors is elucidated by the data in Table 3. Potent activity among the 4'-derivatives is maintained even when a nonpolar substituent (e.g., methyl) is present. While in the case of 2',3'-dideoxynucleosides, in order to maintain activity, the 3'α position must be hydrogen or certain electronegative groups like azido, fluoro, or thiol. For example, 3'-deoxy-3'-methylthymidine[14] and 3'-(propyl-2-ene)-2',3'-dideoxyuridine[15] are known to be inactive against HIV. Furthermore, while 3'-cyano-3'-deoxythymidine is inactive,[16] 4'-cyano-thymidine is among the most potent nucleoside inhibitors known. Thus, the activity elicited by a substituent at the 3'-position of thymidine is not at all predictive of the activity to be expected when the same substituent resides at the 4'-position.

Table 1.   Effect of base substitutions on the inhibition of HIV-1 (LAV-III$_B$) replication in A3.01 cells.

B	IC_{50} (μM)	CC_{25} (μM)	CC_{100} (μM)	S.I.
thymine	0.01	8	200	800
uracil	0.8	200	>200	250
cytosine	0.004	0.21	1.9	52
adenine	0.13	50	>50	385
hypoxanthine	0.65	<22	200	<34
guanine	0.003	0.21	1.9	70
5-Cl-uracil	0.056	1000	>1000	17,900
5-Me-cytosine	0.006	5	>200	833
AZT	0.01	825	3300	82,500
D4T	0.47	133	400	283

Table 2.   Effect of 3'-substitutions on the inhibition of HIV-1 (LAV-III$_B$) replication in A3.01 cells.

R	IC_{50} (μM)	CC_{25} (μM)	CC_{100} (μM)	S.I.
OH	0.01	8	200	800
H	inactive	22	200	---
OCH$_3$	20.4	>200	>200	>9.8
F	19.7	>200	>200	>10
	inactive	2.5	7.4	---

R	IC$_{50}$ (µM)	CC$_{25}$ (µM)	CC$_{100}$ (µM)	S.I.
N$_3$	0.01	8	200	800
CN	0.002	1	15	500
CH$_3$	3.5	111	333	32
CH$_2$OH	12.5	>200	>200	>16
CH$_2$N$_3$	2.1	>333	1000	>159
OCH$_3$	8.49	>200	>200	>24

From the 4'-substituted nucleosides examined thus far, it can be concluded that anti-HIV activity extends over a sizable range of modification if certain positions are left intact. Figure 1 summarizes where changes are tolerated. It appears that in order to retain good activity the 4'-substituted nucleoside must be a 2'-deoxy-β-D-*erythro* derivative.[17] However, it is remarkable that once these criteria were satisfied, none of the changes of the heterocyclic base resulted in a dramatic loss of activity. Furthermore, as mentioned before, the 4'-substituent could be of different types, although azido and cyano were the most effective.

CONFORMATIONAL ANALYSIS

That the antiviral activity of the 4'-substituted nucleosides extends over a variety

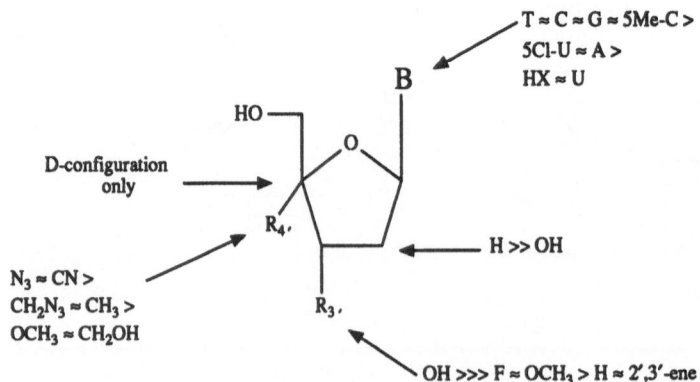

Figure 1.   SAR Summary - In Vitro Inhibition of HIV

of substituent types is suggestive of a common change this modification has on the nucleoside structure. In order to gain insight into how the presence of a group at the 4'-position accounts for the anti-HIV activity, we began by examining the three-dimensional molecular structure of 4'-azidothymidine.

The furanose ring of nucleosides occurs in two ranges of low energy conformations referred to as C-3'-endo, Northern form (N-Type) and C-2'-endo Southern form (S-Type). The preference for either of these forms depends upon the substitution pattern of the nucleoside. Thymidine, for instance, favors a C-2'-endo conformation in its crystalline state,[18] although in solution NMR studies indicate that an equilibrium mixture of C-2'-endo and C-3'-endo forms are present with the former being somewhat more favored.[19] The substitution of a fluorine atom for the 3'-hydroxyl group of thymidine, however, forces the furanose ring to adopt an extreme C-3'-exo (S-Type) conformation.[19,20]

It would be expected that substituting the 4'-hydrogen of a nucleoside for a larger and/or electronegative group would also have an effect on the furanose conformation. A clue as to this effect was provided during its chemical synthesis. During the reaction in which the 5'-iodo group of 3'-O-anisoyl-4'-azido-5'-deoxy-5'-iodothymidine was displaced, an intermediate was formed in which the 3' and 5' oxygens were bridged with a cyclic benzoxonium ion. This sort of bridge, joining oxygens trans disposed on the furanose ring, could only form if the ring is able to easily adopt a C-3'-endo or C-4'-exo (N-Type) conformation (Fig. 2 - only C-3'-endo shown). Such a conformation would position both the 3'-oxygen and 5'-carbon pseudoequitorially on the ring and in close enough proximity to allow joining via the benzoxonium intermediate. The adoption of this proposed conformation may be rationalized on theoretical grounds. The introduction of an electronegative substituent at the 4'-position would have a strong influence on the furanose conformation. Due to the anomeric effect,[21] the 4'-substituent would prefer an axial orientation. Then the 3'-hydroxyl group (or derivatized hydroxyl) would be forced into a pseudoequitorial position to avoid eclipsing the 4'-substituent. The result would be the presence of both the 3'-hydroxyl group and the 5'-carbon pseudoequitorial with the furanose ring in a C-3'-endo/C-4'-exo conformation. As stated above, this is the ideal conformation amenable to the formation of the benzoxonium bridge structure.

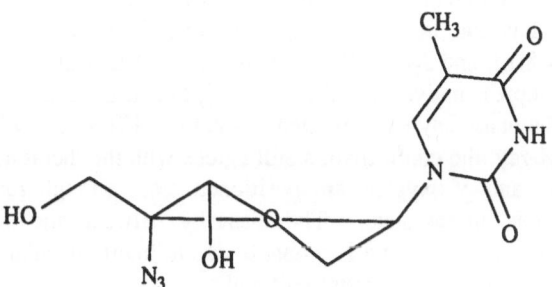

Figure 2.   C3'-endo (N-Type) conformation of 4'-azidothymidine.

While the empirical and theoretical arguments coincidentally pointed to an N-Type furanose conformation, further insight was provided by the X-ray crystallographic structure of 4'-azidothymidine. Two slightly different molecular forms are present in the crystal structure and are depicted in Figure 3. The difference between the two structures is in the glycosidic torsion angle which differs by 17° and the C-4'—C-5' torsion angle which differs by 23°. The furanose ring conformations are identical and are clearly of the N-Type. A quantitative measure of the furanose conformation can be derived using the methodology of Sundaralingam.[18] From the furanose ring torsion angles obtained from the crystal structure, the position of the out-of-plane ring twist or envelope is determined. This "phase angle" is computed to be equal to 14° for both crystallographic forms of 4'-azidothymidine and corresponds to a C-3'-endo (N-Type) conformation. Thus, the crystal structure is in accord with the other conformational evidence and predictions.

We also examined the solution conformation of 4'-azidothymidine by NMR. From the proton-proton coupling constants, $J_{1',2'}$, $J_{1',2''}$, $J_{2',3'}$, $J_{2'',3}$ the torsion angles between the 1', 2', and 3' protons were determined using the method of Slessor and Tracey.[22] This

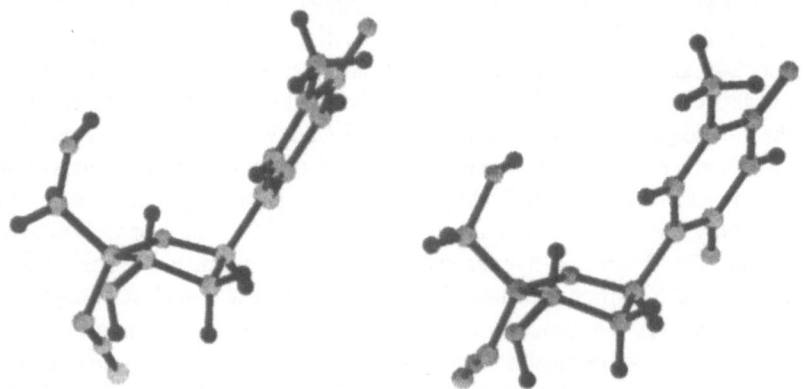

Figure 3.   Crystal structure conformations of the two independent molecules of 4'-azidothymidine.

derivation of dihedral angles is a variation of the Karplus relationships in which the influences of electronegativity and ring strain are minimized through the use of the ratio of Karplus constants, rather than the constants themselves. From the derived proton torsion angles, the torsion angles of all the furanose ring carbons can be obtained. It is these five endocyclic torsion angles which are used to determine the phase angle of pseudorotation.[18] Thus, the proton coupling constants, obtained in D_2O, of $J_{1',2'}$ = 3.79, $J_{1',2''}$ = 8.17, $J_{2',3'}$ = 8.13, and $J_{2'',3'}$ = 8.32 Hz, are associated with a phase angle equal to 63°. This angle is representative of a 4'-exo (N-Type) conformation.

Although the phase angle determined in solution differs somewhat from the phase angle obtained by X-ray, the conformation still agrees with the theoretical arguments and is such that the 3'- and 5'-oxygens are positioned close enough for bridging by the benzoxonium reaction intermediate. The disparity between the solid and solution conformations is not unexpected and is probably due to intermolecular associations (e.g., hydrogen bonding) present in the crystal unit cell.

It is worth noting that while the C-3'-endo (N-Type) conformation is preferred by 4'-azidothymidine, most other potent nucleoside inhibitors of HIV prefer an S-Type conformation or maintain an equilibrium mixture of S- and N-Type. For example, X-ray crystallographic studies have shown that AZT,[15] AZddU,[15] FddT,[20a] 5Cl-FddU,[12] ddA[15,23] and ddC[15] all maintain a C-3'-exo (S-Type) conformation in the solid state. In solution, the conformation is sometimes different. From the analysis of NMR spectra, AZT[20b,24] and AZddU[22] are predicted to exist as a nearly equal mixture of the C-3'-exo (S-Type) and C-3'-endo (N-Type) forms. ddA[24] and ddC[24] slightly favor an N-Type conformation in solution, while FddT[20b,24] remains entirely C-3'-exo (S-Type).

The seemingly strong preference for 4'-azidothymidine to reside in an N-Type conformation (3'-endo or 4'-exo) may account for its ability to act as a chain terminator in spite of the presence of a 3'-hydroxyl group. We have observed that the elongation of a DNA chain on a DNA template slows after the incorporation of a single 4'-azido-thymidine and ceases if two consecutive 4'-azidothymidines are inserted.[3] Double-stranded DNA exists in the B-DNA form in which the furanose moieties are of the S-Type conformation. This orients the 3'-substituent pseudoaxially. 4'-Azidothymidine, on the other hand, maintains its 3'-hydroxyl group in a pseudoequatorial position. Thus, after 4'-azidothymidine joins the growing DNA chain, its 3'-hydroxyl group may be out of position for ideal linking to the next incoming nucleotide. This would impede elongation. Two consecutive insertions of 4'-azidothymidine might be expected to further exaggerate the mispositioning of the 3'-hydroxyl group and lead to complete cessation of polymerization.

CONCLUSIONS

The 4'-position of nucleosides has previously been ignored as a site of modification when seeking inhibitors of HIV. Yet many substitutions at this position are capable of inducing potent anti-HIV activity. Furthermore, in the case of 4'-azido-2'-deoxynucleosides, this activity is maintained throughout a broad series of both purine and pyrimidine bases. Such scope of activity in a single nucleoside series is very rare.

The presence of an azido group at the 4'-position (and this may be true for other substituents as well) induces a strong preference for an unusual N-Type conformation of the furanose ring. It has yet to be determined what role, if any, this conformation may play in the inhibition of HIV, but it may be a factor in the mechanism of DNA chain termination. In any case, the modification of the 4'-position of nucleosides remains a largely unexplored path to potential chemotherapeutics.

ACKNOWLEDGEMENTS

Thanks are due to Mary Jane McRoberts for viral screening, Janis T. Nelson for NMR coupling determinations, and to Nicole Grinder for manuscript preparation.

REFERENCES

1. M. Nasr, C. Litterest, J. McGowan, Computer-assisted sructure-activity correlations of dideoxynucleoside analogs as potential anti-HIV drugs, *Antiviral Res.* 14:125 (1990).

2. H. Maag, R.M. Rydzewski, M.J. McRoberts, D. Crawford-Ruth, J.P.H. Verheyden, and E.J. Prisbe, Synthesis and anti-HIV activity of 4'-azido- and 4'-methoxy-nucleosides, *J. Med. Chem.* 35:1440 (1992).

3. M.S. Chen, R. Suttmann, C. Bach, J.C. Wu, E. Prisbe, M.J. McRoberts, and D. Crawford-Ruth, Mechanism of the inhibitory effect of 4'-azidothymidine (ADRT) on the replication of human immunodeficiency virus in vitro, Fourth International Conference on Antiviral Research, New Orleans, LA, Poster #87 (1991).

4. K.J. Divaker and C.B. Reese, 4-(1,2,4-Triazol-1-yl)- and 4-(3-nitro-1,2,4-triazol-1-yl)-1-(β-D-2,3,5-tri-O-acetylarabinofuranosyl)pyrimidin-2(1H)-ones. Valuable intermediates in the synthesis of derivatives of 1-(β-D-arabinofuranosyl)cytosine (ARA C), *J. Chem. Soc., Perkin Trans. 1*, 1171 (1982).

5.   P.A.M. Herdewijn, Anchimeric assistance of a 5'-O-carbonyl function for inversion of configuration at the 3'-carbon atom of 2'-deoxyadenosine. Synthesis of 3'-azido-2',3'-dideoxyadenosine and 3'-azido-2',3'-dideoxyinosine, *J. Org. Chem.* 53:5050 (1988).

6.   E.J. Prisbe and J.C. Martin, A novel and efficient preparation of 2',3'-dideoxy-nucleosides, *Synth. Commun.* 15:401 (1985).

7.   P. Herdewijn, J. Balzarini, E. DeClercq, R. Pauwels, S. Broder, and H. Vanderhaeghe, 3'-Substituted 2',3'-dideoxynucleoside analogues as potential anti-HIV (HTLV-III/LAV) agents, *J. Med. Chem.* 30:1270 (1987).

8.   D.J. Morgans and H.H. Chapman, unpublished results.

9.   (a) R.D. Youssefyeh, J.P.H. Verheyden, and J.G. Moffatt, 4'-Substituted nucleosides. 4. Synthesis of some 4'-hydroxymethyl nucleosides, *J. Org. Chem.* 44:1301 (1979). (b) G.H. Jones, M. Taniguchi, D. Tegg, and J.G. Moffatt, 4'-Substituted nucleosides. 5. Hydroxymethylation of nucleoside 5'-aldehydes, *J. Org. Chem.* 44:1309 (1979).

10.   C. O-Yang, W. Kurz, E.M. Eugui, M.J. McRoberts, J.P.H. Verheyden, L.J. Kurz, and K.A.M. Walker, 4'-Substituted nucleosides as inhibitors of HIV: An unusual oxetane derivative, *Tetrahedron Lett.* 33:41 (1992).

11.   C. O-Yang, H.Y. Wu, E.B. Fraser-Smith and K.A.M. Walker, Synthesis of 4'-cyanothymidine and analogs as potent inhibitors of HIV, *Tetrahedron Lett.* 33:37 (1992).

12.   A. Van Aerschot, D. Everaert, J. Balzarini, K. Augustyns, L. Jie, G. Janssen, O. Peeters, N. Blaton, C. DeRanter, E. DeClercq, and P. Herdewijn, Synthesis and anti-HIV evaluation of 2',3'-dideoxyribo-5-chloropyrimidine analogues: Reduced toxicity of 5-chlorinated 2',3'-dideoxynucleosides, *J. Med. Chem.* 33:1833 (1990).

13.   M.S. Chen, R.T. Suttmann, J.C. Wu, and E.J. Prisbe, Metabolism of 4'-azido-thymidine, *J. Biol. Chem.* 267:257 (1992).

14.   K. Agyei-Aye and D.C. Baker, Synthesis and evaluation of a series of 1-(3-alkyl-2,3-dideoxy-α,β-D-*erythro*-pentofuranosyl)thymines, *Carbohydr. Res.* 183:261 (1988).

15.   P. Van Roey, J.M. Salerno, C.K. Chu, and R.F. Schinazi, Correlation between preferred sugar ring conformation and activity of nucleoside analogues against human immunodeficiency virus, *Proc. Natl. Acad. Sci., U.S.A.* 86:3929 (1989).

16.   C.W. Greengrass, D.W.T. Hoople, S.D.A. Street, F. Hamilton, M.S. Marriot, J. Bordner, A.G. Dalgleish, H. Mitsuya, and S. Broder, 1-(3-Cyano-2,3-dideoxy-β-, D-*erythro*-pentofuranosyl)thymidine: Synthesis and antiviral evaluation against human immunodeficiency virus, *J. Med. Chem.* 32:618 (1989).

17.   Several 4'-substituted-α-L-*threo*-nucleosides were tested and all were found to be inactive against HIV.

18. C. Altona and M. Sundaralingam, Conformational analysis of the sugar ring in nucleosides and nucleotides. A new description using the concept of pseudorotation, *J. Am. Chem. Soc.* 94:8205 (1972).

19. N. Hicks, O.W. Howarth, and D.W. Hutchinson, N.M.R. studies of the flexibility of the glycosyl ring in thymidine and uridine nucleosides, *Carbohydr. Res.* 216:1 (1991).

20. (a) N. Camerman, D. Mastropaolo, and A. Camerman, Structure of the anti-human immunodeficiency virus agent 3'-fluoro-3'-deoxythymidine and electronic charge calculations for 3'-deoxythymidines, *Proc. Natl. Acad. Sci. U.S.A.* 87:3534 (1990). (b) J. Plavec, L.H. Koole, A. Sandström and J. Chattopadhyaya, Structural studies of anti-HIV 3'-α-fluorothymidine and 3'-α-azidothymidine by 500 MHz ^{1}H-NMR spectroscopy and molecular mechanics (MM2) calculations, *Tetrahedron* 47:7363 (1991).

21. (a) A.J. Kirby. "The Anomeric Effect and Related Stereoelectronic Effects at Oxygen," Springer Verlag, Berlin (1983). (b) P. Deslongchamps. "Stereoelectronic Effects in Organic Chemistry," Pergamon Press, Oxford (1983).

22. K.N. Slessor, and A.S. Tracey, Couplings into methylene groups: A new nuclear magnetic resonance approach to stereochemistry, *Can. J. Chem.* 49:2874 (1971).

23. C.K. Chu, V.S. Bhadti, B. Doboszewski, Z.P. Gu, Y. Kusugi, K.C. Pullaiah, and P. Van Roey, General synthesis of 2',3'-dideoxynucleosides and 2',3'-didehydro-2',3'-dideoxynucleosides, *J. Org. Chem.* 54:2217 (1989).

24. B. Jagannadh, D.V. Reddy, and A.C. Kunwar, ^{1}H NMR study of the sugar pucker of 2',3'-dideoxynucleosides with anti-human immunodeficiency virus (HIV) activity, *Biochem. Biophys. Res. Commun.* 179:386 (1991).

ADENOSINE-DERIVED 5'-α-HALO THIOETHER, SULFOXIDE, SULFONE, AND (5'-HALO)METHYLENE ANALOGUES. INHIBITION OF S-ADENOSYL-L-HOMOCYSTEINE HYDROLASE

Morris J. Robins,[*1] Stanislaw F. Wnuk,[1] Khairuzzaman B. Mullah,[1] N. Kent Dalley,[1] Ronald T. Borchardt,[2] Younha Lee,[2] and Chong-Sheng Yuan[2]

[1]Department of Chemistry, Brigham Young University, Provo, Utah 84602 and [2]Departments of Biochemistry and Medicinal Chemistry, University of Kansas, Lawrence, Kansas 66045

INTRODUCTION

S-Adenosyl-L-methionine is the principal methyl group donor in a variety of metabolic pathways. The byproduct of this process, S-adenosyl-L-homocysteine (AdoHcy), is a potent feedback inhibitor of methyltransferase enzymes.[1] S-Adenosyl-L-homocysteine hydrolase (AdoHcy hydrolase) catalyzes the hydrolysis of AdoHcy (**1**) to adenosine (Ado, **4**) and L-homocysteine (Hcy) via a reversible oxidation, elimination, Michael addition, reduction mechanism[1-3] (see Scheme 1). Since it is crucial for continuing metabolism to remove intracellular AdoHcy in order for enzymatic methylation to proceed, inhibition of AdoHcy hydrolase presents a rational target for anticancer and antiviral chemotherapy.[3b] We hypothesized that adenosine 5'-α-halo thioethers or their oxidized analogues might function as mechanism-based inhibitors of AdoHcy hydrolase if they were bound and converted into inhibitory species by the oxidation/elimination processes.[4] It had been demonstrated that 4',5'-didehydro-5'-deoxyadenosine [9-(5-deoxy-β-D-*erythro*-pent-4-enofuranosyl)adenine, **6**] was accepted as an alternative substrate by AdoHcy hydrolase and converted into Ado and AdoHcy.[2,5] Therefore, we also targeted 5'-halo analogues of **6** as potential mechanism-based inhibitors with adenosine 5'-α-halosulfoxides as precursors. We now summarize our syntheses of 5'-α-halo thioethers, sulfoxides, and sulfones from 5'-S-(alkyl or aryl)-5'-thioadenosines; thermal *syn* conversions of diastereomeric sulfoxides into 5'-halomethylene analogues of **6**; stereochemical determinations by X-ray crystallography and ^{19}F and ^{1}H NMR spectroscopy; and time-dependent inactivation of AdoHcy hydrolase.

5'-(α-FLUORO THIOETHER AND FLUOROMETHYLENE) ANALOGUES

Treatment of thioethers with xenon difluoride,[6] or α-chloro thioethers with potassium fluoride/crown ether,[7] had been reported to give α-fluoro thioethers. McCarthy and

Nucleosides and Nucleotides as Antitumor and Antiviral Agents,
Edited by C.K. Chu and D.C. Baker, Plenum Press, New York, 1993

115

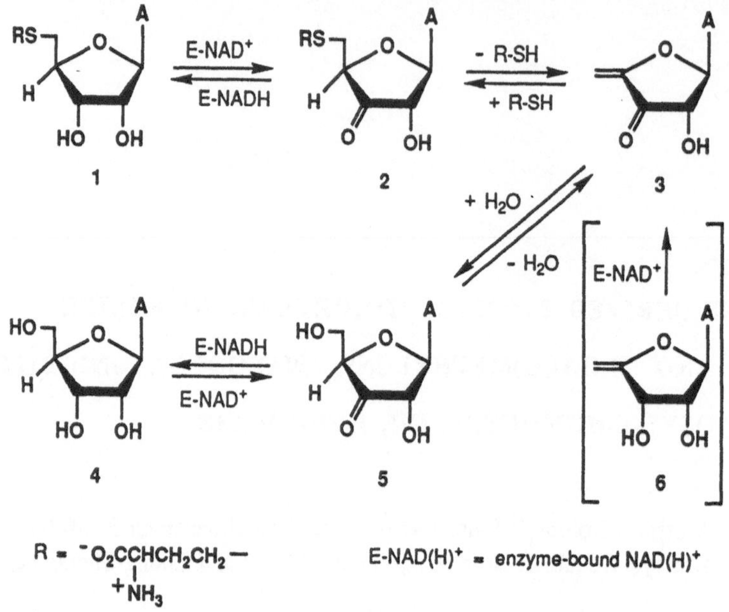

Scheme 1. Mechanism proposed for S-adenosyl-L-homocysteine hydrolase

coworkers reported conversions of alkyl aryl sulfoxides into α-fluoro thioethers with DAST (diethylaminosulfur trifluoride).[8] N-Fluoropyridinium triflates have since been used,[9] and an electrolysis method was reported recently.[10] Alkylation of thiosugars with fluoromethyl reagents to give mono-, di-, and [trifluoro(methyl)]thio compounds also has been noted.[11-13]

We applied the McCarthy DAST procedure[8] to the sulfoxides **9a** (see Scheme 2) derived by selective oxidation of 2',3'-di-O-acetyl-5'-S-phenyl-5'-thioadenosine (**8a**) [~1 equivalent MCPBA (3-chloroperoxybenzoic acid)/-40 °C], but observed minimal conversion to the desired α-fluoro thioethers.[4] The major product was the deoxygenated thioether precursor **8a**. Addition of zinc(II) iodide as catalyst[8] resulted in rapid deoxygenation of **9a** to give **8a**. Facile deoxygenation of sulfoxides to thioethers had been reported with sodium iodide and boron trifluoride etherate,[14] so we investigated other Lewis acid systems that did not contain iodide.

We found that antimony(III) chloride and antimony(V) fluoride catalyzed these α-fluorination reactions efficiently with minimal discoloration and formation of byproducts.[4] Thus, treatment of the **9a** mixture with DAST/SbCl$_3$/CH$_2$Cl$_2$ gave smooth conversion to the **11a** diastereomers. Oxidation of **7a** and fractional crystallization gave the R (at sulfur, S$_R$) sulfoxide (see Figure 1). Acetylation of this isomer and subjection of the resulting **9a**(S$_R$) to the same reaction conditions resulted in formation of the same diastereomeric mixture of **11a**, indicating no dependence on the sulfoxide stereochemistry. The ease of handling of the colorless crystalline SbCl$_3$ led to its choice for general use.[15] McCarthy had noted[8] that 4-methoxyphenyl sulfoxides were much more reactive than their phenyl analogues, and we observed a marked enhancement with 3',5'-di-O-acetyl-5'-[(4-methoxyphenyl)sulfinyl]-5'-deoxyadenosine[4] (**9b**). In fact, treatment of **9b** with DAST/ZnI$_2$ also gave good yields of the α-fluoro thioether diastereomers **11b**, but our catalysis by SbCl$_3$ gave **11b** more rapidly and with less coloration.

It is noteworthy that this SbCl$_3$-catalyzed α-fluorination process is almost as efficient with the 5'-S-phenyl compounds as with the much more expensive 5'-S-(4-methoxyphenyl) analogues.[4,15] The 5'-S-(4-chlorophenyl) nucleoside 9c was converted into diastereomeric α-fluoro thioethers 11c, although this reaction was sluggish and some discoloration occurred. Treatment of sulfoxide derivative 16 (or its phenyl analogue) with DAST, with or without SbCl$_3$, failed to effect a second α-fluorination to give the 5',5'-difluoro-5'-S-aryl-5'-thioadenosine derivatives. We then prepared a uridine 5'-S-(4-hydroxyphenyl) derivative to evaluate the possibility of greater activation of the phenyl ring with a phenolic (vs. alkoxy) substituent. However, *no* α-fluorination was observed with the phenolic compound. Acetylation gave the [4-(acetyloxy)phenyl]thio nucleoside, which underwent mono but not difluorination analogous to the 4-methoxyphenyl compounds.[16]

Fluorination of protected nucleoside thioethers with xenon difluoride was found to proceed smoothly.[16] Treatment of 8a-c with XeF$_2$/CH$_2$Cl$_2$ at low temperature gave the 11 diastereomers. Yields from the two processes were comparable, and diastereomeric ratios were not remarkably different in most cases. The cost and manipulations required for oxidation of nucleoside thioethers to their sulfoxides, plus the ratios of excess DAST noted,[8] sometimes balance the cost of the stoichiometric XeF$_2$. Treatment of the 3',5'-di-O-acetyl

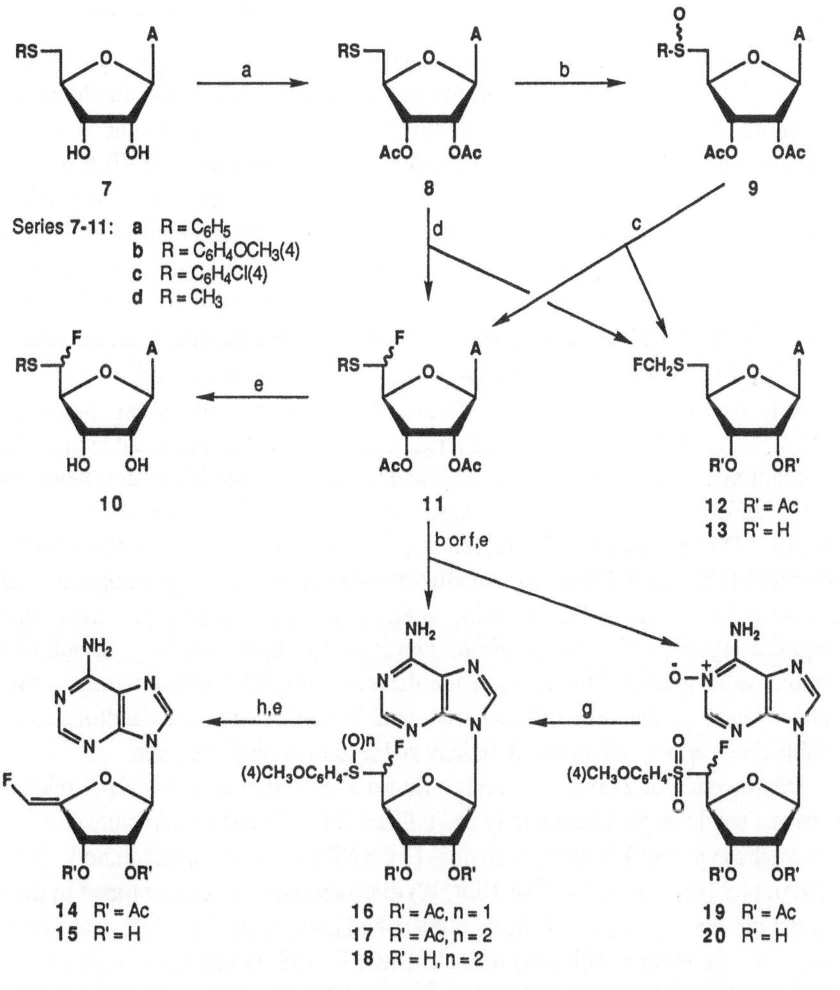

Series 7-11: a R = C$_6$H$_5$
b R = C$_6$H$_4$OCH$_3$(4)
c R = C$_6$H$_4$Cl(4)
d R = CH$_3$

12 R' = Ac
13 R' = H

14 R' = Ac
15 R' = H

16 R' = Ac, n = 1
17 R' = Ac, n = 2
18 R' = H, n = 2

19 R' = Ac
20 R' = H

(a) Ac$_2$O/pyridine. (b) MCPBA/CH$_2$Cl$_2$/-40 °C. (c) DAST/SbCl$_3$/CH$_2$Cl$_2$. (d) XeF$_2$/CH$_2$Cl$_2$/-25 °C to ambient. (e) NH$_3$/MeOH. (f) MCPBA/CH$_2$Cl$_2$/ambient. (g) Si$_2$Cl$_6$/CH$_2$Cl$_2$. (h) Diglyme/i-Pr$_2$NEt/145 °C.

Scheme 2. Synthesis of fluoro analogues from adenosine 5'-thioethers

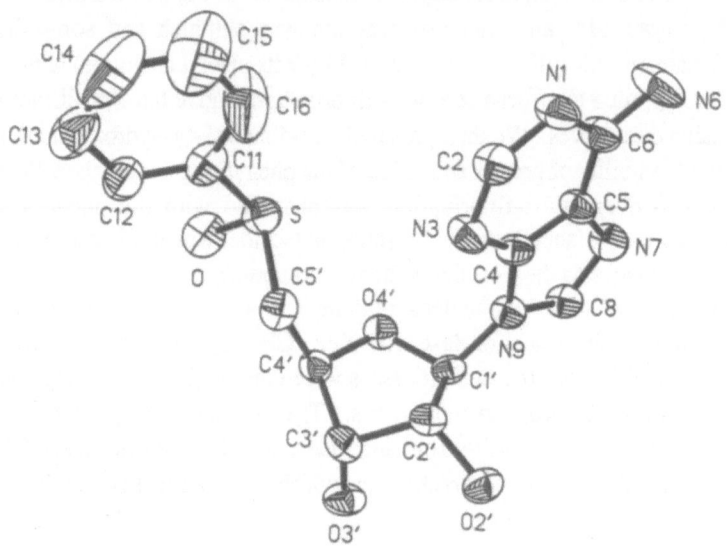

Figure 1. X-Ray crystal structure of 5'-deoxy-5'-phenylsulfinyladenosine(S_R)

derivative (**8d**) of 5'-S-methyl-5'-thioadenosine (**7d**, the metabolic byproduct of enzymatic decarboxylation, followed by aminopropyl transfer from S-adenosyl-L-methionine[17]) with XeF$_2$, or its sulfoxide **9d** with DAST/SbCl$_3$,[4] or DAST,[18] gave the diastereo- and regioisomeric mixture of 5'-fluoro-5'-S-methyl-5'-thio- (**11d**) and 5'-S-(fluoromethyl)-5'-thioadenosine (**12**) derivatives. The sensitive fluoromethyl thioether **12** was stable enough for chromatographic purification, deprotection to **13**, and gentle manipulation. However, the 5'-fluoro-5'-methylthio diastereomers **11d** were labile and decomposed significantly on silica gel in methanol-containing solvents to give mixed methoxy/methylthio acetals.[4] Deprotection (NH$_3$/MeOH) and careful chromatography with basic systems resulted in isolation of the very sensitive 5'-fluoro-5'-S-methyl-5'-thioadenosine (**10d**) diastereomers.[4,18]

Selective oxidation of the α-fluoro thioethers to sulfoxides (~1 equivalent MCPBA/CH$_2$Cl$_2$/-40 °C) was readily effected. Excess MCPBA at ambient temperature gave α-fluoro sulfones. The α-fluoro sulfoxides and sulfones were stable against neutral ambient solvolysis and other processes that caused decomposition of their α-fluoro thioether precursors. This enhanced stability was especially pronounced with the oxidized 5'-S-fluoromethyl (**12**) and 5'-fluoro-5'-S-methyl-5'-thio (**11d**) adenosine compounds, and also with 2'-fluoro-2'-S-methyl-2'-thiouridine analogues.[19] The methyl α-fluoro thioethers decomposed significantly during chromatography and upon standing, in solution or as amorphous solid glasses. In contrast, the derived sulfoxides (diastereomers often were amorphous) and especially the sulfones were well-behaved compounds in protic solvents for reasonable time periods and were indefinitely stable in the crystalline state.

Deprotection of 2',3'-di-O-acetyl-5'-fluoro-5'-S-(4-methoxyphenyl)-5'-thioadenosine (**11b**) gave a solid that was fractionally crystallized (MeOH) and the structure of 5'(S)-fluoro-5'-S-(4-methoxyphenyl)-5'-thioadenosine [**10b**(5'S)] was established by X-ray crystallography (see Figure 2). The **10b**(5'R) diastereomer was concentrated in the mother liquors but did not crystallize. R diastereomers had lower field ^{19}F NMR peaks than their S counterparts. Compound **10b**(5'R) had ^{19}F NMR δ -155.05 (dd $^2J_{F-5'}$ = 53.5 Hz, $^3J_{F-4'}$ = 11.3 Hz) and **10b**(5'S) had δ -160.24 (dd $^2J_{F-5'}$ = 53.5 Hz, $^3J_{F-4'}$ = 18.8 Hz) (upfield from external CCl$_3$F). Acetylation of **10b**(5'S) gave **11b**(5'S) that was oxidized to give the sulfoxide diastereomers, 2',3'-di-O-acetyl-5'-deoxy-5'(S)-fluoro-5'-[(4-methoxyphenyl)-sulfinyl($S_{R/S}$)]adenosine [**16**(5'S, $S_{R/S}$)], with ^{19}F NMR δ -195.16 (dd, $^2J_{F-5'}$ = 46.5 Hz,

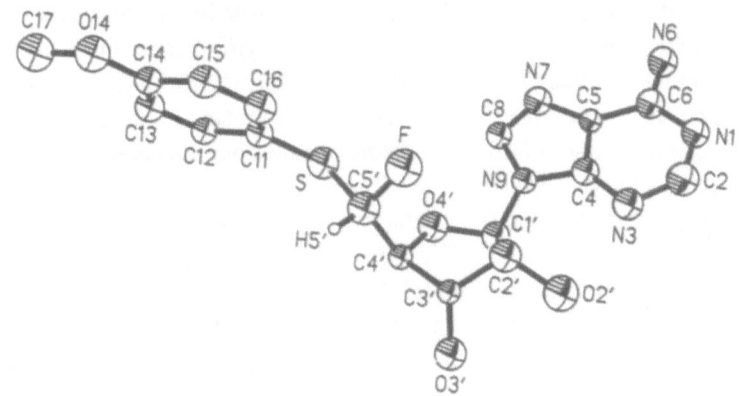

Figure 2. X-Ray crystal structure of 5'(*S*)-fluoro-5'-*S*-(4-methoxyphenyl)-5'-thioadenosine [**10b**(5'*S*)]

$^3J_{F-4'} = 27.5$ Hz, 0.77 F; 5'*S*, S$_S$), -198.23 (dd, $^2J_{F-5'} = 46.5$ Hz, $^3J_{F-4'} = 21.0$ Hz, 0.23 F; 5'*S*, S$_R$). Oxidation of **11b**(5'*R/S*) under similar conditions gave a mixture of four sulfoxide diastereomers plus ~5% of two sulfones.

Oxidation of **11b**(5'*R/S*) with excess MCPBA gave the diastereomeric sulfones **17** (39%) and sulfone 1-N-oxide **19** (51%) that were separated by silica gel chromatography. Deprotection of **17** or **19** gave the corresponding 5'-deoxy-5'-fluoro-5'-[(4-methoxyphenyl)sulfonyl]adenosines (**18**) or 1-N-oxides (**20**). Fractional crystallization of **20** (MeOH) gave 5'-deoxy-5'(*S*)-fluoro-5'-[(4-methoxyphenyl)sulfonyl]adenosine-1-*N*-oxide [**20**(5'*S*)] with ^{19}F NMR δ -192.85 [dd, $^2J_{F-5'} = 45.0$ Hz, $^3J_{F-4'} = 28.5$ Hz, F(5'*S*)]. Hexachlorodisilane effected smooth 1-N-deoxygenation[20] of **19** to give **17**.

Syn 1,2-elimination of the sulfenic acid occurred upon thermolysis of the nucleoside α-fluoro sulfoxides.[21] Attempts to employ milder thermal eliminations of selenoxides were unsuccessful. Attempted syntheses of α-fluoro selenoethers from nucleoside 5'-selenoxides with DAST or DAST/SbCl$_3$ resulted in simple deoxygenation, and selenoethers formed the relatively stable tetracoordinate difluorides[22] upon treatment with XeF$_2$. Thermolysis of **16**(5'*S*, S$_{R/S}$) at ~145 °C for 48 h in ethyldiisopropylamine/diglyme gave 9-[2,3-di-*O*-acetyl-5-deoxy-5(Z)-fluoro-β-D-*erythro*-pent-4-enofuranosyl]adenine[21] (**14**, 71%) by syn elimination of H4' and the sulfoxide(S$_{R/S}$) moiety. Deprotection gave 9-[5-deoxy-5(Z)-fluoro-β-D-*erythro*-pent-4-enofuranosyl]adenine (**15**, 88%). The less inhibitory 5'(*E*) fluoromethylene analogue was prepared similarly.[21]

5'-(α-CHLORO SULFOXIDE AND CHLOROMETHYLENE) ADENOSINES

Several procedures exist for the synthesis of α-chloro thioethers and sulfoxides. The solid iodobenzene dichloride [phenyliodine(III) dichloride] reagent is a convenient source of positive chlorine, and the lipophilic iodobenzene by-product can be recycled. Colonna and coworkers have reported reactions of sulfoxides with iodobenzene dichloride and investigated the stereochemistry of α-chloro sulfoxides obtained under various conditions.[23]

Chromatographic separation of **9a**(S$_{R/S}$) gave amorphous 2',3'-di-*O*-acetyl-5'-deoxy-5'-[(4-methoxyphenyl)sulfinyl(S$_R$)]adenosine [**21a**(S$_R$)] and **22a**(S$_S$). Deacetylation gave the high-melting sulfoxide diastereomers **21b**(S$_R$) and **22b**(S$_S$) (see Scheme 3). Treatment of **21a** with iodobenzene dichloride and potassium carbonate in acetonitrile gave 2',3'-di-*O*-acetyl-5'(*S*)-chloro-5'-deoxy-5'-[(4-methoxyphenyl)sulfinyl(S_*S*)]adenosine (**23a**), 2',3'-di-*O*-acetyl-5'(*R*)-chloro-5'-deoxy-5'-[(4-methoxyphenyl)sulfinyl(S_*R*)]adenosine (**24a**), and minor diastereomers in ratios of ~15:4:1. Analogous treatment of **22a** gave **23a**, **24a**, and minor diastereomers in ratios of ~5.5:8:1. [Note that the absolute configuration at sulfur in sulfoxide **21** and α-chloro sulfoxide **23** (or **22** and **24**) is the same but the *R/S* configuration descriptors change due to the change in Cahn-Ingold-Prelog priority of C5' bearing a chloro substituent.] The diastereomers **23a**(5'*S*, S_*S*) and **24a**(5'*R*, S_*R*) were separated by column chromatography. Deacetylation of **23a**(5'*S*, S_*S*) afforded **23b**(5'*S*, S_*S*) which crystallized from methanol as needles suitable for X-ray analysis (see Figure 3). Analogous treatment of **24a**(5'*R*, S_*R*) (NH$_3$/MeOH) resulted in its decomposition with release of adenine.

Radical-mediated reductive dechlorination (Bu$_3$SnH/AIBN/benzene/Δ) of purified **23a**(5'*S*, S_*S*) and **24a**(5'*R*, S_*R*) gave the sulfoxide diastereomers **21b**(S_*R*) and **22b**(S_*S*), respectively, with retention of stereochemistry at sulfur (^{1}H NMR). No changes were observed upon control treatment of **21b** or **22b** (Bu$_3$SnH/AIBN/benzene/Δ). Thus, the stereochemistry at sulfur in the two major α-chloro sulfoxide diastereomers (as well as in

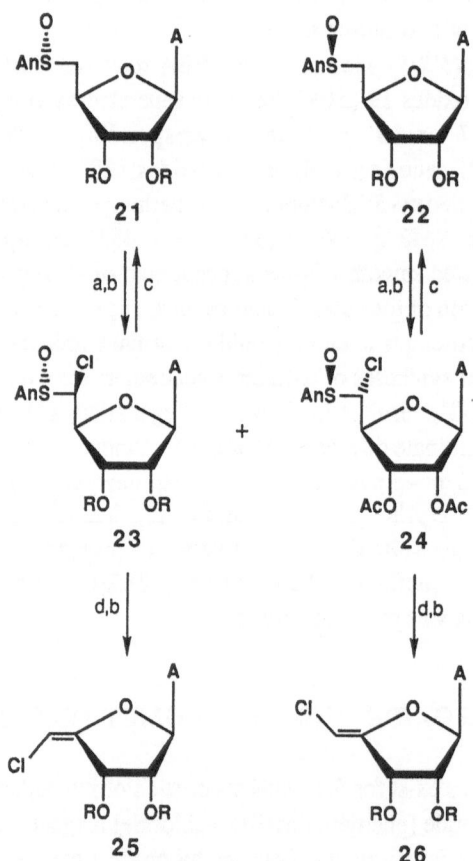

An = (*p*)CH$_3$OC$_6$H$_4$; Series 21-26: **a** R = Ac, **b** R = H.

(a) PhICl$_2$ (1.25 equiv.)/K$_2$CO$_3$/MeCN. (b) NH$_3$/MeOH. (c) Bu$_3$SnH/AIBN/C$_6$H$_6$/Δ. (d) *i*-Pr$_2$NEt/(diglyme or Me$_2$SO)/Δ.

Scheme 3. Synthesis of chloro analogues from adenosine 5'-sulfoxides

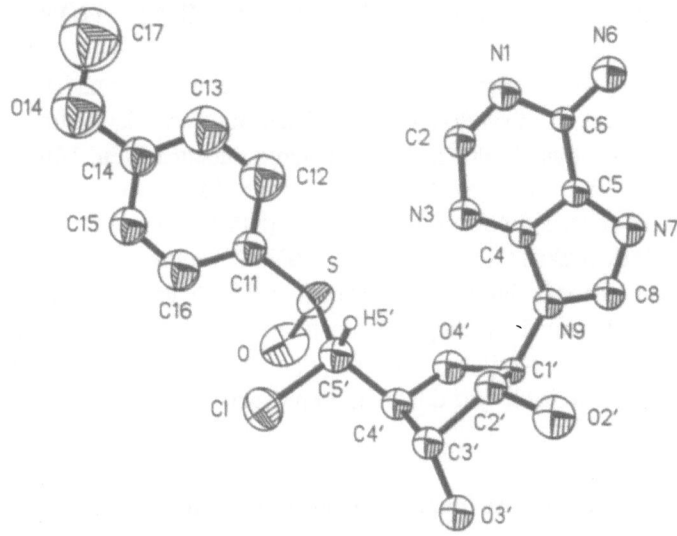

Figure 3. X-Ray crystal structure of 5'(*S*)-chloro-5'-deoxy-5'-[(4-methoxyphenyl)sulfinyl(S$_S$)]adenosine (**23b**)

their precursor sulfoxides) was established by the dechlorination experiments in conjunction with the X-ray crystal structure of **23b**(5'*S*, S$_S$) (Figure 3).

Thermolysis of **23a**(5'*S*, S$_S$) at ~150 °C for 36 h in diglyme containing excess Hünig's base followed by chromatography gave 9-[2,3-di-*O*-acetyl-5(*E*)-chloro-5-deoxy-β-D-*erythro*-pent-4-enofuranosyl]adenine (**25a**, 41%) and recovered **23a**(5'*S*, S$_S$) (11%). Longer thermolysis periods or the use of DMSO as solvent resulted in lower yields of **25a**. The thermolysis of **24a**(5'*R*, S$_R$) proceeded much more readily and was complete within 5 h at ~145 °C (excess *i*-Pr$_2$NEt/diglyme) to give 9-[2,3-di-*O*-acetyl-5(*Z*)-chloro-5-deoxy-β-D-*erythro*-pent-4-enofuranosyl]adenine **26a** (58%).

The more facile thermolysis of **24a**(5'*R*, S$_R$) and relative instability of **26a** at elevated temperatures allowed selective preparations of **25a** or **26a** from mixtures of the α-chloro sulfoxides. When such mixtures were heated at ~145 °C for 4.5 h and chromatographed, **26a/25a** (21%, ~5.7:1) and **23a**(5'*S*, S$_S$) (56%, suitable for further thermolysis to give **25a**) were isolated. Thermolysis of mixtures at ~150 °C for ≥36 h resulted in isolation of **25a** plus unchanged **23a**(5'*S*, S$_S$). This differential stability of the **25**(*E*) and **26**(*Z*) isomers might have resulted in the loss of **26** in an earlier study in which the stereochemistry of their 5'-chloro product was misassigned.[21b] The syn-stereospecificity of sulfoxide thermolysis reactions coupled with our X-ray structure of **23b**(5'*S*, S$_S$) and the reductive dechlorination results which identifed the starting sulfoxide **21b**(S$_R$) and **22b**(S$_S$) structures provided strong evidence for the configuration of **24a**(5'*R*, S$_R$). The stereochemistry of these α-chlorinations also is in harmony with the prior studies of Colonna and coworkers.[23]

Deprotection of **25a** and **26a**, diastereomeric resolution by HPLC, and "diffusion crystallization" gave 9-[5(*E*)-chloro-5-deoxy-β-D-*erythro*-pent-4-enofuranosyl]adenine (**25b**, 80%) and 9-[5(*Z*)-chloro-5-deoxy-β-D-*erythro*-pent-4-enofuranosyl]adenine (**26b**, 83%) Irradiation of H3' with the Z-isomer **26b** resulted in ~5% enhancements of the H5' peak (δ 5.60) in ^{1}H NMR difference NOE experiments, whereas low to negligible analogous effects on the H5' signal (δ 5.90) of the *E*-isomer **25b** were observed. This correlates with our above stereochemical assignments, and parallel effects were observed with the *E* and *Z* 5'-fluoro analogues. Relative H5' peak shifts, vinyl proton couplings, and C5' ^{13}C NMR shifts also correlate with our assigned structures.

Several compounds were examined for concentration-dependent inactivation of AdoHcy hydrolase from beef liver (Table 1), and candidates were selected for evaluation[24] of time-dependent inactivation (Table 2). The 5'-α-fluoro thioethers **10a**, **10b**, and **10c** were quite potent inactivators. However, it was observed that **10a-c** were unstable in the buffer test solutions, and other compounds were formed (HPLC) soon after the samples were dissolved. ^{1}H NMR peaks at δ ~9.8 and ~11.7 (DMSO-d_6) were present after **10a** was allowed to stand in aqueous solution. The first peak was stable upon addition of D_2O, but the second peak at δ ~11.7 rapidly disappeared upon D_2O exchange. This is consistent with chemical hydrolysis of the 5'-α-fluoro thioethers **10** to give the epimeric 4'-carbaldehyde **28**, hydroxy enol ether **29**, and aldehyde dihydrate **30** products (see Scheme 4).

Table 1. Qualitative inhibition of bovine liver AdoHcy hydrolase

compound	concentration-dependent inactivation
7b	–
7b-sulfoxides	–
7b-sulfone	–
10a	++
10b	++
10c	+++
uridine analogue of **10b**	–
18	–
20	–
25	±
26	+++

Table 2. Time-dependent inactivation of bovine liver AdoHcy hydrolase

compound	K_i(nM)	k_2(min^{-1})	k_2/K_i(M^{-1} min^{-1})
10b(5'R/S)	~300	~4	133×10^5
10c(5'R/S ~1.6)	21.4	0.24	112×10^5
10c(5'R/S ~0.6)	28.1	0.32	114×10^5
15(Z)	22	0.042	19.1×10^5
E-isomer of **15**	48	0.046	9.58×10^5
26(Z)	54.5	0.046	8.44×10^5

The possibility that a component of this "adenosine 5'-aldehyde" mixture is the actual inhibitor was supported by the following observations: (1) the chemically stable **7b** was not an inhibitor of AdoHcy hydrolase; (2) Ado and Hcy were rapidly converted into AdoHcy by AdoHcy hydrolase, but **7b** and Hcy gave no detected AdoHcy; (3) the chemically stable sulfones **18** and **20** derived from **10b** were not inhibitors; (4) the uridine analogues of **10b** [5'(R/S)-fluoro-5'-S-(4-methoxyphenyl)-5'-thiouridine] were not inhibitors.

Since **7b** was not an inhibitor and apparently was not oxidized to its 3'-keto intermediate, it is unlikely that 5'-fluoro derivatives of **7b** (i.e., the active **10b**) would be alternative substrates. It is more probable that the **10b** diastereomers undergo chemical hydrolysis to an active species. This is consistent with the facile hydrolysis of **10b** to give an active species and the absence of formation of this species from the chemically stable **18**. Specific inhibition by a species derived from adenosine 5'-α-fluoro thioethers rather than nonspecific inhibition by fluoride, thiophenol, or generic nucleoside 5'-aldehyde products is supported by lack of inhibition with the uridine analogue of **10b**.

Moffatt et al.[25] had reported the lability of nucleoside 5'-aldehydes and isomerizations via enolized species. Deprotection of their 6-N-benzoyl-2',3'-O-isopropylideneadenosine 5'-aldehyde 1,3-diphenylimidazolidine derivative[25b] **27** gave a mixture with ^{1}H NMR peaks similar to those observed from aqueous solutions of **10a** (see above). The 5'-α-fluoro thioethers **10** are thioacetal analogues, and methanolysis of the methylthio diastereomers **11d** gave methoxy-methylthio mixed acetals on silica gel columns as noted above.[4] Thus, the diastereomeric hydroxy enol ethers **29** (Scheme 4) are plausible analogues of 4',5'-didehydro-5'-deoxyadenosine (**6**, Scheme 1), which is oxidized to enone **3** (which in turn undergoes Michael addition of Hcy to give **2** and then AdoHcy, or water to give **5** and then Ado).

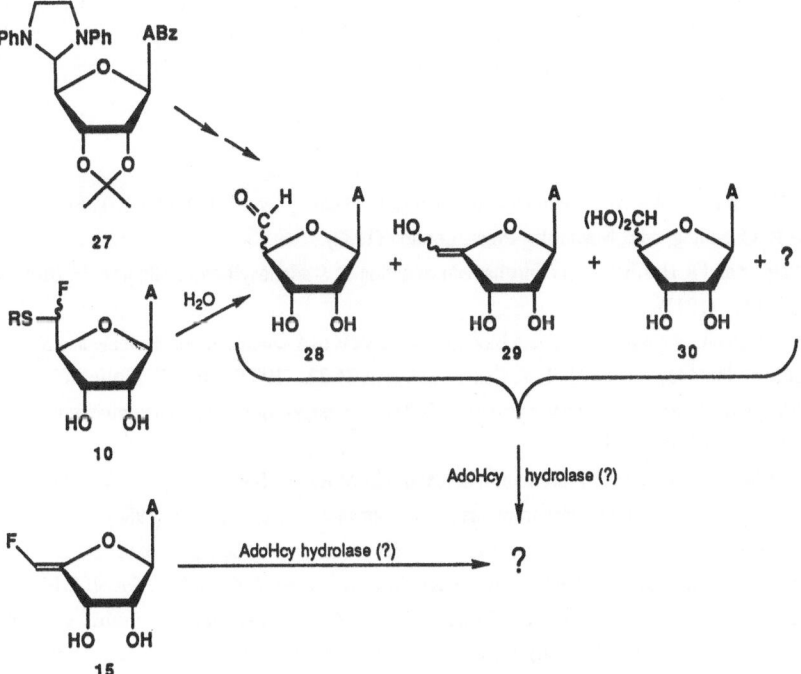

Scheme 4. Possible inhibitors of AdoHcy hydrolase derived from adenosine 5'-aldehyde

In order to examine this hypothesis directly, "adenosine 5'-aldehyde" was generated from intermediate 27[25b] and also from the adenosine mixed 5'-methoxy-5'-methylthioacetal.[4] We were gratified to find that the purified mixture of aldehyde-derived species exhibited potent AdoHcy inhibitory activity (Borchardt, R. T. et al., unpublished results).

Slight differences in inhibitory concentrations and rates of inactivation of AdoHcy hydrolase were observed between diastereomer-enriched samples of the 4-chlorophenylthio compounds 10c (Table 2) and larger differences relative to the 4-methoxyphenylthio analogues 10b were found. Current studies indicate that different rates and/or mechanisms of chemical hydrolysis of the α-fluoro thioethers can account for these differences.

It is intriguing to consider whether the inhibitory 5'E and 5'Z (15) fluoro analogues of 4',5'-didehydro-5'-deoxyadenosine (6) have intrinsic activity per se, or if they also function as prodrugs and are converted by AdoHcy hydrolase into the same/similar inhibitory species as the 5'-α-fluoro thioethers 10. McCarthy and coworkers[26] have postulated inhibitory pathways involving enzymatic oxidation at C-3' of 15 followed by conjugate addition/elimination processes. Both the direct enzyme-binding (covalent inactivation) and hydrolytic possiblilites were noted. Oxidation of 15 at C-3' followed by Michael addition of water and elimination of hydrogen fluoride could produce the same 3'-keto-5'-aldehyde mixture as oxidation of adenosine 5'-aldehyde at C-3'. Enzyme-mediated hydrolysis of fluoride from 15 to give the hydroxy enol ethers 29 might produce the same species as hydrolysis of 10. Analogous considerations with respect to the potent 5'(Z)-chloro analogue 26 are possible. Studies to explore such possibilities are in progress.

It is noteworthy that qualitative differences exist between the 25(E) and 26(Z) 5'-chloromethylene isomers with respect to inhibition of AdoHcy hydrolase. The authentic Z-isomer 26 is a time-dependent inactivator with potency comparable to that of its 5'-fluoromethylene analogue 15 (Table 2). In contrast, the E-isomer 25 showed little activity in our preliminary screen (Table 1) and was reported to be a *reversible* inhibitor in another study (in which the Z-configuration had been tentatively assigned).[21b]

Acknowledgment. We thank the American Cancer Society (Grant No. DHP-34) and the National Institutes of Health (Grant No. GM 29332) for support and Mrs. Kathryn M. Rollins for assistance with the manuscript.

REFERENCES

1. "The Biochemistry of S-Adenosylmethionine and Related Compounds," E. Usdin, R.T. Borchardt, and C.R. Creveling, eds., Macmillan Press, London (1982).

2. J.L. Palmer and R.H. Abeles, The mechanism of action of S-adenosylhomocysteinase, *J. Biol Chem.* 254:1217 (1979).

3. (a) P.M. Ueland, Pharmacological and biochemical aspects of S-adenosylhomocysteine and S-adenosylhomocysteine hydrolase, *Pharmacol. Rev.* 34:223 (1982); (b) M.S. Wolfe and R.T. Borchardt, S-Adenosyl-L-homocysteine hydrolase as a target for antiviral chemotherapy, *J. Med. Chem.* 34:1521 (1991).

4. M.J. Robins and S.F. Wnuk, Fluorination at C5' of Nucleosides. Synthesis of the new class of 5'-fluoro-5'-S-aryl(alkyl)thionucleosides from adenosine, *Tetrahedron Lett.* 29:5729 (1988).

5. R.J. Parry and L.J. Askonas, Studies of Enzyme Stereochemistry. Elucidation of the stereochemistry of the reaction catalyzed by S-adenosylhomocysteine hydrolase, *J. Am. Chem. Soc.* 107:1417 (1985).

6. (a) M. Zupan, Fluorination with xenon difluoride: Part IX. Reaction with phenylsubstituted sulphides, *J. Fluorine Chem.* 8:305 (1976); (b) R.K. Marat and A.F. Janzen, Reaction of xenon difluoride.

Part III. Oxidative-fluorination and α-fluorination of sulfur(II) compounds, *Can. J. Chem.* 55:3031 (1977).

7. K.M. More and J. Wemple, The synthesis of aryl fluoromethyl sulfoxides, *Synthesis* 791 (1977).

8. J.R. McCarthy, N.P. Peet, M.E. LeTourneau, and M. Inbasekaran, (Diethylamino)sulfur trifluoride in organic synthesis. 2. The transformation of sulfoxides to α-fluoro thioethers, *J. Am. Chem. Soc.* 107:735 (1985).

9. T. Umemoto and G. Tomizawa, α-Fluorination of sulfides with *N*-fluoropyridinium triflates, *Bull. Chem. Soc. Jpn.* 59:3625 (1986).

10. (a) T. Fuchigami, M. Shimojo, A. Konno, and K. Nakagawa, Electrolytic partial fluorination of organic compounds. 1. Regioselective anodic monofluorination of organosulfur compounds, *J. Org. Chem.* 55:6074 (1990); (b) T. Brigaud and E. Laurent, Oxidative fluorination of sulfides in presence of Et₃N•3HF, *Tetrahedron Lett.* 31:2287 (1990).

11. (a) S. Nishikawa, A. Ueno, H. Inoue, and Y. Takeda, Effect of 5'-difluoromethylthioadenosine, an inhibitor of methylthioadenosine phosphorylase, on proliferation of cultured cells, *J. Cell. Physiol.* 133:372 (1987); (b) Y. Takeda, T. Mizutani, A. Ueno, K. Hirose, E. Tanahashi, and S. Nishikawa, Preparation of 5'-deoxy-5'-fluoromethylthio ribonucleosides and antitumor agents containing them, Jpn. Pat. Appl. 87/45,495 (1987), *Chem. Abstr.* 110:58012m (1989); (c) Y. Takeda, A. Mizutani, A. Ueno, K. Hirose, E. Tanahashi, and S. Nishikawa, Preparation, testing, and formulation of 5-deoxy-5-fluoromethylthioribose derivatives as antitumor agents, Jpn. Pat. Appl. 86/70,889 (1986), *Chem. Abstr.* 109:38178w (1988).

12. A.J. Gianotti, P.A. Tower, J.H. Sheley, P.A. Conte, C. Spiro, A.J. Ferro, J.H. Fitchen, and M.K. Riscoe, Selective killing of *Klebsiella pneumoniae* by 5'-trifluoromethylthioribose, *J. Biol. Chem.* 265:831 (1990).

13. M.E. Houston Jr., D.L. Vander Jagt, and J.F. Honek, Synthesis and biological activity of fluorinated intermediates of the methionine salvage pathway, *Bioorg. Med. Chem. Lett.* 1:623 (1991).

14. Y.D. Vankar and C.T. Rao, Sodium iodide/boron trifluoride etherate: A mild reagent system for the conversion of allylic and benzylic alcohols into corresponding iodides and sulfoxides into sulfides, *Tetrahedron Lett.* 26:2717 (1985).

15. S.F. Wnuk and M.J. Robins, Antimony(III) chloride exerts potent catalysis of the conversion of sulfoxides to α-fluoro thioethers with (diethylamino)sulfur trifluoride, *J. Org. Chem.* 55:4757 (1990).

16. M.J. Robins, S.F. Wnuk, K.B. Mullah, and N.K. Dalley, Nucleic acid related compounds. 68. Fluorination at C5' of nucleoside 5'-thioethers with DAST/antimony(III) chloride or xenon difluoride to give 5'-*S*-aryl-5'-fluoro-5'-thiouridines, *J. Org. Chem.* 56:6878 (1991).

17. F. Schlenk, Methylthioadenosine, *Adv. Enzymol. Related Areas Mol. Biol.* 54:195 (1983).

18. J.R. Sufrin, A.J. Spiess, D.L. Kramer, P.R. Libby and C.W. Porter, Synthesis and antiproliferative effects of novel 5'-fluorinated analogues of 5'-deoxy-5'-(methylthio)adenosine, *J. Med. Chem.* 32:997 (1989).

19. M.J. Robins, K.B. Mullah, S.F. Wnuk, and N.K. Dalley, Nucleic acid related compounds. 73. Fluorination of uridine 2'-thioethers with xenon difluoride or (diethylamino)sulfur trifluoride. Synthesis of stable 2'-[alkyl(or aryl)sulfonyl]-2'-deoxy-2'-fluorouridines, *J. Org. Chem.* 57:2357 (1992).

20. M. MacCoss, E.K. Ryu, R.S. White, and R.L. Last, A new synthetic use of nucleoside N^1-oxides, *J. Org. Chem.* 45:788 (1980).

21. (a) J.R. McCarthy, E.T. Jarvi, D.P. Matthews, M.L. Edwards, N.J. Prakash, T.L. Bowlin, S. Mehdi, P.S. Sunkara, and P. Bey, 4',5'-Unsaturated 5'-fluoroadenosine nucleosides: Potent mechanism-based inhibitors of *S*-adenosyl-L-homocysteine hydrolase, *J. Am. Chem. Soc.* 111:1127 (1989); (b) E.T. Jarvi, J.R. McCarthy, S. Mehdi, D.P. Matthews, M.L. Edwards, N.J. Prakash, T.L. Bowlin, P.S.

    Sunkara, and P. Bey, 4',5'-Unsaturated 5'-halogenated nucleosides. Mechanism-based and competitive inhibitors of S-adenosyl-L-homocysteine hydrolase, *J. Med. Chem.* 34:647 (1991).

22. K.J. Wynne, The preparation and properties of diorganoselenium difluorides, *Inorg. Chem.* 9:299 (1970).

23. (a) M. Cinquini and S. Colonna, Synthesis of α-halogeno-sulphoxides, *J. Chem. Soc., Perkin Trans. 1* 1883 (1972); (b) M. Cinquini, S. Colonna, R. Fornasier, and F. Montanari, Stereochemistry of α-halogeno-sulphoxides. Part I. Inversion of chirality at the sulphinyl sulphur atom in a reaction not involving the breaking of the sulphinyl bonds at the chiral sulphur atom, *J. Chem. Soc., Perkin Trans. 1* 1886 (1972); (c) P. Calzavara, M. Cinquini, S. Colonna, R. Fornasier, and F. Montanari, Stereochemistry of α-halo sulfoxides. II. Interdependent stereochemistry at sulfur and α-carbon in the α-halogenation of sulfoxides, *J. Am. Chem. Soc.* 95:7431 (1973); (d) R. Annunziata and S. Colonna, Stereochemistry of α-halogeno-sulphoxides. Part 5. Absolute stereochemistry of α-chlorination of benzyl *t*-butyl sulphoxide, *J. Chem. Soc., Perkin Trans. 1* 1052 (1977).

24. M.S. Wolfe, Y. Lee, W.J. Bartlett, D.R. Borcherding, and R.T. Borchardt, 4'-Modified analogues of aristeromycin and neplanocin A: Synthesis and inhibitory activity toward S-Adenosyl-L-homocysteine hydrolase, *J. Med. Chem.* 35:1782 (1992).

25. (a) J.G. Moffatt, Chemical transformations of the sugar moiety of nucleosides, *in:* "Nucleoside Analogues: Chemistry, Biology, and Medical Applications," R.T. Walker, E. DeClercq, and F. Eckstein, eds., Plenum Press, New York (1979); (b) R.S. Ranganathan, G.H. Jones, and J.G. Moffatt, Novel analogs of nucleoside 3',5'-cyclic phosphates. I. 5'-Mono- and dimethyl analogs of adenosine 3',5'-cyclic phosphate, *J. Org. Chem.* 39:290 (1974).

26. (a) S. Mehdi, E.T. Jarvi, J.R. Koehl, J.R. McCarthy, and P. Bey, The mechanism of inhibition of S-adenosyl-L-homocysteine hydrolase by fluorine-containing adenosine analogs, *J. Enzyme Inhib.* 4:1 (1990); (b) P. Bey, J.R. McCarthy, and I.A. McDonald, Terminal fluoroolefins: Synthesis and application to mechanism-based enzyme inhibition, *in:* "Effects of selective fluorination and reactivity," J.T. Welch, ed., ACS Symposium Series 456, American Chemical Society, Washington, D. C. (1991).

APPROACHES TO NOVEL ISOMERIC NUCLEOSIDES AS ANTIVIRAL AGENTS

Vasu Nair

Department of Chemistry, The University of Iowa

Iowa City, Iowa 52242, U.S.A.

INTRODUCTION

The finding that the human immunodeficiency virus (HIV) is the etiologic agent of acquired immunodeficiency syndrome (AIDS) has focused considerable attention on the design and synthesis of compounds that would inhibit the replication of this and related viruses. A key enzyme encoded by HIV and involved in its replication is the multi-functional HIV reverse transcriptase (HIV RT) which has provided a molecular target for the design of potential anti-HIV therapeutic agents. Among the HIV RT targeted inhibitors are the dideoxynucleosides, 2',3'-dideoxyadenosine (ddA), 2',3'-dideoxyinosine (ddI), 2',3'-dideoxyguanosine (ddG), 2',3'-dideoxycytidine (ddC), and dideoxynucleoside derivatives such as 3'-α-azido-3'-deoxythymidine (AZT) (Scheme 1).[1-17]
The mechanism of action of these compounds appear to be through their triphosphates (produced inside the cytoplasm of the target cell) which act as enzyme inhibitors of HIV RT through incorporation and termination of the growing viral DNA chain.[1, 2, 5, 13, 18-20]

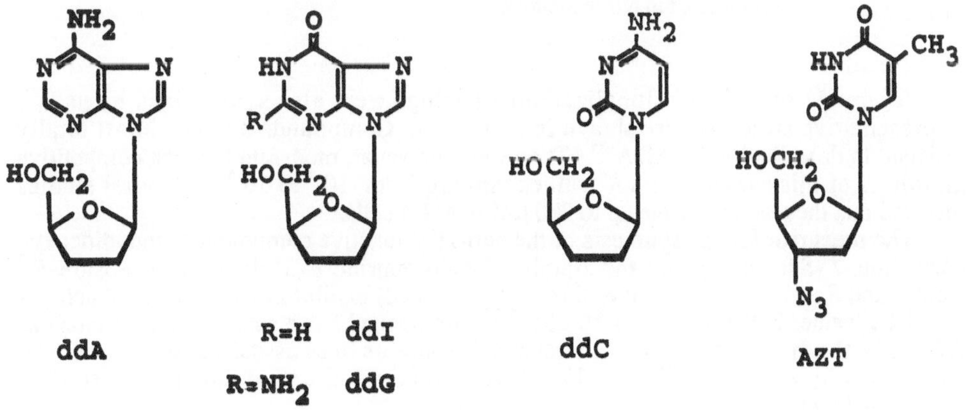

Scheme 1. Structures of some anti-HIV active dideoxynucleosides.

Nucleosides and Nucleotides as Antitumor and Antiviral Agents,
Edited by C.K. Chu and D.C. Baker, Plenum Press, New York, 1993

Dideoxynucleosides, particularly those of the purine family, are very unstable with respect to hydrolytic cleavage of the glycosidic bond. This inherent chemical property, which results from both the absence of the 2'- and 3'-hydroxyl groups (-I effect) and the presence and involvement of the proximal ring oxygen, limits the usefulness of these compounds as antiviral agents and biological probes. In addition, 2',3'-dideoxyadenosine is a substrate for mammalian adenosine deaminase (ADA) and is easily converted by enzyme-catalyzed hydrolytic deamination to 2',3'-dideoxyinosine.[18, 21] The design and synthesis of compounds that would be more stable with respect to glycosidic bond cleavage and enzymatic deamination would be of considerable significance in this area. In a program in our laboratory directed at the discovery of novel, stable nucleosides and nucleotides with anti-HIV potential, we have investigated a number of strategically modified dideoxynucleosides, their phosphorylated analogs, their prodrugs, and structures regioisomeric with the known active dideoxynucleosides. This chapter will discuss the synthesis, structural studies, enzymology, stability data, and *in vitro* anti-HIV studies of some selected novel dideoxynucleosides from our laboratory.

RESULTS AND DISCUSSION

2',3'-Dideoxynebularine (2',3'-dideoxypurine nucleoside, ddPN) 1[22] was one of a number of C-6 modified purine dideoxynucleosides synthesized in our laboratory (Scheme 2). It was designed to be a prodrug of ddI and was synthesized from 5'-protected ddA by reductive deamination using a procedure previously developed by us,[23] followed by deprotection. Compound 1 was found to be inactive against the cytopathic effect of HIV-1 and HIV-2 in MT-4 cells. It was not toxic to MT-4 cells at 200 μM. It was not a substrate for xanthine oxidase.[24] In contrast, C-6 substituted purine dideoxynucleosides that are substrates for mammalian ADA, show anti-HIV activity [e.g., 6-iodo ddPN, 2, synthesized from 5'-protected ddA by radical deamination/halogenation (n-pentyl nitrite, CH_2I_2, CH_3CN, Δ), followed by deprotection].

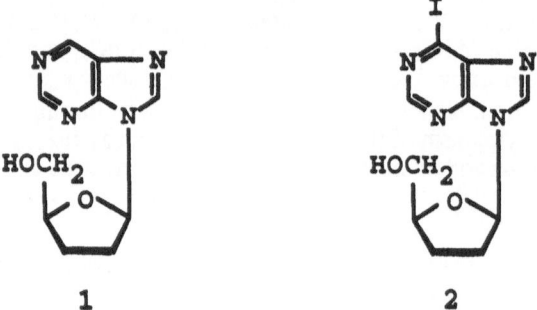

Scheme 2. Two representative examples of C-6 modified purine dideoxynucleosides synthesized in our laboratory.

A number of C-2 functionalized ddA analogs were also synthesized by us.[7,22] Representative structures are shown in Scheme 3. Compounds 3 were almost totally resistant to deamination by ADA.[25] They were, however, moderate to weak competitive inhibitors of this enzyme with K_i values ranging from 10^{-4} to 10^{-5}. Antiviral studies revealed that they were inactive up to 200 μM in MT-4 cells.

The precursor for the synthesis of the anti-HIV inactive compound, 2-iododideoxy-adenosine, 7 (Scheme 4), was the 2-amino-6-chloropurine 2',3'-dideoxynucleoside, 5.[7] Compound 5 and its 5'-derivatives 6 (e.g., 5'-acylated) exhibit *in vitro* anti-HIV activity with EC_{50} values in the range of 5-10 μM.[26, 27] Compound 5 is a substrate for the enzyme, ADA, and the mechanism of its antiviral activity appears to be associated with its cellular conversion by ADA to ddG (Fig. 1). Several other prodrugs of ddG have also been synthesized by us.

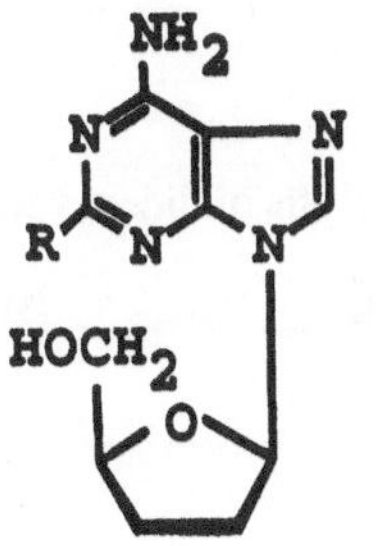

3

$$R = I, \; CN, \; C_2H_5, \; CF_3, \; SCH_3$$

Scheme 3. Representative examples of C-2 substituted dideoxyadenosines synthesized.

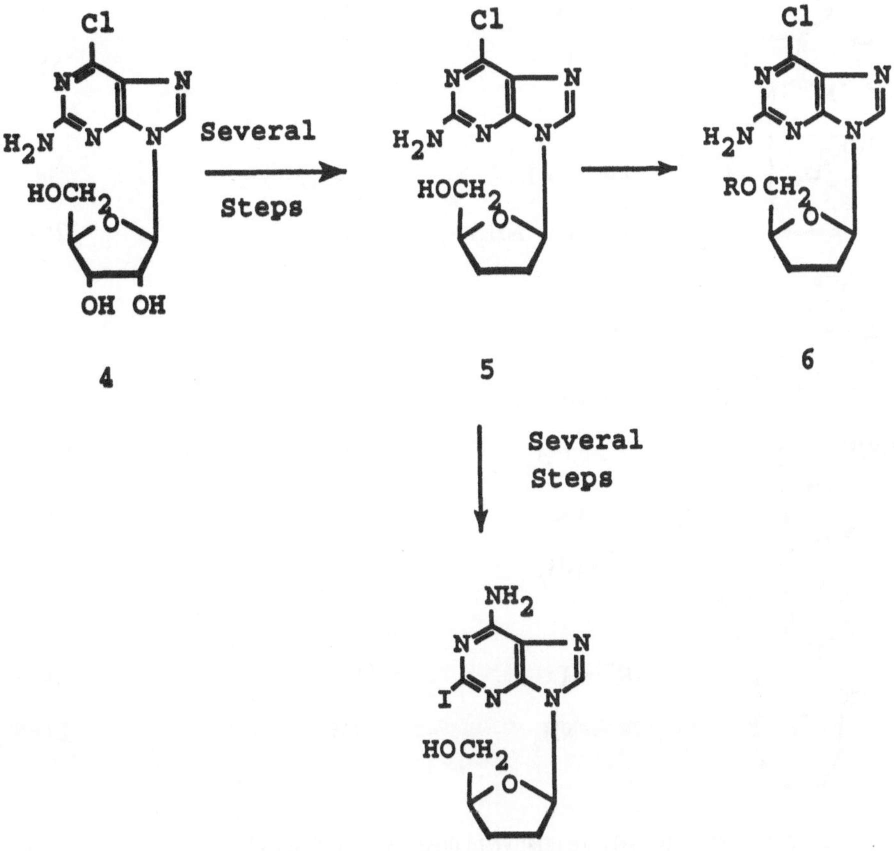

Scheme 4. Synthesis of 2-amino-6-chloropurine dideoxynucleoside and 2-iodo ddA.

The glycosidic bond stability of 2, 6, and 8-substituted purine dideoxynucleosides was also examined by us.[28] While substitutions at the 2 and 6 positions result in small to moderate effects on the stability of the glycosidic bond of these dideoxynucleosides, the most dramatic effect is seen with substitution at the 8-position (Table 1). For example, one of the anti-HIV active compounds synthesized in our program was 8-hydroxy ddA.[29] It is resistant to hydrolysis even at pH 1!

Table 1. Relative rates of hydrolysis of dideoxypurine nucleosides at pH 3 and 22 °C.

Compound	Relative Rate at pH 3[a]	λ (nm)[b]
R = H	100	254.5
R = NH$_2$	20	256.5
R = CN	47	259
R = I	55	259
R = SCH$_3$	64	264
R = CH$_2$CH$_3$	75	259
R = CF$_3$	79	255
ddI (R' = H)	74	234
ddG (R' = NH$_2$)	84	254
R" = OH	0[c]	–
R" = OCH$_2$Ph	39	254.5
R" = SCH$_3$	40	269
R" = OCH$_3$	61	253
R" = NH$_2$	2050	263
R"' = H	177	244.5
R"' = NH$_2$	110	244.5

a. Rates of hydrolysis are relative to dideoxyadenosine (rate = 100). The apparent first-order rate constant for the hydrolysis of ddA at pH 3 is 8.23×10^{-4} min.$^{-1}$

b. Rate of change in absorbance monitored at this wavelength by differential UV spectroscopy. The monitoring wavelength represents the wavelength of maximum difference at pH 13 between the intact dideoxynucleoside and its cleaved heterocyclic base anion. The reference bases were synthesized in each case where they were unknown or not commercially available.

c. No detectable hydrolysis even at pH 1.

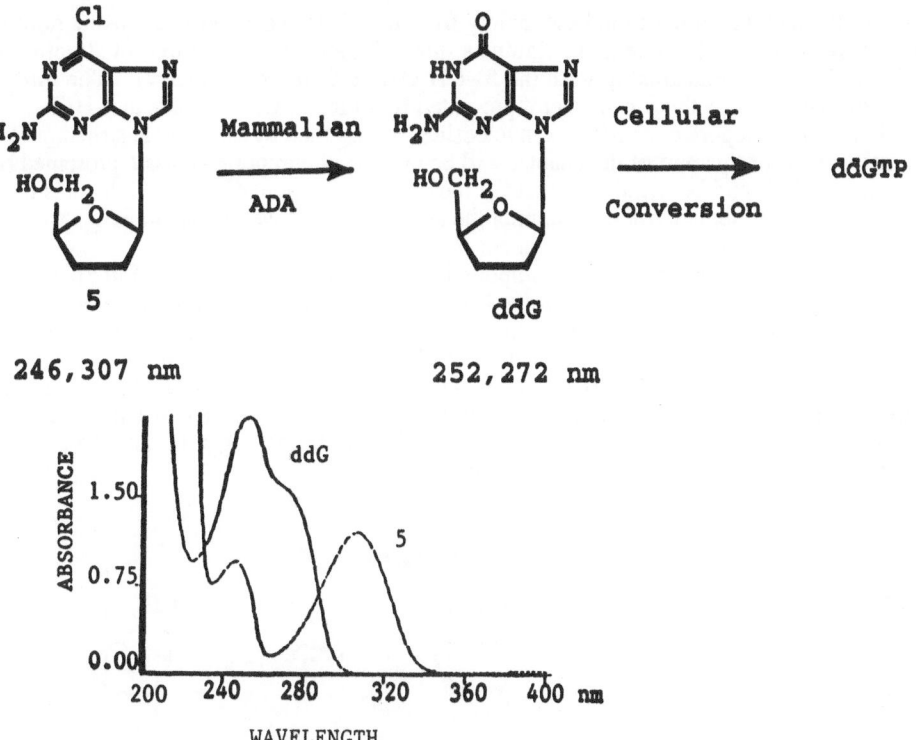

Figure 1. Conversion of 2-amino-6-chloro ddPN to ddG by mammalian ADA.

Although numerous analogs of dideoxynucleosides derived from the natural nucleosides have been synthesized since the discovery of their antiviral activity, the synthesis of structures regioisomeric with the known active "natural" dideoxynucleosides (Scheme 5, Structure A) have received much less attention. One such class of compounds

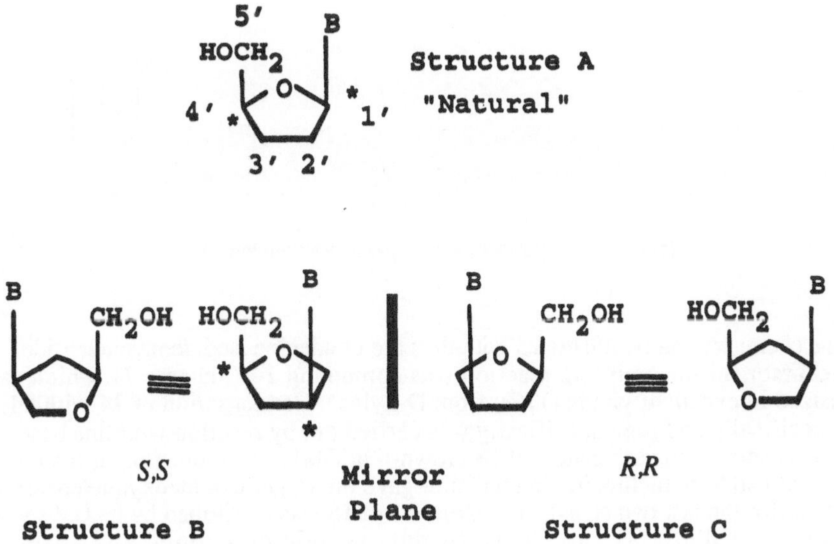

Scheme 5. Relationship of the "natural" and regioisomeric dideoxynucleosides.

involves the transposition of the base moiety from the 1'- to the 2'-position (using normal nucleoside numbering), while maintaining as shown in both representations of structure B (Scheme 5) its *cis* relationship with the 5'-CH$_2$OH (*S,S* stereochemistry). The mirror image of structure B is the *R,R*-isomer represented by structure C. As some anti-HIV work has already been reported on dideoxynucleosides represented by the general structure C,[30] the major focus of this part of the chapter will be on the yet unreported class represented by structure B. It should be mentioned that the most active compound of type C that has been synthesized is the isonucleoside containing the adenine base,[30] which had an EC$_{50}$ in ATH8 cells of the order of 10 μM with CC$_{50}$ values > 200 μM.

Approaches to the synthesis of compounds of structure B are illustrated in Scheme 6. The key steps in the methodology involved synthesis of appropriately tailored precursors **8** from natural D-sugars. Coupling of the base to **8** with inversion of stereochemistry followed by elaboration and/ or deprotection would furnish the target isodideoxynucleoside **10**. The approach would be different for those bases (e.g., cytosine) where there are regiochemical problems in the direct coupling procedure. In these cases the base was constructed on a pre-existing stereochemically defined β-amino sugar **11**, and the product **12** then was elaborated and finally deprotected.

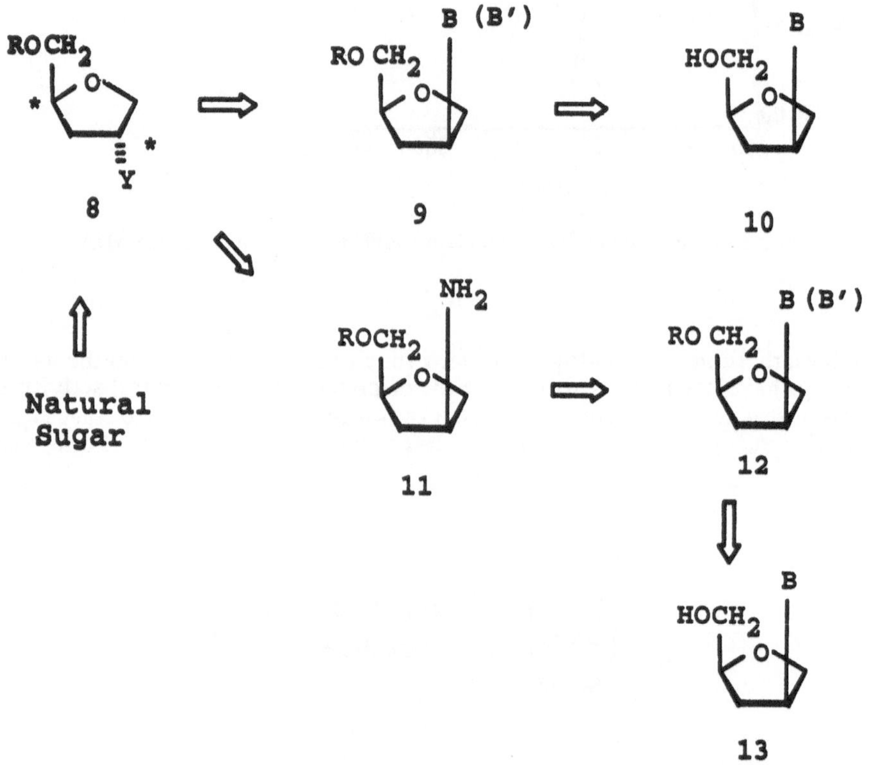

Scheme 6. Approaches to 2'-isodideoxynucleosides.

The chemistry can be illustrated with the case of adenine isodideoxynucleoside. The key precursor for the coupling reaction was compound **14** (Scheme 7), which can be synthesized in excellent yields (17%) from D-xylose. Condensation of **14** with adenine stereospecifically and regiospecifically was carried out by reaction with this base in the presence of potassium carbonate and 18-crown-6 in DMF. Deprotection of the resulting product with sodium methoxide in methanol gave the target isodideoxynucleoside **15** in 55% yield (for the last two steps). The structure of **15** was confirmed by its UV spectrum (260 nm, ε 14,000), ^{1}H and ^{13}C NMR data (single compound and absence of diastereoisomer), and optical rotation ([α]$_D$ = -26.6°). This magnitude of levorotation is the

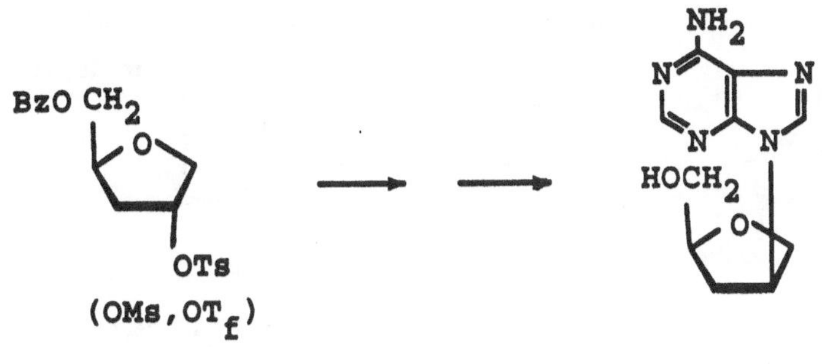

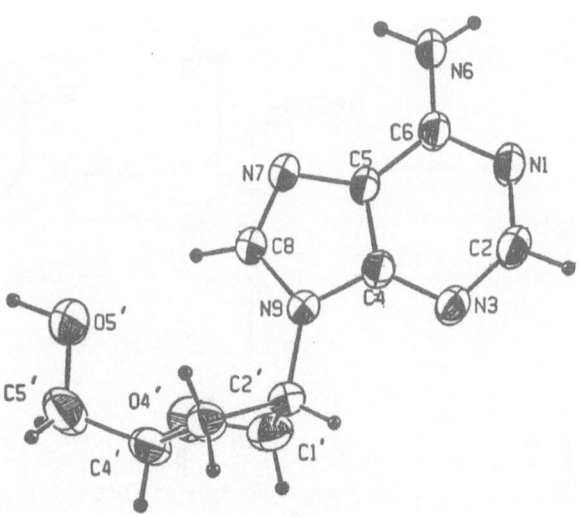

Scheme 7. Direct coupling methodology to (-) isoddA.

Figure 2. ORTEP plot showing the structure, configuration and conformation of compound **15**.

same as that for ddA. Total structure, regiochemistry, conformation and absolute configuration were all confirmed by single-crystal X-ray data (Fig. 2). Crystallographic data showed the base to be in the preferred *anti* conformation, and the carbohydrate moiety showed C-1' *exo*/ O-4' endo conformation [31,32]

Coupling of **14** with 6-chloropurine and 2-amino-6-chloropurine and subsequent deprotection of the 5'-benzoate group and regeneration of the lactam functionality on the base with aqueous NaOH gave, respectively, isoddI (30%) and isoddG (37%). The thymine analog was synthesized by the direct coupling procedure, although the yield was low and the conversion was accompanied by *bis-O*-alkylated and *bis-N,O*-alkylated products. The structure of the desired N^1-alkylated compound was evident from its UV spectrum (271.5 nm in MeOH) and from its high-field ^{1}H NMR data. Uracil gave similar synthetic results as thymine but with slightly higher yields (Scheme 8).

The U and T analogs could be synthesized more efficiently by construction of the base on the pre-existing β-amino sugar **17** (Scheme 9). The latter was prepared almost quantitatively from **14** by azide displacement followed by catalytic reduction. Treatment of **17** with 3-ethoxy-2-propenoyl isocyanate (generated *in situ* from 3-ethoxy-2-propenoyl chloride and silver isocyanate[33]), and subsequent acid-catalyzed cyclization of the resulting adduct **18**, gave isoddU **19**. The overall yield for this sequence of reactions from the tosylate **14** was about 50%. IsoddT **(20)** was synthesized using a similar approach.

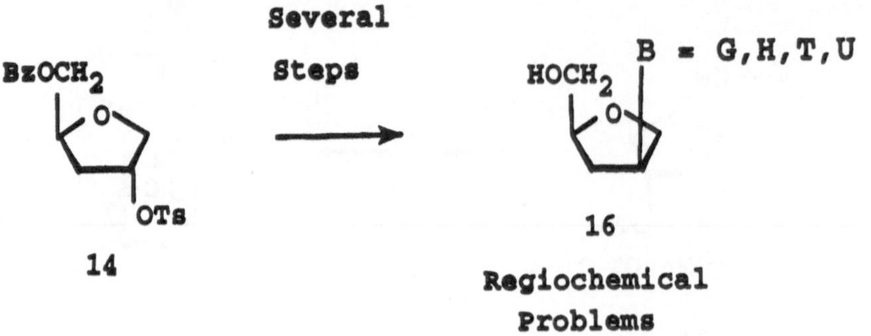

Several
Steps

B = G,H,T,U

Regiochemical
Problems

Scheme 8. Direct coupling and elaboration for lactam bearing isodideoxynucleosides.

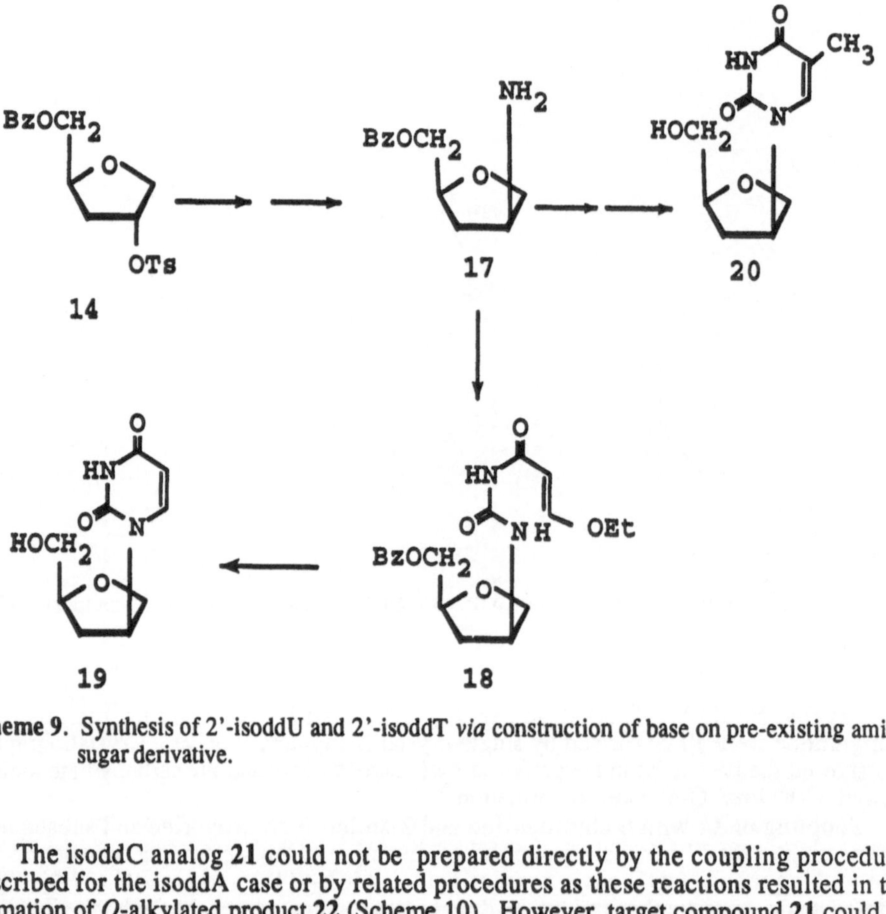

Scheme 9. Synthesis of 2'-isoddU and 2'-isoddT *via* construction of base on pre-existing amino sugar derivative.

The isoddC analog **21** could not be prepared directly by the coupling procedure described for the isoddA case or by related procedures as these reactions resulted in the formation of O-alkylated product **22** (Scheme 10). However, target compound **21** could be prepared through the protected isoddU (**23**) by conversion to the 4-triazolo derivative,[34] followed by treatment with ammonium hydroxide and deprotection with sodium methoxide in methanol.

Unequivocal differentiation between the N^1-alkylated product **21** and the O^2-alkylated product **22** could not be ascertained from their UV spectra or their high-field ^{1}H and ^{13}C NMR chemical shift data. However, J-modulated selective INEPT ^{1}H-^{13}C NMR correlations[35] clearly established the position of linkage to the base for each of these compounds. Representative INEPT spectra are shown in Fig. 3 (normal nucleoside numbering used). Selective INEPT ^{1}H-^{13}C NMR correlations are potentially very useful for structure determination work in nucleoside chemistry, but this NMR technique has seen little use in this area.

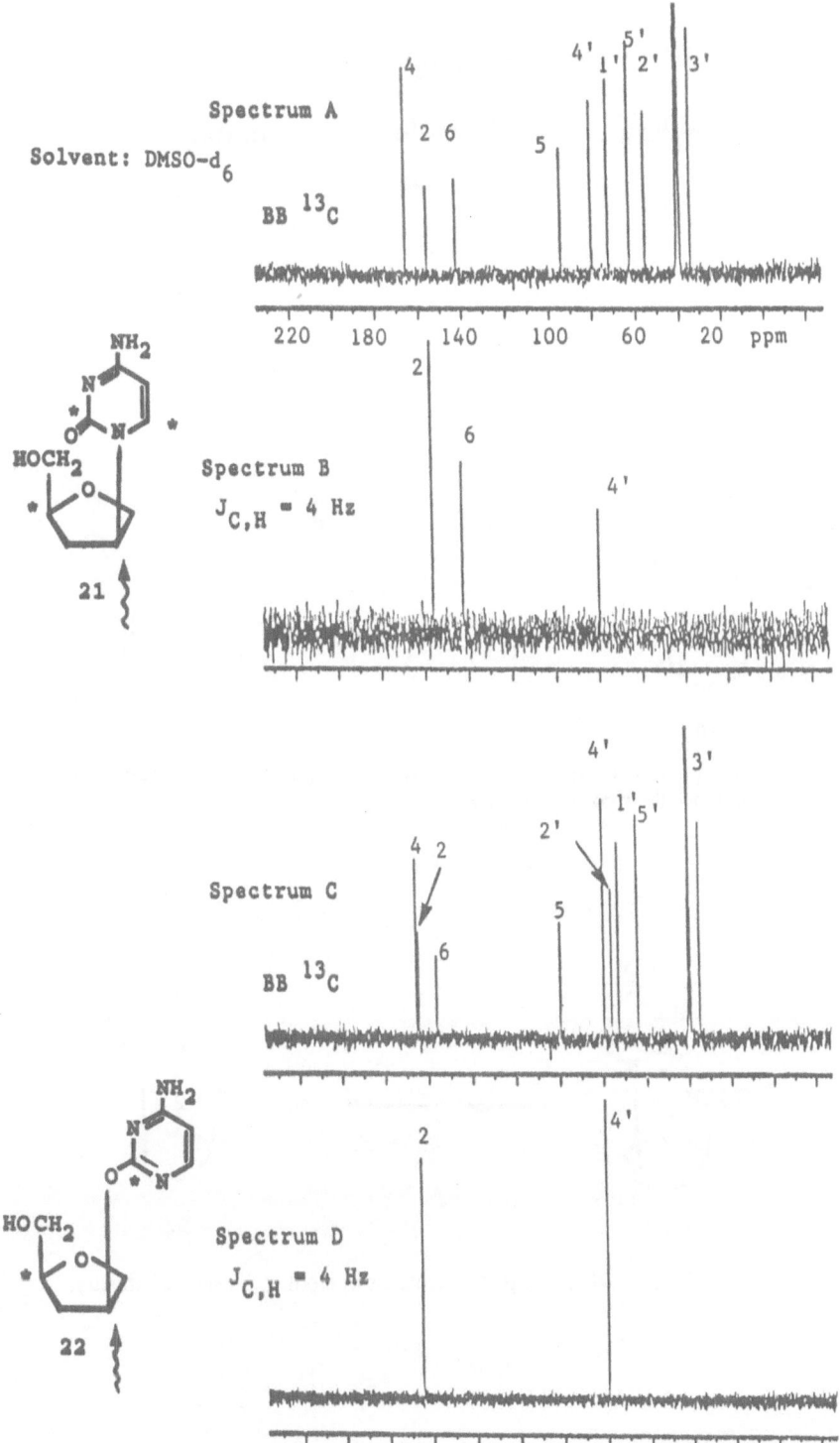

Figure 3. One-dimensional *J*-modulated selective INEPT ^{1}H-^{13}C NMR correlations that differentiate between structures **21** and **22**. A = Broad-band ^{1}H decoupled ^{13}C NMR spectrum for **21**; B = INEPT spectrum for **21** (H-2' irradiated); C = broad-band ^{1}H decoupled ^{13}C NMR spectrum for **22**; D = INEPT spectrum for **22** (H-2' irradiated).

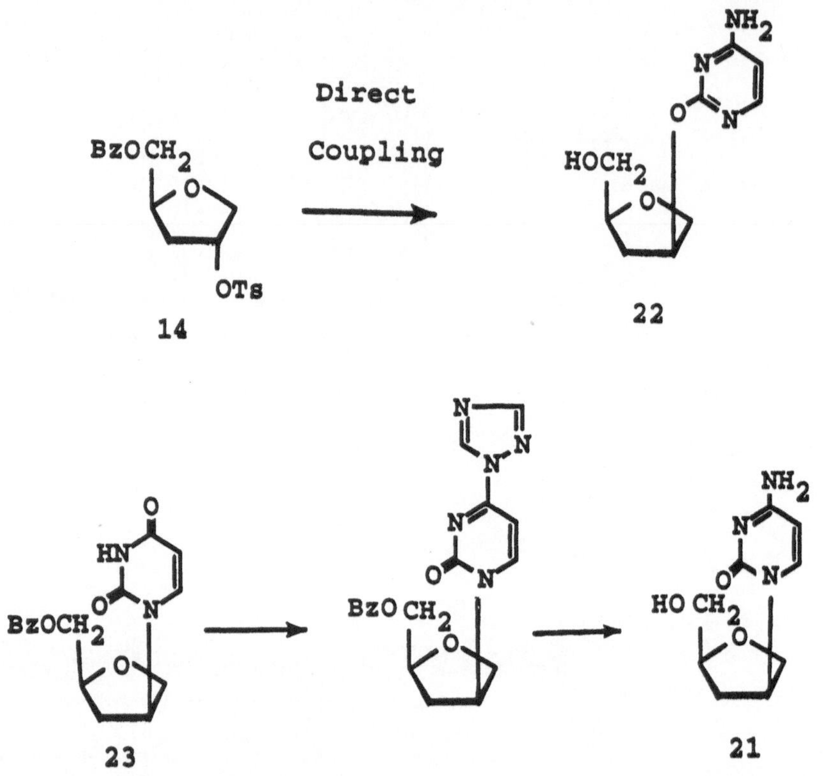

Scheme 10. Attempted synthesis of isoddC by direct coupling and successful synthesis from isoddU *via* the triazolo derivative.

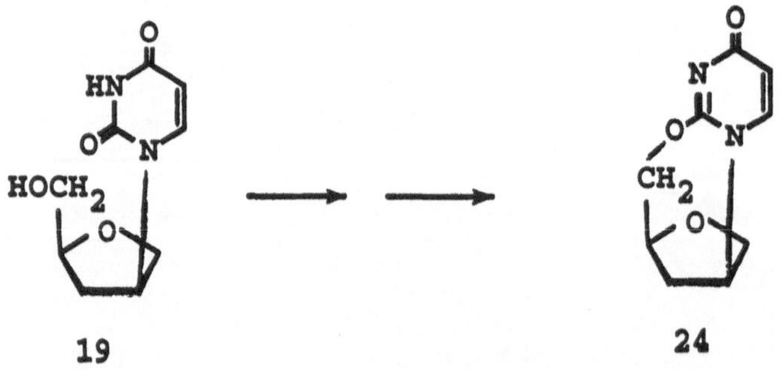

Scheme 11. Formation of cyclonucleosides as proof of *cis* stereochemistry.

Chemical proof for the *cis* relationship of the base and 5'-CH$_2$OH (normal nucleoside numbering) in these compounds came from the formation of cyclonucleosides through the treatment of isoddN 5'-tosylates with DBU (Scheme 11).

Glycosidic bond stabilities of these dideoxynucleosides were studied using differential UV spectroscopy by procedures previously described by us.[28] Both the purine and pyrimidine isodideoxynucleosides were found to be very stable under both acidic and basic conditions. For example, while "natural" ddA is cleaved rapidly at pH 3 ($t_{1/2}$ < 1/2 hour), its regioisomer, isoddA has a half-life of greater than 16 days even at pH 1.

The stability of the isomeric ddNs towards mammalian enzymes of hydrolytic deamination was also studied. As the 3'-α-hydroxyl group is a necessary requirement for substrate activity against cytidine deaminase,[36] the isoddC analog would not be a substrate for cytidine deaminase. The behavior with respect to mammalian adenosine deaminase was much more difficult to predict. The consequences of deamination with ADA would be detrimental because the isoddI produced would have to traverse a very inefficient salvage pathway to reach the potentially anti-HIV active isoddATP (Scheme 12). Even for "natural" ddI, this pathway is inefficient.[37] Consistent with this assessment was the observation that isoddI did not exhibit any anti-HIV activity. IsoddA was found to be highly resistant to deamination by mammalian ADA (Scheme 13). Using large excesses of enzyme and prolonged reaction times, we were able to determine the rate of deamination of this compound to be 0.0017% that of ddA. Inhibition studies with the natural substrate, adenosine, showed that isoddA was a moderate to weak competitive inhibitor with a K_i of 0.84×10^{-4} M.

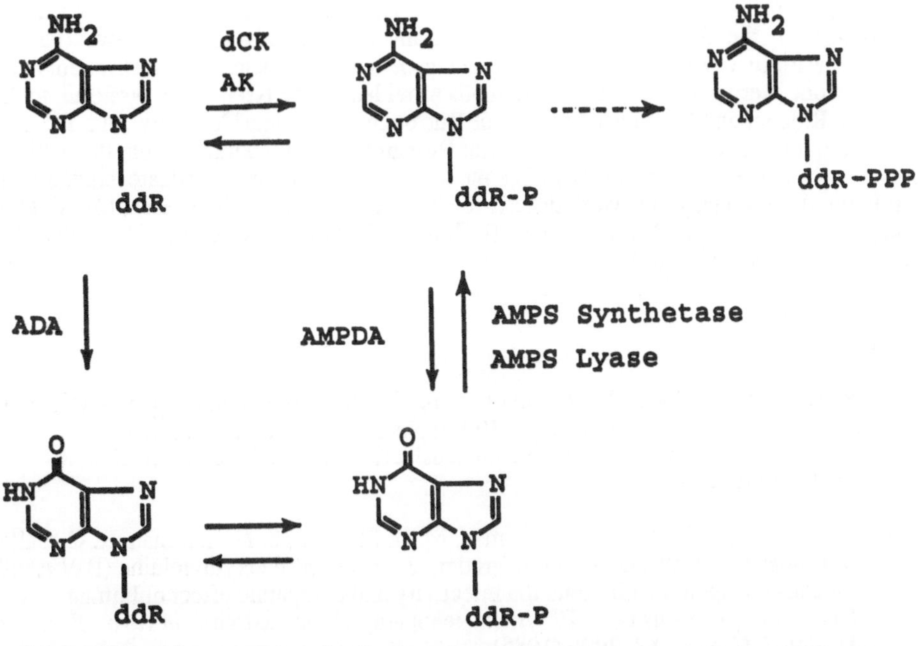

Scheme 12. Salvage pathway for the conversion of ADA substrates to ddATP compounds.

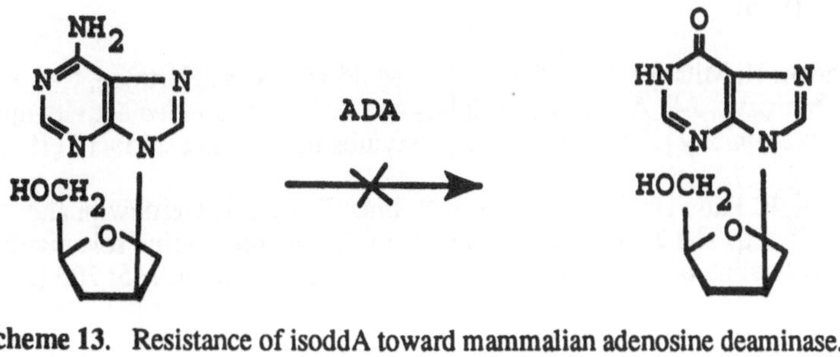

Scheme 13. Resistance of isoddA toward mammalian adenosine deaminase.

The *in vitro* anti-HIV studies of these isodideoxynucleosides showed that optically active *S,S*-isodideoxyadenosine **15** and its derivatives (acylated and phosphorylated) were strongly active. This anti-HIV activity of **15** appears to be comparable to those observed for its mirror image, the corresponding *R,R*-isomer.[30] Both isoddG and isoddI were inactive. Antiviral evaluation of the pyrimidine isodideoxynucleosides and various other isomeric nucleosides are currently in progress and will be reported elsewhere.

SUMMARY

Anti-HIV active 2',3'-dideoxynucleosides that are derived from the natural nucleosides have disadvantages in terms of glycosidic bond hydrolytic stability and stability in terms of certain cellular enzymes. The design, synthesis, structural and enzymological studies and preliminary antiviral results of a new class of anti-HIV agents that overcome these inherent problems of instability are described. In these optically active isodideoxynucleosides with *S,S* absolute stereochemistry, the purine and pyrimidine bases have been transposed from the natural 1'-position to the 2'-position.

ACKNOWLEDGMENTS

Support of this research investigation by the National Institutes of Health (AI29842 and AI32851), the Burroughs Wellcome Company, and the University of Iowa (Faculty Scholar Award to V.N.) is gratefully acknowledged. The high-field NMR instruments used in this work were purchased in part by funds provided by the NIH. I am grateful to Dr. Janet L. Rideout and Dr. Marty St. Clair of Burroughs Wellcome Company and Dr. Erik De Clercq of the Rega Institute of Medical Research for collaborations on the antiviral work. It is a pleasure to acknowledge the contributions of my graduate students and postdoctoral associates to the work described. They are: Ms. Zoraida Nuesca, Mr. Todd B. Sells, Mr. Pascal Bolon, Mr. Lawrence B. Zintek, Dr. Greg S. Buenger, Dr. Arthur G. Lyons and Dr. David F. Purdy.

REFERENCES

1. H. Mitsuya and S. Broder, Inhibition of the *in vitro* infectivity and cytopathic effect of human T-lymphotropic virus type III/ lymphadenopathy-associated virus (HTLV-III/ LAV) by 2',3'-dideoxynucleosides, *Proc. Natl. Acad. Sci. U. S. A.* 83: 1911 (1986).

2. H. Mitsuya, K. J. Weinhold, P. A. Furman, M. H. St. Clair, S. N. Lehrman, R. C. Gallo, D. Bolognesi, D. W. Barry and S. Broder, 3'-Azido-3'-deoxythymidine (BW A509U): An antiviral agent that inhibits the infectivity and cytopathic effect of human T-lymphotropic virus type III/ lymphadenopathy-associated virus *in vitro*, *Proc. Natl. Acad. Sci. U. S. A.* 82: 7096 (1985).

3. R. Yarchoan, R. W. Klecker, K. J. Weinhold, P. D. Markham, H. K. Lyerly, D. T. Durack, E. Gelman, S. N. Lehrman, R. M. Blum, D. W. Barry, G. M. Shearer, M. A. Fischl, H. Mitsuya, R. C. Gallo, J. M. Collins, D. P. Bolognesi, C. E. Meyers and S. Broder, Administration of 3'-azido-3'-deoxythymidine, an inhibitor of HTLV-III/ LAV replication, to patients with AIDS or AIDS-related complex, *Lancet* i: 575 (1986).

4. R. Yarchoan, H. Mitsuya, R. V. Thomas, J. M. Pluda, N. R. Hartman, C-F. Perno, K. S. Marczyk, J-P. Allain, D. G. Johns and S. Broder, *In vivo* activity against HIV and favorable toxicity profile of 2',3'-dideoxyinosine, *Science* 245:412 (1989).

5. J. Balzarini, R. Pauwels, M. Baba, M. J. Robins, R. Zou, P. Herdewijn and E. De Clercq, The 2',3'-dideoxyriboside of 2,6-diaminopurine selectively inhibits HIV replication *in vitro*, *Biochem. Biophys. Res. Commun.* 145: 269 (1987).

6. H. Mitsuya, R. Yarchoan and S. Broder, Molecular targets for AIDS therapy, *Science* 249: 1533 (1990).

7. V. Nair and G. S. Buenger, Novel, stable congeners of the antiretroviral compound, 2',3'-dideoxyadenosine, *J. Am. Chem. Soc.* 111: 8502 (1989).

8. C. K. Chu, G. V. Ullas, L. S. Jeong, S. K. Ahn, B. Doboszewski, Z. X. Lin, J. W. Beach, R. F. Schinazi, Synthesis and structure-activity relationships of 6-substituted 2',3'-dideoxypurine nucleosides as potential anti-human immunodeficiency virus agents, *J. Med. Chem.* 33: 1553 (1990).

9. M. M. Mansuri, J. E. Starrett, Jr., I. Ghazzouli, M. J. M. Hitchcock, R. Z. Sterzycki, V. Brankovan, T. S. Lin, E. M. August, W. H. Prusoff, J-P. Sommadossi and J. C. Martin, 1-(2,3-dideoxy-β-D-glycero-pent-2-enofuranosyl)thymine. A highly potent and selective anti-HIV agent, *J. Med. Chem.* 32: 461 (1989).

10. M. Nasr, C. Litterst, J. McGowan, Computer-assisted structure-activity correlations of dideoxynucleoside analogs as potential anti-HIV drugs, *Antiviral Res.* 14: 125 (1990).

11. R. Vince, M. Hua, J. Brownell, S. Daluge, F. Lee, W. M. Shannon, G. C. Lavelle, J. Qualls, O. S. Weislow, R. Kiser, P. G. Canonico, R. H. Schultz, V. L. Narayanan, J. G. Mayo, R. H. Shoemaker and M. R. Boyd, Potent and selective activity of a new carbocyclic nucleoside analog (carbovir: NSC 614846) against human immuno-deficiency virus *in vitro*, *Biochem. Biophys. Res. Commun.* 156: 1046 (1988).

12. V. E. Marquez, C. K-H. Tseng, H. Mitsuya, S. Aoki, J. A. Kelley, H. Ford, Jr., J. S. Roth, S. Broder, D. G. Johns, J. S. Driscoll, Acid-stable 2'-fluoro purine dideoxynucleosides as active agents against HIV, *J. Med. Chem.* 33: 978 (1990).

13. E. De Clercq, HIV inhibitors targeted at the reverse transcriptase, *AIDS Res. Human Retroviruses* 8: 119 (1992).

14. G. W. Koszalka, T. A. Krenitsky, J. L. Rideout and C. L. Burns, Therapeutic nucleosides, European Patent No. 286 425 (1988).

15. J. W. Beach, L. S. Jeong, A. J. Alves, D. Pohl, H. O. Kim, C-N. Chang, S-L. Doong, R. F. Schinazi, Y-C. Cheng and C. K. Chu, Synthesis of enantiomerically pure (2'R, 5'S)-(-)-1-[2-(hydroxymethyl) oxathiolan-5-y1] cytosine as a potent antiviral agent against HBV and HIV, *J. Org. Chem.* 57: 2217 (1992).

16. H. Maag, R. M. Rydzewski, M. J. McRoberts, D. Crawford-Ruth, J. P. H. Verheyden and E. J. Prisbe, Synthesis and anti-HIV activity of 4'-azido and 4'-methoxy-nucleosides, *J. Med. Chem.* 35: 1440 (1992).

17. J. A. Secrist III, R. M. Riggs, K. N. Tiwari and J. A. Montgomery, Synthesis and anti-HIV activity of 4'-thio-2',3'-dideoxynucleosides, *J. Med. Chem.* 35: 533 (1992).

18. D. A. Cooney, G. Ahluwalia, H. Mitsuya, A. Fridland, M. Johnson, Z. Hao, M. Dalal, J. Balzarini, S. Broder and D. G. Johns, Initial studies on the cellular pharmacology of 2',3'-dideoxyadenosine, an inhibitor of HTLV-III infectivity, *Biochem. Pharmacol.* 36: 1765 (1987).

19. G. Ahluwalia, D. A. Cooney, H. Mitsuya, A. Fridland, K. P. Flora, Z. Hao, M. Dalal, S. Broder and D. G. Johns, Initial studies on the cellular pharmacology of 2',3'-dideoxyinosine, an inhibitor of HIV infectivity, *Biochem. Pharmacol.* 36: 3797 (1987).

20. J. E. Dahlberg, H. Mitsuya, S. B. Blam, S. Broder and S. A. Aaronson, Broad spectrum antiretroviral activity of 2',3'- dideoxynucleosides, *Proc. Natl. Acad. Sci. U.S.A.* 84: 2469 (1987).

21. A. Bloch, M. J. Robins, J. R. McCarthy, Jr., The role of the 5'-hydroxyl groups of adenosine in determining substrate specificity for adenosine deaminase, *J. Med. Chem.* 10: 908 (1967).

22. V. Nair, Development of methodologies for the strategic modification of purine ribonucleoside systems, *Nucleosides Nucleotides* 8: 699 (1989).

23. V. Nair and S.D. Chamberlain, Reductive deamination of aminopurine nucleosides, *Synthesis* 401 (1984).

24. T. A. Krenitsky, W. W. Hall, P. deMiranda, L. M. Beauchamp, H. J. Schaeffer and P. D. Whiteman, 6-Deoxyacyclovir: A xanthine oxidase-activated prodrug of acyclovir, *Proc. Natl. Acad. Sci. U.S.A.* 81: 3209 (1984).

25. V. Nair, G. S. Buenger and T. B. Sells, Inhibition of mammalian adenosine deaminase by novel functionalized 2',3'-dideoxyadenosines, *Biochim. Biophys. Acta* 1078: 121 (1991).

26. T. Shirasaka, K. Murakami, H. Ford, Jr., J. A. Kelley, H. Yoshioka, E. Kojima, S. Aoki, S. Broder and H. Mitsuya, Lipophilic halogenated congeners of 2'-3'-dideoxypurine nucleosides active against human immunodeficiency virus *in vitro*, *Proc. Natl. Acad. Sci. U.S.A.* 87: 9426 (1990).

27. V. Nair and G. S. Buenger, Approaches to new dideoxynucleosides, *Nucleosides Nucleotides* 10: 307 (1991); V. Nair, T. B. Sells, A. G. Lyons and D. F. Purdy, Novel modified nucleosides and their phosphorylated analogues as potential anti-HIV agents, *Antiviral Res.* Suppl. I: 44 (1991).

28. V. Nair and G.S. Buenger, Hydrolysis of dideoxygenated purine nucleosides: Effect of modification of the base moiety, *J. Org. Chem.* 55: 3695 (1990).

29. G. S. Buenger and V. Nair, Deoxygenated purine nucleosides substituted at the 8-position: Chemical synthesis and stability, *Synthesis* 962 (1990).

30. K. B. Frank, E. V. Connell, M. J. Holman, D. M. Huryn, B. C. Sluboski, S. Y. Tam, L. J. Todaro, M. Weigele, D. D. Richman, H. Mitsuya, Anabolism and mechanism of action of Ro24-5098, an isomer of 2',3'-dideoxyadenosine (ddA) with anti-HIV activity, *Ann. N.Y. Acad. Sci.* 616: 408 (1990).

31. M. Sundaralingam, Structure and conformation of nucleosides and nucleotides and their analogs as determined by X-ray diffraction, *Ann. N.Y. Acad. Sci.* 255: 3 (1975).

32. E. W. Taylor, P. Van Roey, R. F. Schinazi and C. K. Chu, A stereochemical rationale for the activity of anti-HIV nucleosides, *Antiviral Chem. Chemother.* 1: 163 (1990).

33. Y. F. Shealy and C. A. O'Dell, Synthesis of the carbocyclic analogs of uracil nucleosides, *J. Heterocycl. Chem.* 13: 1015 (1976).

34. W. L. Sung, Chemical conversion of thymidine into 5-methyl-2'-deoxycytidine, *J. Chem. Soc., Chem. Commun.* 11: 1089 (1981).

35. A. Bax, Structure determination and spectral assignment by pulsed polarization transfer *via* long-range ^{1}H-^{13}C couplings, *J. Magn. Reson.* 57:314 (1984).

36. D. A. Cooney, M. Dalal, H. Mitsuya, J. B. McMahon, M. Nadkarni, J. Balzarini, S. Broder and D. G. Johns, Initial studies on the cellular pharmacology of 2',3'-dideoxycytidine, an inhibitor of HTLV-III infectivity, *Biochem. Pharmacol.* 35: 2065 (1986).

37. V. Nair and T.B. Sells, Interpretation of the roles of adenylosuccinate lyase and of AMP deaminase in the anti-HIV activity of 2',3'-dideoxyadenosine and 2',3'-dideoxyinosine, *Biochim. Biophys. Acta* 1119: 201 (1992).

CARBOCYCLIC 7-DEAZAGUANOSINE NUCLEOSIDES
AS ANTIVIRAL AGENTS

Stewart W. Schneller*, Xing Chen, and Suhaib M. Siddiqi

Department of Chemistry, University of South Florida

Tampa, FL 33620-5250

INTRODUCTION

Nucleosides in which the ribofuranosyl oxygen is replaced by a methylene to result in a cyclopentane ring are referred to as carbocyclic nucleosides.[1] The first such compound was carbocyclic adenosine (**1**, aristeromycin), which was synthesized in its racemic form[2] prior to isolation of the (-)-enantiomer from *Streptomyces citricolor*.[3] The antiviral activity of **1**[4] has stimulated the search for other carbocyclic nucleosides that would display a more favorable therapeutic index.[5]

A consistent structural unit in recent antiviral drug development is the presence of the guanine base with a wide variety of modified ribofuranosyl substituents at the N-9 position.[6] In the latter regard, carbocyclic 7-deazaguanosine (**2**) has been reported[7] to have potentially meaningful antiherpetic activity. However, further study of carbocyclic 7-deazaguanosine as a prototype structure for potential antiviral agents has not been described. Support for the expectation that an extended investigation of the carbocyclic 7-deazaguanosine series will lead to fruitful results can be found in the work of Townsend, Drach and their co-workers,[8] who have reported that 7-deazaadenosine-derived agents have significant antiviral potential. Thus,

Nucleosides and Nucleotides as Antitumor and Antiviral Agents,
Edited by C.K. Chu and D.C. Baker, Plenum Press, New York, 1993

several years ago, we sought to evaluate carbocyclic nucleosides derived from 7-deazaguanosine as antiviral agents. Several examples of this study are presented in this report.

DISCUSSION

Rotationally Restricted Analogue of Buciclovir

Buciclovir (3, (R)-DHBG),[9] which shows antiherpetic activity,[10] can be viewed as a carba analogue of the antiviral agent DHPG (4)[6b] lacking a methylene group. Carbocyclic 2',3'-dideoxy-3'-oxa-7-deazaguanosine (5) was conceived as a rotationally restricted derivative of 3 still capable of triphosphate formation at the C-5' center, a likely requirement[11,12] for antiviral activity. In addition to its relationship to 3, compound 5 can be considered an analogue of carbocyclic 2',3'-dideoxy-3'-oxaadenosine (6, isoddA),[13] which shows anti-HIV activity, and carbocyclic 2',3'-dideoxy-3'-azapurine nucleosides (for example, 7),[14] and carbocyclic 2',3'-dideoxy-3'-thia- and -2'-oxapurine and pyrimidine nucleosides,[15,16] which were inactive.

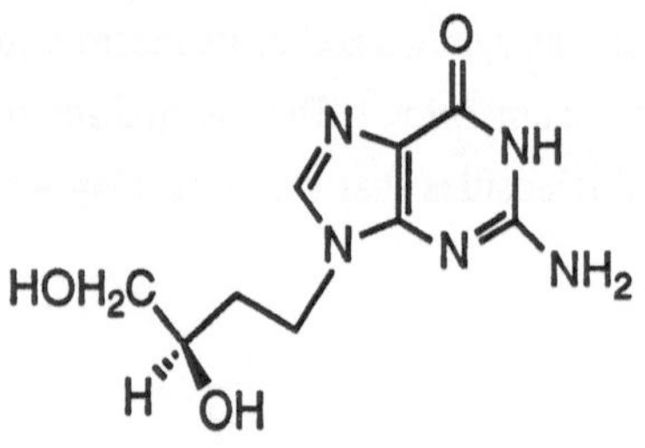

3

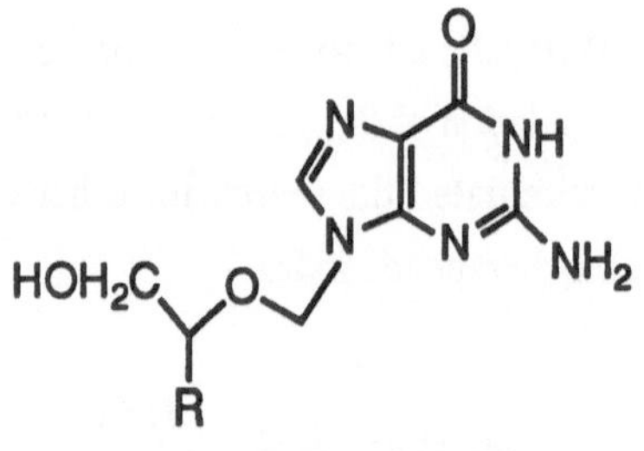

4, R = CH$_2$OH

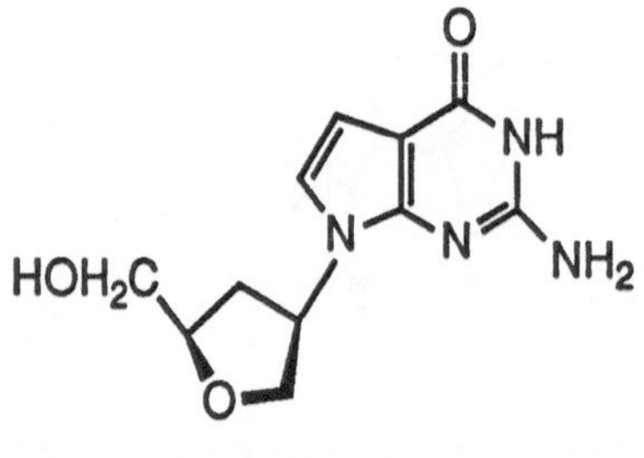

5

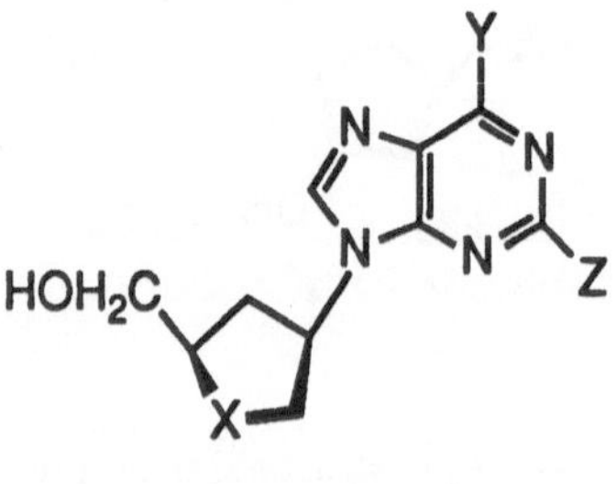

6, X = O; Y = NH$_2$; Z = H
7, X = NH; Y = OH; Z = NH$_2$

The synthesis of 5 (Scheme 1) was easily accomplished by reaction of 8 with 2-(2-amino-4,6-dichloropyrimidin-5-yl)acetaldehyde dimethyl acetal[17] to give 9, which was not fully characterized but readily cyclized to 10 upon treatment with acid. Acidic hydrolysis of 10 to the desired 2-amino-7-[(1R,4R)-4-hydroxymethyl-3-oxa-1-cyclopentyl]-7H-pyrrolo-

[2,3-*d*]pyrimidin-4(3*H*)-one (**5**) was then accomplished. Attempts were unsuccessful for preparing **10** by reaction of 2-amino-4-chloro-7*H*-pyrrolo[2,3-*d*]pyrimidine (**11**)[18] with **15** (Scheme 2) (or **16**, which would have required desilylation) in the presence of potassium carbonate.

In order to employ Scheme 1 for preparing **5**, it was first necessary to prepare (1*R*,4*R*)-4-(*t*-butyldimethylsilyloxy)methyl-3-oxa-1-cyclopentylamine (**8**). Thus, by modification of procedures that led to the preparation of **6**,[13] the synthesis of **8** began with

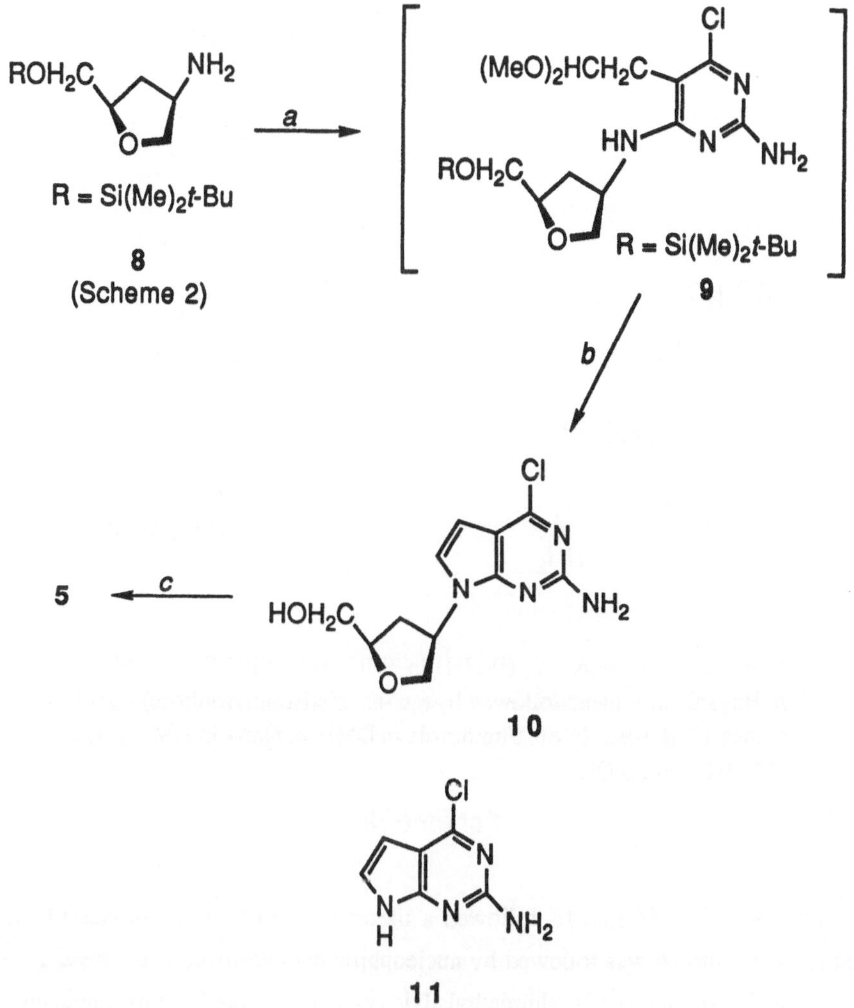

Reaction conditions: *a*, 2-(2-amino-4,6-dichloropyrimidin-5-yl)acetaldehyde dimethyl acetal and Et₃N in BuOH; *b*, 2 N HCl in dioxane, room temperature; *c*, 2 N HCl in MeOH, reflux.

Scheme 1

the conversion of 5-O-acetyl-1,2-O-isopropylidene-α-D-xylofuranose (**12**)[19] into its 3-O-[(methylthio)thiocarbonyl] derivative **13** (Scheme 2). It should be noted that the 3-O-(imidazolylthiocarbonyl) derivative of **12** has been reported,[19] but we found (i) that the use of carbon disulfide and methyl iodide for preparing **13** to be a more convenient preparation and (ii) that the subsequent Barton reaction[19,20] on **13** was easily worked-up to give 5-O-acetyl-3-deoxy-1,2-O-isopropylidene-α-D-*erythro*-pentofuranose (**14**).

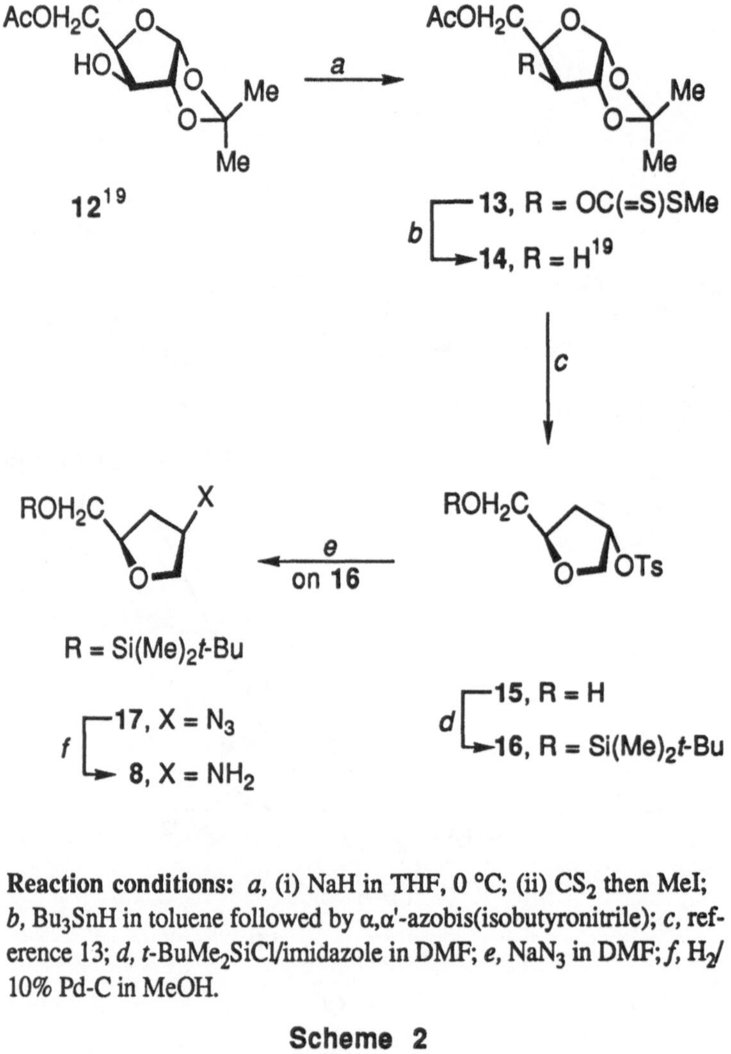

Reaction conditions: *a*, (i) NaH in THF, 0 °C; (ii) CS$_2$ then MeI; *b*, Bu$_3$SnH in toluene followed by α,α'-azobis(isobutyronitrile); *c*, reference 13; *d*, *t*-BuMe$_2$SiCl/imidazole in DMF; *e*, NaN$_3$ in DMF; *f*, H$_2$/ 10% Pd-C in MeOH.

Scheme 2

Conversion of **14** into **15** followed a literature route[13] with comparable yields. Silylation of **15** into **16** was followed by nucleophilic displacement of the tosyl group by azide to provide (1R,4R)-4-(t-butyldimethylsilyloxy)methyl-3-oxa-1-cyclopentylazide (**17**). Reduction of **17** then gave the requisite **8**.

Compound **5** was evaluated against herpes simplex virus types 1 and 2, human cytomegalovirus, vaccinia, vesicular stomatitis, coxsackie B4, polio, parainfluenza 3, reo 1, sindbis, and semliki forest viruses and human immunodeficiency viruses 1 and 2 and found to lack activity towards all of these viruses. Analogue 5 was also non-cytotoxic.

(-)-Carbocyclic 7-Deazaoxetanocin G

The naturally occurring oxetanocin A (18)[21] and synthetic oxetanocin G (19)[6d] represent a novel class of nucleosides possessing antiviral properties.[22] Modification of the unique oxetanosyl-*N*-glycoside structural feature of 18 and 19 into the carbocyclic nucleoside framework led to the synthesis of 20 and 21 as racemates[23] and enantiomers,[24] which have shown activity against herpesviruses and HIV.[22b,c,24,25] In view of the potent and selective anti-HCMV properties of the *R*-enantiomer of 21,[24b] the synthesis of the *R* stereoisomer 22 was undertaken and accomplished.

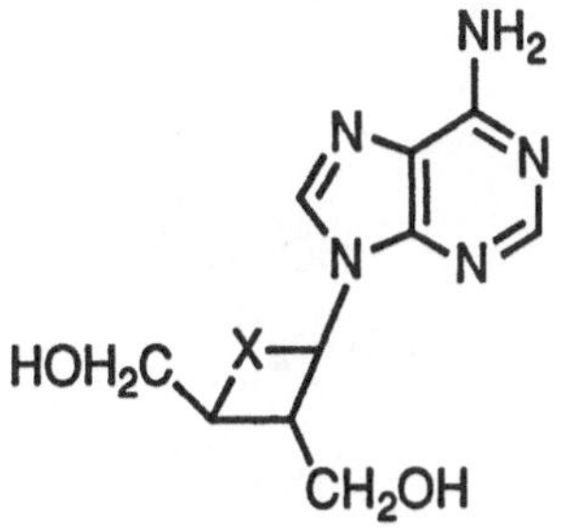

(-)-18, X = O
20, X = CH$_2$

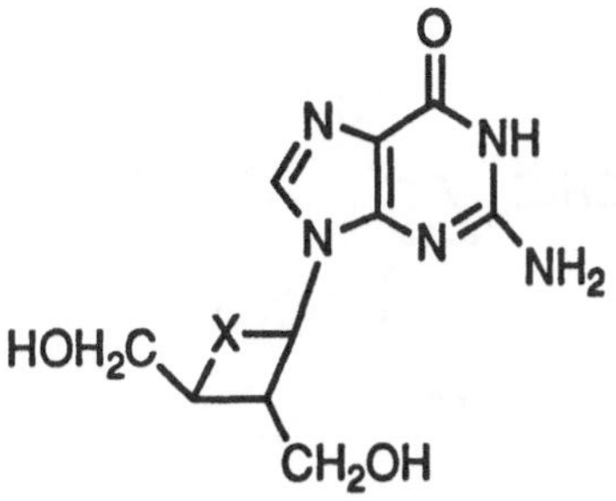

(-)-19, X = O
21, X = CH$_2$

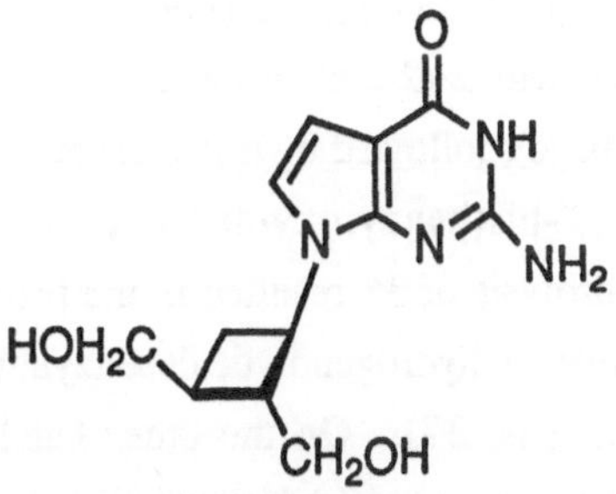

(*R*)-22

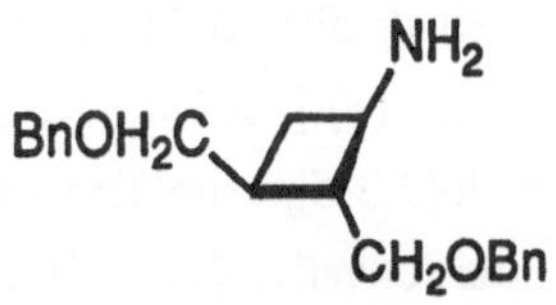

23

The previously described 7-deazapurine carbocyclic nucleoside synthesis (see Scheme 1) indicated that the most efficient route to 22 would be via reaction of the protected chiral amine 23 with the dimethylacetal of 2-(2-amino-4,6-dichloropyrimidin-5-yl)acetaldehyde followed by ring closure, hydrolysis and deprotection. A review of the literature revealed two enantioselective routes to cyclobutyl derivatives that had been used in the chiral synthesis of carbocyclic oxetanocins[24] and could be employed for preparing a precursor to 23. In one case,[24a] however, the initial step involved a [2+2]-cycloaddition reaction of not easily obtainable reagents in the presence of a chiral titanium compound as catalyst. In the other case,[24b] resolution of a chiral mixture of cycloadducts was done through diastereomeric amides that added steps to the synthesis. In the search for a more convenient means to **23**, our attention turned to using enzymes for optical resolution.[26] As a consequence, we chose to use *Pseudomonas cepacia* lipase[27] on the (±)-acetate **28** (Scheme 3) as an efficient means to **23**.

Thus, benzylation of *trans*-3,3-diethoxy-1,2-bis(hydroxymethyl)cyclobutane (**24**)[23a,b] to **25** (Scheme 3) was followed by hydrolysis to *trans*-2,3-bis(benzyloxymethyl)-cyclobutanone (**26**).[28] Reduction of **26** with LS-Selectride[29] to (±)-(1α,2α,3β)-2,3-bis(benzyloxymethyl)cyclobutanol (**27**) occurred with no indication (by ^{1}H and ^{13}C NMR) of the diastereomeric alcohol **29** having been formed. Acetylation of **27** gave **28**, which, when subjected to treatment with *Pseudomonas cepacia* lipase, yielded (+)-**28** (>99% ee[30]) and (-)-**27** (>99% ee[30]). The structural assignments for (+)-**28** and (-)-**27** were accomplished by conversion into the enantiomeric alcohols **30** and **31** and comparing the optical rotation data for each product with the reported[24b] values for these latter compounds.

Saponification of (+)-**28** to (+)-**27** was followed by tosylation to (+)-[1*S*,2*S*,3*R*]-2,3-bis(benzyloxymethyl)-1-cyclobutyl tosylate (**32**). Treatment of **32** with sodium azide (to **33**), followed by reduction, provided (+)-[1*R*,2*S*,3*R*]-2,3-bis(benzyloxymethyl)-cyclobutylamine (**23**), which, in turn, upon reaction with 2-(2-amino-4,6-dichloropyrimidin-5-yl)acetaldehyde dimethyl acetal (to **34** on next page), followed by ring closure with dilute acid, produced (+)-[1'*R*,2'*S*,3'*R*]-2-amino-7-[2',3'-bis(benzyloxymethyl)cyclobutyl]-4-chloropyrrolo[2,3-*d*]pyrimidine (**35**). Basic hydrolysis of **35** resulted in the protected 7-deazaguanine derivative (+)-**36**.[31] Attempts to carry out hydrogenolytic debenzylation of **36** produced the 5,6-dihydro derivative of **22** (that is, **37**). On the other hand, use of ethanethiol/boron trifluoride on **36** led to the target compound (-)-[1'*R*,2'*S*,3'*R*]-2-amino-7-[2',3'-bis(hydroxymethyl)cyclobutyl]pyrrolo[2,3-*d*]pyrimidin-4(3*H*)-one (**22**, 70% ee,[30] {[α]$_D^{25}$ -12.29° (*c* 0.358, DMSO)}). It should be noted that reaction of **32** (Scheme 3) with 2-amino-4-chloro-7*H*-pyrrolo[2,3-*d*]pyrimidine (**11**)[18] using sodium hydride as the base provided another route to **35**.

Compound (-)-**22** has been evaluated against herpes simplex virus types 1 and 2, vaccinia virus, vesicular stomatitis virus, and HIV-1 and -2 with potent activity found towards herpes simplex type 1 (MIC = 0.01 μg/mL) and type 2 (MIC = 0.04 μg/mL). Some activity towards HIV-1 (EC$_{50}$ = 231±11 μM) and HIV-2 (EC$_{50}$ = 221±30 μM) was also observed. This analogue was also found to be non-cytotoxic.

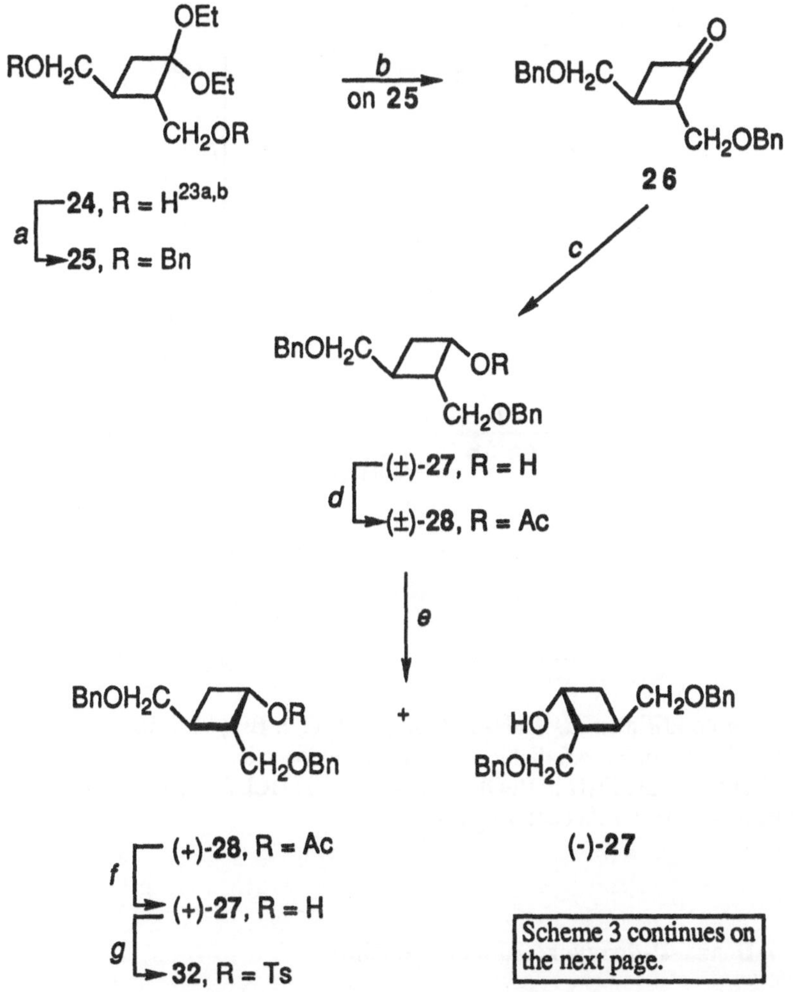

Scheme 3 continues on the next page.

Reaction conditions: *a,* (i) NaH in DMF; (ii) BnBr; *b,* 0.5% H_2SO_4 in MeCN; *c,* LS-Selectride in THF, -78 °C; *d,* Ac_2O/pyridine; *e, Pseudomonas cepacia* lipase in phosphate buffer; *f,* KOH in MeOH; *g,* TsCl/pyridine.

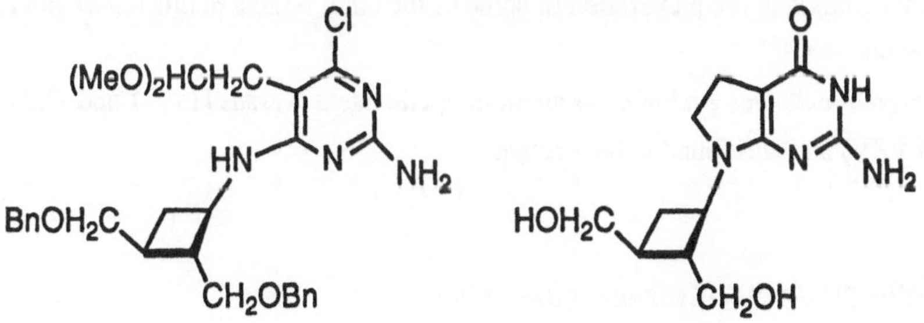

34 **37**

Scheme 3

Scheme 3 (Continued)

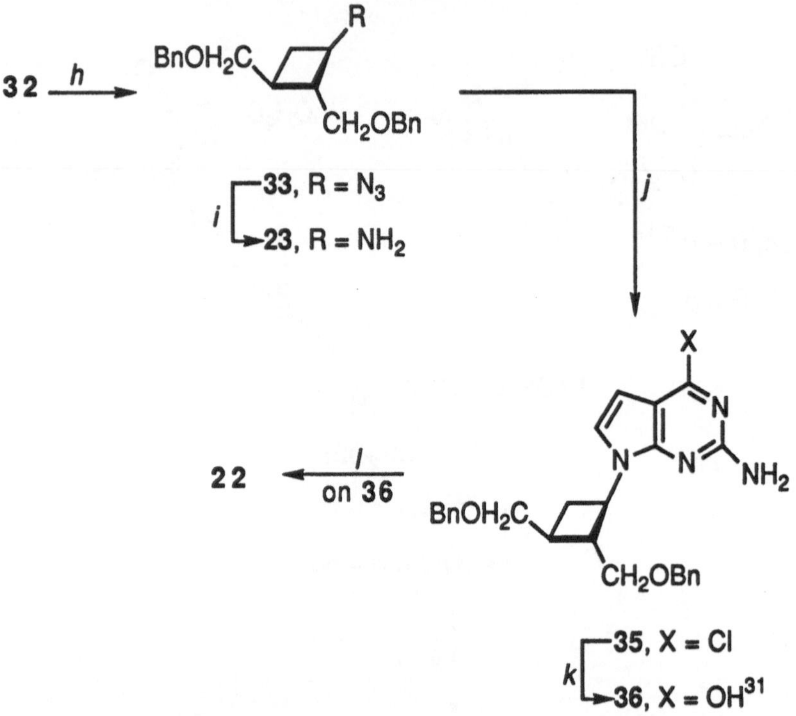

Reaction conditions: *a*, NaN$_3$ in DMF, 100 °C; *i*, BH$_3$•THF in THF; *j*, (i) 2-(2-amino-4,6-dichloropyrimidin-5-yl)acetaldehyde dimethyl acetal and Et$_3$N in BuOH, 100 °C; (ii) 2 N HCl; *k*, 5% aq. NaOH in MeOH; *l*, EtSH and BF$_3$•Et$_2$O.

Carbocyclic 7-Deaza-2'-deoxy*ara*fluoroguanosine

Borthwick and colleagues reported[32] that carbocyclic 2'-deoxy*ara*fluoroguanosine (**38**) displayed potent anti-HSV-1 and -2 properties. In view of this, we sought the 7-deaza analogue (**39**) whose synthesis is shown in Scheme 4. The reactions employed in that Scheme are common to the preparation of some of the other targets in this report and require no elaboration.

Derivative **39** was evaluated as an antiherpetic agent (versus HSV-1 and -2, HCMV, EBV, and VZV) and was found to be inactive.

Carbocyclic N^2-Aryl-7-deaza*ara*guanosine

Wright and his co-workers[35] have found that placement of an arylamino side-chain at

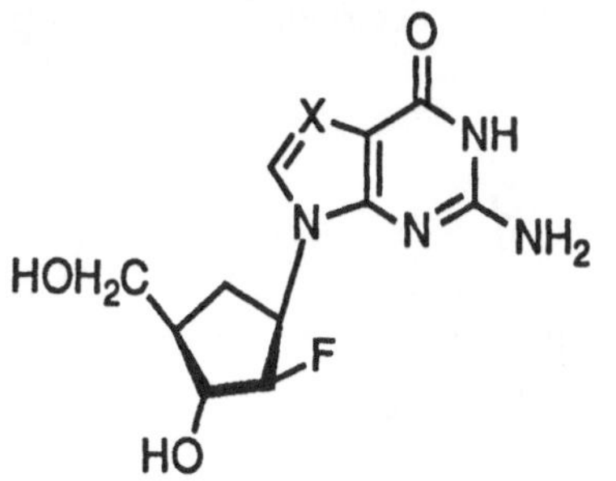

38, X = N
39, X = CH

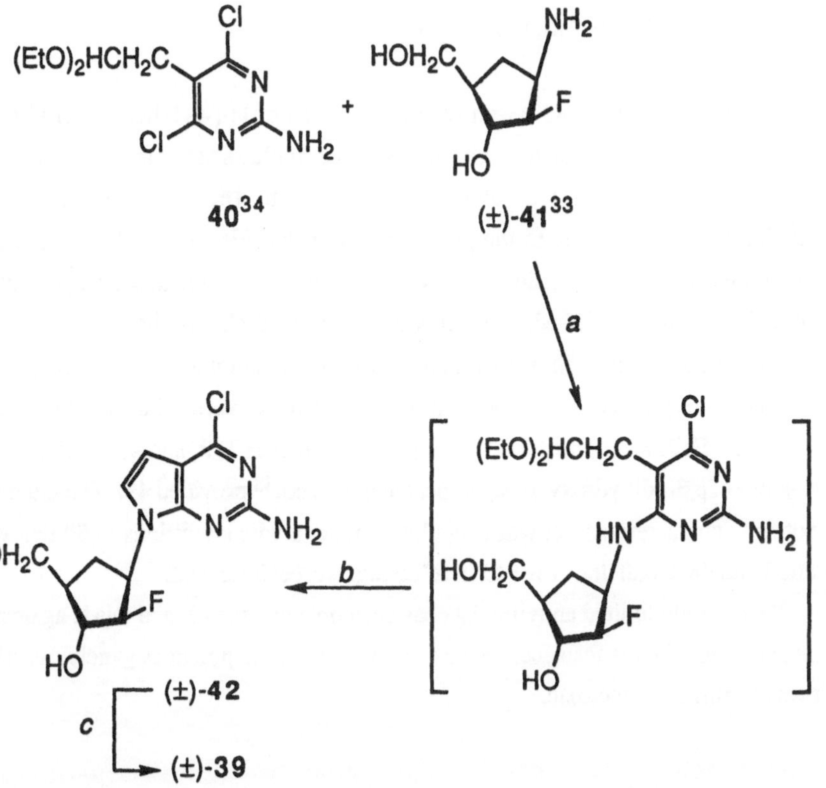

40[34]   (±)-**41**[33]

a

(±)-**42**

c

(±)-**39**

Reaction conditions: *a*, Et₃N in BuOH, reflux; *b*, 1 N HCl, 2 days; *c*, 2 N HCl/MeOH, reflux.

Scheme 4

the C-2 position of purine nucleosides induces their nucleotide derivatives to show DNA polymerase inhibitory activity including polymerases from viral sources (HSV, EBV, and vaccinia viruses).[36] In view of the ability of *ara* analogues to act as 2'-deoxy alternatives in the presence of DNA polymerase,[35] the carbocyclic 7-deaza*ara*guanosine derivative **43** was identified as a target compound to determine if Wright's prototype structure could be extended to the carbocyclic series.

(±)-**43**
Ar = 4-(1-butyl)phenyl[36]

45

The synthesis of (±)-**43** began with the reaction of N-[4-(1-butyl)phenyl]guanidine nitrate with diethyl allylmalonate to give **44** (Scheme 5) plus a small amount of the isomeric **45**.[37] To achieve successful chlorination of **44** to **46**, it was necessary to add tetraethylammonium chloride to the phosphorus oxychloride. In our hands, the standard ozonolysis approach for converting the allyl side chain of **46** to the acetaldehyde substituent of **47** could not be accomplished, apparently due to susceptibility of the arylamino side chain to the reaction conditions. Thus, treatment of **46** with osmium tetroxide in the presence of sodium periodate yielded **47** that was, in turn, readily converted into its dimethyl acetal derivative **48**. Following procedures analogous to Schemes 1, 3, and 4, reaction of **48** with (±)-4α-amino-2β,3α-dihydroxy-1α-cyclopentanemethanol[38] provided **49**. The latter product was not fully characterized but was ring closed under acidic conditions to **50** that was then converted into the target derivative **43** in refluxing hydrochloric acid.

To date only limited antiviral data has been obtained for **43**, and this is against human cytomegalovirus. In this instance, **43** was about 3.5 times as potent as ganciclovir (DHPG), but it was 13 times as cytotoxic.

(±)-7-Deazacarbovir

(-)-Carbovir (carbocyclic 2',3'-didehydro-2',3'-dideoxyguanosine, **51**)[6c,39,40] has been identified as a potent and selective inhibitor of human immunodeficiency virus (HIV)[41] and, as a consequence, has been the focus of considerable recent synthetic attention.[42] Following the theme of the research being described herein, it was desirable to prepare racemic 7-deazacarbovir ((±)-**52**) to determine its anti-HIV potential prior to synthesis of the enantiomer that is configurationally equivalent to (-)-**51**. The preparation of (±)-**52** (Scheme 6), which was based on the preparation of **51**,[39] began with the basic hydrolysis of (±)-(3α,5α)-3-acetamido-5-(acetoxymethyl)cyclopentene (**53**)[43] to (±)-(3α,5α)-3-amino-5-

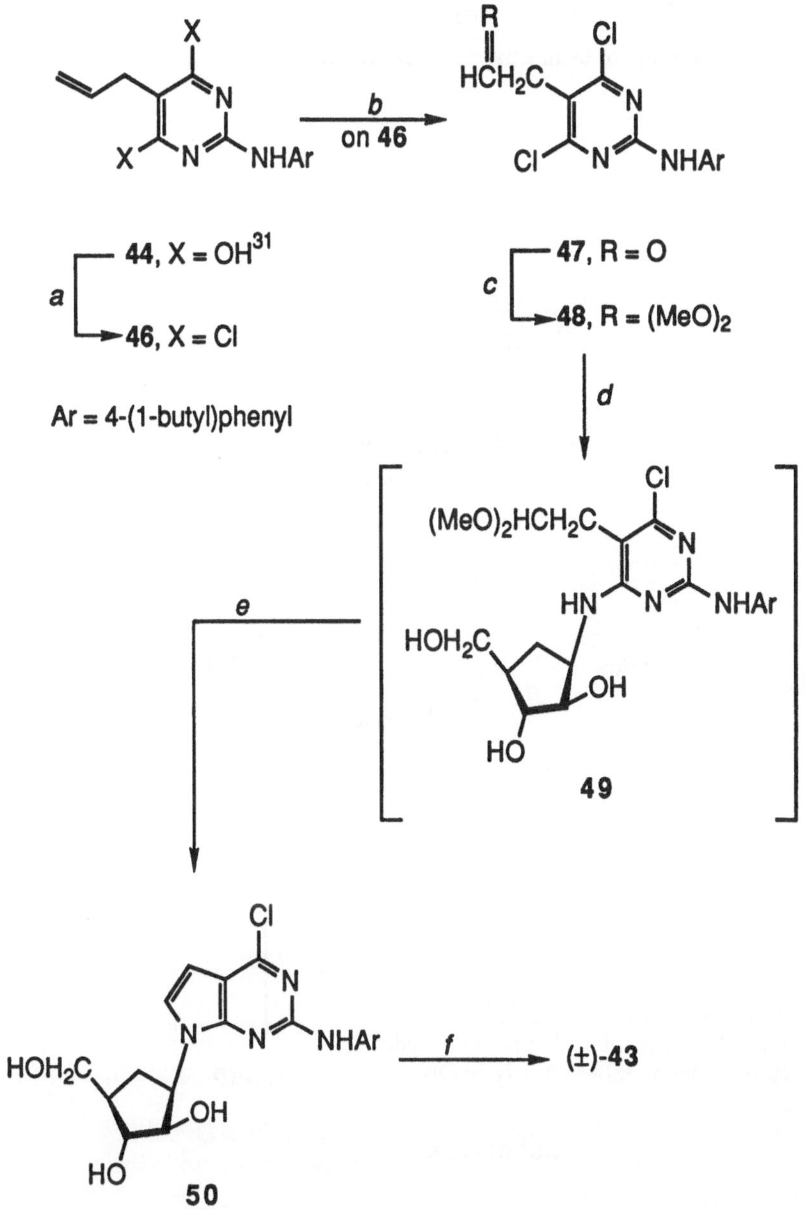

Reaction conditions: a, POCl$_3$/Et$_4$NCl/N,N-diethylaniline in MeCN, 70 °C then 100 °C; b, OsO$_4$/NaIO$_4$ in MeOH/acetone/H$_2$O; c, NH$_4$Cl/pyridinium p-toluenesulfonate in absolute MeOH; d, (±)-4α-amino-2β,3α-dihydroxy-1α-cyclo-pentanemethanol[38] in BuOH containing Et$_3$N, heat; e, 2 N HCl in 1,4-dioxane; f, 2 N aqueous HCl in EtOH, reflux.

Scheme 5

(hydroxymethyl)cyclopentene. Reaction of the latter compound with 2-(2-amino-4,6-di-chloropyrimidin-5-yl)acetaldehyde[7] resulted in **54** without the need for a separate pyrrole ring closure step. Basic hydrolysis of **54** yielded **52**.

Compound **52** was found to be inactive towards HIV *in vitro*.

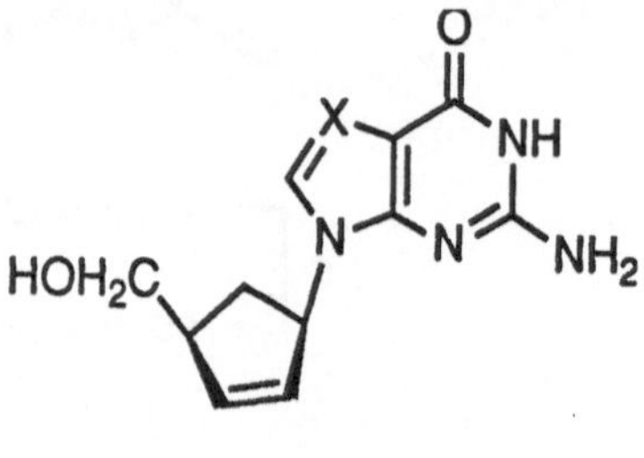

(-)-**51**, X = N
(±)-**52**, X = CH

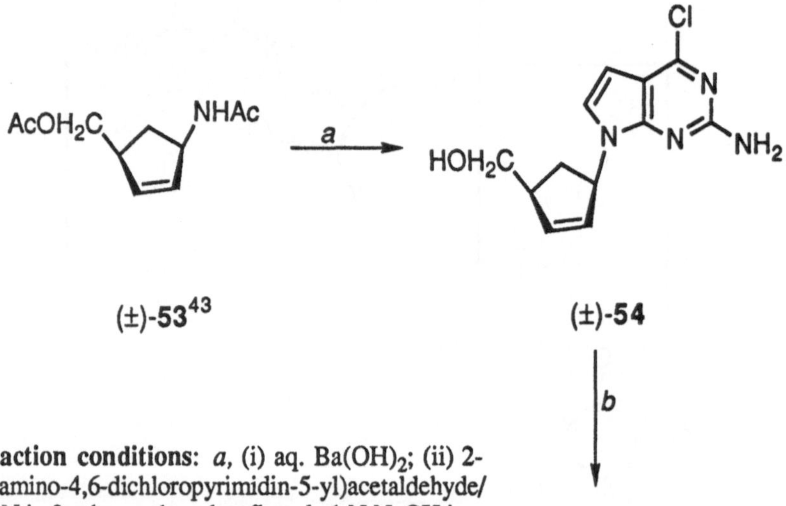

(±)-**53**[43]

(±)-**54**

b

(±)-**52**

Reaction conditions: *a*, (i) aq. Ba(OH)$_2$; (ii) 2-(2-amino-4,6-dichloropyrimidin-5-yl)acetaldehyde/Et$_3$N in 2-ethoxyethanol, reflux; *b*, 1 N NaOH in EtOH, reflux.

Scheme 6

CONCLUSIONS

Except for the oxetanocin G analogue **22**, carbocyclic 7-deazaguanosines do not seem to provide a fruitful source of potential antiviral agents. This observation agrees with others[44] who have noted decreases in antiviral activity when carbocyclic 7-deazapurine nucleosides are compared to their purine parent systems. This correlation carries over into the cytotoxicity results in that the carbocyclic 7-deazaguanosines reported herein showed no cytotoxicity, which is in contrast to what is often seen in the 7-deazapurine nucleoside series.

ACKNOWLEDGEMENTS

This project was supported by funds from the Department of Health and Human Services (NO1-AI-72645) and this is appreciated. The assistance of Drs. Erik De Clercq, Peter Medveczky, and Christopher Tseng in obtaining the antiviral data for the compounds described in this report is also gratefully acknowledged.

REFERENCES

1. (a) V.E. Marquez and M.-I.Lim, Carbocyclic nucleosides, *Med. Res. Rev.* 6: 1 (1986). (b) J.A. Montgomery, Approaches to antiviral chemotherapy, *Antiviral Res.* 12: 113 (1989).

2. (a) Y.F. Shealy and J.D. Clayton, 9-[β-DL-2α,3α-Dihydroxy-4β-(hydroxymethyl)-cyclopentyl]adenine, the carbocyclic analog of adenosine, *J. Am. Chem. Soc.* 88: 3885 (1966). (b) Y.F. Shealy and J.D. Clayton, Synthesis of carbocyclic analogs of purine ribonucleosides, *J. Am. Chem. Soc.* 91: 3075 (1969).

3. (a) T. Kusaka, H. Yamamoto, M. Shibata, M. Muroi, T. Kishi, and K. Mizuno, *Streptomyces citricolor* nov. sp. and a new antibiotic, aristeromycin, *J. Antibiot. (Tokyo)* 21: 255 (1968). (b) T. Kishi, M. Muroi, T. Kusaka, M. Nishikawa, K. Kamiya, and K. Mizuno, The structure of aristeromycin, *Chem. Pharm. Bull.* 20: 940 (1972). (c) P. Herdewijn, J. Balzarini, E. De Clercq, and H. Vanderhaeghe, Resolution of aristeromycin enantiomers, *J. Med. Chem.* 28: 1385 (1985).

4. E. De Clercq, S-Adenosylhomocysteine hydrolase inhibitors as broad-spectrum antiviral agents, *Biochem. Pharmacol.* 36: 2567 (1987).

5. For example, see M.S. Wolfe and R.T. Borchardt, S-Adenosylhomocysteine hydrolase as a target for antiviral chemotherapy, *J. Med. Chem.* 34: 1521 (1991).

6. For example, (a) acyclovir: G.B. Elion, P.A. Furman, J.A. Fyfe, P. De Miranda, L. Beauchamp, and H.J. Schaeffer, Selectivity of action of an antiherpetic agent, 9-(2-hydroxyethoxymethyl)guanine, *Proc. Natl. Acad. Sci. U.S.A.* 74: 5716 (1977). (b) 9-[(1,3-dihydroxy-2-propoxy)methyl]guanine (ganciclovir, DHPG): (i) J.C. Martin, C.A. Dvorak, D.F. Smee, T.R. Matthews, and J.P.H. Verheyden, 9-[(-1,3-Dihydroxy-2-propoxy)methyl]guanine: a new potent and selective antiherpes agent, *J. Med. Chem.* 26: 759 (1983); (ii) A.K. Field, M.E. Davies, C. DeWitt, H.C. Perry, R. Liou, J. Germershausen, J.D. Karkas, W.T. Ashton, D.B.R. Johnston, and R.L. Tolman, 9-{[2-Hydroxy-1-(hydroxymethyl)ethoxy]methyl}guanine: a selective inhibitor of herpes group virus replication, *Proc. Natl. Acad. Sci. U.S.A.* 80: 4139 (1983); and, (iii) K.O. Smith, K.S. Galloway, W.L. Kennell, K.K. Ogilvie, and B.K. Radatus, A new nucleoside analog, 9-[[2-hydroxy-1-(hydroxymethyl)ethoxy]methyl]guanine, highly active in vitro against herpes simplex

virus types 1 and 2, *Antimicrob. Agents Chemother.* 22: 55 (1982). (c) carbovir: L.L. Bondoc, Jr., W.M. Shannon, J.A. Secrist, III, R. Vince, and A. Fridland, Metabolism of the carbocyclic nucleoside analogue carbovir, an inhibitor of human immunodeficiency virus, in human lymphoid cells, *Biochemistry* 29: 9839 (1990). (d) oxetanocin G: N. Shimada, S. Hasegawa, S. Saito, T. Nishikiori, A. Fujii, and T. Takita, Derivatives of oxetanocin: oxetanocins H, X and G, and 2-aminooxetanocin A, *J. Antibiot.* 40: 1788 (1987). (e) carbocyclic 2'-deoxy*ara*fluoroguanosine: A.D. Borthwick, B.E. Kirk, K. Biggadike, A.M. Exall, S. Butt, S.M. Roberts, D.J. Knight, J.A.V. Coates, and D.M. Ryan, Fluorocarbocyclic nucleosides: synthesis and antiviral activity of 2'- and 6'-fluorocarbocyclic 2'-deoxyguanosines, *J. Med. Chem.* 34: 907 (1991).

7. M. Legraverend, R.-M.N. Ngongo-Tekam, E. Bisagni, and A. Zerial, (±)-2-Amino-3,4-dihydro-7-[2,3-dihydroxy-4-(hydroxymethyl)-1-cyclopentyl]-7*H*-pyrrolo[2,3-*d*]pyrimidin-4-ones: new carbocyclic analogues of 7-deazaguanosine with antiviral activity, *J. Med. Chem.* 28: 1477 (1985).

8. (a) S.R. Turk, C. Shipman, Jr., R. Nassiri, G. Genzlinger, S.H. Krawczyk, L.B. Townesnd, and J.C. Drach, Pyrrolo[2,3-*d*]pyrimidine nucleosides as inhibitors of human cytomegalovirus, *Antimicrob. Agents Chemother.* 31: 544 (1987). (b) J.S. Pudlo, N.K. Saxena, M.R. Nassiri, S.R. Turk, J.C. Drach, and L.B. Townsend, Synthesis and antiviral activity of certain 4- and 4,5-disubstituted 7-[(2-hydroxy-ethoxy)methyl]pyrrolo[2,3-*d*]pyrimidines, *J. Med. Chem.* 31: 2086 (1988). (c) J.S. Pudlo, M.R. Nassiri, E.R. Kern, L.L. Wotring, J.C. Drach, and L.B. Townsend, Synthesis, antiproliferative, and antiviral activity of certain 4-substituted and 4,5-disubstituted 7-[(1,3-dihydroxy-2-propoxy)methyl]pyrrolo[2,3-*d*]pyrimi-dines, *J. Med. Chem.* 33: 1984 (1990).

9. A. Larsson, B. Öberg, S. Alenius, C.-E. Hagberg, N.-G. Johansson, B. Lindborg, and G. Stening, 9-(3,4-Dihydroxybutyl)guanine, a new inhibitor of herpesvirus multiplication, *Antimicrob. Agents Chemother.* 23: 664 (1983).

10. (a) A.-C. Ericson, A. Larsson, F.Y. Aoki, W.-A. Yisak, N.-G. Johansson, B. Öberg, and R. Datema, Antiherpes effects and pharmacokinetic properties of 9-(4-hydroxybutyl)guanine and the (*R*) and (*S*) enantiomers of 9-(3,4-dihydroxy-butyl)guanine, *Antimicrob. Agents Chemother.* 27: 753 (1985). (b) R. Datema, N,G, Johansson, and B. Öberg, On the antiviral efficacy of some drugs interfering with nucleic acid synthesis in virus-infected cells, *Chem. Scripta* 26: 49 (1986).

11. A.K. Field, M.E. Davies, C.M. DeWitt, H.C. Perry, R. Liou, J. Germershausen, J.D. Karkas, W.T. Ashton, D.B.R. Johnston, and R.L. Tolman, 9-{[1-Hydroxy-1-(hydroxymethyl)ethoxy]methyl}guanine: a selective inhibitor of herpes group virus replication, *Proc. Natl. Acad. Sci. USA* 80: 4139 (1983).

12. (a) J.E. Reardon and T. Spector, Herpes simplex virus type 1 DNA polymerase, *J. Biol. Chem.* 264: 7405 (1989). (b) D.F. Smee, J.C. Martin, J.P.H. Verheyden, and

T.R. Matthews, Anti-herpesvirus activity of the acyclic nucleoside 9-(1,3-dihydroxy-2-propoxymethyl)guanine, *Antimicrob. Agents Chemother.* 23: 676 (1983). (c) A. Larsson, K. Stenberg, A.-C. Ericson, U. Haglund, W.-A. Yisak, N.-G. Johansson, B. Öberg, and R. Datema, Mode of action, toxicity, pharmacokinetics, and efficacy of some new antiherpesvirus guanosine analogs related to buciclovir, *Antimicrob. Agents Chemother.* 30: 598 (1986). (d) R. Datema, A.-C. Ericson, H.J. Field, A. Larsson, and K. Stenberg, Critical determinants of antiherpes efficacy of buciclovir and related acyclic guanosine analogs, *Antiviral Res.* 7: 303 (1987). (e) J.D. Karkas, W.T. Ashton, L.F. Canning, R. Liou, J. Germershausen, R. Bostedor, B. Arison, A.K. Field, and R.L. Tolman, Enzymatic phosphorylation of the antiherpetic agent 9-[(2,3-dihydroxy-1-propoxy)methyl]guanine, *J. Med. Chem.* 29: 842 (1986).

13. D.M. Huryn, B.C. Sluboski, S.Y. Tam, L.J. Todaro, and M. Weigele, Synthesis of *iso*-DDA, member of a novel class of anti-HIV agents, *Tetrahedron Lett.* 30: 6259 (1989).

14. (a) M.L. Peterson and R. Vince, Synthesis and biological evaluation of 4-purinylpyrrolidine nucleosides, *J. Med. Chem.* 34: 2787 (1991). (b) K.E. Ng and L.E. Orgel, Replacement of the 3'-CH group by nitrogen in the carbocyclic analogue of thymidine, *J. Med. Chem.* 32: 1754 (1989).

15. M.J. Bamford, D.C. Humber, and R. Storer, Synthesis of (±)-2'-oxa-carbocyclic-2',3'-dideoxynucleosides as potential anti-HIV agents, *Tetrahedron Lett.* 32: 271 (1991).

16. M.F. Jones, S.A. Noble, C.A. Robertson, and R. Storer, Tetrahydrothiophene nucleosides as potential anti-HIV agents, *Tetrahedron Lett.* 32: 247 (1991).

17. S.M. Siddiqi and S.W. Schneller, unpublished results.

18. F. Seela, A. Kehne, and H.-D. Winkeler, Synthese von acyclo-7-desazaguanosin durch regiospezifische phasentransferalkylierung von 2-amino-4-methoxy-7*H*-pyrrolo[2,3-*d*]pyrimidin, *Liebigs Ann. Chem.* 137 (1983).

19. S. De Bernado, J.P. Tengi, G.J. Sasso, and M. Weigele, Clavalanine (Ro 22-5417), a new clavam antibiotic from *Streptomyces clavuligerus*. 4. a stereorotational synthesis, *J. Org. Chem.* 50: 3457 (1985).

20. D.H.R. Barton and S.W. McCombie, A new method for the deoxygenation of secondary alcohols, *J. Chem. Soc., Perkin Trans.* 1 1574 (1975).

21. (a) N. Shimada, S. Hasegawa, T. Harada, T. Tomisawa, A. Fujii, and T. Takita, Oxetanocin, a novel nucleoside from bacteria, *J. Antibiot.* 39: 1623 (1986). (b) H. Nakamura, S. Hasegawa, N. Shimada, A. Fujii, T. Takita, and Y. Iitaka, The X-ray structure determination of oxetanocin, *J. Antibiot.* 39: 1626 (1986).

22. (a) H. Hoshino, N. Shimizu, N. Shimada, T. Takita, and T. Takeuchi, Inhibition of infectivity of human immunodeficiency virus by oxetanocin, *J. Antibiot.* 40: 1077 (1987). (b) Y. Nishiyama, N. Yamamoto, K. Takahashi, and N. Shimada, Selective inhibition of human cytomegalovirus replication by a novel nucleoside, oxetanocin G,

Antimicrob. Agents Chemother. 32: 1053 (1988). (c) Y. Nishiyama, N. Yamamoto, Y. Yamada, H. Fujioka, N. Shimada, and K. Takahashi, Efficacy of oxetanocin G against herpes simplex virus type 2 and murine cytomegalovirus infections in mice, *J. Antibiot.* 42: 1308 (1989).

23.   (a) M. Honjo, T. Maruyama, Y. Sato, and T. Horii, Synthesis of the carbocyclic analogue of oxetanocin A, *Chem. Pharm. Bull.* 37: 1413 (1989). (b) W.A. Slusarchyk, M.G. Young, G.S. Bisacchi, D.R. Hockstein, and R. Zahler, Synthesis of SQ-33,054, a novel cyclobutane nucleoside with potent antiviral activity, *Tetrahedron Lett.* 30: 6453 (1989). (c) D.W. Norbeck, E. Kern. S. Hayashi, W. Rosenbrook, H. Sham, T. Herrin, J.J. Plattner, J. Erickson, J. Clement, R. Swanson, N. Shipkowitz, D. Hardy, K. Marsh, G. Arnett, W. Shannon, S. Broder, and H. Mitsuya, Cyclobut-A and cyclobut-G: broad-spectrum antiviral agents with potential utility for the therapy of AIDS, *J. Med. Chem.* 33: 1281 (1990).

24.   (a) Y.-i. Ichikawa, A. Narita, A. Shiozawa, Y. Hayashi, and K. Narasaka, Enantio- and diastereo-selective synthesis of carbocyclic oxetanocin analogues, *J. Chem. Soc., Chem. Commun.* 1919 (1989). (b) G.S. Bisacchi, A. Braitman, C.W. Cianci, J.M. Clark, A.K. Field, M.E. Hagen, D.R. Hockstein, M.F. Malley, T. Mitt, W.A. Slusarchyk, J.E. Sundeen, B.J. Terry, A.V. Tuomari, E.R. Weaver, M.G. Young, and R. Zahler, Synthesis and antiviral activity of enantiomeric forms of cyclobutyl nucloside analogues, *J. Med. Chem.* 34: 1415 (1991).

25.   (a) S. Hayashi, D.W. Norbeck, W. Rosenbrook, R.L. Fine, M. Matsukura, J.J. Plattner, S. Broder, and H. Mitsuya, Cyclobut-A and cyclobut-G, carbocyclic oxetanocin analogs that inhibit the replication of human immunodeficiency virus in T cells and monocytes and macrophages *in vitro*, *Antimicrob. Agents Chemother.* 34: 287 (1990). (b) A.K. Field, A.V. Tuomari, B. McGeever-Rubin, B.J. Terry, K.E. Mazina, M.L. Haffey, M.E. Hagen, J.M. Clark, A. Braitman, W.A. Slusarchyk, M.G. Young, and R. Zahler, (±)-(1α,2β,3α)-9-[2,3-Bis(hydroxymethyl)cyclobutyl]-guanine [(±)-BHCG or SQ 33054]: a potent and selective inhibitor of herpesviruses, *Antiviral Res.* 13: 41 (1990).

26.   H.G. Davies, R.H. Green, D.R. Kelly, and S.M. Roberts. Biotransformations in Preparative Organic Chemistry. The Use of Isolated Enzymes and Whole Cell Systems in Synthesis, Academic Press, San Diego, CA, 1989.

27.   (a) D.L. Hughes, J.J. Bergan, J.S. Amato, M. Bhupathy, J.L. Leazer, J.M. McNamara, D.R. Sidler, P.J. Reider, and E.J.J. Grabowski, Lipase-catalyzed asymmetric hydrolysis of esters having remote chiral/prochiral centers, *J. Org. Chem.* 55: 6252 (1990). (b) A recent review on *Pseudomonas fluorescens* lipase (Z.-F. Xie, *Pseudomonas fluorescens* lipase in asymmetric synthesis, *Tetrahedron: Asymmetry* 2: 733 (1991)) may actually be a review on *Pseudomonas cepacia* lipase (see footnote 7 of reference 27a).

28. C.-N. Hsiao and S.M. Hannick, Efficient synthesis of protected (2S,3S)-2,3-bis(hydroxymethyl)cyclobutanone, key intermediates for the synthesis of chiral carbocyclic analogues of oxetanocin, *Tetrahedron Lett.* 31: 6609 (1990) reports the synthesis of the (+)-enantiomer of **26** in a rather lengthy process.

29. Reference 23b reports LS-Selectride provided the greatest configurational selectivity at the hydroxyl bearing carbon when reducing *trans*-2,3-bis[(benzoyloxy)methyl]-cyclobutanone.

30. Enantiomeric purity was determined by HPLC using a CHIRALCEL OJ column (0.46 cm x 25 cm) and eluting with 15% 2-PrOH-hexane.

31. Compounds **36** and **44** are drawn in the enol form for the sake of convenience in structural representation, but no studies were conducted to determine their tautomeric preferences.

32. A.D. Borthwick, S. Butt, K. Biggadike, A.M. Exall, S.M. Roberts, P.M. Youds, B.E. Kirk, B.R. Booth, J.M. Cameron, S.W. Cox, C.L.P. Marr, and M.D. Shill, Synthesis and enzymatic resolution of carbocyclic 2'-*ara*fluoroguanosine: a potent new anti-herpetic agent, *J. Chem. Soc., Chem. Commun.* 656 (1988).

33. (a) A.D. Borthwick, D.N. Evans, B.E. Kirk, K. Biggadike, A.M. Exall, P. Youds, S.M. Roberts, D.J. Knight, and J.A.V. Coates, Fluoro carbocyclic nucleosides: synthesis and antiviral activity of 2'- and 6'-fluoro carbocyclic pyrimidine nucleosides including carbocyclic 1-(2-deoxy-2-fluoro-β-D-arabinofuranosyl)-5-methyluracil and carbocyclic 1-(2-deoxy-2-fluoro-β-D-arabinofuranosyl)-5-iodouracil, *J. Med. Chem.* 33: 179 (1990). (b) K. Biggadike, A.D. Borthwick, D. Evans, A.M. Exall, B.E. Kirk, S.M. Roberts, L. Stephenson, and P. Youds, Use of diethylaminosulfur trifluoride (DAST) in the preparation of synthons of carbocyclic nucleosides, *J. Chem. Soc., Perkin Trans.* 1 549 (1988).

34. Compound **40** was prepared by a modification of a procedure described in reference 7.

35. G.E. Wright and N.C. Brown, Deoxyribonucleotide analogs as inhibitors and substrates of DNA polymerase, *Pharmacol. Ther.* 47: 447 (1990).

36. N.N. Khan, G.E. Wright, L.W. Dudycz, and N.C. Brown, Butylphenyl dGTP: a selective and potent inhibitor of mammalian DNA polymerase alpha, *Nucleic Acid Res.* 12: 3695 (1984).

37. F. Seela, W. Bussmann, A. Götze, and H. Rosemeyer, Isomeric N-methyl-7-deazaguanosines: synthesis, structural assignment, and inhibitory activity on xanthine oxidase, *J. Med. Chem.* 27: 981 (1984).

38. R. Vince, J. Brownell, and S. Daluge, Carbocyclic analogues of xylofuranosylpurine nucleosides. synthesis and antitumor activity, *J. Med. Chem.* 27: 1358 (1984).

39. R. Vince and M. Hua, Synthesis and anti-HIV activity of carbocyclic 2',3'-didehydro-2',3'-dideoxy 2,6-disubstituted purine nucleosides, *J. Med. Chem.* 33: 17 (1990).

40. R. Vince and J. Brownell, Resolution of racemic carbovir and selective inhibition of human immunodeficiency virus by the (-)-enantiomer, *Biochem. Biophys. Res. Commun.* 168: 912 (1990).

41. R. Vince, M. Hua, J. Brownell, S. Daluge, F. Lee, W.M. Shannon, G.C. Lavelle, J. Qualls, O.S. Weislow, R. Kiser, P.G. Canonico, R.H. Schultz, V.L. Narayanan, J.G. Mayo, R.H. Shoemaker, and M.R. Boyd, Potent and selective activity of a new carbocyclic nucleoside analog (carbovir: NSC 614846) against human immunodeficiency virus *in vitro*, *Biochem. Biophys. Res. Commun.* 156: 1046 (1988).

42. (a) M.R. Peel, D.D. Sternbach, and M.R. Johnson, A short, enantioselective synthesis of the carbocyclic nucleoside carbovir, *J. Org. Chem.* 56: 4990 (1991). (b) A.M. Exall, M.F. Jones, C.-L. Mo, P.L. Myers, I.L. Paternoster, H. Singh, R. Storer, G.G. Weingarten, C. Williamson, A.C. Brodie, J. Cook, J.D.E. Lake, C.A. Meerholz, P.J. Turnbull, and R.M. Highcock, Synthesis of (-)-aristeromycin and X-ray structure of (-)-carbovir, *J. Chem. Soc., Perkin Trans.* 1 2467 (1991). (c) M.F. Jones, P.L. Myers, C.A. Robertson, R. Storer, and C. Williamson, Total synthesis of (-)-carbovir, *J. Chem. Soc., Perkin Trans.* 1 2479 (1991). (d) L.-L. Gundersen, T. Benneche, and K. Undheim, Pd(O)-catalyzed allylic alkylation in the synthesis of (±)-carbovir, *Tetrahedron Lett.* 33: 1085 (1992). (e) C.T. Evans, S.M. Roberts, K.A. Shoberu, and A.G. Sutherland, Potential use of carbocyclic nucleosides for the treatment of AIDS: chemo-enzymatic syntheses of the enantiomers of carbovir, *J. Chem. Soc., Perkin Trans.* 1 589 (1992).

43. S. Daluge and R. Vince, Synthesis of carbocyclic aminonucleosides, *J. Org. Chem.* 43: 2311 (1978).

44. J.A. Secrist, III, S.J. Clayton, and J.A. Montgomery, (±)-3-(4-Amino-1*H*-pyrrolo[2,3-*d*]pyrimidin-1-yl)-5-(hydroxymethyl)-(1α,2α,3β,5β)-1,2-cyclopentane-diol, the carbocyclic analogue of tubercidin, *J. Med. Chem.* 27: 534 (1984).

L-ASCORBIC AND D-ISOASCORBIC ACIDS: CHIRON SOURCES FOR 1',2'-*SECO*-NUCLEOSIDES/TIDES, PHOSPHONATES, AND OTHER MOLECULES OF BIOLOGICAL INTEREST

Elie Abushanab

Departments of Medicinal Chemistry and Chemistry
University of Rhode Island
Kingston, RI 02881

The introduction of acyclovir (ACV, **1a**) as a clinically useful anti-herpetic drug has spurred interest in other acyclic nucleosides as potential chemotherapeutic agents. Ganciclovir (DHPG, **1b**), a closely related analog, has found clinical utility in the treatment of infections caused by human cytomegalovirus (HCMV). Among several acyclic nucleosides, a few have shown promising anti-HIV activity. Adenallene (**2a**) and cytallene (**2b**) are two such compounds that inhibit the expression of HIV-1 gag-encoded protein and suppress DNA synthesis at concentrations that do not affect normal T-cells *in vitro*.[1] 9-[(2-Phosphonylmethoxy)ethyl]adenine (PMEA, **3a**) and *S*-9-[(3-hydroxy-2-phosphonyl-methoxy)propyl]adenine (HPMPA, **3b**) and -cytosine (HPMPC, **3c**) are phosphonic acid derivatives that have been found to be active against a series of DNA[2] as well as HIV viruses.[3]

ACV and DHPG can be considered as *bis-nor-* and *nor*-guanosine, respectively, and must be phosphorylated to exert antiviral activity. While ACV has no chiral center, DHPG has a prochiral carbon and phosphorylation of only the *pro-S*-hydroxyl group gives the active species.[4] As the carbon framework of the acyclic side chain approaches that of the naturally occurring pentose, the number of chiral centers increases and, consequently, the preparation of optically pure isomers becomes more difficult. 1',2'-*seco*-Nucleosides have two chiral centers and, to the best of our knowledge, only one report has appeared in the literature that described in detail the synthesis of all four isomers of 1',2'-*seco*-guanosine,[5,6] which had no antiherpetic activity.[4] The reported synthetic route to these compounds could not be easily applied to prepare a variety of 1',2'-*seco*-nucleosides for a

Nucleosides and Nucleotides as Antitumor and Antiviral Agents,
Edited by C.K. Chu and D.C. Baker, Plenum Press, New York, 1993

more systematic evaluation of their antiviral activity. Thus, there was a need for a more efficient synthesis to prepare an acyclic chiron for further elaboration into 1',2'-*seco*-purine and -pyrimidine nucleosides.

1	**2**	**3**
a: R = H	a: B = A	a: R = H, B = A
b: R = CH₂OH	b: B = C	b: R = CH₂OH, B = A
		c: R = CH₂OH, B = C

Retrosynthetic analysis demonstrated the need for a 1,2,3-butanetriol and a 1,2,3,4-butanetetrol with 2*R*, 3*S* chirality, which corresponds to that found in 2-deoxy-D-ribose. Our initial involvement with L-ascorbic acid (*vide infra*) led us to examine an equally accessible diastereomer, D-isoascorbic acid. While the former has been extensively studied, the same is not true for the latter. This could be due to early difficulties encountered in the preparation of 5,6-*O*-protected derivatives.[7] Unlike their counterparts from L-ascorbic acid, which are equally accessible from *R*-tartrates, chirons derived from D-isoascorbic acid are not easily obtained from other natural sources. They can be available from *meso*-tartrates only when differentiation between the two carboxyl groups is feasible.

Degradation of D-isoascorbic acid has been successfully completed to furnish a variety of selectively protected chirons,[8,9] including those mentioned earlier for the synthesis of 1',2'-*seco*-nucleosides. The synthetic route is depicted in Scheme 1. Starting with D-isoascorbic acid (**4**), the key epoxide **7** was obtained in 40% overall yield. Regiospecific opening of **7** with a variety of nucleophiles was possible. Among those used were the hydride and benzylate ions to give **8** (**a** and **b**), respectively, which were converted to **8** (**c** and **d**). Acid hydrolysis of the isopropylidene ring was followed by epoxidation using the Mitsunobu reaction to provide **10** (**a** and **b**). These, in turn, were opened with benzylate anion to give **11**. Chloromethylation with HCl/formaldehyde, followed by treatment of ethers **12** with either silylated pyrimidines or the sodium salt of purines furnished benzylated 1',2'-*seco*-nucleosides. Deblocking with Pearlman's catalyst [Pd(OH)₂/C] gave the target compounds **13**. In this manner, 1',2'-*seco*- and 1',2'-*seco*-2'-deoxyguanosine, uridine, thymidine, cytosine, and triazole-3-carboxamide were prepared.[10,11] Additionally, 1',2'-*seco*-analogues of nucleosides with known antiviral activity such as **14**, **15**, and **16** were also synthesized following similar procedures.[12-14]

The 1',2'-*seco*-nucleosides prepared were tested against a number of DNA, RNA, and HIV viruses and were found devoid of significant activity. This was presumed to be due to their inability to serve as substrates for phosphorylation by a kinase. Consequently,

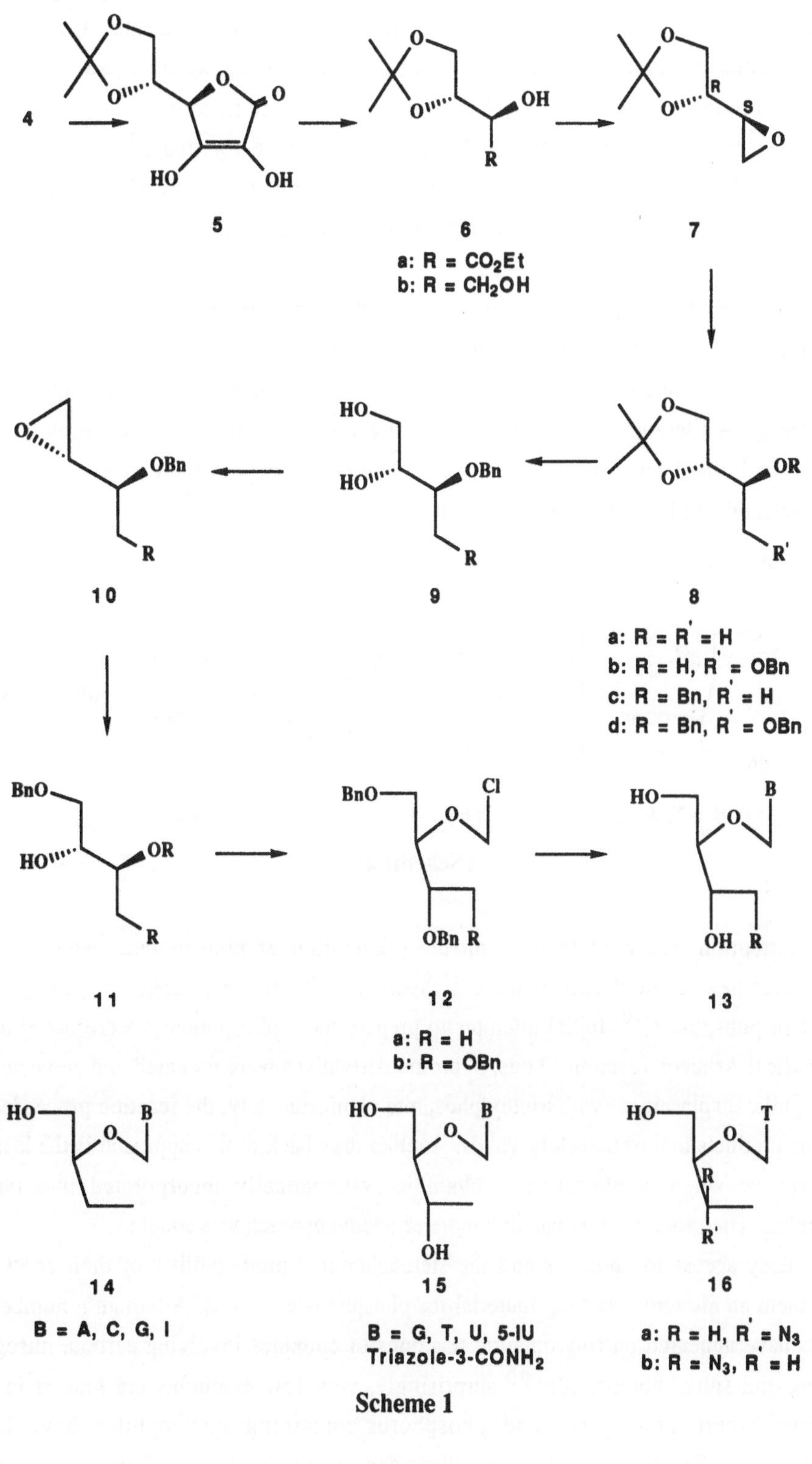

Scheme 1

it was decided to prepare phosphate and phosphonate esters in the hope of bypassing the requisite phosphorylation at C-5'. Using standard procedures, a number of phosphate and 3',5'-cyclic phosphate esters were prepared and tested as the ammonium salts. It was hoped that the reduced number of charges in the cyclic phosphate would allow better cell penetration to be followed by intracellular phosphodiesterase hydrolysis to give the 1',2'-seco-nucleotide. This approach has literature precedence for ara-C[15] and 6-MP riboside.[16] Unfortunately, no measurable improvement in biolgocial activity was observed.

During the course of our work on phosphate esters, an interesting dephosphorylation reaction was observed. While diphenylphosphate esters in the 2'-deoxy series could be easily prepared, the same was not true for the 2'-oxy analogues **17**. Upon debenzylation, the 2'-OH group participated in an *intra*molecular nucleophilic attack at C-5' resulting in the formation of a tetrahydrofuran derivative (**19**) and the elimination of diphenylphosphate (Scheme 2). A literature search did not reveal a precedent for this observation.[17] The reaction took place under neutral and mild conditions through the intermediacy of 1',2'-seco-nucleotide diphenylphosphate **18** to give the product in 80% yield.[18]

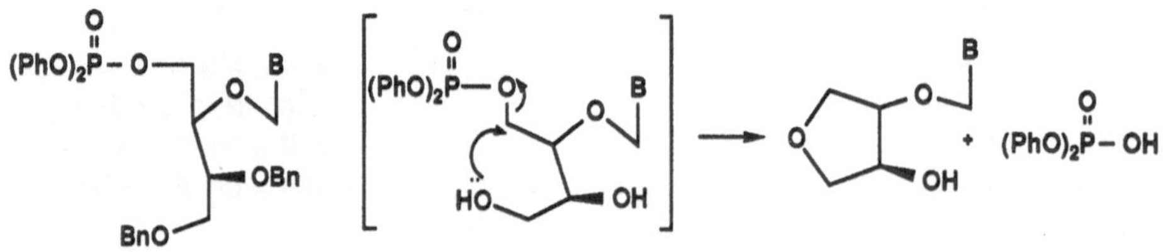

17: B = T, A   18   19

Scheme 2

Attention was next focused on the preparation of phosphonate esters. Such compounds have been shown to serve as hydrolytically stable isosteres of biologically important phosphates.[19] Initial attempts to prepare these phosphonates were based upon the classical Arbuzov reaction. Thus, 5'-bromo derivatives were prepared and subjected to nucleophilic displacement with triethylphosphite. Unfortunately, the reaction proceeded to give the products in unsatisfactory yields. Another drawback to this approach is the lack of convergence where a phosphonate chain is systematically incorporated into target molecules. Therefore, an alternate and more efficient approach was sought.

Easy access to epoxides and the stereochemical predictability of their reactions made them an attractive starting material for phosphonate chirons. Although a number of reports have appeared on ring-opening reactions of epoxides involving carbon, nitrogen, oxygen, and sulfur nucleophiles,[20] surprisingly, very few examples are known in the literature where phosphorus and phosphorus-containing nucleophiles have been described.[21-24] The success of these reactions depended upon the use of excess epoxide, a procedure that is of little use when one equivalent is used.

The report on the synthesis of the phosphonate isostere of AZT 5'-phosphate by a BF3·Et2O-catalyzed nucleophilic addition of diethyl methanephosphonate to an oxetane[25] prompted us to apply the same reaction conditions to the opening of epoxides. Indeed, when the anion of the same phosphonate was generated with BuLi in THF, followed by the sequential addition of epoxide **10a** or **10b** and BF3·Et2O at -78°, the corresponding β-hydroxyphosphonates **20** (**a** and **c**) were obtained in quantitative yield.[26] Extension of this reaction to a phosphorus nucleophile proved to be equally effective. Thus, when diethyl phosphite was condensed with the same epoxides under identical reaction conditions, the corresponding β-hydroxyphosphonates **20** (**b** and **d**) were isolated in

Table 1 - Synthesis of β- and γ-hydroxyphosphonates

Epoxide	Nucleophile*	n	Yield %
R = (OBn)	a	2	100
	b	1	100
(OBn, OBn)	a	2	100
	b	1	77
(dioxolane)	a	2	65
	b	1	64
CH2Ph	a	2	76
	b	1	77
CH2OTs	a	2	79
(OBn, OTs)	b	1	65
(OBn, OBz)	a*	2	90
	b	1	72
(OBn, Br)	a	2	84
	b	1	65

*a = CH3-P(=O)-(EtO)2 ; b = H-P(=O)-(EtO)2   * Ester cleavage took place

comparable yield. Methyl and benzylphosphonates behaved similarly giving the respective esters in good yield. A summary of ring opening reactions is shown in Table 1.

obtained chirons **20**, the synthesis of 1',2'-*seco*-nucleoside phosphonates was undertaken. The synthetic route followed was identical to that described earlier for the preparation of 1',2'-*seco*-nucleosides (Scheme 3). The hydroxy groups in **20** were converted to the corresponding chloromethyl ethers (**21**) and were used to alkylate purines and pyrimidines. The phosphonic acids (**22**) were obtained by treatment of their esters with trimethylsilyl bromide.[27] Like 1',2'-*seco*-nucleosides and -nucleotides, these compounds were also inactive as antiviral agents.

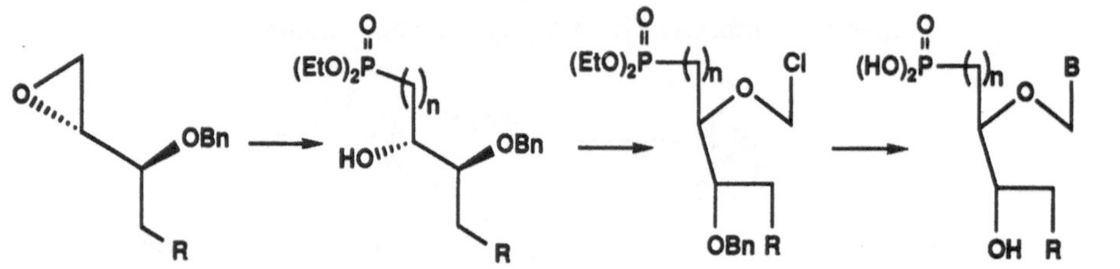

a: R = H, n = 2
b: R = H, n = 1
c: R = OBn, n = 2
d: R = OBn, n = 1

B = A, G, C, T, U

a: R = H
b: R = OH

Scheme 3

In an attempt to impart rigidity to the chain and, hopefully, improve chances for phosphorylation by a kinase, it was decided to prepare some 2,3'- and 2,2'-anhydronucleosides. Consequently, the suitably protected *seco*-thymidine tosylates **23** and **24** were prepared and subjected to base-catalyzed cyclization reactions (Scheme 4). Tosylate **23**, using NaH/DMF, furnished the desired compound **25** along with other elimination products. However, using DBU as the base resulted in exclusive formation of **25** in 85% yield. Attempted cyclizations of **24** did not proceed as expected. Instead, an unusual *inter*molecular dimerization reaction took place to form **26** whose structure was assigned based on mass spectral and NMR analyses.[28] The preferential *intra* and *inter*molecular cyclization of **23** on one hand, and **24** on the other, is attributable to the difference in the ring size of the desired product. This is not surprising since seven-membered rings are more readily formed than eight-membered ones.[29] This reaction does not appear to have been previously reported. The closest analogy was found in the area of aza-crown ethers.[30]

In addition to providing easy access to 1',2'-*seco*-nucleosides, -nucleotides, and phosphonate isosteres, L-ascorbic and D-isoascorbic acids proved to be useful in the preparation of equally attractive chirons that can be incorporated into other molecules of biological interest. Diol **6b** is particularly attractive since it represents an easy entry into selectively protected erythritols. Having the *R*- and *S*-chiralities, access to both *R*- and *S*-

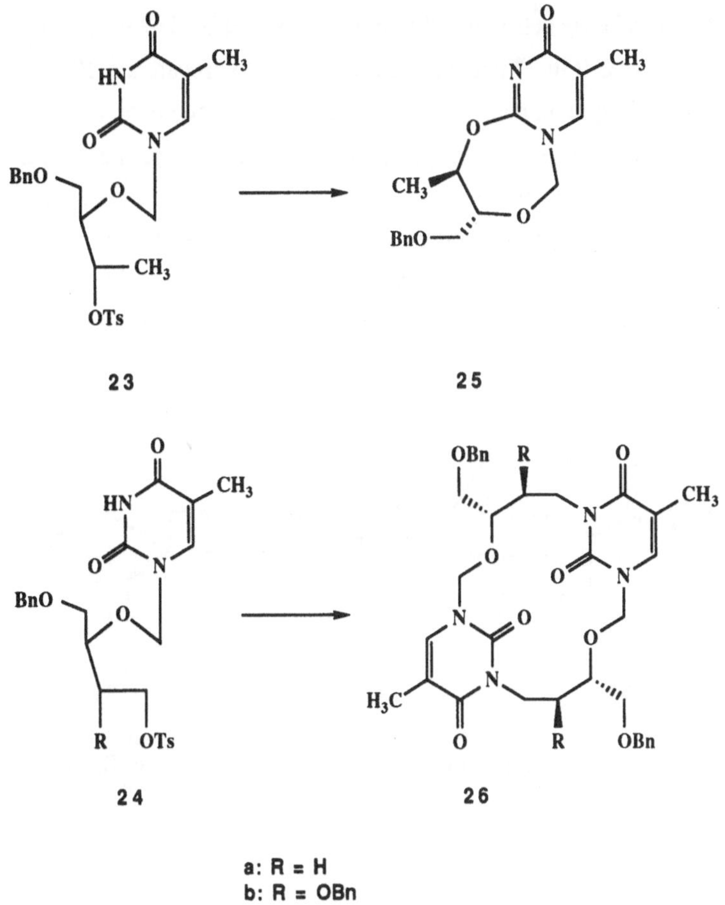

23　　　　　　　　　**25**

24　　　　　　　　　**26**

a: R = H
b: R = OBn

Scheme 4

glyceraldehydes from a common precursor became possible. Thus, periodate cleavage of
6b and **9b** furnished the corresponding *R*- and *S*-glyceraldehydes.[31]

An efficient synthesis of pure *R*- and *S*-1,2-*O*-isopropylidene-1,2,4-butanetriols (**28**) was
also developed Scheme 5. These compounds have been used in asymmetric syntheses of
various natural products.[32] However, the reported routes to **28** suffer from certain
drawbacks, such as difficulty in separation or the use of costly starting material, that detract
from their utility. Starting with tosylate **27**, obtained from **6a** (Scheme 1), the *S*-isomer was
prepared in two steps in excellent yield. The *R*-isomer was similarly obtained from the 2*R*
3*S*-diastereomer of **6a**.[33]

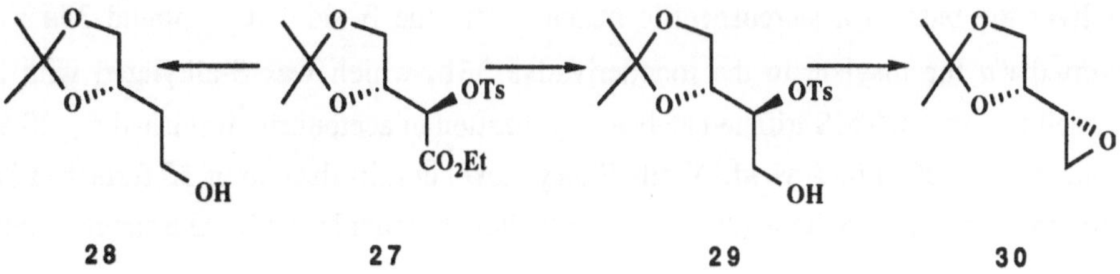

28　　　　　　**27**　　　　　　**29**　　　　　　**30**

Scheme 5

A practical and easy entry into all stereoisomers of epoxide **7** became possible when the stereochemistry at C-2 in **27** could be easily inverted to prepare the 2*R*, 3*R*-epoxide **30**. These compounds, along with their 2*S*, 3*S*-, and 2*S*, 3*R*-diastereomers, were converted into all four stereoisomers of methyl 2-deoxypentofuranosides,[34] compounds that are attractive targets for the synthesis of other molecules of biological interest. Earlier syntheses of these compounds suffer from either one of two drawbacks: low yields or the separation of enantiomers and/or diastereomers.[35] Conversion of the 2*R*,3*S*-epoxide **7** to a 2-deoxy-D-ribose derivative is used as an example to illustrate the preparation of all four stereoisomers (Scheme 6).

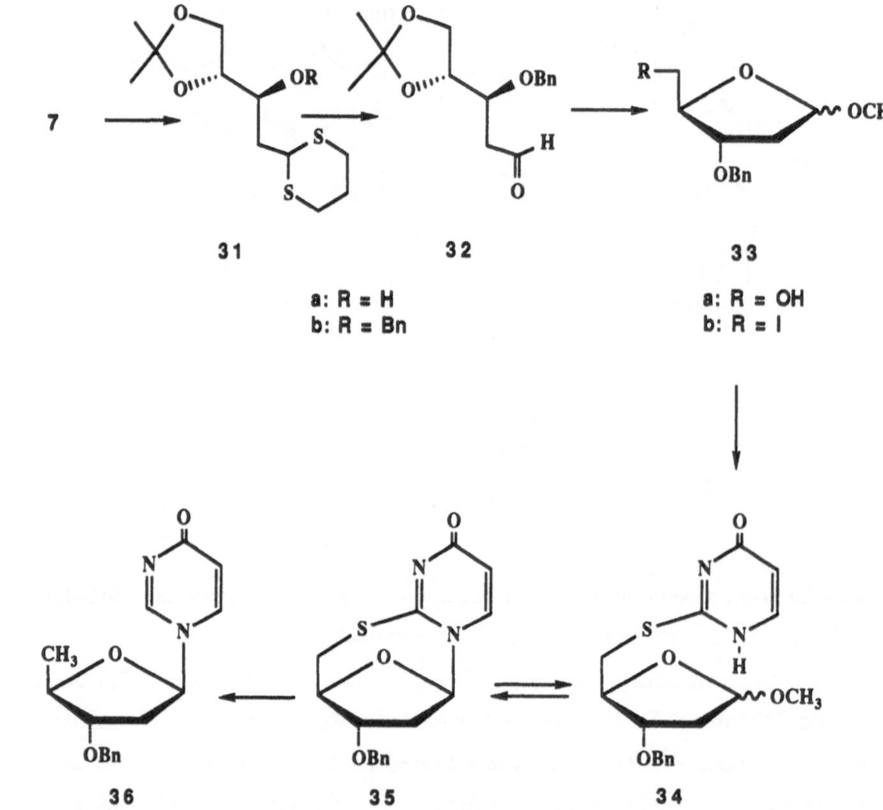

Scheme 6

There are several reports that describe attempts to favor the formation of β- over α-D-2'-deoxynucleosides.[36-39] The availability of **33a** prompted us to use the chirality at C-4 to deliver the base in a stereospecific manner from the β side. Compound **33a** was converted *via* the tosylate to the iododerivative **33b**, which was S-alkylated with 2-thiouracil to give **34**. TMS triflate-catalyzed cyclization in acetonitrile furnished the 2S,5'-cyclonucleoside **35** in 65% yield. While Raney nickel desulfurization of **35** furnished **36**, an unprecedented reaction took place where palladium hydroxide catalyzed a stereospecific alcoholysis of the nucleoside bond resulting in formation of the a-anomer of **34**. Extensive ^{1}H and ^{13}C NMR studies were supportive of the assigned structures.[40]

Another application of sugar chirality for the stereospecific formation of a nucleoside bond is demonstrated in the synthesis of 2',3'-didehydro-3'-deoxythymidine (D4T, Scheme 7). The synthesis is patterned after the work of Shannanhoff and Sanchez, who used the chirality at C-2 in D-arabinose for the exclusive formation of β-D-nucleosides.[41] Application of this approach to 3-deoxy-D-arabinose does not seem to have been reported. The latter compound was prepared from D-glucose *via* D-arabinonolactone following a literature procedure.[42] Catalytic reduction of **37a** furnished the 3-deoxy derivative **37b**.[43] This was converted to **38** by a sequence of reactions involving hydrolysis, silylation, and DIBAL reduction. Construction of the thymine ring, phenyl thiolate opening of the anhydro bond, followed by esterification with pivaloyl chloride gave **40**, from which the target compound D4T was obtained by sulfoxide formation, followed by DBU elimination, in an overall yield of 21% starting from **37a**.[44]

A recent application of our 2-deoxypentofuranose chemistry has resulted in the synthesis of rigid analogs of diacylglycerols as inhibitors of protein kinase C (PK-C). The four isomers of 2-deoxypentofuranonolactones were prepared as illustrated for the 2*S*, 3*R*-

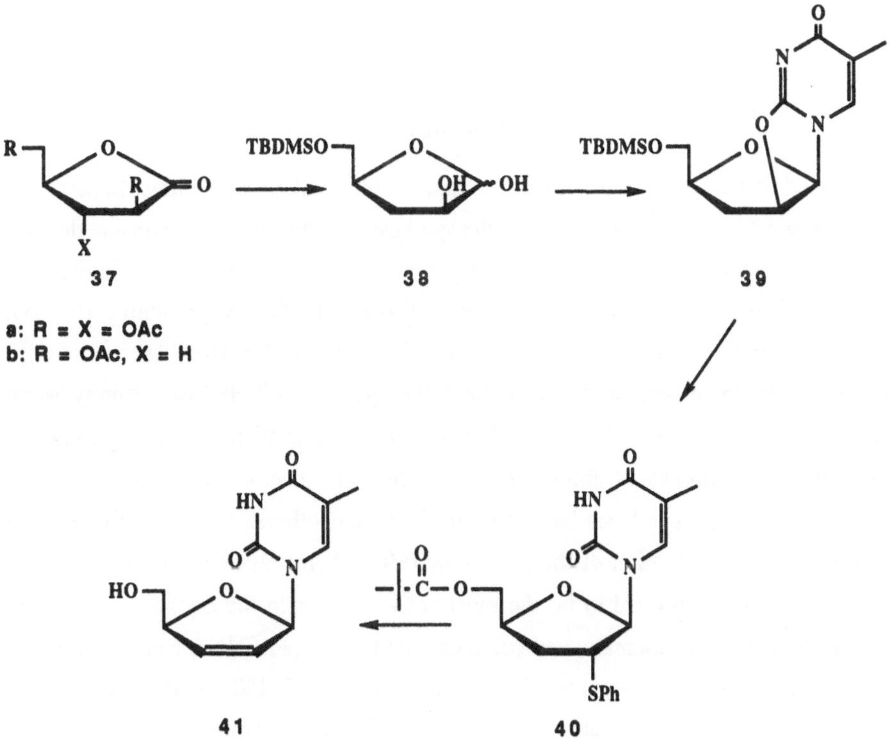

Scheme 7

isomer **45** (Scheme 8). Eight isomers resulting from monoesterification at either C-3 or C-5 with myristic acid were prepared. Two enantiomers **45** and **46** were shown to be competitive inhibitors of phorbol dibutyrate binding to PK-C with K_i values of 5.3 and 2.5 µM, respectively. The value obtained for **46** is five times greater than that of *rac*-glycerol-1-myristate-2-acetate (GMA), providing an approximation of the conformation of GMA when it binds to PK-C.[45]

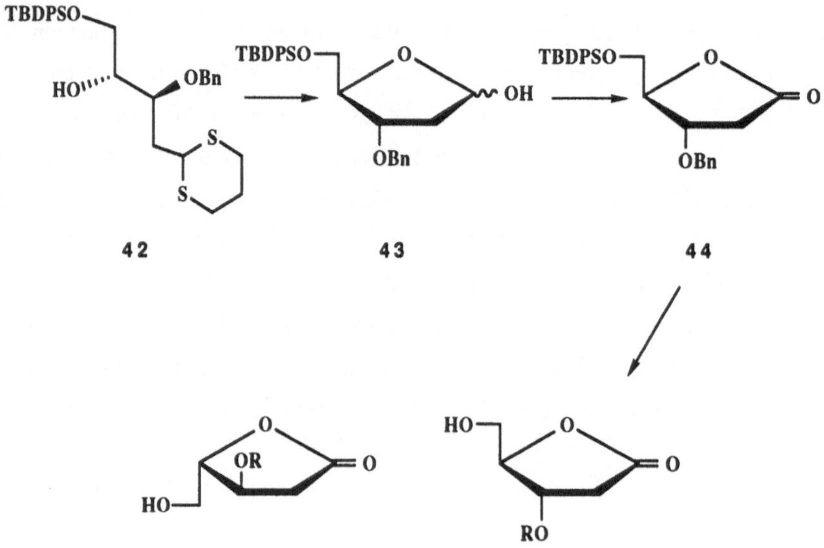

Scheme 8

In addition to complimenting D-isoascorbic acid in providing diastereomeric chirons, L-ascorbic acid has been used in the synthesis of inhibitors of adenosine deaminase (ADA, adenosine aminohydrolase). ADA has a broad substrate specificity and deaminates a variety of adenosine analogs such as 9-β-D-arabinofuranosyladenine (ara-A), 8-azaadenosine, and 2',3'-dideoxyadenosine (ddA).[46] This may inactivate these analogs and limit their utility in antiviral and cancer chemotherapy. (±)-9(2-Hydroxy-3-nonyl)adenine (EHNA) was designed by Schaeffer and co-workers as a semi-tight binding inhibitor of ADA[47] and was used in early biological studies where enhancement of the activity of ara-A was observed.[48] The synthesis by our group,[49,50] and others,[51,52] of chiral isomers of EHNA allowed identification of the (+)-2*S*,3*R*-(**50**, Scheme 9) as the most potent ADA inhibitor. (+)-EHNA proved to be the most active isomer in the enhancement of ara-A activity against human pancreatic and colon carcinomas.[53] (+)-EHNA has a short duration of action and is believed to be metabolized by a cytochrome P-450-mediated hydroxylation in the nonyl chain.[54] Thus, an attractive goal would be the development of derivatives that are one to two orders of magnitude more potent than (+)-EHNA; such derivatives might produce enhanced, but still reversible, ADA inhibition.

Our early synthesis starting from L-rhamnose was rather lengthy and a more efficient and practical route was developed using chirons derived from L-ascorbic acid.[55] The method is versatile and allows the preparation of (+)-EHNA derivatives. The synthesis of the 9'-OH derivative **53** is depicted in Scheme 9.

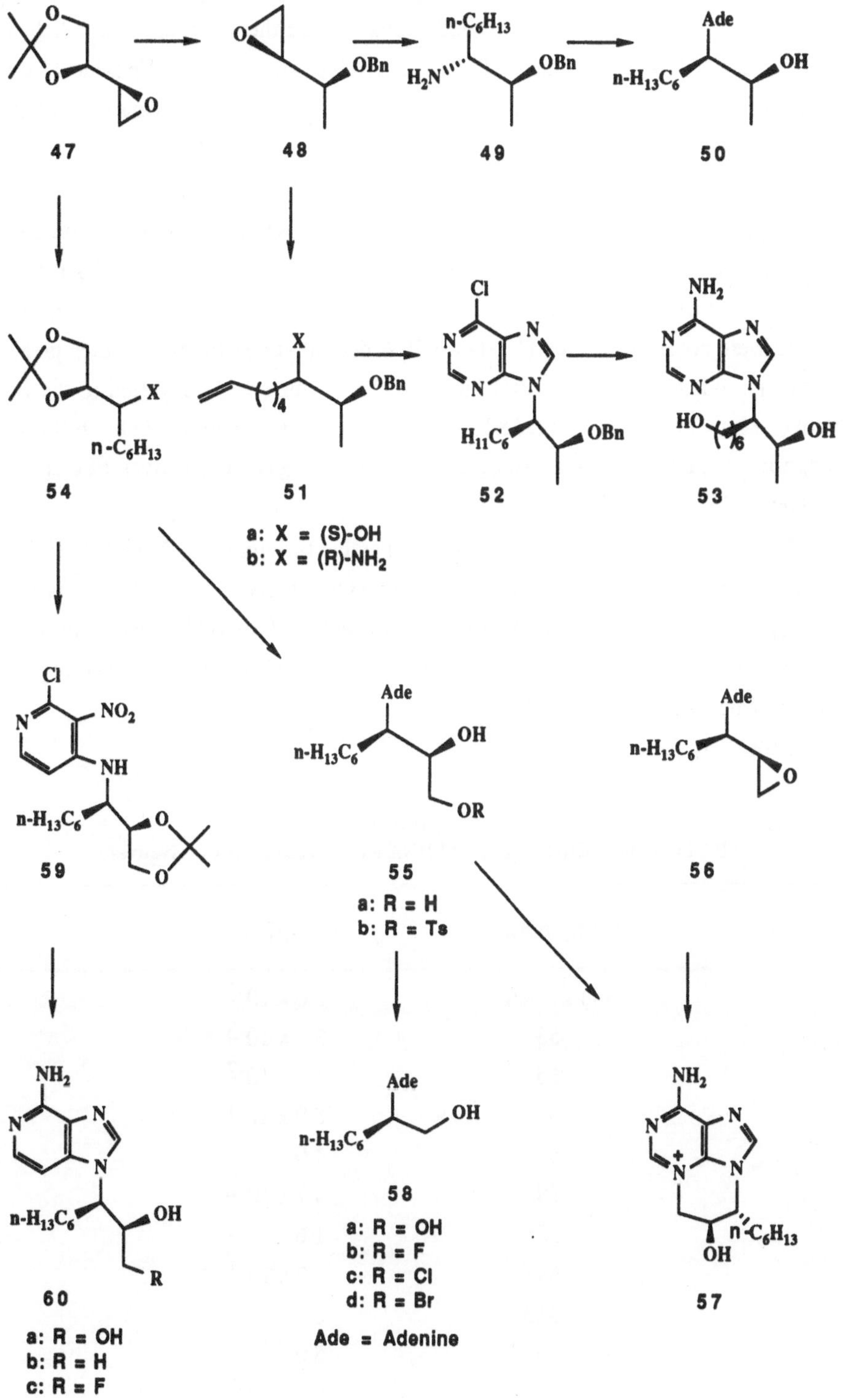

47 **48** **49** **50**

54 **51** **52** **53**

a: X = (S)-OH
b: X = (R)-NH₂

59 **55** **56**

a: R = H
b: R = Ts

60 **58** **57**

a: R = OH
b: R = H
c: R = F

a: R = OH
b: R = F
c: R = Cl
d: R = Br

Ade = Adenine

Scheme 9

The addition of 1-pentenyl-5-magnesium bromide to epoxide **48** furnished alcohol **51a**. This was converted to amine **51b** by reduction of an azide, obtained by the Mitsunobu reaction,[55] and conversion to **52** by known methods.[47] The final product was secured by a three-step sequence involving amination, hydroboration and debenzylation.

The versatility of the L-ascorbic acid approach is further demonstrated in Scheme 9 where C-1', *nor*-C-1', and 3-deaza (+)-EHNA derivatives are prepared. Epoxide **47** served as the starting material to all three classes of compounds. When treated with pentylmagnesium chloride, alcohol **54a** was obtained and was converted to amine **54b** by mesylation, azide displacement, and reduction. C-1' Hydroxy (+)-EHNA (**55a**) was then obtained as described earlier. Tosylation of **55a** and subsequent treatment with sodium ethoxide furnished the unstable epoxide **56**. Both compounds **55b** and **56** led, upon standing, to the tricyclic derivative **57**. Efforts to prepare C-1' derivatives from either compound were not fruitful. *Nor*-C-1'-(+)-EHNA derivatives **58** (**a-d**) were then prepared. Reductive periodate cleavage of diol **55a** gave **58a**, which was converted to the fluoro derivative **58b** by treatment with DAST. On the other hand, tosylation of **58a**, followed by nucleophilic displacement with Bu4NCl and Bu4NBr gave derivatives **58c** and **58d**, respectively.

Reports that 3-deaza (±)-EHNA had comparable activity to (±)-EHNA,[56,57] and the fact that N-3 cyclization reactions in C-1' substituted derivatives are not possible prompted us to prepare the optically active molecule as well as other C-1'-derivatives. The preparation of target compounds **60** (**a-c**) followed published procedures,[57] and proceeded through intermediate **59**.

Table 2
ADA inhibitory activities of C-1' EHNA and 3-deaza EHNA derivatives

Compound	$K_i(\mu M)$
(+)-EHNA	2.0×10^{-3}
53	3.4×10^{-3}
55a	5.6×10^{-2}
56	8.9×10^{-1}
57	1.0
58a	4.7×10^{-2}
58b	1.6
58c	3.2×10^{-1}
58d	6.0×10^{-1}
60a	5.0×10^{-1}
60b	6.2×10^{-3}
60c	4.7×10^{-3}

As shown in Table 2, diminished potencies were evident in *nor*-(+)-EHNA derivatives. The deletion of the 1'-methyl group caused a 23-fold decrease in activity for **58a**, which is comparable to what has been observed for the racemate.[58] Derivatives of *nor*-(+)-EHNA, compounds **58 (b-d)**, had inhibitory activities 7- to 34-fold lower than *nor*-(+)-EHNA (**58a**) and 160-800-fold lower than (+)-EHNA itself. This seems to further substantiate the existence of an auxiliary methyl pocket on the enzyme[60] and the importance of the hydroxyl group to binding. The C-1' hydroxy derivative **55a** showed a 28-fold decrease in inhibitory activity. Interestingly, the rigid cyclic analog **57** showed higher affinity for the enzyme than the natural nucleoside substrates. The activity of **57** was also comparable to that of epoxide **56**, which converted slowly to **57** during testing procedures and caused no detectable irreversible inhibition of the enzyme, suggesting a lack of alkylation at the active site.[59]

As expected, 3-deaza (+)-EHNA (**60b**) was found to be almost twice as active as the racemate.[57] The fluoro derivative **60c** showed slightly higher affinity for the enzyme than the parent compound **60b**. Since there is little change in size, the improved activity might be due to increased lipophilicity at C-1'. On the other hand, the C-1' hydroxy derivative **60a** proved to be 80-times less active than **60b**. This is possibly due to the large change in size and/or increased hydrophilicity of the substitutent. This appears to suggest that the binding to the methyl pocket as proposed by Schaeffer,[60] may be affected by both steric and electronic factors. The activity of 9'-hydroxy-(+)-EHNA (**50**) is rather interesting. Should **50** prove to be an active metabolite of (+)-EHNA, it can be used for the attachment of ligands to help purify adenosine deaminase. Current efforts are aimed at preparing other chain hydroxylated derivatives to further probe the requirements for binding at the hydrophobic site of ADA.

In this chapter, an attempt has been made to illustrate the utility of L-ascorbic and D-isoascorbic as sources of chirons in organic synthesis. Although the 1',2'-*seco*-nucleosides and their derivatives have not shown significant activity, our goal is to develop methods that bypass requirements for phosphorylation and to generate 1',2'-*seco*- and other nucleotides intracellularly.

Acknowledgments: The cooperation of students, colleagues, and especially that of Dr. Raymond P. Panzica, is gratefully acknowledged. Special thanks are due to Mrs. Rena Fullerton for her patience and skilled typing, and to Dr. Chandra Vargeese for drawing the structures.

References

1. S. Hayashi, S. Phadatre, J. Zemlicka, M. Matsukura, H. Matsuya, and S. Broder, Adenallene and cytallene: Acyclic nucleoside analogs that inhibit replication and cytopathic effect of human immunodeficiency virus *in vitro*, *Proc. Natl. Acad. Sci. USA*, 85:6127 (1988).

2. E. De Clercq, T. Sakuma, M. Baba, R. Pauwels, J. Balzarini, I. Rosenberg, and A. Holy, Antiviral activity of phosphonylmethoxyalkyl derivatives of purines and pyrimidines, *Antiviral Res.* 8:261 (1987).

3. R. Pauwels, J. Balzarini, D. Schols, M. Baba, J. Desmyter, I. Rosenberg, A. Holy, and E. De Clercq, Phosphonylmethoxyethylpurine derivatives. A new class of anti-human immunodeficiency virus agents, *Antimicrob. Agents Chemother.* 32:1025 (1988).

4. M. MacCoss, R.L. Tolman, W.T. Ashton, A.F. Wagner, J. Hannah, A.K. Field, J.D. Karkas, and J.I. Germershausen, Synthetic, biochemical and antiviral aspects of selected acyclonucleosides and their derivatives, *Chem. Script.* 26:113 (1986).

5. M. MacCoss, A. Chen, and R. Tolman, Synthesis of the chiral acyclonucleoside antiherpetic agent (S)-9-(2,3-dihydroxy-1-propoxymethyl)guanine, *Tetrahedron Lett.*, 26:1815 (1985).

6. M. MacCoss, A. Chen, and R. Tolman, Syntheses of all 4 possible diastereomers of the acyclonucleoside 9-(1,2,4-trihydroxy-2-butoxymethyl)guanine from carbohydrate precursors, *Tetrahedron Lett.* 26:4287 (1985).

7. A. Tanaka and K. Yamashita, A novel synthesis of (R)- and (S)-4-hydroxytetrahydro-furan-2-ones, *Synthesis*, 570 (1987).

8. E. Abushanab, P. Vemishetti, R.W. Leiby, H.K. Singh, A.B. Mikkilineni, D.C.-J. Wu, R. Saibaba, and R.P. Panzica, The Chemistry of L-ascorbic and D-isoascorbic acids. 1. The preparation of chiral butanetriols and -tetrols, *J. Org. Chem.* 53:2598 (1988).

9. R. Saibaba, M.S.P. Sarma, and E. Abushanab, The Chemistry of L-ascorbic and D-isoascorbic acids. 3: Efficient syntheses of pure R- and S- 1,2-O-isopropylidene-1,2,4-butanetriols, *Synth. Commun.* 19:3077 (1989).

10. P. Vemishetti, R.W. Leiby, E. Abushanab, and R.P. Panzica, A practical synthesis of ethyl 1,2,4-triazole-3-carboxylate and its use in the formation of chiral 1',2'-*seco*-nucleosides of ribovirin, *J. Heterocycl. Chem.* 25:651 (1988).

11. P. Vemishetti, E. Abushanab, R.W. Leiby, and R. P. Panzica, Synthesis of chiral 1',2'-*seco*-nucleosides of guanine and uracil, *Nucleosides /Nucleotides* 8:201 (1989).

12. P. Vemishetti, R. Saibaba, R.P. Panzica, and E. Abushanab, The preparation of 2'-deoxy-2'-fluoro-1',2'-*seco*-nucleosides as potential antiviral agents, *J. Med. Chem.* 33:681 (1990).

13. P. Vemishetti, H.I. El-Subbagh, E. Abushanab, and R.P. Panzica, Synthesis of 1',2'-*seco*-nucleoside analogues of AZT, *Nucleosides /Nucleotides* 11:739 (1992).

14. E. Abushanab and M.S.P. Sarma, 1',2'-*seco*-Dideoxynucleosides as potential anti-HIV agents, *J. Med. Chem.* 32:76 (1989).

15. R.A. Long, G.L. Szekeres, T.A. Khwaja, R.W. Sidwell, L.N. Simon, and R.K. Robins, Synthesis and antitumor and antiviral activities of 1-β-D-arabinofuranosylpyrimidine 3',5'-cyclic phosphate, *J. Med. Chem.* 15:1215 (1972).

16. R.B. Meyer, T.E. Stone, and B. Ulman, 2'-O-Acyl-6-thioinosine cyclic 3',5'-phosphate as prodrugs of thioinosinic acid, *J. Med. Chem.* 22:811 (1977).

17. The author wishes to thank Dr. J. G. Moffat who pointed out a footnote (no. 41) in the following reference describing a similar observation. H.G. Khorana, G.M. Terner, R.W. Wright, and J. G. Moffat, Cyclic phosphates. III. Some general observations on the formation and properties of five-, six- and seven-membered cyclic phosphates, *J. Am. Chem. Soc.* 79:430 (1957).

18. A.F. Cichy, P. Vemishetti, and E. Abushanab, An unusual dephosphorylation reaction, *Nucleosides /Nucleotides* 8:957 (1989).

19. R. Engel in "The role of phosphonates in living systems" R.L. Hilderband, Ed., CRC Press, Inc. Boca Raton, FL, 1988.

20. J. Gorzynski Smith, Synthetically useful reactions of epoxides, *Synthesis* 629 (1984), and references cited therein.

21. G.V. Chelintsev and V.K. Kuskov, Diacid tautomerism, *J. Gen. Chem.* U.S.S.R. 16:1481 (1946); *Chem. Abstr.* 5441b (1947).

22. J. Gasteiger and C. Herzig, β-Ketophosphonates from α-chlorooxiranes, *Tetrahedron Lett.*, 21:2687 (1980).

23. T. Azuhata and Y. Okamoto, Synthesis of diethyl 2-(trimethylsiloxy)-alkane-phosphonates from epoxides and diethyltrimethylsilylphosphite, *Synthesis* 916 (1983).

24. T. Azuhata and Y. Okamoto, Synthesis of dialkyl 2-(dialkoxyphosphinyloxy)-alkanephosphonates, *Synthesis* 417 (1984).

25. H. Tanaka, M. Fukui, K. Haraguchi, M. Masaki, and T. Miyasaka, Cleavage of a nucleosidic oxetane with carbanions: Synthesis of a highly promising candidate for anti-HIV agents. A phosphonate isostere of AZT-5'-phoshate, *Tetrahedron Lett.* 30:2567 (1989).

26. S. Racha, Z. Li, H.I. El-Subbagh, and E. Abushanab, A facile synthesis of β- and γ-hydroxyphosphonates from epoxides, *Tedrahedron Lett.* 33:5491 (1992).

27. S. Racha, H.I. El-Subbagh, C. Vargeese, R.P. Panzica, and E. Abushanab, unpublished results.

28. A.F. Cichy, R. Saibaba, H.I. El-Subbagh, R.P. Panzica, and E. Abushanab, 1',2'-*seco*-Thymidines. The preparation of 2,3'-anhydro derivatives and the formation of two unusual dimeric products, *J. Org. Chem.* 56:4653 (1991).

29. F.A. Carey and R.J. Sundberg, *Advanced Organic Chemistry*, Part A, Plenum Press, New York (1984), p. 147.

30. K.E. Krakowiak, J.S. Bradshaw, and D. J. Zamecka-Krakowiak, Synthesis of *aza*-crown ethers, *Chem. Rev.* 89:929 (1989).

31. A.B. Mikkilineni, P. Kumar, and E. Abushanab, The Chemistry of L-ascorbic and D-isoascorbic acids. 2. *R* and *S* Glyceraldehydes from a common intermediate, *J. Org. Chem.* 53:6005 (1988).

32. K. C. Luk and C. C. Wei, Preparation of derivatives of (*R*)-1,2,3-butanetriol from L-ascorbic acid, *Synthesis*, 226 (1988).

33. R. Saibaba, M.S.P. Sarma, and E. Abushanab, The chemistry of L-ascorbic and D-isoascorbic acids. 3: Efficient synthesis of pure R- and S-1,2-O-isopropylidene-1,2,4-butanetriols, *Synth. Commun.* 19:3077 (1989).

34. C. Vargeese and E. Abushanab, Chemistry of L-ascorbic and D-isoascorbic acids. 4. An efficient synthesis of 2-deoxypentofuranoses, *J. Org. Chem.* 55:4400 (1990).

35. Ref. 34 and references cited therein.

36. J. N. Freskos, Synthesis of 2'-deoxypyrimidine nucleosides *via* copper(I) iodide catalysis, *Nucleosides /Nucleotides* 8:549 (1989).

37. T. Okauchi, H. Kubota, and K. Narasaka, Stereoselective syntheses of β-2-deoxy-ribonucleosides from 1-O-acetyl-3-O-[2-(methylsulfinyl)ethyl]-2-deoxyribose, *Chem. Lett.* 801 (1989).

38. L. Wilson and D. Liotta, A general method for controlling glycosylation stereochemistry in the synthesis of 2'-deoxyribose nucleosides, *Tetrahedron Lett.* 31:1815 (1990).

39. C.K. Chu, J.R. Babu, J.W. Beach, S.K. Ahn, H. Huang, L.S. Jeong, and S.J. Lee, A highly stereoselective glycosylation of 2-(phenylselenyl)-2,3-dideoxy-ribose derivative with thymine: Synthesis of 3'-deoxy-2',3'-didehydrothymidine and 3'-deoxythmidine, *J. Org. Chem.* 55:1418 (1990).

40. H.I. El-Subbagh, L.J. Ping and E. Abushanab, A stereospecific synthesis of pyrimidine b-D-2'-deoxyribonucleosides, *Nucleosides /Nucleotides* 11:603 (1992).

41. D.H. Shannanhoff and R.A. Sanchez, 2,2'-Anhydropyrimidine nucleosides. Novel syntheses and reactions, *J. Org. Chem.* 38:893 (1973).

42. W.J. Humphlett, Synthesis of some esters and lactones of aldonic acids, *Carbohydr. Res.* 4:157 (1967).

43. K. Bock, I. Lundt, and C. Pedersen, Preparation of 3-deoxyaldonolactones by hydrogenolysis of acetylated aldonolactones, *Acta Chem. Scand. Ser. B* 155 (1981).

44. C. Vargeese and E. Abushanab, A practical and stereospecific approach to the synthesis of 3'-deoxy-2',3'-didehydrothymidine (D4T), *Nucleosides /Nucleotides* 11:1549 (1992).

45. K. Teng, V.E. Marquez, G.W.A. Milne, J.J. Barchi, Jr., M.G. Kazanietz, N.E. Lewin, P.M. Blumberg, and E. Abushanab, Conformationally constrained analogues of diacylglycerol. Interaction of γ-lactones with the phorbol ester receptor of protein kinase C, *J. Am. Chem. Soc.* 144:1061 (1992).

46. T. Haertle, C.J. Carrera, D.B. Wasson, L.C. Sowers, D.D. Richman, D.A. Carson, Metabolism and anti-human immunodeficienty virus-1 activity of 2-halo-2',3'-dideoxyadenosine derivatives, *J. Biol. Chem.* 263:5870 (1988).

47. H.J. Schaeffer, C.F. Schwender, Enzyme Inhibitors: XXVI. Bridging hydrophobic and hydrophilic regions on adenosine deaminase with some 9-(2-hydroxy-3-alkyl)-adenines, *J. Med. Chem.* 17:6 (1974).

48. W.J. Suling, L.S. Rice, W.M. Shannon, Effects of 2'-deoxycoformycin and *erythro*-9-(2-hydroxy-3-nonyl)adenine on plasma levels and urinary excretion of 9-β-D-arabino-furanosyladenine in the mouse, *Cancer Treat. Rep.* 62:369 (1978).

49. G. Bastian, M. Bessodes, R.P. Panzica, E. Abushanab, S.F. Chen, J.D. Stoeckler, and R.E. Parks, Jr., Adenosine deaminase inhibitors. Conversion of a single chiral synthon into *erythro-* and *threo*-9-(2-hydroxy-3-nonyl)adenines, *J. Med. Chem.* 24:1383 (1981).

50. M. Bessodes, G. Bastian, E. Abushanab, R.P. Panzica, S.F. Berman, E.J. Marcaccio, Jr., S.-F. Chen, J.D. Stoeckler, and R.E. Parks, Jr., Effect of chirality in *erythro*-9-2-hydroxy-3-nonyl)adenine (EHNA) on adenosine deaminase inhibition, *Biochem. Pharmacol.* 31:879 (1982).

51. D.C. Baker, J.C. Hanvey, L.D. Hawkins, and J. Murphy, Identification of the bioactive enantiomer of *erythro*-3-(adenin-9-yl)-2-nonanol (EHNA). A semi-tight binding inhibitor of adenosine deaminase, *Biochem. Pharmacol.* 30:1159 (1981).

52. D.C. Baker and L.D. Hawkins, Synthesis of inhibitors of adenosine deaminase. A total synthesis of *erythro*-3-(adenin-9-yl)-2-nonanol and its isomers from chiral precursors, *J. Org. Chem.* 47:2179 (1982).

53. M.Y. Chu, E. Chu, E. Abushanab, R. P. Panzica, and P.C. Calabresi, Effects of the chiral isomers of erythro-9-(2-hydroxy-3-nonyl)adenine on the anti-neoplastic activity of adenosine analogs against human pancreatic DAN and human colon HCT-8 carcinomas. *Proc. Am. Assoc. Cancer Res.* 29:1394 (1988).

54. W.R. McConnell, S.M. El Dareer, and D.L. Hill, Metabolism and disposition of *erythro*-9-(2-hydroxy-3-nonyl)[^{14}C]adenine in the rhesus monkey, *Drug Metab. Disp.* 8:5 (1980).

55. E. Abushanab, M. Bessodes, and K. Antonakis, Practical enantiospecific synthesis of (+)-*erythro*-9(2S-hydroxy-3R-nonyl)adenine, *Tetrahedron Lett.* 25: 3841 (1984).

56. I. Antonini, G. Cristalli, P. Franchetti, M. Grifantini, S. Martelli, G. Lupidi, and F. Riva, Adenosine deaminase inhibitors. Synthesis of deaza analogues of erythro-9-(2-hydroxy-3-nonyl)adenine, *J. Med. Chem.* 27:274 (1984).

57. G. Cristalli, P. Franchetti, M. Grifantini, S. Vittori, G. Lupidi, F. Riva, T. Bordoni, C. Geroni, M.A. Verini, Adenosine deaminase inhibitors. Synthesis and biological activity of deaza analogues of *erythro*-9-(2-hydroxy-3-nonyl)adenine, *J. Med. Chem.* 31:390 (1988).

58. H. J. Schaeffer and C.F. Schwender, Enzyme inhibitors XXIV. Bridging hydrophobic and hydrophilic regions on adenosine deaminase, *J. Pharm. Sci.* 60:1204 (1971).

59. G.C.B. Harriman, A.F. Poirot, E. Abushanab, R.M. Midgett, and J.D. Stoeckler, Synthesis and biological evaluation of C1' derivatives of (+)-*erythro*-9-(2S-hydroxy-3R-nonyl)adenine, *J. Med. Chem.*, in press.

60. H.J. Schaeffer, Design and evaluation of enzyme inhibitors, especially of adenosine deaminase, Topics in medicinal chemistry, Rabinowitz and Myerson, 3:1 (1972).

SYNTHESIS AND ANTICANCER AND ANTIVIRAL ACTIVITY OF CERTAIN PYRIMIDINE NUCLEOSIDE ANALOGUES

Tai-Shun Lin and Mao-Chin Liu

Department of Pharmacology and The Comprehensive Cancer Center
Yale University School of Medicine
New Haven, Connecticut 06510

3'-AZIDO, 3'-AMINO, 2',3'-UNSATURATED, 2',3'-DIDEOXY, AND 5-SUBSTITUTED ANALOGUES OF PYRIMIDINE NUCLEOSIDES

3'-Azido-3'-deoxythymidine (AZT) was first synthesized by Horwitz et al.[1] and subsequently was found by Mitsura et al.[2] to be a potent inhibitor of the replication of the human immunodeficiency virus (HIV) that is responsible for the acquired immunodeficiency syndrome (AIDS). AZT has been reported to be of marked benefit in the therapy of AIDS patients, but the usefulness of AZT is limited by its bone marrow toxicity.[3] Various 3'-azido, 3'-amino, 2',3'-unsaturated, 2',3'-dideoxy, and 5-substituted analogues of pyrimidine deoxyribonucleosides have been synthesized and evaluated against Moloney-murine leukemia virus (M-MuLV), a mammalian T-lymphotropic retrovirus, *in vitro* in our laboratory.[4] Among these 3'-azido and 3'-amino derivatives, 3'-azido-3'-deoxythymidine (**2**, AZT) was the most active against M-MuLV *in vitro* with an EC_{50} value of 0.02 μM. The 3'-azido analogues of 5-bromo- and 5-iodo-2'-deoxyuridine, compounds **5** and **6**, also showed significant antiviral activity with EC_{50} values of 1.5 and 3.0 μM, respectively. The 3'-azido derivative of 2'-deoxyuridine (**1**, AZDU, CS-87), and the 3',5'-diazido and 3'-amino derivatives of thymidine, **11** and **12**, demonstrated moderate antiviral activity yielding the corresponding EC_{50} values of 52, 17, and 42 μM. Conversely, the 3'-azido analogue of 2'-deoxycytidine, compound **7**, only showed moderate inhibition against M-MuLV with an EC_{50} value of 58 μM. The other 3'-azido and 3'-amino derivatives in this group were found to be practically inactive. Among the 2',3'-unsaturated, 2',3'-dideoxy derivatives of pyrimidine deoxyribonucleosides, the 2',3'-unsaturated analogue of thymidine (**17**, D4T) and 2'-deoxycytidine (**18**, D4C) and the 3'-deoxy analogue of 2'-deoxycytidine (**24**, DDC) produced significant antiviral activity with EC_{50} values of 2.5, 3.7, and 4.0 μM, respectively. The results are summarized in Table 1.

Nucleosides and Nucleotides as Antitumor and Antiviral Agents,
Edited by C.K. Chu and D.C. Baker, Plenum Press, New York, 1993

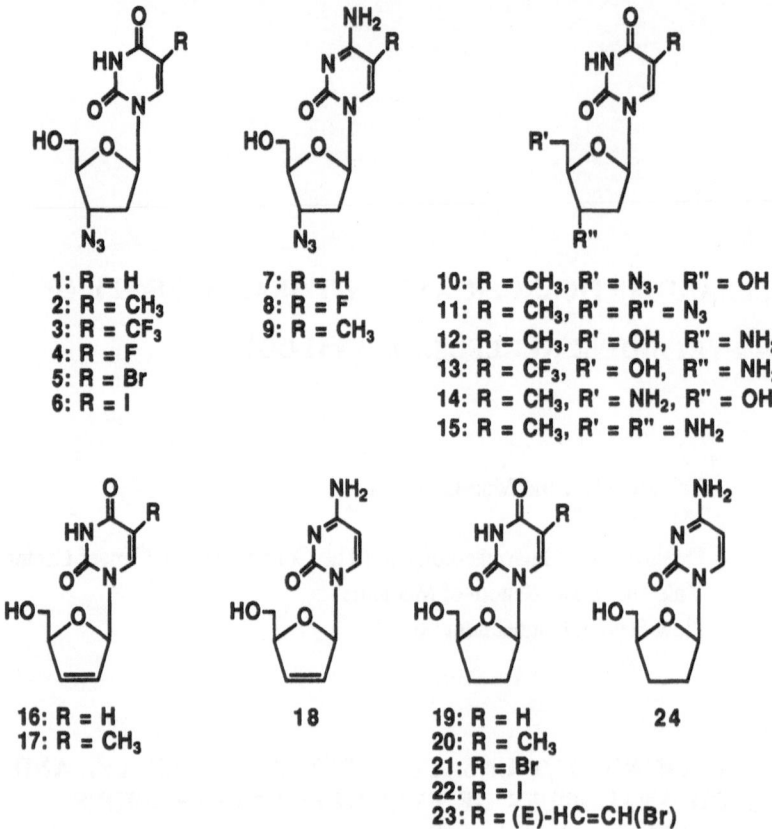

Figure 1. Structures of various azido, amino, 2',3'-unstaturated, and 3'-deoxy analogues of pyrimidine 2'-deoxynucleosides.

Table 1. Antiviral activity of various azido, amino, 2',3'-unsaturated, and 2',3'-dideoxy analogues of pyrimidine deoxyribonucleosides on the replication of M-MuLV in 3T3 mouse cells.

Compd	EC_{50}[1] (µM)	Compd	EC_{50}[1] (µM)
1	52	13	>100
2	0.02	14	>100
3	>100	15	>100
4	>100	16	>100
5	1.5	17	2.5
6	3.0	18	3.7
7	58	19	>100
8	>100	20	>100
9	>100	21	>100
10	>100	22	>100
11	17	23	>100
12	42	24	4.0

[1]The EC_{50} values were estimated from dose-response curves and represent the drug concentration (µM) required to inhibit 50% of the syncytial forming units of the virus.

There appears to be a relationship between the antiviral activity and the electron-withdrawing capacity of the substituents in the 5-position of the nucleoside analogues as evidenced by the decrease in the antiviral activity when the methyl moiety was replaced (CF_3, $F > I > Br > CH_3$). These differences may be related to their substrate activity for thymidine kinase, which is required for activation, or to the differences in metabolic conversion to the di- or triphosphate, or to the relative affinity of the nucleoside triphosphate analogue for the reverse transcriptase.

Replacement of the uracil moiety of the 3'-azido nucleoside analogue **1** with the cytosine to form **7** did not affect its antiviral activity; however, when the substituent on carbon-5 of the cytosine moiety was either fluoro **8** or methyl **9**, the antiviral activity was markedly reduced. This could be explained if these 3'-azido-2',3'-dideoxycytidine analogues were required to be deaminated by deoxycytidine deaminase for which the 5-methyl or 5-fluoro analogues are not substrates. Other possibilities include differences in metabolic conversion to the di- and triphosphates analogues, as well as the relative affinities of the triphosphate analogues for the reverse transcriptase.

The azido group in the 3'-position of the deoxyribose moiety of the thymidine (**2**, AZT) is critical since transfer to the 5'-position (e.g., in **10**) resulted in marked decrease in activity. However, retention of the 3'-azido group with addition of an azido group to the 5'-position (as in **11**) increased the antiviral activity relative to **10**, but decreased its activity relative to **2**. Thus a primary hydroxyl group in the 5'-position is beneficial. Although one may presume the 5'-hydroxyl moiety is required for substrate activity for thymidine kinase, this can not be a prerequisite for the antiviral activity of the 3',5'-diazido analogue **11**, unless the 5'-azido moiety were hydrolytically cleaved, either chemically or enzymatically, to a hydroxyl group, a reaction which is most unlikely.

Reduction of the 3'-azido moiety of **2** to an amino group (**12**) markedly decreased activity; however, moderate antiviral activity was retained. Whether the activity of **12** is related to the bonafide antiviral activity of this compound or to the cytotoxic properties[5] of **12** is not clear.

Replacement of the 3'-azido moiety of **2** with hydrogen (**20**) resulted in loss of activity; however, subsequent removal of a hydrogen atom from both the 2'- and 3'-carbon produced the active compound **17** (D4T); thus the azido moiety per se is not an absolute requirement for the antiviral activity of the thymidine analogues. However, replacement of the 5-methyl group of **17** with a hydrogen (**16**) yielded a loss of activity. The azido moiety conferred moderate antiviral activity to 2'-deoxycytidine (**7**), but this moiety also is not an absolute requirement since its replacement with a hydrogen (**24**) resulted in a marked increase in activity. Subsequent removal of a hydrogen from the 2'- and 3'-carbon of **24** to produce **18** did not affect the increase in antiviral activity from that of **7**. Deamination of **24** produced **19** with a concomitant loss of antiviral activity, which could not be recovered by insertion of a variety of substituents on the carbon-5 of the uracil moiety (**20-23**).

Compounds **1-9**, as well as other 3'-azido analogues of pyrimidine deoxyribonucleosides (compounds **25-37**), which were also synthesized in our laboratory, were evaluated against HIV-1 *in vitro*, and the antiviral activity was expressed by the concentration (μM) that inhibits 50% of viral replication. The results are shown in Table 2.[6]

Among the 3'-azido analogues of pyrimidine deoxyribonucleosides, 3'-azido-3'-deoxythymidine (**2**, AZT) was the most active against HIV-1 *in vitro* with an EC_{50} value of 0.002 μM. Conversely, 3'-azido-3'-deoxy-6-azathymidine (**37**) was practically inactive (EC_{50} >100 μM). The 3'-azido derivatives of 3'-deoxy-3-(3-oxo-1-propenyl)thymidine (**36**), 2'-deoxyuridine (**1**), 5-bromo-2'-deoxyuridine (**5**), 2'-deoxy-5-fluorocytidine (**8**), 2'-deoxy-5-iodouridine (**6**), 2'-deoxycytidine (**7**), 2'-deoxy-5-fluorouridine (**4**), 2'-deoxy-5-thio-

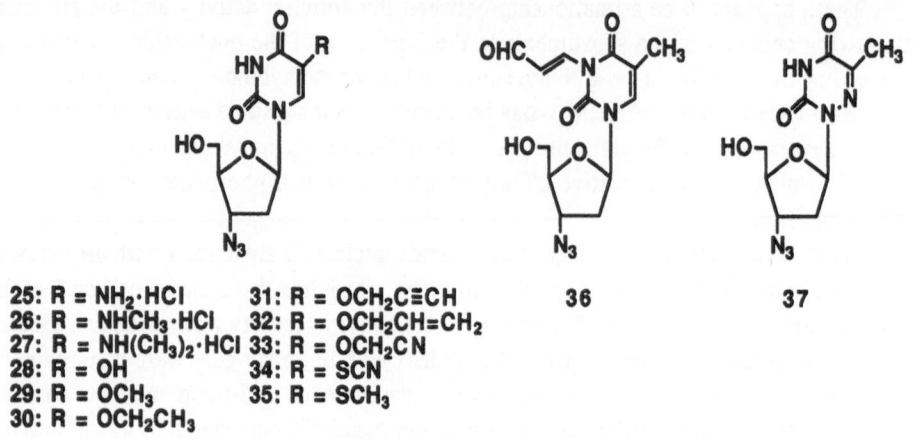

25: R = NH₂·HCl   **31:** R = OCH₂C≡CH
26: R = NHCH₃·HCl   **32:** R = OCH₂CH=CH₂
27: R = NH(CH₃)₂·HCl **33:** R = OCH₂CN
28: R = OH   **34:** R = SCN
29: R = OCH₃   **35:** R = SCH₃
30: R = OCH₂CH₃

Figure 2. Structures of various 3'-azido analogues of pyrimidine 2'-deoxynucleosides.

Table 2. Antiviral activity of various 3'-azido analogues of pyrimidine deoxyribonucleosides on the replication of human immunodeficiency virus (HIV-1) in human peripheral blood mononuclear cells.

Compd	EC_{50}[1] (µM)	Compd	EC_{50} (µM)
1	0.2	27	>100
2	0.002	28	10
3	>100	29	70
4	4.8	30	54
5	1.0	31	38
6	1.1	32	>100
7	1.2	33	16
8	1.0	34	5.1
9	5.1	35	>100
25	6.2	36	0.01
26	>100	37	>100

[1]The EC_{50} values were estimated from dose-response curves and represent the drug concentration (µM) required to inhibit 50% of viral replication.

cyanatouridine (**34**), 2'-deoxy-5-methylcytidine (**9**), 2'-deoxy-5-aminouridine (**25**), and 2'-deoxy-5-hydroxyuridine (**28**) also demonstrated significant antiviral activity with EC_{50} values of 0.01, 0.2, 1.0, 1.0, 1.1, 1.2, 4.8, 5.1, 5.1, 6.2, and 10 µM, respectively. However, the 3'-azido derivatives of 5-[(cyanomethyl)oxy]-2'-deoxyuridine (**33**), 2'-deoxy-5-(2-propynyloxy)uridine (**31**), 2'-deoxy-5-ethoxyuridine (**30**), and 2'-deoxy-5-methoxyuridine (**29**) only showed moderate antiviral activity with respective EC_{50} values of 16, 38,

54, and 70 µM. The other 3'-azido derivatives in this series, compounds **3, 26, 27, 32**, and **35**, were found to be practically inactive (EC_{50} >100 µM). These compounds were not toxic to the host human peripheral blood mononuclear cell (PBM) at >100 µM except for compound **36**, which was toxic at 10 µM. However, it is clear that the data may vary considerably from murine to human retroviruses. For example, compound **1** (AZDU) was first reported by Schinazi, Chu, and co-workers[7-9] to have significant anti-HIV-1 activity in PBM cells with an EC_{50} value of 0.2 µM, while it has an EC_{50} value of 52 µM against Moloney murine leukemia virus.[4] AZDU (**1**) was reported to have low bone marrow toxicity[9] and is currently undergoing clinical trial. The synthesis of AZDU was first reported by our laboratory,[10,11] which started with tritylation of uridine with trityl chloride in pyridine, followed by mesylation of the 5'-O-trityl derivative with methanesulfonyl chloride in pyridine. By treatment of the resulting 2',3'-dideoxy-3'-O-mesyl-5'-O-trityluridine with sodium hydroxide in aqueous ethanol solution, the 3'-lyxo nucleoside was formed via the 2,3'-anhydronucleoside. Mesylation of the 3'-lyxo nucleoside, followed by treatment of the mesylate with lithium azide in DMF, gave the 3'-azido derivative, which was then detritylated with 80% acetic acid to afford AZDU. Subsequently, the synthesis of AZDU was also reported by Krenitsky et al.[12] and Chu et al.[9]

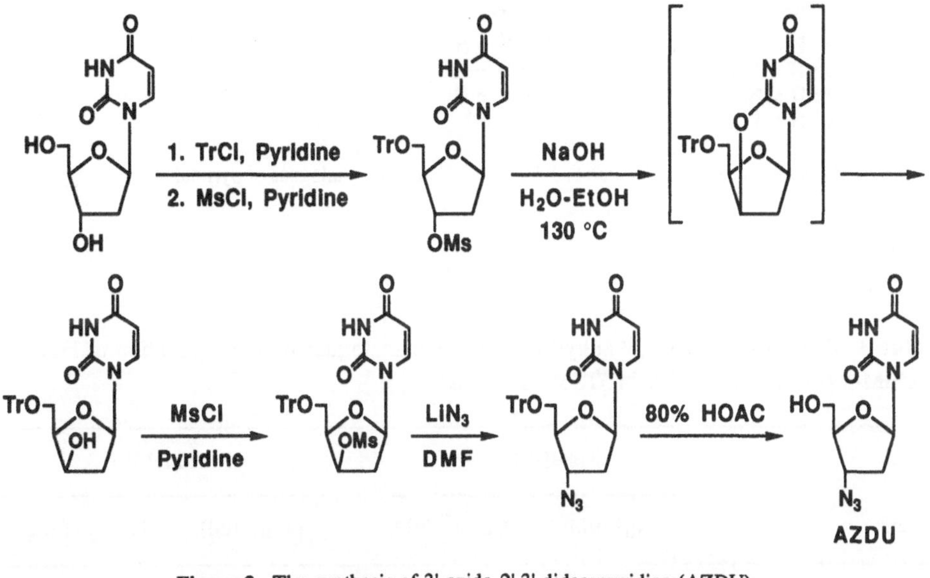

Figure 3. The synthesis of 3'-azido-2',3'-dideoxyuridine (AZDU).

Several 2,5'-anhydro analogues of 3'-azido-2',3'-dideoxyuridine (**1**, AZDU), 3'-azido-3'-deoxythymidine (**2**, AZT), 3'-azido-5-bromo-2',3'-dideoxyuridine (**5**), 3'-azido-2',3'-dideoxy-5-iodouridine (**6**), and the 3'-azido derivative of 2'-deoxy-5-methylisocytidine, compounds **38-42**, have been synthesized and evaluated against human immunodeficiency virus (HIV-1) and Rauscher-murine leukemia virus (R-MuLV).[13] In general, the parent compounds **1, 2, 5**, and **6** were approximately 1.8 to 6 times more active against HIV-1 than their corresponding 2,5'-anhydro derivatives (compounds **38-41**, Table 3). Compounds **38-41** have significant antiviral activity with respective EC_{50} values of 4.95, 0.56, 28, and 27.1 µM. 3'-Azido-2',3'-dideoxy-5-methylisocytidine (**42**) also demonstrated anti-HIV-1 activity, with an EC$_{50}$ value of 12 µM. The host cell cytotoxicity of anhydro-AZT (**39**) had a TCID$_{50}$ value of >100 µM, whereas the parent compound, AZT, was cytotoxic with a TCID$_{50}$ value of 29 µM. The anhydro derivatives, compounds **40** and **41**, had TCID$_{50}$ values of 168 and 194 µM, respectively, and hence were slightly less cytotoxic than their parent compounds (**5** and **6**), which had TCID$_{50}$ values of 127 and 153 µM. These compounds were also evaluated against R-MuLV *in vitro*. Among them, AZT, 3'-azido-2',3'-dideoxy-5-iodouridine (**6**), 3'-azido-2',3'-dideoxy-5-bromouridine (**5**), and anhydro-AZT (**39**) were found to be most active, with EC$_{50}$ values of 0.023, 0.21, 0.23, and 0.27 µM, respectively (Table 3).

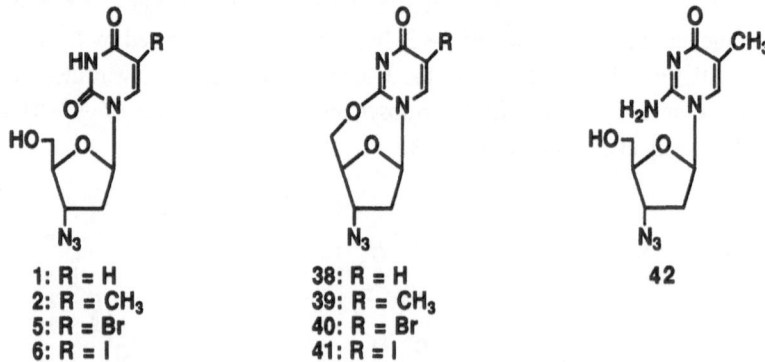

1: R = H
2: R = CH$_3$
5: R = Br
6: R = I

38: R = H
39: R = CH$_3$
40: R = Br
41: R = I

42

Figure 4. Structures of various 3'-azido and 3'-azido-2,5'-anhydro analogues of pyrimidine 2'-deoxyyribo-nucleosides.

Table 3. Effect of various 2,5'-anhydro nucleoside analogues on the replication of HIV-1 in CEM-F cells and R-MuLV in 3T3 mouse cells.

Compd	HIV-1		R-MuLV	
	EC$_{50}$[1] (µM)	TCID$_{50}$[2] (µM)	EC$_{50}$ (µM)	TCID$_{50}$ (µM)
1	2.76	166	138	>395
2	0.1	29	0.023	>400
5	8.73	127	0.23	>301
6	12.6	153	0.21	>264
38	4.95	>100	ND[3]	ND
39	0.56	>100	0.27	>400
40	28	168	16.1	>319
41	27.1	194	17.7	>277
42	12	470	114	>376

[1]The EC$_{50}$ values represent the drug concentration (µM) required to inhibit 50% of viral replication.
[2]The TCID$_{50}$ values represent the drug concentration (µM) required to inhibit 50% of host cell replication.
[3]ND: not determined.

The stability of these 2,5'-anhydro nucleoside derivatives (38-41) which displayed antiviral activity, as well as compound 42, were determined at pH 7.5 and 2.0 (Table 4).[13] This was performed to determine if the biological activity of this class of compounds was due to the anhydro compound itself, or whether hydrolysis of the 2,5'-anhydro linkage occurred in solution to yield the parent compound which, as previously shown, had antiviral activity. All of the compounds tested are stable at neutral pH, with half-life values ranging from 60.5 h for 2,5'-anhydro-3'-azido-5-bromo-2',3'-dideoxyuridine (40) to >168 h for 3'-azido-2',3'-dideoxy-5-methylisocytidine (42). At low pH, however, the compounds displayed shorter and more variable half-lives, ranging from 11.3 min for 2,5'-anhydro-3'-azido-2',3'-dideoxyuridine (38) to 140 min for 3'-azido-2',3'-dideoxy-5-methylisocytidine (42). Pyrimidine base formation was not detected from any of these compounds. Moreover, the compounds decomposed to yield 7.5% or less of the parent compound after one half-life at either neutral or acidic pH, except for 18% of 2,5'-anhydro-3'-azido-2',3'-dideoxyuridine (38), which was recovered as AZDU (1).

Table 4. Stability of various 2,5'-anhydro nucleoside analogues at pH 7.4 and 2.0.

| Compd | Half-life ($t_{1/2}$, 37 °C) | |
	pH 7.4	pH 2.0
38	120 h (ND)[1]	11.3 min (18)[2]
39	>75 h (0)	29.5 min (7.5)
40	60.5 h (3.5)	14.3 min (3.5)
41	95 h (3.8)	12 min (3.8)
42	>168 h (ND)	140 min (ND)

[1]ND refers to not detectable, when the limits of detection correspond to a peak size that is 0.5% of that of the anhydro compound initially present.

[2]The number in the parentheses indicates amount of parent compound formed at $t_{1/2}$ and is expressed as the percentage of the initial anhydro compound present.

Further evidence to suggest the anhydro nucleosides have their unique biological activity was provided by Simpson et al.[14] who investigated the effect of these compounds on mitochondrial DNA synthesis. At 25 µM, AZT inhibited the uptake of [^{3}H]dATP into mitochondrial DNA by 51%, whereas anhydro-AZT inhibited DNA synthesis by only 5%. Conversely, at a concentration of 25 µM, 2,5'-anhydro-3'-azido-2',3'-dideoxy-5-iodouridine (anhydro-5-I-AZDU) inhibited mitochondrial DNA synthesis more strongly than did the parent compound, 3'-azido-2',3'-dideoxy-5-iodouridine, with a ratio of 100% versus 12%. Anhydro-5-I-AZDU would not be expected to show greater biological activity if such activity were solely dependent upon conversion of an inactive prodrug to an active (parent) species.

These data suggest that the basis for antiviral activity of the 2,5'-anhydro nucleoside analogues does not depend on a conversion of "inactive" (anhydro) prodrug to yield an active (parent) species, but rather the anhydro compounds appear to be active per se. However, an intracellular cleavage of the 2,5'-anhydro linkage or other metabolic conversions cannot be ruled out.

In order to study the molecular basis of the antiviral activity, as well as the inhibitory activity to mitochondrial DNA synthesis concerning these anhydro nucleoside derivatives, [2-^{14}C]2,5'-anhydro-3'-azido-3'-deoxythymidine ([2-^{14}C]anhydro-AZT) and [2-^{14}C]2,5'-anhydro-3'-azido-2',3'-dideoxy-5-iodouridine ([2-^{14}C]anhydo-5-I-AZDU) have also been synthesized.[15]

Various 2',3'-dideoxynucleosides showed significant activity against HIV viruses.[16] 3'-Deoxy-2',3'-didehydrothymidine (D4T) and 2',3'-dideoxy-2',3'-didehydrocytidine (D4C) were found by us[17-19] and other laboratories[20-25] to have potent inhibitory activities against HIV. D4T is a more potent inhibitor of HIV than DDT in human peripheral blood mononuclear cells *in vitro*. Although DDT was at least 10-fold less toxic than D4T to uninfected cells, the unsaturated analogue (D4T) was about 19 times more potent as an antiviral agent. D4T (0.1 μM) was markedly more effective in inhibiting viral replication than DDT (1.0 μM) when treatment was delayed by one or two days after infection.[17] Bone marrow toxicity studies have shown that D4T is approximately 100 times less toxic than AZT to normal human granulocyte macrophage cells (ID_{50} = 100 and 1 μM, respectively). Whereas, for erythrocyte progenitor cells, D4T is only slightly less toxic than AZT (ID_{50} = 10 μM and 6.7 μM, respectively).[18] D4T is metabolized in cells to the mono-, di-, and triphosphate nucleotides. It was suggested that the initial conversion to the monophosphate is catalyzed by thymidine kinase. This enzyme has a 600-fold lower affinity for D4T than for thymidine and catalyzes the rate-limiting step in production of the triphosphate.[26] Nevertheless, intracellular concentrations of the triphosphate approximately equal to the reported K_i for human immunodeficiency virus reverse transcriptase are attained with extracellular concentrations of free drug as low as 0.05 μM. The pattern of phosphorylation is different from that of AZT, which has an affinity for thymidine kinase equivalent to that of thymidine and is easily phosphorylated. The rate-limiting step in formation of AZT triphosphate is the conversion of mono- to the diphosphate, and thus the monophosphate accumulates. On removal of D4T or AZT from the media, both triphosphates have an intracellular half-life of about 200 min, and this rate ultimately controls the rate of elimination of the drugs from cells. The differences in metabolism of D4T and AZT observed *in vitro* may be responsible for the differences in toxicity seen *in vitro* and *in vivo* and support the exploration of the clinical utility of D4T as a human anti-immunodeficiency viral agent.[26] The therapeutic index of D4T appears to be close to 5000,[17] and D4T now is under clinical phase III trial.

An X-ray crystallographic analysis of DDT and D4T was recently reported.[27] Crystals of D4T belong to space group $P2_1$ with cell dimensions a = 11.662 (1), b = 5.422 (1), and c = 16.233 (3) Å. The results show that the X-ray structures of DDT and D4T are similar in both their glycosyl geometries and hydrogen bonding behavior. The biological differences of the two compounds are presumably due to their differences in the activation or pharmacokinetics. This suggests that recognition of nucleosides by the relevant kinases is essential for biological activity. It seems more likely that it is the differences in conformational flexibility of these two molecules that contributes to the differences in biological efficacy.

D4T and DDT were first synthesized by Horwitz et al.,[28] and [2-^{14}C]D4T was recently synthesized[29] by a modification of Horwitz's methodology for the study of the metabolism and mechanism of action of D4T.

DDC and D4C were also first synthesized by Horwitz et al.[30] D4C is about as equally potent as DDC; however, Balzarini et al.[31] found that D4C is markedly less stable, due to spontaneous degradation, than DDC. The structural features and conformation of 2',3'-

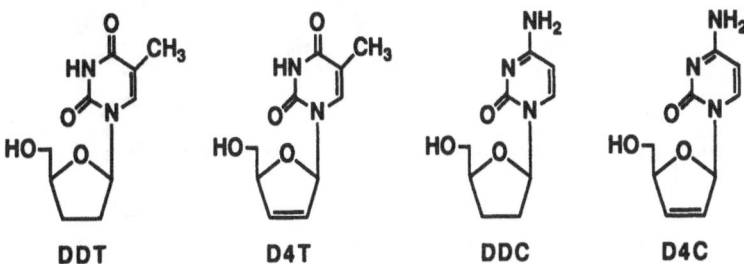

Figure 5. Structures of DDT, D4T, DDC, and D4C.

dideoxycytidine (DDC) and 2',3'-didehydro-2',3'-dideoxycytidine (D4C) were determined by X-ray crystallography.[32,33] DDC crystallizes in the tetragonal space group $P4_12_12$ with cell dimensions $a = b = 8.698$ (4) and $c = 26.155$ (9) Å.[32] The conformation of the furanose ring corresponds to the unusual C-3' exo/C-4' endo ($_3T^4$) pucker, similar to that found in one of the molecules of AZT. The fact that both AZT and DDC have similar unusual conformations is remarkable, but whether these structural features can be correlated with anti-HIV activity cannot be assessed at the present time.[32] D4C crystallizes in the orthorhombic space group $P2_12_12_1$ with cell dimensions $a = 8.603$ (1), $b = 9.038$ (1), and $c = 25.831$ (2) Å and with two independent molecules in the asymmetric unit ($Z = 8$).[33] In both molecules, the temperature parameters of the atoms in the five-membered ring are substantially higher than those of the cytosine atoms. The largest displacements are exhibited by C-2' and C-3', a surprising result in view of these atoms being double-bonded.[33] It is possible that the increased mobility of these atoms reflects the previously mentioned lability of 2',3'-didehydro-2',3'-dideoxynucleosides.[31] A mechanism which would explain the observed release of free base is shown in Figure 6.

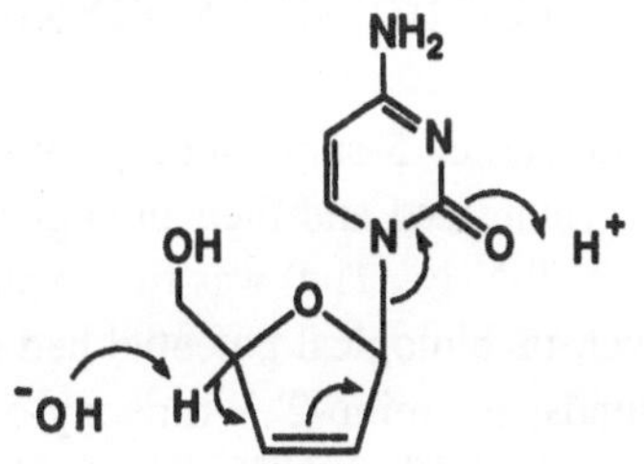

Figure 6. Postulated mechanism for the degradation of D4C.

2',3'-Didehydro-3'-deoxy-6-azathymidine (**43**, 6-aza-D4T), 3'-deoxy-6-azathymidine (**44**, 6-aza-DDT), and 2',3'-dideoxy-6-azacytidine (**45**, 6-aza-DDC) have also been synthesized and evaluated against HIV-1 *in vitro* in the CEM-F cell line and found to be inactive.[34] It appears that the replacement of the 6-carbon in the pyrimidine base with a nitrogen resulted in the loss of antiviral activity. These compounds probably are not substrates for the appropriate kinases.

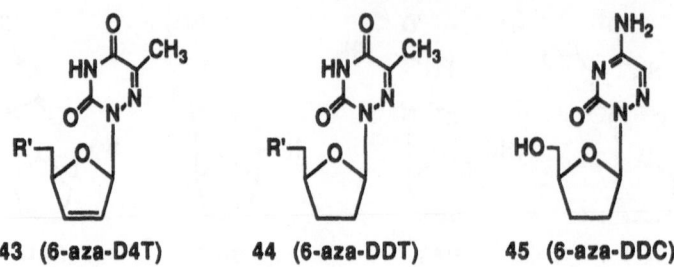

43 (6-aza-D4T)   **44 (6-aza-DDT)**   **45 (6-aza-DDC)**

Figure 7. Structures of 6-aza-D4T, 6-aza-DDT, and 6-aza-DDC.

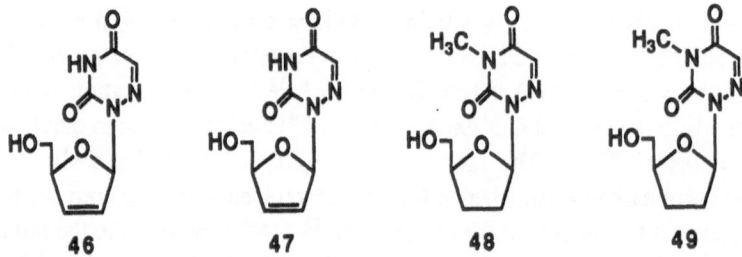

46   **47**   **48**   **49**

Figure 8. Structures of 2',3'-didehydro-2',3'-dideoxy and 2',3'-dideoxy derivatives of 6-azauridine and the corresponding 3-*N*-methyl derivatives.

Recently, Rosowsky and Pai[35] reported the synthesis of 2',3'-didehydro-2',3'-dideoxy and 2',3'-dideoxy derivatives of 6-azauridine (compounds **46** and **48**) and the corresponding 3-*N*-methyl derivatives (compounds **47** and **49**).

AMINO ANALOGUES OF PYRIMIDINE DEOXYRIBONUCLEOSIDES

Various 3'-amino analogues of 5-substituted pyrimidine deoxyribonucleosides, compounds **50-61**, have been synthesized and their biological activities evaluated.[5,11,36] 3'-Amino-3'-deoxythymidine (**53**, 3'-NH$_2$-dThd) was first synthesized by Miller and Fox;[37] and by Horwitz et al.[1] However, its biological potential had not been extensively investigated. Among these compounds, 3'-amino-2',3'-dideoxy-5-fluorouridine (**51**, 3'-NH$_2$-FdUrd), 3'-amino-3'-deoxythymidine (**53**, 3'-NH$_2$-dThd), 3'-amino-2',3'-dideoxycytidine (**59**, 3'-NH$_2$-dCyd), and 3'-amino-2',3'-dideoxy-5-fluorocytidine (**60**) were found to have significant activity against both murine L1210 and sarcoma 180 (S-180) neoplastic cells, with EC$_{50}$ values of 15, 1, 0.7, and 10 μM; and 1, 5, 4, and 1 μM, respectively. 3'-Amino-2',3'-dideoxyuridine (**50**) and 3'-amino-2',3'-dideoxy-5-methylcytidine (**61**) also inhibited the replication of L1210 and sarcoma 180 cells, with respective EC$_{50}$ values of 18 and 50 μM; and 25 and 20 μM. The 3'-azido derivatives demonstrated either less activity (**3**, **4**, and **8**) or loss of activity (**1**, **2**, and **7**) in comparison to their 3'-amino analogues. The 3'-azido-5'-fluoro and 3'-amino-5'-fluoro derivatives of thymidine, **57** and **58**, and the 5'-amino and 3',5'-diamino derivatives of thymidine, **55** and **56**, were found to be not active against either L1210 or S-180 neoplastic cells *in vitro*. The results are summarized in Table 5.

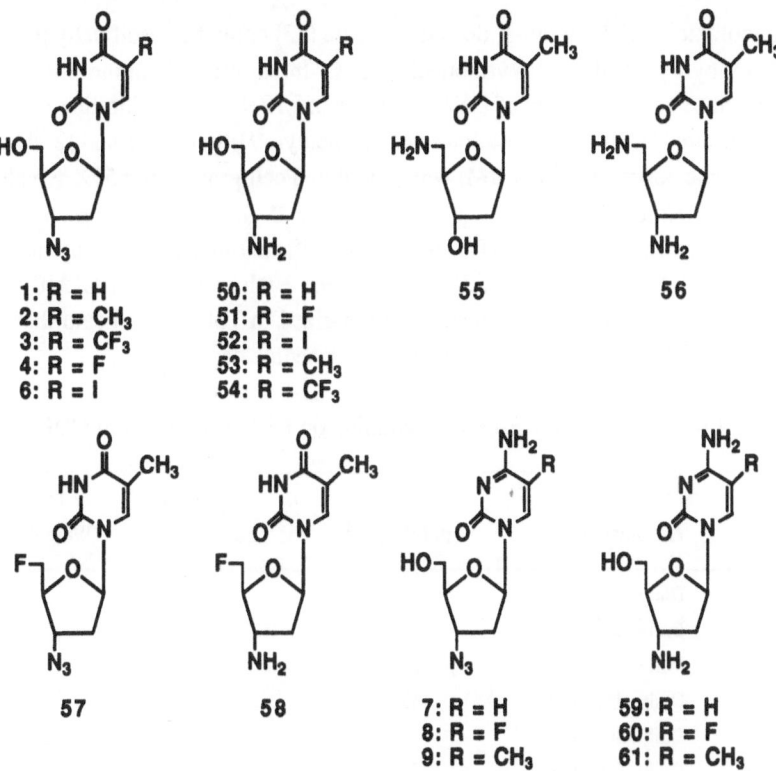

Figure 9. Structures of various azido and amino analogues of pyrimidine 2'-deoxyribonucleosides.

Table 5. EC$_{50}$ values of various 3'-azido and 3'-amino analogues of pyrimidine deoxy-ribonucleoside on the replication of L1210 and sarcoma 180 cells *in vitro*.

Compd	EC$_{50}$[1] (μM) L1210	S-180	Compd	EC$_{50}$ (μM) L1210	S-180
1	NA[2]	NA	52	400	ND
2	NA	NA	53	1	5
3	400	100	54	150	100
4	20	40	55	NA	NA
6	NA	NA	56	NA	NA
7	NA	NA	57	NA	NA
8	25	ND[3]	58	NA	NA
9	NA	NA	59	0.7	4
50	18	50	60	10	1
51	15	1	61	25	20

[1]The EC$_{50}$ values were estimated from dose-response curves compiled from at least two independent experiments and represent the drug concentration (μM) required to inhibit replication of L1210 or S-180 cells by 50%.

[2]NA: not active.

[3]ND: not determined.

The cytotoxicity of 3'-amino-3'-deoxythymidine (53) could be specifically prevented or reversed only by pyrimidine deoxyribonucleosides. In addition, 3'-amino-3'-deoxythymidine is a potent selected inhibitor of DNA synthesis in L1210 cells. Inhibition of DNA biosynthesis, measured by the incorporation rates of [methyl-^{3}H]thymidine and [2-^{3}H]adenine, was notable at the concentration (1 μM) which inhibited cell growth by 65%. No alterations were detected to either RNA or protein synthesis at a 25-fold greater drug concentration. These findings suggest that 3'-NH$_2$-dThd acts as a specific antimetabolite of thymidine.[38]

3'-Amino-3'-deoxythymidine (53) also has been evaluated against L1210 leukemia-bearing CDF$_1$ female mice.[39] The results, summarized in Table 6, indicate that the antileukemic effect of 3'-NH$_2$-dThd was both dose and schedule dependent.

Table 6. Effect of 3'-amino-3'-deoxythymidine on L1210 leukemia in CDF$_1$ mice.

Dose (mg/kg)	Schedule of Drug Administration	% Increase in Lifespan[1]	Long-term Survivors[2]	Maximum Weight Loss[3] (%)
20	Daily, days 1-6	0	0/5	2.5
40	Daily, days 1-6	11	0/5	2.0
80	Daily, days 1-6	22	0/5	4.0
160[4]	Daily, days 1-6	42	1/11	4.5
240	Daily, days 1-6	44	0/6	6.0
320	Daily, days 1-6	72	0/6	4.0
160[5]	2 x Daily, days 1-6	147	1/18	5.7
320[6]	2 x Daily, days 1-6	144	5/12	8.0
640[6]	2 x Daily, days 1-6	128	5/12	8.0
160	3 x on day 1 only	12	0/10	4.0
320	3 x on day 1 only	19	0/10	5.0

[1]Percent increase in lifespan was calculated using the median days of death. In all experiments the median day of death for control animals was 8-9 days.

[2]Animals surviving >60 days.

[3]The maximum weight loss is expressed as the maximum percent decrease in body weight post drug injection (days 3-4).

[4]Average of two experiments.

[5]Average of three experiments.

[6]Average of two experiments (all long-term survivors were seen in one experiment).

There was a progressive increase in the lifespan of the animals treated daily for 6 days as the dose of 3'-NH$_2$-dThd was raised from 20 mg/kg to 320 mg/kg. Although there was one 60-day survivor and marked increases in lifespans, this was not a curative schedule, based on the estimated number of cells surviving therapy. If the regimen was changed to twice a day for three days, the antileukemic effect was dramatically improved. On this schedule, increase in the median lifespan of approximately 140% was achieved, and in one experiment there were a number of long-term survivors. Most importantly, this therapeutic effect was accomplished with very little drug-induced toxicity. The maximum weight loss by the animals was only 8%.

The antitumor efficacy of 1-β-D-arabinofuranosylcytosine (ara-C) and its various derivatives is diminished by their susceptibility to deamination. However, biological and biochemical studies have shown that 3'-amino-2',3'-dideoxycytidine (**59**) was not only resistant to deamination by partially purified cytidine/deoxycytidine deaminase derived from human KB cells,[40] but also produced 80% long-term survivors (>60 days) when administered twice daily at a optimal dosage of 20 mg/kg for a total of 9 days to L1210 leukemia-bearing CDF_1 female mice.[41] 3'-NH_2-dThd was found to produce an S-phase-specific block in exponentially growing L1210 leukemia cells. The mono- and triphosphate forms of the drug were detected within a few hours after the treatment of the intact cells with 3'-amino-2',3'-dideoxycytidine (**59**). No significant change in the deoxynucleoside triphosphate levels was observed during the early stages of treatment. However, within 24 h, a 2-fold increase in the amount of the deoxynucleoside triphosphate was seen. The triphosphate form of the drug competitively inhibited dCTP incorporation into calf thymus DNA using highly purified DNA polymerase α. The K_i was determined to be 9.6 μM with respect to dCTP. Incorporation of the analogue into DNA was not detected. On the other hand, sucrose gradient analysis suggested that incorporation of the analogue into actively synthesized DNA may account for the biological activity of this compound. Treatment with 3'-NH_2-dCyd induced single-strand breaks in actively synthesized DNA, but no double-strand breaks were observed in the presence of the analogue. The data indicate that 3'-amino-2',3'-dideoxycytidine (**59**) specifically interferes with DNA replication at level of DNA polymerase by inhibiting chain elongation.[41]

Several platinum(II) complexes of 3',5'-diamino-3',5'-dideoxythymidine (**62**), 5'-amino-5'-deoxythymidine (**64**), and 3'-amino-3'-deoxythymidine (**66**) and the respective 2'-deoxyuridine amino nucleoside complexes, (**63, 65,** and **67**), have been synthesized and biologically evaluated.[42] Whereas compounds **62–65** and **67** had no inhibitory effect on the replication of murine L1210 cells in culture, compound **66** inhibited these cells yielding an EC_{50} value of 0.8 μM. Compound **66** was therefore selected for study in mice bearing the L1210 leukemia. The results indicated that compound **66** had a dose-dependent effect on the survival of mice, with a T/C x 100 value of 175 at a dose of 320 mg/kg. However, compound **66** is considerably less potent than cisplatin, cis-[$PtCl_2(NH_3)_2$], based on the report of Macquet and Butow.[43] These investigators reported a T/C x 100 value of 205 when a single injection of cisplatin was administered at a dose of 8 mg/kg.

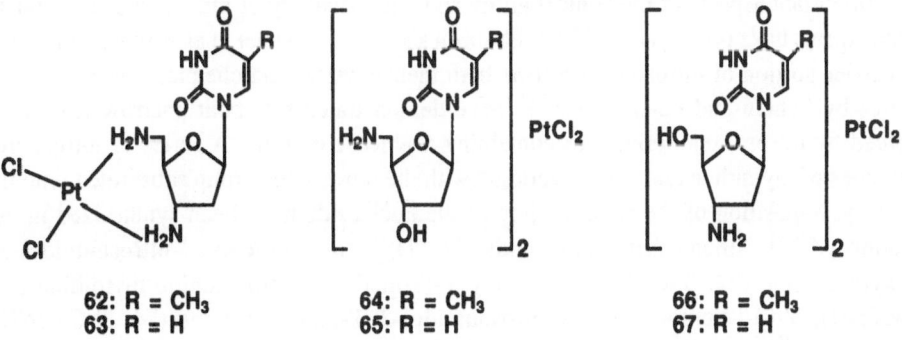

Figure 10. Structures of 3'-amino-3'-deoxy-, 5'-amino-5'-deoxy-, and 3',5'-diamino-3',5'-dideoxythymidine platinum(II) complexes.

5-TRIFLUOROMETHYL- AND 5-PENTAFLUOROETHYL-PYRIMIDINE NUCLEOSIDE ANALOGUES

In view of the biological activity of 1-β-D-arabinofuranosylthymine (ara-T) and other analogues of thymidine, a number of analogues of ara-T with various substituents replacing the 5-methyl group in the pyrimidine moiety, such as 1-β-D-arabinofuranosyl-5-ethyluracil, 1-β-D-arabinofuranosyl-5-vinyluracil, and 1-β-D-arabinofuranosyl-(E)-5-(2-bromovinyl)-uracil, have been synthesized. Because of the potent antiviral activity of 5-(trifluoromethyl)- and 2'-deoxy-5-ethyluridine, several 5-(perfluoroalkyl)pyridine arabinofuranosyl and 2'-deoxyribofuranosyl nucleoside analogues (68-71) have been synthesized and their biological activities evaluated.[44] The 5-trifluoromethyl-substituted derivatives, 68 and 70, demonstrated significant antiviral activity against herpes simplex virus type 1 (HSV-1), with EC_{50} values of 7 and 5 µM, respectively. The acetate 68 was about equally cytotoxic to L1210, S-180, and Vero cells. At 100 µM, compound 68 inhibited the growth of L1210, S-180, and Vero cells by 95 to 100%; however, at the same concentration, the unblocked nucleoside 70 was about 25-fold less toxic to Vero cells (4% inhibition). Thus, compound 70 gave a favorable therapeutic index of 64 against HSV-1 *in vitro*. Conversely, the 5-pentafluoroethyl derivatives, 69 and 71, were not active.

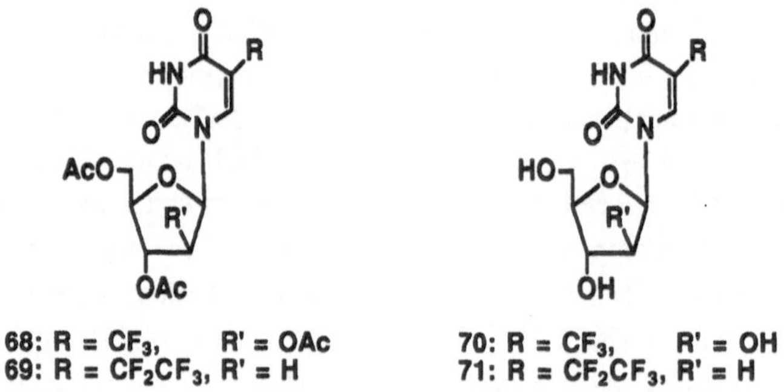

68: R = CF₃,   R' = OAc
69: R = CF₂CF₃, R' = H

70: R = CF₃,   R' = OH
71: R = CF₂CF₃, R' = H

Figure 11. Structures of several 5-trifluoromethyl- and 5-pentafluoroethyl-pyrimidine nucleoside analogues.

NUCLEOSIDE NITROSOUREA ANALOGUES

Important aspects of the clinical utility and pharmacology of nitrosoureas as antineo-plastic agents have been reviewed.[45] It has been shown by Wheeler et al.[46] that alteration of the carrier portion of nitrosoureas affects both their physical and chemical properties, and studies by Schein and co-workers[47,48] have demonstrated that bone marrow toxicity is reduced by nitrosourea derivatives containing a glucose carrier. A series of nitrosourea analogues of thymidine and 2'-deoxyuridine with the nitrosourea group substituted onto the 3'- or the 5'-position of the sugar moiety of the nucleoside have been synthesized in our laboratory.[49-51] Among these compounds (72-77), 3'-(3-chloroethyl-3-nitrosoureido)-3'-deoxythymidine (72, 3'-CTNU) , 3'-deoxy-3'-(3-methyl-3-nitrosoureido)thymidine (73, 3'-MTNU), 3'-[3-(2-chloroethyl)-3-nitrosoureido]-2',3'-dideoxyuridine (74, 3'-CdUNU), and 3'-[3-(2-chloroethyl)-3-nitrosoureido]-2',3'-dideoxy-5-fluorouridine (75, 3'-CFdUNU) exhibited significant antitumor activity both *in vitro* and *in vivo*.

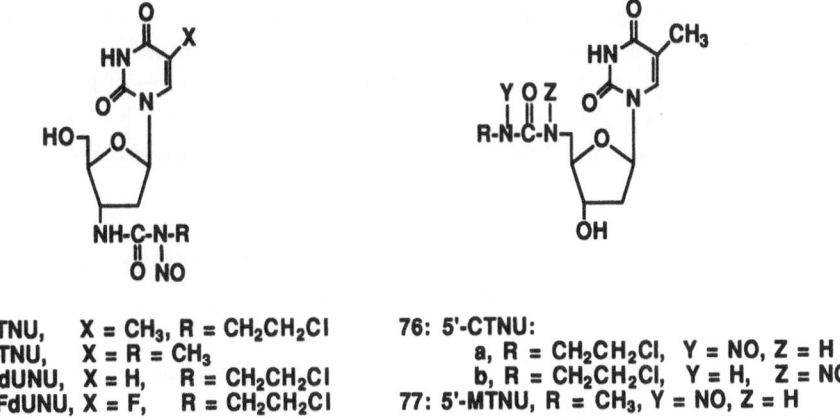

72: 3'-CTNU,   X = CH₃, R = CH₂CH₂Cl
73: 3'-MTNU,   X = R = CH₃
74: 3'-CdUNU,   X = H,   R = CH₂CH₂Cl
75: 3'-CFdUNU, X = F,   R = CH₂CH₂Cl

76: 5'-CTNU:
  a, R = CH₂CH₂Cl, Y = NO, Z = H
  b, R = CH₂CH₂Cl, Y = H, Z = NO
77: 5'-MTNU, R = CH₃, Y = NO, Z = H

Figure 12. Structures of various nucleoside nitrosourea analogues.

The synthesis of the nucleoside nitrosoureas is depicted in Figure 13. Conversion of the amino nucleosides to the corresponding urea derivatives was achieved by reaction with either chloroethylisocyanate or methylisocyanate. These urea derivatives were then treated with sodium nitrite in 80% acetic acid solution at 0 °C to afford the various nucleoside nitrosoureas.[49-51]

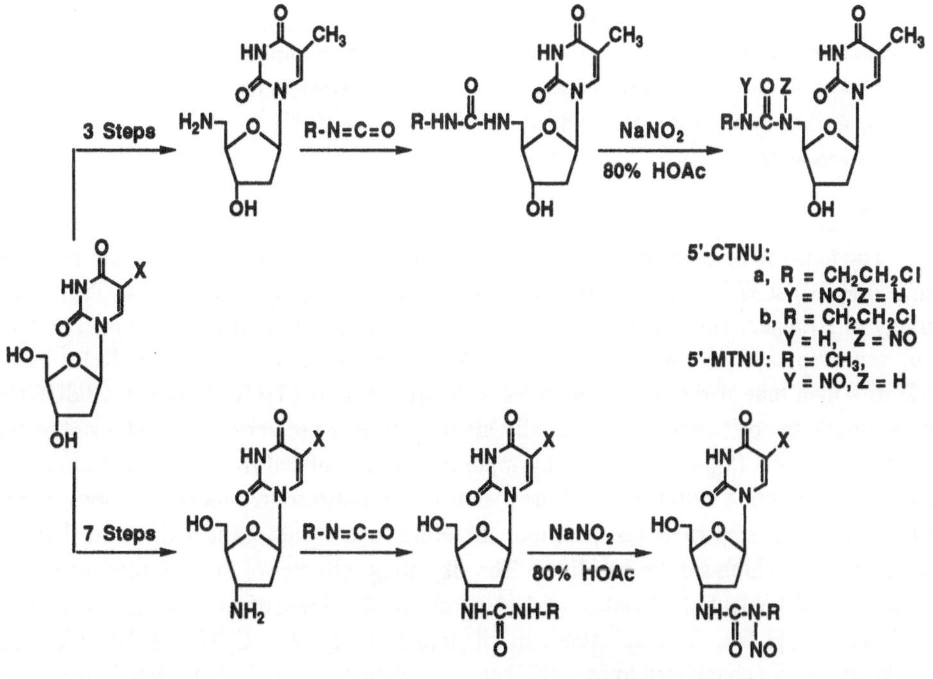

3 Steps   H₂N   R-N=C=O   R-HN-C-HN   NaNO₂   R-N-C-N
                                      80% HOAc

5'-CTNU:
  a, R = CH₂CH₂Cl
     Y = NO, Z = H
  b, R = CH₂CH₂Cl
     Y = H, Z = NO
5'-MTNU: R = CH₃,
         Y = NO, Z = H

7 Steps   NH₂   R-N=C=O   NH-C-NH-R   NaNO₂   NH-C-N-R
                                      80% HOAc

3'-CTNU,   X = CH₃, R = CH₂CH₂Cl
3'-MTNU,   X = R = CH₃
3'-CdUNU,   X = H,   R = CH₂CH₂Cl
3'-CFdUNU, X = F,   R = CH₂CH₂Cl

Figure 13. Synthesis of various nucleoside nitrsourea analogues.

191

It has been suggested that therapeutic effectiveness might be maximized in drugs possessing high alkylating and low carbamoylating activities.[46] The half-lives, the alkylating and carbamoylating activities, as well as the EC_{50} values of the most active nucleoside nitrosoureas 3'-CTNU, 3'-CdUNU, 3'-CFdUNU, 3'-MTNU, and 1,3-bis(2-chloroethyl)-1-nitrosourea (BCNU) against L1210 leukemia cells in culture, are listed in Table 7.[49,51]

Table 7. Alkylating and carbamoylating activity and cytotoxicity against L1210 cells *in vitro* of various nucleoside nitrosoureas.

Compd	Half-life[1] (min)	Alkylating activity[2]	Carbamoylating activity[3]	EC_{50}[4] (μM)
BCNU	50.0	1.00	1.00	4.0
72 (3'-CTNU)	37.3	2.00	0.85	1.5
73 (3'-MTNU)	ND[5]	0.05	0.90	1.0
74 (3'-CdUNU)	27.3	1.65	0.85	12.5
75 (3'-CFdUNU)	36.1	1.78	0.77	2.5
76 (5'-CTNU)	ND	0.47	1.60	6.6

[1]Half-life determined by incubation at 37 °C in PBS (pH 7.4).
[2]Alkylating activity of all compounds is compared to BCNU; for BCNU, $\Delta A_{540}/120 = 0.42$. At least three determinations were made for each compound, and the standard deviation was less than 10% in all cases.
[3]Carbamoylating activity of all compounds is compared to BCNU; for 3'-CTNU, 78% of 5'-amino-5'-deoxythymidine was present as products other than unreacted 5'-amino-5'-deoxythymidine. Three determinations were made for each compound with the standard deviation less than 10% of the mean in all cases.
[4]EC_{50} values were estimated from dose-response curves compiled from at least three separate experiments and represent the drug concentration (μM) needed to inhibit cell growth by 50%.
[5]ND: not determined.

The half-life of each compound was determined by incubation in phosphate-buffered saline (pH 7.4) at 37 °C, followed by reversed-phase chromatographic analysis by the methodology of Brubaker and Prusoff.[52] A plot of log [nitrosourea] vs. time was linear in each case, indicating pseudo-first-order kinetics. The half-life (27.3 min) of 3'-CdUNU is about 25% less than that of the other two nucleoside nitrosoureas (3'-CTNU and 3'-CFdUNU), and although the difference is statistically significant, it is not very large. The site of the structure variation (5-position of the pyrimidine ring) is isolated from inductive effects of resonance interaction with the site of the 3'-(chloroethyl)nitrosourea moiety involved in the rate-determining step of decomposition; therefore, one would expect these nucleoside nitrosoureas to exhibit similar half-lives. The alkylating activities of these nitrosoureas were measured by the extent of alkylation of 4-(4-nitrobenzyl)pyridine that occurs within 120 min at 37 °C and pH 6.4. The difference in alkylating ability of 3'-CTNU, 3'-CdUNU, and 3'-CFdUNU is not considered to be significant. Alkylating activity is dependent in part upon the rate of decomposition of the nitrosoureas, since decomposition is a prerequisite for the formation of the alkylating species. There is no apparent correlation between the alkylating activities and the EC_{50} values. For example, 3'-CdUNU, which is the least cytotoxic, has an alkylating activity 65% greater than that of BCNU. It is possible that alkylating activity measured at pH 6 does not necessarily correspond to the degree of alkylation that occurs at physiological pH. Carbamoylating activity was determined by measurement of the extent of carbamoylation of 5'-amino-5'-deoxythymidine upon incubation with the nitrosourea at 37 °C for 6 h in phosphate-buffered saline.[53] The carbamoylating activities of the 3'-nitrosourea analogues fall within a relatively narrow range from 0.77 to 1.0 (Table 7). There is no clear relationship between half-life, alkylating activity, or carbamoylating activity and the EC_{50} value. Furthermore, the cytotoxicity is clearly not related to the formation of the corresponding 3'-aminonucleoside analogues formed during decomposition (Table 8). The EC_{50} values for the 3'-aminodeoxyribonucleosides derived from 3'-CdUNU and 3'-CFdUNU are significantly greater than that of the nitrosourea analogues from which they could have derived. The reverse relationship would have been expected if the activity were due to the aminonucleoside formed as a byproduct of decomposition of the nitrosourea nucleoside. Additional evidence for the active component not being the 3'-aminonucleosides is supported by the finding that no increase in survival of the treated mice relative to the untreated mice was found, when 3'-amino-2',3'-dideoxyuridine or 3'-amino-2',3'-dideoxy-5-fluorouridine were administered to mice bearing the L1210 leukemia.[51]

Table 8. Comparison of the EC_{50} values of several 3'-amino analogues of pyrimidine deoxyribonucleosides and their corresponding nitrosourea derivatives on the replication of L1210 cells *in vitro*.

Compd	$EC_{50}{}^1$ (µM)
3'-CTNU	1.5
3'-NH$_2$-dThd	0.8
3'-CdUNU	12.5
3'-NH$_2$-dUrd	18.0
3'-CFdUNU	2.5
3'-NH$_2$-FdUrd	15.0

[1]The EC_{50} values were estimated from dose-response curves compiled from at least two independent experiments and represents the drug concentration (µM) required to inhibit replication of L1210 cells by 50%.

The nucleoside nitrosoureas, 3'-CTNU, 3'-CdUNU, and 3'-CFdUrd, exhibited marked anticancer activity against the L1210 leukemia in tumor-bearing mice, when given as a single intraperitoneal injection, 24 h after tumor inoculation. Whereas 160 mg/kg of 3'-CTNU was lethal, 80 mg/kg produced 100% long-term (>60 days) survivors, 40 mg/kg produced 50 to 83% long-term survivors, and 20 mg/kg produced 33% long-term survivors.[50,51,54] At an optimum dosage level of 40 mg/kg, 3'-CdUNU and 3'-CFdUrd produced 90% and 60% long-term survivors, respectively.[51] Although 3'-CdUNU has a relatively low alkylating activity and a relatively high EC_{50} value *in vitro* in comparison to that of the other two nucleoside nitrosoureas, 3'-CdUNU has equally good anticancer activity *in vivo*. Clearly, the anticancer activity is a complex function of these and other variables.

The coadministration of thymidine (2 g/kg) did not prevent the initial loss of weight by 3'-CTNU, but it did prevent the lethality otherwise produced in nontumor-bearing mice. This same dose of thymidine when administered alone to L1210- or P388-bearing mice demonstrated no anticancer activity. However, when thymidine was administered in combination with 3'-CTNU, it enhanced the antitumor activity of 3'-CTNU.[55] This enhancement of the anticancer activity of 3'-CTNU by thymidine was not limited to this compound alone. Furthermore, the coadministration of thymidine with 1,3-bis(2-chloroethyl)-1-nitrosourea (BCNU) produced an even greater enhancement of antitumor activity than that observed when thymidine was coadministered with 3'-CTNU.[55] The biochemical basis for the enhancement of the antitumor activity of BCNU and 3'-CTNU has been studied.[56] In the presence of a 5- and 25-fold excess of thymidine, a 1.3- and 1.5-fold respective increase in the uptake of radioactivity from 0.1 mM [chloroethyl-^{14}C]BCNU was observed. Similarly, an enhancement of DNA alkylation was observed upon treatment of L1210 cells for up to 3 h with 0.1 mM [chloroethyl-^{14}C]BCNU from 70 pmol ^{14}C/mg DNA in control to 85, 95, and 120 pmol ^{14}C/mg DNA with equimolar 5- and 25-fold excess thymidine, respectively. No effect of thymidine on the uptake of 0.1 mM [chloroethyl-^{14}C]-3'-CTNU was observed, although a small increase in DNA alkylation at 3 h was evident. DNA repair, as measured by the amount of radioactivity remaining associated with the DNA after an initial 2-h treatment with labelled BCNU, was largely unaffected by thymidine. Although thymidine appears to enhance the cellular uptake of BCNU and the alkylation of DNA by both BCNU and 3'-CTNU, dealkylative repair, in the presence of thymidine, proceeds unhindered.[56] Recently,[57] 3'-CTNU was found to dramatically enhance the expression of the c-myb proto-oncogene in both a concentration- and time-dependent manner in L1210 leukemia cells, where the expression of the c-myc proto-oncogene was suppressed.

Montgomery and Thomas[58] also reported the synthesis of a series of methylnitrosourea analogues of the nucleosides adenosine, uridine, and cytidine as potential active site-directed irreversible enzyme inhibitors, but only 5'-deoxy-5'-(3-methyl-3-nitrosoureido)adenosine (78) was found to have weak cytotoxcity against H.Ep-2 cells in culture, and 3'-deoxy-3'-(3-methyl-3-nitrosoureido)adenosine (79) and 3'-deoxy-3'-(3-methyl-3-nitrosoureido)uridine (80) produced slight activity against the L1210 leukemia *in vivo*. Montgomery et al.[59] also synthesized nitrosoureido-nucleosides with the nitrosoureido function spaced an appropriate distance from C-5' as potential inhibitors of enzymes that phosphorylate purine and pyrimidine nucleotides, but only 1-(3,4-didehydro-2,4-dioxo-1(2H)-pyrimidinyl)-1-deoxy-N-[2-(3-methyl-3-nitrosoureido)ethyl]-β-D-ribofuranuronamide (81) was marginally active against the P388 leukemia *in vivo*, producing a 28% ILS at 200 mg/kg when given daily for 9 days.

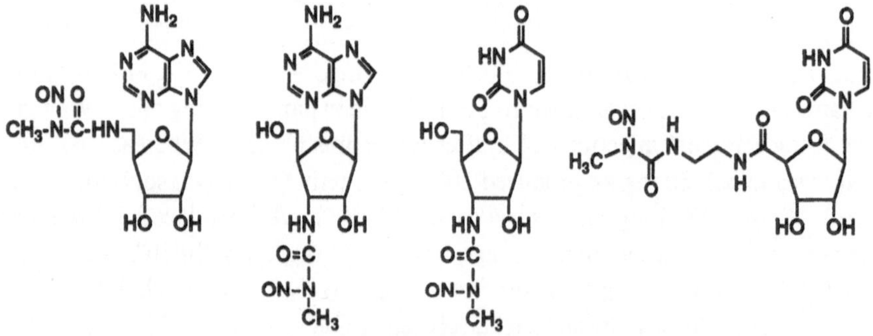

Figure 14. Structures of several nitrosourea analogues of adenosine and uridine.

NUCLEOSIDE PHOSPHORAZIRIDINE ANALOGUES

A variety of compounds possessing the bis(aziridinyl)phosphinoyl moiety have been synthesized as possible alkylating agents. Compounds of this type have been shown to have anticancer activity in several experimental systems.[60,61] Some of these bis(aziridinyl)phosphinoyl compounds appeared to potentiate the effect of X-irradiation against animal and human neoplasms and, in addition, have been reported to exhibit lower hematological toxicity in comparison to other kinds of alkylating agents.[62,63] To determine the impact of including a bis(aziridinyl)phosphinoyl group on thymidine antimetabolite, which could direct the alkylating activity of this moiety to the level of DNA on cytotoxcity, PP-bis(aziridinyl)phosphinic N-3'-thymidinylamide (**82**) and PP-bis(aziridinyl)phosphinic N-5'-thymidinylamide (**83**) have been synthesized in our laboratory.[64] Both compounds demonstrated significant anticancer activity against cultured L1210 leukemia cells. The 3'-bis(aziridinyl)phosphinic amide derivative of thymidine (**82**) was found to be about 11 times more potent than its 5'-counterpart (**83**), with estimated EC_{50} values of 0.6 and 7 μM, respectively.

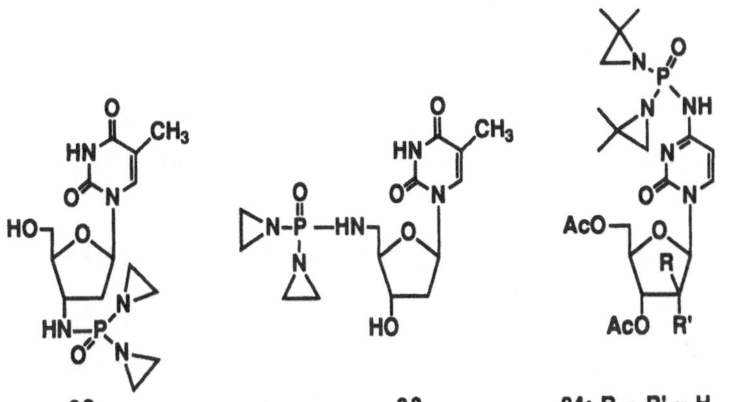

Figure 15. Structures of various nucleoside phosphoraziridine analogues.

Breiner et al.[65] also reported the syntheses of a series of 2,2-dimethylphosphoraziridine-type antitumor agents, in which the reactive bis(2,2-dimethyl-1-aziridinyl)phosphinyl group was linked through a carbamate or amide linkage to thymidine or cytosine nucleoside moieties. The cytosine derivatives, compounds **84** and **85**, showed significant activity against the P388 leukemia in mice, producing % T/C values of 188 and 206, respectively, at a dosage level of 256 mg/kg, injected ip on day 1 following implant of 10^6 P388 cells in a group of six CDF$_1$ female mice.

NUCLEOSIDE CYCLOPHOSPHAMIDE ANALOGUES

A number of cyclophosphamide derivatives of purine and pyrimidine nucleosides have been synthesized by Okruszek and Verkade[66] and our laboratory,[67] respectively. Among these derivatives, compounds **86a**, and **86b**, which are isomeric at phosphorus, were active against KB tumor cells in culture with EC_{50} values of 1.1 and 1.2 μg/mL, respectively.[66]

Compounds **87, 88**, and **90** also have significant inhibitory effects on the replication of L1210 cells *in vitro* with the respective EC$_{50}$ values of 1.2, 1.5, and 1.4 x 10^{-5} M. Compound **89**, which has a -NH- linkage instead of an oxygen at the 3'-position in the sugar moiety, resulted in the loss of anticancer activity. On the contrary, cyclophosphamide (cytoxan) has no cytotoxicity *in vitro*.[67] These findings illustrate the unusual biological properties of this new series of nucleoside cyclophosphamide analogues and suggest an additional or completely different mode of activation.

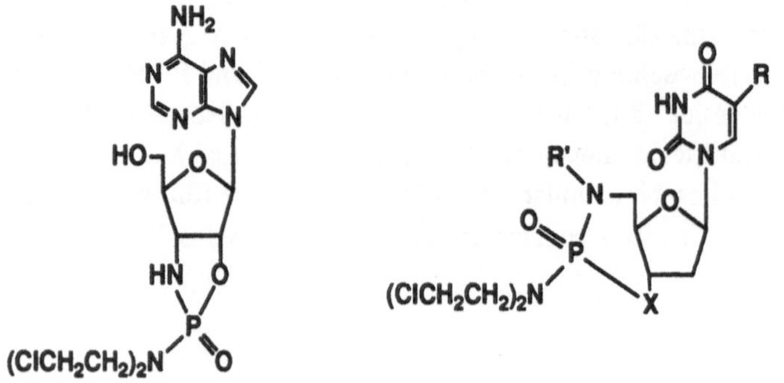

86a, b

87: R = CH$_3$, R' = H,   X = O
88: R = CH$_3$, R' = CH$_3$, X = O
89: R = CH$_3$, R' = H,   X = NH
90: R = I,   R' = H,   X = O

Figure 16. Structures of various nucleoside cyclophosphoramide analogues.

N-(3-OXO-1-PROPENYL)-SUBSTITUTED NUCLEOSIDE ANALOGUES

Bleomycin, when chelated with iron, reacts with DNA in the presence of molecular oxygen to form *N*-(3-oxo-1-propenyl)-substituted pyrimidines and purines, which are cytotoxic to a variety of tumor cells in culture. On the basis of these findings, Johnson et al.[68] reported the synthesis of a series of 3-(3-oxo-1-propenyl)-substituted derivatives of pyrimidine and purine bases, and pyrimidine nucleosides. 3-(3-Oxo-1-propenyl)thymidine (**91**), which was one of the most interesting compounds in this series, has shown significant cytotoxic activity against L1210 leukemia, Lewis lung carcinoma, B16 melanoma, DLD-1 human carcinoma, and HeLa cells in culture, as well as anticancer activity *in vivo* against the L1210 murine leukemia. 3-(3-Oxo-1-propenyl)thymidine (**91**) selectively blocks DNA synthesis in HeLa cells and inhibits the activities of thymidine kinase and DNA polymerase-α. Available evidence also suggests that 3-(3-oxo-1-propenyl)thymidine (**91**) readily undergoes addition-elimination in the presence of nucleophiles with the thymidine moiety acting as the leaving group.[69] Based on these concepts, a variety of 3-(3-oxo-1-propenyl) derivatives of 2',3'-dideoxy and carbocyclic pyrimidine nucleosides have been synthesized in our laboratory[70] as potential anticancer agents. In addition, 3-(3-oxo-1-propenyl) analogues of nucleosides with known chemotherapeutic efficacy, such as carbocyclic thymidine and 3'-deoxy-3'-fluorothymidine were also synthesized. The resulting products are visualized to have the potential to act in a synergistic manner to produce cytotoxicity by complementary inhibition. The 3-(3-oxo-1-propenyl) moiety presumably causes damage to DNA, thereby interfering with its function as a template in replication. The nucleoside antimetabolite inhibits the appropriate enzyme(s) after its conversion to the respective nucleotide, which serves in the envisioned action to minimize the repair of DNA lesions. Among the compounds evaluated, 3'-deoxy-3'-fluoro-3-(3-oxo-1-propenyl)thymidine (**92**), 3'-azido-3'-deoxy-3-(3-oxo-1-propenyl)thymidine (**36**), and (±)-1-[(1α,3β,4α)-3-hydroxy-4-(hydroxymethyl)cyclopentyl]-5-methyl-3-(3-oxo-1-propenyl)-2,4 (1*H*,3*H*)pyrimidinedione (**93**) were found to be the most active compounds against murine L1210, P388, S-180, and human CCRF-CEM lymphoblastic leukemia cell lines *in vitro* with EC$_{50}$ values of 0.5, 0.2, 0.1, and 0.3 µM; 1.2, 0.5, 1.0, and 1.0 µM; and 0.8, 0.7, 1.5, and 3.0 µM, respectively. Of these 3-(3-oxo-1-propenyl) derivatives, compound **92** and its parent compound, 3'-deoxy-3'-fluorothymidine, had similar EC$_{50}$ values: 0.2 and 0.2 µM (P388); 0.1 and 0.1 µM (S-180); 0.3 and 0.3 µM (CCRF-CEM), except for 0.5 and 2.0 µM (L1210), respectively. However, a comparison of their cytotoxicity at higher concentrations showed significant differences. For example, the percent of cell inhibition at 10 µM of 3-(3-oxo-1-propenyl) derivative **92** was 100% in all the given cell lines, whereas, the percent of cell inhibition at 100 µM of the parent compound, 3'-deoxy-3'-fluorothymidine, was only 84% (L1210), 94% (P388), 86% (S-180), and 75% (CCRF-CEM), respectively. These findings indicate that at 10 µM concentration, the 3-(3-oxo-1-propenyl)-derivative of 3'-deoxy-3'-fluorothymidine, compound **92**, is approximately 10 times more inhibitory on the growth of these cells line than the parent compound.

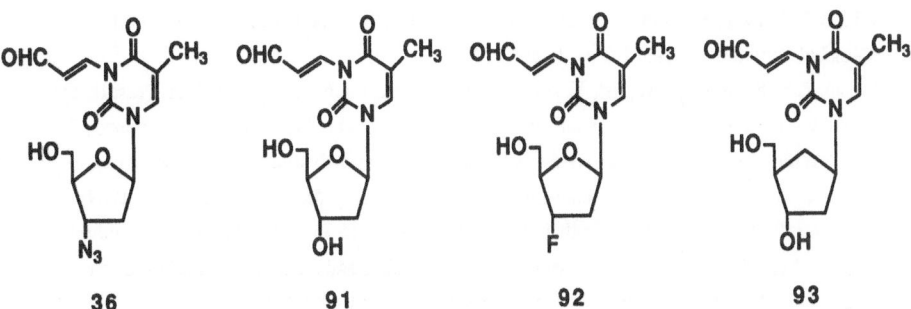

Figure 17. Structures of several *N*-(3-oxo-1-propenyl)-substituted nucleoside analogues.

REFERENCES

1. J.P. Horwitz, J. Chua, and M. Noel, Nucleosides. V. The monomesylates of 1-(2-deoxy-β-D-lyxofuranosyl)thymidine, *J. Org. Chem.* 29:2076 (1964).

2. H. Mitsuya, K.J. Weinhold, P.A. Furman, M.H. St. Clair, S. Nusinoff-Lehrman, R.C. Gallo, D. Bolognesi, D.W. Barry, and S. Broder, 3'-Azido-3'-deoxythymidine (BW A509U): an antiviral agent that inhibits the infectivity and cytopathic effect of human T-lymphotropic virus type III/lymphadenopathy-associated virus *in vitro*, *Proc. Natl. Acad. Sci. USA* 82:7096 (1985).

3. D.D. Richman, M.A. Fischl, M.H. Grieco, M.S. Gottlieb, P.A. Volberding, O.L. Laskin, J.M. Leedom, J.E. Groopman, D. Mildvan, M.S. Hirsch, G.G. Jackson, D.T. Durack, and S. Nusinoff-Lehrman, The toxicity of azidothymidine (AZT) in the treatment of patients with AIDS and AIDS-related complex, *N. Engl. J. Med.* 317:192 (1987).

4. T.S. Lin, M.S. Chen, C. McLaren, Y.S. Gao, I. Ghazzouli, and W.H. Prusoff, Synthesis and antiviral activity of various 3'-azido, 3'-amino, 2',3'-unsaturated, and 2',3'-dideoxy analogues of pyrimidine deoxyribonucleosides against retroviruses, *J. Med. Chem.* 30:440 (1987).

5. T.S. Lin and W.H. Prusoff, Synthesis and biological activity of several amino analogues of thymidine, *J. Med. Chem.* 21:109 (1978).

6. T.S. Lin, J.Y. Guo, R.F. Schinazi, C.K. Chu, J.N. Xiang, and W.H. Prusoff, Synthesis and antiviral activity of various 3'-azido analogues of pyrimidine deoxyribonucleosides against human immunodeficiency virus (HIV-1, HTLV-III/LAV), *J. Med. Chem.* 31:336 (1988).

7. R.F. Schinazi, C.K. Chu, M.-K. Ahn, J.-P. Sommadossi, and H. McClure, Cancer and Aids: approaches to prevention and therapy, in "Abbott-UCLA Symposium-Human Retroviruses," Keystone, CO, April 1-6, 1987.

8. R.F. Schinazi, C.K. Chu, M.-K. Ahn, J.-P. Sommadossi, and H. McClure, Selective *in vitro* inhibition of human immunodeficiency virus (HIV) replication by 3'-azido-2',3'-dideoxyuridine (CS-87), *J. Cell. Biochem.* 1987, 11D:74 (1987).

9. C.K. Chu, R.F. Schinazi, M.-K. Ahn, G.V. Ullas, Z.P. Gu, Structure-activity relationships of pyrimidine nucleosides as antiviral agents for human immunodeficiency virus type 1 in peripheral blood mononuclear cells, *J. Med. Chem.* 32:612 (1989).

10. T.S. Lin and W.R. Mancini, "Synthesis and antineoplastic activity of several 3'-azido- and 3'-amino pyrimidine deoxyribonucleosides", Abstracts, 184th ACS National Meeting, Kansas City, Missouri, September, 1982, MEDI 31.

11. T.S. Lin and W.R. Mancini, Synthesis and antineoplastic activity of 3'-azido and 3'-amino analogues of pyrimidine deoxyribonucleosides, *J. Med. Chem.* 26:544 (1983).

12. T.A. Krenitsky, G.A. Freeman, S.R. Shaver, L.M. Beacham III, S. Hurlbert, N.K. Cohn, L.P. Elwell, and J.W.T. Selway, 3'-Amino-2',3'-dideoxyribonucleosides of some pyrimidines: synthesis and biological activities, *J. Med. Chem.* 26:891 (1983).

13. T.S. Lin, Z.Y. Shen, E.M. August, V. Brankovan, H. Yang, I. Ghazzouli, and W.H. Prusoff, Synthesis and antiviral activity of several 2,5'-anhydro analogues of 3'-azido-3'-deoxythymidine, 3'-azido-2',3'-dideoxyuridine, 3'-azido-2',3'-dideoxy-5-halouridine, and 3'-deoxythymidine against human immunodeficiency virus and Rauscher-murine leukemia virus, *J. Med. Chem.* 32:1891 (1989).

14. M.V. Simpson, C.D. Chin, S.A. Keilbaugh, T.S. Lin, and W.H. Prusoff, Studies on the inhibition of mitochondrial DNA replication by 3'-azido-3'-deoxythymidine and other dideoxynucleoside analogs which inhibit HIV-1 replication, *Biochem. Pharmacol.* 38(7):1033 (1989).

15. T.S. Lin, M.C. Liu, E.M. August, E.M. Birks, and W.H. Prusoff, Syntheses of [2-^{14}C]2,5'-anhydro-3'-azido-3'-deoxythymidine and [2-^{14}C]2,5'-anhydro-3'-azido-2',3'-dideoxy-5-iodouridine: inhibitors of human immunodeficiency virus (HIV-1), *J. Labelled Compd. Radiopharm.* 29(12):1315 (1991).

16. H. Mitsuya and S. Broder, Inhibition of the *in vitro* infectivity and cytopathic effect of human T-lymphotropic virus type III/lymphadenopathy-associated virus (HTLV-III/LAV) by 2',3'-dideoxynucleosides, *Proc. Natl. Acad. Sci. USA* 83:1911 (1986).

17. T.S. Lin, R.F. Schinazi, and W.H. Prusoff, Potent and selective *in vitro* activity of 3'-deoxythymidin-2'-ene (3'-deoxy-2',3'-didehydrothymidine) against human immunodeficiency virus, *Biochem. Pharmacol.* 36(17):2713 (1987).

18. M.M. Mansuri, J.E. Starrett, I. Ghazzouli, M.J.M. Hitchcock, R.Z. Sterzycki, V. Brankovan, T.S. Lin, E.M. August, W.H. Prusoff, J.-P. Sommadossi, and J.C. Martin, 1-(2,3-Dideoxy-β-D-*glycero*-pent-2-enofuranosyl)thymine. A highly potent and selective anti-HIV agent, *J. Med. Chem.* 32:461 (1989).

19. T.S. Lin, R.F. Schinazi, M.S. Chen, E. Kinney-Thomas, and W.H. Prusoff, Antiviral activity of 2',3'-dideoxycytidin-2'-ene (2',3'-dideoxy-2',3'-didehydrocytidine) against human immunodeficiency virus *in vitro*, *Biochem. Pharmacol.* 36(3):311 (1987).

20. M. Baba, R. Pauwels, P. Herdewijn, E. De Clercq, J. Desmyter, and M. Vandeputte, Both 2',3'-dideoxythymidne and its 2',3'-unsaturated derivative (2',3'-dideoxythymidinene) are potent and selective inhibitors of human immunodeficiency virus replication *in vitro*, *Biochem. Biophys. Res. Commun.* 142(1):128 (1987).

21. C.K. Chu, R.F. Schinazi, B.H. Arnold, D.L. Cannon, B. Doboszewski, V.B. Bhadit, and Z. Gu, Comparative activity of 2',3'-saturated and unsaturated pyrimidine and purine nucleosides against human immunodeficiency virus type 1 in peripheral blood mononuclear cells, *Biochem. Pharmacol.* 37(19):3543 (1988).

22. J. Balzarini, P. Herdewijn, and E. De Clercq, Differential patterns of intracellular metabolism of 2',3'-didehydro-2',3'-dideoxythymidine and 3'-azido-2',3'-dideoxythymidine, two potent anti-human immunodeficiency virus compounds, *J. Biol. Chem.* 264(11):6127 (1989).

23. H. Hartmann, M.W. Vogt, A.G. Durno, M.S. Hirsch, G. Hunsmann, and F. Eckstein, Enhanced *in vitro* inhibition of HIV-1 replication by 3'-fluoro-3'-deoxythymidine compared to several other nucleoside analogs, *AIDS Res. Human Retrovir.* 4(6):457 (1988).

24. Y. Hamamoto, H. Nakashima, T. Matsui, A. Matasuda, T. Ueda, and N. Yamamoto, Inhibitory effect of 2',3'-didehydro-2',3'-dideoxynucleosides on infectivity, cytopathic effects, and replication of human immunodeficiency virus, *Antimicrob. Agents Chemother.* 31(6):907 (1987).

25. J. Balzarini, R. Pauwels, P. Herdewijn, E. De Clercq, D.A. Cooney, G.-L. Kang, M. Dalal, D.G. Johns, and S. Broder, Potent and selective anti-HTLV-III/LAV activity of 2',3'-dideoxycytidinene, the 2',3'-unsaturated derivative of 2',3'-dideoxycytidine, *Biochem. Biophys. Res. Commun.* 140(2):735 (1986).

26. H.T. Ho and M.J.M. Hitchcock, Cellular pharmacology of 2',3'-dideoxy-2',3'-didehydrothymidine, a nucleoside analog active against human immunodeficiency virus, *Antimicrob. Agents Chemother.* 33(6):844 (1989).

27. W.E. Harte, Jr, J.E. Starrett, Jr, J.C. Martin, M.M. Mansuri, Structural studies of the anti-HIV agent 2',3'-didehydro-2',3'-dideoxythymidine (D4T), *Biochem. Biophys. Res. Commun.* 175(1):298 (1991).

28. J.P. Horwitz, J. Chua, M.A. Da Rooge, M. Noel, and I.L. Klundt, Nucleosides. IX. The formation of 2',3'-unsaturated pyrimidine nucleosides *via* a novel β-elimination reaction, *J. Org. Chem.* 31:205 (1966).

29. T.S. Lin, Y.S. Gao, E.M. August, H.Y. Qian, and W.H. Prusoff, Synthesis of [2-^{14}C]3'-deoxythymidin-2'-ene (d4T) and [5-^{125}I]3'-azido-2',3'-dideoxy-5-iodouridine: potent inhibitors of human immunodeficiency virus (HIV-1), *J. Labelled Compd. Radiopharm.* 27(6):669 (1989).

30. J.P. Horwitz, J. Chua, M. Noel, and J.T. Donatti, Nucleosides. XI. 2',3'-Dideoxycytidine, *J. Org. Chem.* 32:817 (1967).

31. J. Balzarini, G.-J. Kang, M. Dalal, P. Herdewijn, E. De Clercq, S. Broder, and D.G. Johns, The anti-HTLV-III (anti-HIV) and cytotoxic activity of 2',3'-didehydro-2',3'-dideoxyribonucleosides: a comparison with their parental 2',3'-dideoxyribonucleosides, *Mol. Pharmacol.* 32:162 (1987).

32. G.I. Birnbaum, T.S. Lin, and W.H. Prusoff, Unusual structural features of 2',3'-dideoxycytidine, an inhibitor of the HIV (AIDS) virus, *Biochem. Biophys. Res. Commun.* 151(1):608 (1988).

33. G.I. Birnbaum, J. Giziewicz, T.S. Lin, and W.H. Prusoff, Structural features of 2',3'-dideoxy-2',3'-didehydrocytidine, a potent inhibitor of the HIV (AIDS) virus, *Nucleosides Nucleotides* 8(7):1259 (1989).

34. T.S. Lin, J.H. Yang, and Y.S. Gao, Synthesis of 2',3'-unsaturated and 2',3'-dideoxy analogs of 6-azapyrimidine nucleosides as potential anti-HIV agents, *Nucleosides Nucleotides* 9(1):97 (1990).

35. A. Rosowsky and N.N. Pai, Synthesis of the 2',3'-didehydro-2',3'-dideoxy and 2',3'-dideoxy derivatives of 6-azauridine and a new route to 2',3'-didehydro-2',3'-dideoxy-5-chlorouridine, *Nucleosides Nucleotides* 10(4):837 (1991).

36. T.S. Lin, Y.S. Gao, and W.R. Mancini, Synthesis and biological activity of various 3'-azido and 3'-amino analogues of 5-substituted pyrimidine deoxyribonucleosides, *J. Med. Chem.* 26:1691 (1983).

37. N. Miller and J.J. Fox, Nucleosides. XXI. Synthesis of some 3'-substituted 2',3'-dideoxyribonucleosides of thymine and 5-methylcytosine, *J. Org. Chem.* 29:1772 (1964).

38. P.H. Fischer, T.S. Lin, and W.H. Prusoff, Reversal of the cytotoxicity of 3'-amino-3'-deoxythymidine by pyrimidine deoxyribonucleosides, *Biochem. Pharmacol.* 28:991 (1979).

39. T.S. Lin, P.H. Fischer, and W.H. Prusoff, Effect of 3'-amino-3'-deoxythymidine on L1210 and P388 leukemias in mice, *Biochem. Pharmacol.* 31(1):125 (1982).

40. W.R. Mancini and T.S. Lin, Ribo- and deoxyribonucleoside effect on 3'-amino-2',3'-dideoxycytidine-induced cytotoxicity in cultured L1210 cells, *Biochem. Pharmacol.* 32(16):2427 (1983).

41. W.R. Mancini, M.S. Williams, and T.S. Lin, Specific inhibition of DNA biosynthesis induced by 3'-amino-2',3'-dideoxycytidine, *Biochemistry* 27:8832 (1988).

42. T.S. Lin, R.X. Zhou, K.J. Scanlon, W.F. Brubaker, Jr., J.J. Lee, K. Woods, C. Humphreys, and W.H. Prusoff, Synthesis and biological activity of several amino nucleoside-platinum(II) complexes, *J. Med. Chem.* 29:681 (1986).

43. J.-P. Macquet and J.-L. Butour, Platinum-amine compounds: importance of the labile and inert ligands for   their pharmacological activities toward L1210 leukemia cells, *J. Natl. Cancer Inst..* 70:899 (1983).

44. T.S. Lin and Y.S. Gao, Synthesis and biological activity of 5-(trifluoromethyl)- and 5-(pentafluoroethyl)-pyrimidine nucleoside analogues, *J. Med. Chem.* 26:598 (1983).

45. E.P. Mitchell and P.S. Schein, Contributions of nitrosoureas to cancer treatment, *Cancer Treat. Rep.* 70(1):31 (1986).

46. G.P. Wheeler, B.J. Bowdon, J.A. Grimsley, and H.H. Lloyd, Interrelationships of some chemical, physicochemical, and biological activities of several 1-(2-haloethyl)-1-nitroureas, *Cancer Res.* 34:194 (1974).

47. P.S. Schein, M.J. O'Connell, J. Blom, S. Hubbard, I.T. Magrath, P. Bergevin, P.H. Wiernik, J.L. Ziegler, and V.T. DeVita, Clinical antitumor activity and toxicity of streptozotocin (NSC-85998), *Cancer* 34:993 (1974).

48. T. Anderson, M.G. McMenamin, and P.S. Schein, Chlorozotocin, 2-[3-(2-chloroethyl)-3-nitrosureido]-D-glucopyranose, an antitumor agent with modified bone marrow toxicity, *Cancer Res.* 35:761 (1975).

49. T.S. Lin, P.H. Fischer, G.T. Shiau, and W.H. Prusoff, Antineoplastic agents. 1. Synthesis and anti-neoplastic activities of chloroethyl- and methylnitrosourea analogues of thymidine, *J. Med. Chem.* 21:130 (1978).

50. W.H. Prusoff, T.S. Lin, M.S. Chen, P.H. Fischer, W.R. Mancini, W. Brubaker, J.J. Lee, and K. Woods, Development of nitrosourea nucleosides as anticancer agents, in "Development of Target-Oriented Anticancer Drugs", Y.C. Cheng, B. Goz, and M. Minkoff, eds., Raven Press, New York (1983).

51. T.S. Lin, W.F. Brubaker, Jr., Z.H. Wang, S. Park, and W.H. Prusoff, Antineoplastic activity of 3'-(chloroethyl)nitrosourea analogues of 2'-deoxyuridine and 2'-deoxy-5-fluorouridine, *J. Med. Chem.* 29:862 (1986).

52. W.F. Brubaker, Jr. and W.H. Prusoff, Preparative and analytical high-performance liquid chromatographic methods in the synthesis and analysis of decomposition of nitrosourea nucleosides, *J. Chromatogr.* 322:455 (1985).

53. W.F. Brubaker, Jr., H.P. Zhao, and W.H. Prusoff, Measurement of carbamoylating activity of nitrosoureas and isocyanates by a novel high-pressure liquid chromatography assay, *Biochem. Pharmacol.* 35(14):2359 (1986).

54. T.S. Lin, P.H. Fischer, J.C. Marsh, and W.H. Prusoff, Antitumor activity of the 3'-chloroethylnitrosourea analog of thymidine and the prevention by co-administered thymidine of lethality but not of anticancer activity, *Cancer Res.* 42:1624 (1982).

55. T.S. Lin and W.H. Prusoff, Enhancement of the anticancer activity of bis(2-chloroethyl)nitrosourea in mice by coadministration of 2'-deoxyuridine, 2'-deoxycytidine, or thymidine, *Cancer Res.* 47:394 (1987).

56. E.M. August and W.H. Prusoff, Effect of thymidine on uptake, DNA alkylation, and DNA repair in L1210 cells treated with 1,3-bis(2-chloroethyl)-1-nitrosourea or 3'-[3-(2-chloroethyl)-3-nitrosoureido]-3'-deoxythymidine, *Cancer Res.* 48:4272 (1988).

57. X.K. Zhang, M.L Zucker, D.P. Huang, T.S. Lin, W.H. Prusoff, and J.F. Chiu, Alteration of cellular

oncogene expression in L1210 cells by a nitrosourea analog of thymidine, *Cancer Commun.* 3(4):119 (1991).

58. J.A. Montgomery and H.J. Thomas, Nitrosoureidonucleosides, *J. Med. Chem.* 22:1109 (1979).

59. J.A. Montgomery, H.J. Thomas, R.W. Brockman, and G.P. Wheeler, Potential inhibitors of nucleotide biosynthesis. 1. Nitrosoureidonucleosides. 2, *J. Med. Chem.* 24:184 (1981).

60. T.J. Bardos, Z.F. Chmielewicz, and P. Hebborn, Structure-activity relationships of alkylating agents in cancer chemotherapy, *Ann. N.Y. Acad. Sci.* 163:1006 (1969).

61. G.L. Wampler, W. Regelson, and T.J. Bardos, Absence of cross-resistance to alkylating agents in cyclophosphamide-resistant L1210 leukemia, *Eur. J. Cancer* 14:977 (1978).

62. G.L. Wampler, M. Kuperminc, and W. Regelson, Phase I study of ethylbis(2,2-dimethyl-1-aziridinyl)phosphinate (AB-163), *Cancer Chemother. Pharmacol.* 4:49 (1980).

63. G.L. Wampler, J.A. Wasum, and R. Belgrad, Radiation potentiating effect of ethyl bis(2,2-dimethyl-1-aziridinyl) phosphinate (AB-163) *Int. J. Rad. Oncol. Biol. Phys.* 5:1681 (1979).

64. T.S. Lin, G.L. Cai, and A.C. Sartorelli, Synthesis of bis(aziridinyl)phosphinic amide derivatives of thymidine as potential anticancer agents, *Nucleosides Nucleotides* 7(3):403 (1988).

65. R.G. Breiner, W.C. Rose, J.A. Dunn, J.E. MacDiarmid, and T.J. Bardos, Synthesis of new nucleoside phosphoraziridines as potential site-directed antineoplastic agents, *J. Med. Chem.* 33:2596 (1990).

66. A. Okruszek and J.G. Verkade, 2',3'-Bis(2-chloroethyl)aminophosphoryl-3'-amino-3'-deoxyadenosine: a cyclic nucleoside with antitumor activity, *J. Med. Chem.* 22:882 (1979).

67. T.S. Lin, P.H. Fischer, and W.H. Prusoff, Synthesis and antineoplastic activity of a novel series of phosphoramide mustard analogues of pyrimidine deoxyribonucelosides, *J. Med. Chem.* 23:1235 (1980).

68. F. Johnson, K.M.R. Pillai, A.P. Grollman, L. Tseng, and M. Takeshita, Synthesis and biological activity of a new class of cytotoxic agents: *N*-(3-oxoprop-1-enyl)substituted pyrimidines and purines, *J. Med. Chem.* 27:954 (1984).

69. A.P. Grollman, M. Takeshita, K.M.R. Pillai, and F. Johnson, Origin and cytotoxic properties of base propenals derived from DNA, *Cancer Res.* 45:1127 (1985).

70. T.S. Lin, J.Y. Guo, and X.H. Zhang, Synthesis and anticancer activity of 3-(3-oxoprop-1-enyl)-substituted analogues of carbocyclic pyrimidine nucleosides and 2',3'-dideoxy pyrimidine nucleosides, *Nucleosides Nucleotides* 9(7):923 (1990).

STRUCTURE-ACTIVITY CORRELATIONS OF

2',3'-DIDEOXY- AND 2',3'-DIDEHYDRO-

2',3'-DIDEOXYPYRIMIDINE NUCLEOSIDES

AS POTENTIAL ANTI-HIV DRUGS

Mohamed Nasr and Steven R. Turk

Division of AIDS
National Institute of Allergy and Infectious Diseases
National Institutes of Health
Bethesda, Maryland 20892

INTRODUCTION

The first compound approved for the clinical treatment of human immunodeficiency virus (HIV) was the nucleoside analog 3'-azido-3'-deoxythymidine (AZT; zidovudine).[1] Discovery of its antiviral activity prompted extensive evaluation of other nucleosides for anti-HIV efficacy and to date 2',3'-dideoxyinosine (ddI; didanosine) and 2',3'-dideoxycytidine (ddC; zalcitabine) additionally have been approved on a limited basis for clinical treatment of this virus. These nucleosides share a common mode of action, namely phosphorylation to the corresponding 5'-triphosphates which act as inhibitors of the virus-encoded reverse transcriptase.[2,3] Substantial effort also has been directed towards developing non-nucleoside drugs (e.g. protease inhibitors, tat antagonists) which inhibit viral targets other than reverse transcriptase.[4] The pursuit of more effective nucleoside analogs nonetheless remains an area of high interest to many investigators. This review provides detailed structure-activity data for two classes of nucleosides, the 2',3'-dideoxy- and 2',3'-didehydro-2',3'-dideoxypyrimidine nucleosides, in the hopes it will prove useful to investigators in identifying new synthetic target molecules while avoiding unnecessary duplication of previous synthetic efforts.

The present compilation extends our earlier publications reviewing dideoxynucleosides and includes more current and comprehensive information than previously reported.[5,6] The data contained in this review represent the published screening results of numerous investigators evaluating compounds for antiviral activity against HIV-1 and for cytotoxicity. All of the original data were generated using in vitro cell-based assays.

Nucleosides and Nucleotides as Antitumor and Antiviral Agents,
Edited by C.K. Chu and D.C. Baker, Plenum Press, New York, 1993

METHODS

Data for this review were abstracted from chemical databases compiled and maintained by the Division of AIDS. Information contained in our databases is acquired by surveillance of primary literature sources and consists of published data on compounds which have been evaluated pre-clinically against HIV or certain opportunistic infections associated with the acquired immunodeficiency syndrome (AIDS). The databases were established to help members of the Division of AIDS track therapeutic developments in the treatment of AIDS and to serve as an information base available to all researchers. Our HIV database contains entries for over 4000 different chemical entities, including over 800 nucleosides. We recently have established a similar database for compounds which have been evaluated against various AIDS-associated opportunistic infections. The latter database currently contains information on more than 2000 compounds.

Database analysis for this review was accomplished through substructure searching followed by retrieval and sorting of data using the software program ChemBase® (Molecular Design Limited, San Leandro, CA). Biological activity is reported using the parameters of ED_{50} (concentration to inhibit virus replication by 50% of control values), ID_{50} (concentration to decrease viability of uninfected cells by 50% of control values), and SI (selectivity index; ratio of ID_{50} to ED_{50}).

RESULTS

For practical reasons we regrettably were unable to present or acknowledge every dideoxy- and didehydrodideoxypyrimidine nucleoside which has been published. In our desire to focus only on primary screening data for purposes of structure-activity relationship analysis we likewise have presented only a fraction of the biological data available on many of the compounds mentioned in this review. We have tried to present the most potent and selective compounds reported in the literature as well as representatives from particular groups of analogs which showed moderate to little activity. References have been listed within the data tables to help point investigators to structural analogs which may not have been included in the tables.

Results have been tabulated into tables representing two major categories: 2',3'-dideoxynucleosides and 2',3'-didehydro-2',3'-dideoxynucleosides. The former category was subdivided into separate tables detailing thymidine, uridine, and cytidine analogs. Within each data table the entries were subgrouped according to the cell type in which the in vitro activity was evaluated. Data within each subgrouping were listed in descending order of selectivity index. Since a variety of experimental methodologies were used by the individual investigators cited in the tables, caution should be exercised when comparing data between different investigators and different cell types.

2',3'-Dideoxythymidine Nucleosides

Table 1 lists data for 45 dideoxythymidine nucleosides which were evaluated in one or more of six different cell lines. AZT was the most widely studied member of this group of compounds and represents one of the most potent and selective antiviral nucleosides ever identified in vitro. Its ED_{50} ranged between 0.002 and 2.4 μM while its SI varied from 19 to >50,000 among the cell lines reported here. Stereochemical change of the 3'-azido group from the α to β face of the sugar (3'-threo-AZT) diminished antiviral activity significantly in ATH8 cells but much less so in H9 cells.

Introduction of a 2'-hydroxyl group in the β-configuration to AZT (3'-Az-ara-T) led to an inactive compound. Reduction of the azido moiety of AZT to an amino group (3'-NH_2-ddT) also abolished activity. Efforts to replace the azido group with a variety of groups (e.g. pyrrole, cyano, isocyano, chloro, bromo, methoxy) were similarly unsuccessful. The notable exceptions were replacement with hydrogen (ddT) and fluorine (3'-F-ddT). The former compound was less potent than AZT and possessed an ED_{50} of 0.17 to 6 μM when tested in CEM, MT-4, and PBL cells. Surprisingly, the compound showed no significant activity when tested in ATH8 cells. Fluorine substitution led to a compound with high activity (ED_{50} ranged from 0.001 to 1.4 μM) but with lower selectivity (SI varied from 10 to 500) than that of AZT due to increased toxicity. The 2'-fluoro analogs of ddT and AZT (both α- and β-anomers) as well as the 3',3'-difluoro analog of ddT were either much less potent or selective than 3'-F-ddT.

A number of 4'-substituted thymidine nucleosides recently have been synthesized and show unexpected potency. The 4'-azido and 4'-azidomethyl analogs were both potent and selective, however insufficient data was provided in the preliminary publications to fully assess the selectivity of the other analogs. The azido function was not an essential requirement for activity by this series of compounds since 4'-methyl and 4'-hydroxymethyl analogs also showed activity. It is interesting to note that 4'-cyanothymidine was quite active whereas 4'-cyano-3'-deoxythymidine was not.

Modification of the 5'-OH group of AZT has been studied extensively. The 5'-hydrogenphosphonate (AZT-5'-H_2PO_3) and 5'-O-methanephosphonate (AZT-5'-$OCH_2PO_3Na_2$) are isosteres of AZT 5'-monophosphate and showed potent activity, albeit less than AZT itself. Many efforts to develop prodrugs of AZT have been made but only a few examples were included in this review (e.g. 5'-nicotinate and 5'-tyrosinate). In general, these prodrugs had in vitro potencies slightly lower than AZT. Nucleotide homo- and heterodimers, compounds in which two nucleosides are joined at the 5'-positions by a phosphate group (e.g. AZT-P-AZT) or other type of bridge, have met with similar success as prodrugs.

Replacement of the hydroxyl group on the 5'-carbon of thymidine by an azido gave a moderately active compound (5'-Az-5'-deoxy-T) whereas replacement by an amino gave an inactive one (5'-NH_2-5'-deoxy-T). Insertion of a methylene unit between the 4'- and 5'-carbons of AZT (5'-CH_2OH-5'-deoxy-AZT) eliminated the antiviral activity of AZT.

The nucleosides described above all share the normal β-D nucleoside configuration. Inversion of configuration at the 4'-carbon of the sugar leads to the α-L series of nucleosides, depicted in Figure 1. No significant antiviral activity was noted in either CEM or MT-4 cells for α-L-ara-T ($R_1=R_3=$OH; $R_2=$H), α-L-erythro-T ($R_1=$OH,$R_2=R_3=$H), or 3'-azido- and 3'-fluoro-α-L-threo-ddT ($R_1=R_3=$H, $R_2=$Az or F).[7,8]

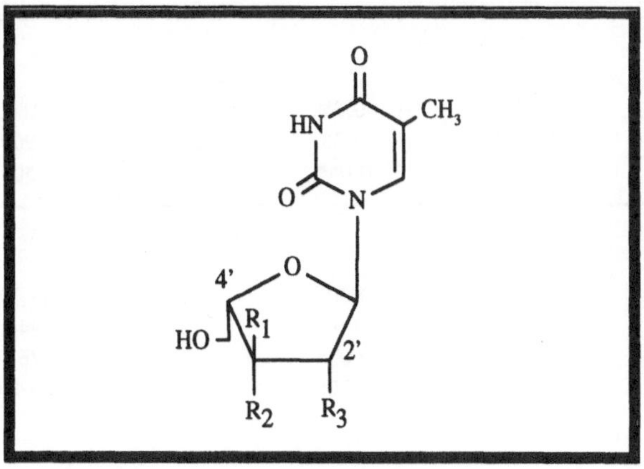

Figure 1. Chemical structure of 2',3'-substituted-α-L-thymidine nucleosides.

Table 1. Antiviral and cytotoxic evaluation of 2',3'-dideoxythymidine nucleosides.[1]

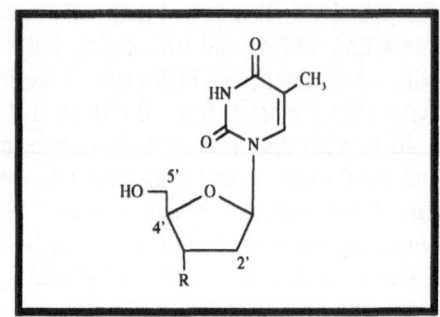

	R	ED_{50}	ID_{50}	SI	Ref
A3.01					
4'-Az-T	OH	0.01	>8	>800	14
4'-CH$_2$N$_3$-T	OH	0.45	>300	>660	15
4'-Me-T	OH	3.5	--	--	15
4'-CH$_2$OH-T	OH	12.5	--	--	15
4'-CN-T	OH	0.002	--	--	16
4'-CN-ddT	H	>200	--	--	16
ATH8					
ddT	H	100	>2000	>20	17
AZT	N$_3$	2.4	45	19	17
3'-F-ddT	F	1.4	15	10	17
3'-threo-AZT	β-N$_3$	>200	>500	2	18
3'-Cl-ddT	Cl	>500	>500	1	17
3'-Br-ddT	Br	500	180	<1	17
3'-MeO-ddT	OMe	100	88	<1	17
3'-Az-ara-T	N$_3$	>100	--	--	19
CEM					
AZT	N$_3$	0.003	>10	>3300	20
ddT	H	0.38	240	630	21
3'-MeS-ddT	SMe	NA	--	--	22
3'-MeSO-ddT	SOMe	NA	--	--	22
3'-MeSO$_2$-ddT	SO$_2$Me	NA	--	--	22
H9					
AZT	N$_3$	0.05	90	1800	23
3'-threo-AZT	β-N$_3$	0.26	250	960	24
3'-F-ddT	F	0.04*	20*	500	23
MT-4					
AZT	N$_3$	0.005	53	11,778	25
AZT-5'-H$_2$PO$_3$	N$_3$	0.03	102	3400	26
AZT-5'-Nicotinate	N$_3$	0.037	99	2676	27
(AZT)$_2$MePO$_3$	N$_3$	0.76*	>336*	>440	28

Table 1 (continued)

3'-F-ddT	F	0.001	0.197	197	29
AZT-5'-OCH$_2$PO$_3$Na$_2$	N$_3$	1.28	161	125	30
ddT	H	6	>625	>104	29
3'-F-5'-OCH$_2$PO$_3$Na$_2$-ddT	F	0.9	17	19	30
2'-F-ddT	H	53	>500	>9	11
3'-(Pyrrol-1-yl)-ddT	NC$_4$H$_4$	201	>500	>2	31
2'-ara-F-AZT	N$_3$	103	212	2	25
3'-CN-ddT	CN	>16	30	2	9
3'-NH$_2$-5'-OCH$_2$PO$_3$Na$_2$-ddT	NH$_2$	>500	>500	1	30
3'-Isocyano-ddT	NC	>0.8	0.88	1	32
3'-Acetylthio-ddT	SCOMe	27	29	1	33
3'-CH$_2$OH-ara-T	CH$_2$OH	>100	>100	1	34
2'-ara-F-ddT	H	>500	>500	1	35
3'-β-CN-ddT	β-CN	>250	>250	1	9
3',3'-F$_2$-ddT	F$_2$	39	48	1	36
3'-Ethynyl-ara-T	C≡C	>500	>500	1	33
3'-CN-ara-T	CN	>500	>500	1	33
5'-CH$_2$OH-5'-deoxy-AZT	N$_3$	NA	--	--	37
3'-Cyanamino-ddT	NHCN	NA	--	--	38

PBL

AZT	N$_3$	0.002	>100	>50,000	10
N^3-(3-Oxo-1-propenyl)-AZT	N$_3$	0.01	>100	>10,000	10
AZT-5'-Tyrosinate	N$_3$	0.11	300	2730	39
AZT-P-AZT	N$_3$	0.010	17	1650	40
AZT-P-ddI	N$_3$	0.11	>100	>900	40
ddT	H	0.17	>100	588	41
AZT-5'-Retinoate	N$_3$	0.20	25	125	39
5'-Az-5'-deoxy-T	OH	8.6	>100	>11	42
3'-I-ddT	I	46	>100	>2	42
5'-NH$_2$-5'-deoxy-T	OH	77	>100	1	42
3'-NH$_2$-ddT	NH$_2$	>100	>100	1	42

[1]The following symbols are used throughout all tables: *=data converted to μM from literature values in μg/ml; --=data not reported in reference; ara=arabinosyl; Az=azido; C=cytidine; d=2'-deoxy; dd=2',3'-dideoxy; d4=2',3'-didehydro-2',3'-dideoxy; Et=ethyl; Me=methyl; NA=not active; Pr=propyl; T=thymidine; U=uridine.

2',3'-Dideoxyuridine Nucleosides

Data for 39 dideoxyuridine analogs is presented in Table 2. The prototypic compound of this series is 2',3'-dideoxyuridine (ddU). In marked contrast to ddT, little antiviral activity was exhibited by this compound (ED$_{50}$ >200 μM). Just as AZT was more potent than ddT, 3'-azido-ddU (AzddU) possessed greater antiviral activity than did ddU. Among the cell lines examined here AzddU had an ED$_{50}$ ranging between 0.36 and 0.5 μM and a SI which varied between >400 and >2170.

In a pattern consistent with the thymidine analogs introduction of a 2'-hydroxyl group in the β-configuration to AzddU (Az-ara-U) or inversion of the stereochemical conformation of the 3'-azido group (3'-threo-AzddU) inactivated the lead compound. Reduction of the azido to an amino group (3'-NH$_2$-ddU) or replacement by hydroxymethyl, isocyano, or cyano[9] moieties similarly abolished activity. The 3'-I-ddU analog possessed moderate activity with an SI of 8, whereas 3'-F-ddU possessed higher potency and selectivity. The latter compound was more cytotoxic and thus less selective than AzddU.

Synthetic attempts to alter the substitution pattern of fluorine in the 2' and 3' positions were unsuccessful in improving activity or selectivity over that of 3'-F-ddU.

A number of 4'-substituted 2'-deoxyuridine nucleosides have been reported which possess potent antiviral activity, albeit of a lesser magnitude than the corresponding thymidine analogs. 4'-Az-dU, for example, had an ED_{50} of 0.8 μM and an $SI > 250$ whereas 4'-Az-T had respective values of 0.01 μM and > 800. Chlorination of the 5-position of 4'-Az-dU gave a compound with both increased potency and selectivity. Further exploration of this class of compounds to better define structure-activity relationships appears warranted.

Modification of the 5'-hydroxyl of ddU to the 5'-O-methanephosphonate (ddU-5'-$OCH_2PO_3Na_2$) did not significantly improve the poor activity of the parent compound. Replacement of the hydroxyl group at the 5'-position of AzddU by an isocyano substituent greatly diminished activity. Compounds in which the 5'-hydroxyl of 2'-deoxyuridine was substituted by amino, azido, or isocyano moieties were inactive.

Different substitution patterns at position 5 of the heterocyclic base of AzddU also have been examined extensively. The 5-methyl derivative is AZT and was a more potent and selective agent than AzddU. Replacement of the methyl group with a trifluoromethyl (5-CF_3-AzddU) abolished activity, as did replacement by methoxy, methoxymethyl, or propyl substituents. The 5-SCN, -NH_2, -OH, and -SMe analogs are mentioned in the references[10] as possessing low activity. 5-Et-AzddU oddly was quite active in PBL cells, as was 5-Et-ddU, whereas neither compound showed significant antiviral activity when assayed in MT-4 cells. Halogenation of the 5-position of AzddU gave compounds which were potent, albeit less so than AzddU itself. The combination of 5-chlorination and 3'-fluorination of ddU (3'-F-5-Cl-ddU) gave a compound that was similar in potency and selectivity to AzddU. Among this series of 5-halogenated derivatives of 3'-F-ddU, the 5-chloro derivative was the most active, the order of potency being $Cl > Br > I > F$ with the 5-F analog not showing appreciable activity.[11]

Table 2. Antiviral and cytotoxic evaluation of 2',3'-dideoxyuridine nucleosides.

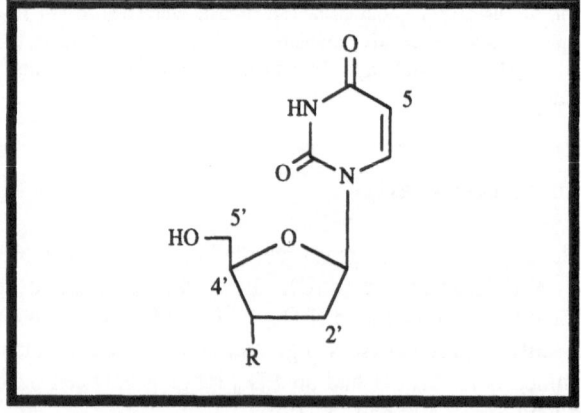

	R	ED_{50}	ID_{50}	SI	Ref
A3.01					
4'-Az-5-Cl-dU	OH	0.056	> 1000	> 17,900	14
4'-Az-dU	OH	0.8	> 200	> 250	14
4'-CN-dU	OH	0.06	--	--	16

Table 2. (continued)

ATH8					
ddU	H	>500	>500	1	17
Az-ara-U	N$_3$	>100	--	--	19
3'-Me-ara-U	Me	>100	--	--	19
C8166					
2'-ara,3'-F$_2$-ddU	F	>100	--	--	43
2'-ara-F-ddU	H	>100	--	--	43
H9					
AzddU	N$_3$	0.5	>200	>400	24
3'-F-ddU	F	0.5	>100	>200	24
3'-threo-AzddU	β-N$_3$	>100	>100	1	24
2'-Az-ddU	H	NA	--	--	44
MT-4					
3'-F-5-Cl-ddU	F	0.38	535	1408	11
AzddU	N$_3$	0.36	244	677	29
3'-F-ddU	F	0.04	16	400	29
5-Et-AzddU	N$_3$	64	418	>6	29
2'-F-AzddU	N$_3$	8.4	46	>5	35
ddU-5'-OCH$_2$PO$_3$Na$_2$	H	90	>500	>5	30
ddU	H	210	>625	>3	29
5'-isocyano-5'-deoxy-AzddU	N$_3$	>40	104	<3	45
3',5-F$_2$-ddU	F	50	50	1	46
3'-CH$_2$OH-ddU	CH$_2$OH	>100	>100	1	34
3'-isocyano-ddU	NC	>500	>500	1	32
5-Et-ddU	H	>625	>625	1	29
2',3'-F$_2$-ddU	F	>500	>500	1	35
5'-isocyano-5'-deoxy-dU	OH	>200	125	<1	45
5-CH$_2$OMe-ddU	H	NA	--	--	47
PBL					
AzddU	N$_3$	0.46	1000	2170	42
5-Et-AzddU	N$_3$	1.0	1000	1000	42
5-Br-AzddU	N$_3$	1.0	>100	>100	10
5-F-AzddU	N$_3$	4.8	>100	>21	10
5-Et-ddU	H	4.9	>100	20	42
3'-I-ddU	I	12.1	>100	8	42
3'-NH$_2$-ddU	NH$_2$	60	100	2	42
3'-NH$_2$-5-Et-ddU	NH$_2$	54.9	>100	2	42
5-Pr-AzddU	N$_3$	63	100	2	42
5-MeO-AzddU	N$_3$	70	>100	>1	10
3'-I-5-Et-ddU	I	86	100	1	42
5'-NH$_2$-5'-deoxy-5-Et-dU	OH	94.3	100	1	42
5'-NH$_2$-5'-deoxy-dU	OH	>100	>100	1	42
5-CF$_3$-AzddU	N$_3$	>100	>100	1	10
5'-Az-5'-deoxy-dU	OH	>100	>100	1	42

2',3'-Dideoxycytidine Nucleosides

Table 3 lists data for 35 different dideoxycytidine nucleosides. 2',3'-Dideoxycytidine (ddC) has been evaluated in five of the seven cell lines listed in the table. It possessed an ED_{50} which ranged between 0.001 and 0.24 μM and an SI which varied between 175 and >9000. In contrast to analogous studies in the thymidine and uridine series introduction of a 3'-azido substituent to ddC gave a compound (AzddC) which was less active (ED_{50} of 0.66 to 7.6 μM) and no more selective (SI of 21 to >600) than its parent.

Introduction of a 2'-hydroxyl group in the β-configuration to either ddC (3'-deoxy-ara-C) or AzddC (3'-Az-ara-C) gave inactive compounds. 3'-Halogenated ddC analogs exhibited less antiviral activity than did either ddC or AzddC. Addition of fluorine to the 2'-position of ddC from its β-face gave a surprisingly potent compound (2'-ara-F-ddC) with an ED_{50} concentration of 0.7 to 10 μM and an SI value of >5 to 710. Other patterns of fluorination at the 2'- and 3'-positions of ddC and AzddC did not improve activity.

As was the case with the thymidine and uridine nucleosides, the addition of an azido or cyano group to the 4'-position of 2'-deoxycytidine led to very potent compounds. 4'-Azidocytidine, in comparison, was inactive. The 2'-deoxycytidine analogs in general were as potent as the thymidine derivatives and more potent than the 2'-deoxyuridine analogs. 4'-Cyano-2'-deoxycytidine was significantly more potent than was its 2',3'-dideoxycytidine counterpart. This trend was also observed in the thymidine series.

The 5'-O-methanephosphonate of ddC (ddC-5'-OCH$_2$PO$_3$Na$_2$) was devoid of activity. The dinucleoside methylphosphonate of ddC [(ddC)$_2$MePO$_3$] showed only moderate activity, much less than exhibited by ddC itself.

Substitution of the 5-position of the pyrimidine ring has been examined, but has given somewhat confusing results. The introduction of a methyl group to the 5-position of ddC (5-Me-ddC) led to a 10-fold decrease in potency in PBL cells (ED_{50}=0.17μM). Substitution of methyl by an ethyl group led to a further reduction in potency. In contrast to findings in PBL cells, 5-Me-ddC was reported to be inactive in ATH8 cells.[12] It is unclear whether the reported lack of activity of this compound in ATH8 cells is due to a true lack of antiviral activity or to enhanced cytotoxicity in this cell line. Even though 5-Me-ddC itself apparently has not been evaluated in MT-4 cells two compounds containing this substitution pattern are listed in the table. Addition of the 5-methyl group to 3'-F-ddC increased antiviral potency (and cytotoxicity), whereas analogous methylation of 2'-ara-F-ddC effectively eliminated activity.

Addition of the 5-methyl group to AzddC increased the antiviral potency in both MT-4 and PBL cells. Neither AzddC[12] nor 5-Me-AzddC were active in ATH8 cells, however. Metabolic studies have suggested that 5-Me-AzddC is converted slowly via intracellular deamination to AZT.[13] Hydroxylation of the nitrogen in position 4 of the heterocycle of 5-Me-AzddC (N^4-OH-5-Me-AzddC) had little effect on potency in MT-4 cells, however methylation of the same N^4-position (N^4,5-Me$_2$-AzddC) diminished activity about 10-fold (ED_{50}=17.3 μM). In contrast, the monomethylated analog N^4-Me-AzddC was inactive.

Fluorination of the 5-position of both ddC and AzddC gave analogs with similar potencies as the parent compounds. Chlorination or bromination at the 5-position of ddC gave inactive compounds, whereas chlorination of AzddC (5-Cl-AzddC) did not significantly alter the antiviral activity but did decrease cytotoxicity somewhat. Similarly, the 5-chloro derivative of 3'-F-ddC (3'-F-5-Cl-ddC) significantly decreased the cytotoxicity exhibited by the parent compound.

2',3'-Didehydro-2',3'-dideoxypyrimidine Nucleosides

Data for 17 didehydrodideoxypyrimidine nucleosides are presented in Table 4.

Table 3. Antiviral and cytotoxic evaluation of 2',3'-dideoxycytidine nucleosides.

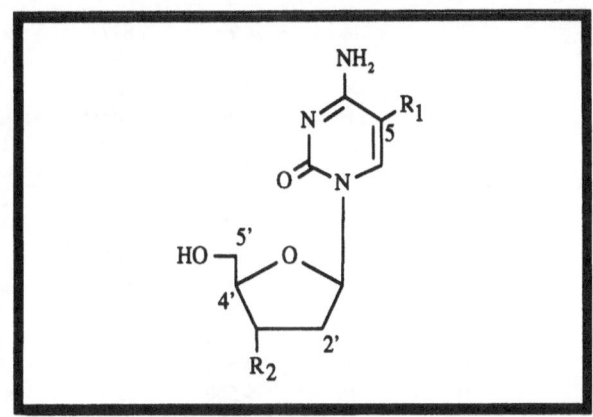

	R_1	R_2	ED_{50}	ID_{50}	SI	Ref
A3.01						
4'-Az-dC	H	OH	0.004	>0.21	>53	14
4'-Az-C	H	OH	>8	>8	1	14
4'-CN-dC	H	OH	0.003	--	--	16
4'-CN-ddC	H	H	1.7	--	--	16
ATH8						
ddC	H	H	0.20	35	175	17
5-F-ddC	F	H	<0.5	>50	>100	48
3'-F-ddC	H	F	8	>250	>31	17
5-Br-ddC	Br	H	>200	>200	1	48
5-Me-AzddC	Me	N_3	>100	>100	1	17
N^4,5-Me$_2$-AzddC	Me	N_3	>100	>100	1	17
3'-Deoxy-ara-C	H	H	>100	NR	NR	19
3'-Az-ara-C	H	N_3	>100	NR	NR	19
C8166						
ddC	H	H	0.125	1000	8000	43
2'-ara-F-ddC	H	H	0.70	500	710	43
2'-ara,3'-F$_2$-ddC	H	F	10	>100	>10	43
2'-ara,5-F$_2$-ddC	F	H	>100	NR	NR	43
CEM						
ddC	H	H	0.033*	6.2*	>180	49
2'-ara-F-ddC	H	H	10*	>100*	>10	49
H9						
2'-ara,3'-F$_2$-5-Me-ddC	Me	F	NA	--	--	50
2'-F-5-Me-ddC	Me	H	NA	--	--	50
MT-4						
ddC	H	H	0.24	313	1304	25
5-Me-AzddC	Me	N_3	1.8	>1000	>555	33
2'-ara-F-ddC	H	H	2.45	872	343	25

Table 3. (continued)

5-Cl-AzddC	Cl	N_3	9	877	97	51
N^4-OH-5-Me-AzddC	Me	N_3	1.5	92	61	33
N^4,5-Me$_2$-AzddC	Me	N_3	17.3	>1000	>58	33
3'-F-5-Cl-ddC	Cl	F	26	>1000	>38	51
AzddC	H	N_3	7.6	160	21	29
(ddC)$_2$MePO$_3$	H	H	26*	>410*	>16	28
3'-F-5-Me-ddC	Me	F	1.7	7.7	5	11
3'-F-ddC	H	F	16	26	2	29
N^4-Me-AzddC	H	N_3	605	>1000	2	33
ddC-5'-OCH$_2$PO$_3$Na$_2$	H	H	>500	>500	1	30
5-Cl-ddC	Cl	H	>500	>500	1	52
2'-ara-F-5-Me-ddC	Me	H	>500	>500	1	35
2',3'-F$_2$-ddC	H	F	>500	>500	1	35
2'-F-AzddC	H	N_3	>500	243	<1	35
5-CH$_2$OMe-ddC	CH$_2$OMe	H	NA	--	--	47
5-CH$_2$OPr-ddC	CH$_2$OPr	H	NA	--	--	47

PBL

ddC	H	H	0.011	>100	>9000	41
5-Me-AzddC	Me	N_3	0.08	>200	>2500	42
AzddC	H	N_3	0.66	>400	>600	42
5-Me-ddC	Me	H	0.17	>100	>580	41
5-F-AzddC	F	N_3	1.0	>100	>100	10
5-Et-ddC	Et	H	4.9	>100	>20	41
3'-I-ddC	H	I	12	>100	>8	41
2'-ara-F-ddC	H	H	<8.7*	48*	>5	49

2',3'-Didehydro-2',3'-dideoxythymidine (d4T) was more active and selective than its saturated counterpart ddT, having an ED$_{50}$ which ranged from 0.009 to 4.1 μM and an SI which varied from 27 to >7700. Replacement of the oxygen at position 4 of the heterocycle by a sulfur (4-thio-d4T) diminished activity 50-fold. The 2'-fluoro derivative of d4T was inactive, whereas 3'-F-d4T retained moderate activity, albeit less in magnitude than that exhibited by 3'-F-ddT. The 4'-hydroxymethyl and 4'-azidomethyl congeners of d4T were inactive, in contrast to the marked activity shown by the analogously substituted thymidines.

No activity was noted for d4U or any of its listed derivatives (e.g. 2'-F-d4U and 5-Cl-d4U). It was somewhat surprising to find that 5-Et-d4U was inactive in PBL cells, given that 5-Et-ddU showed moderate activity (ED$_{50}$=4.9 μM) in the same cell line.

Equally or slightly more active than ddC was its unsaturated counterpart d4C. The latter compound also was more cytotoxic than ddC, however, resulting in d4C having a lower SI than ddC. The 2'- and 3'-fluoro analogs of d4C were less active than the parent compound. Chlorination of the 5-position of d4C gave a compound much less active than d4C whereas methylation of the same position gave a compound nearly as active as d4C. Both substituted compounds were at least 10-fold more active than the saturated ddC analogs.

Table 4. Antiviral and cytotoxic evaluation of 2',3'-didehydro-2',3'-dideoxypyrimidine nucleosides.

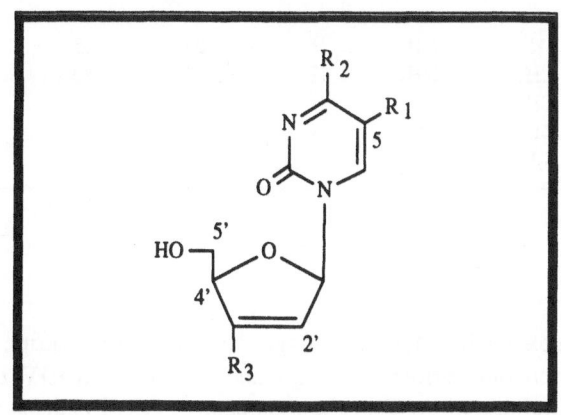

	R_1	R_2	R_3	ED_{50}	ID_{50}	SI	Ref
A3.01							
4'-CH$_2$OH-d4T	CH$_3$	OH	H	>200	--	--	15
4'-CH$_2$N$_3$-d4T	CH$_3$	OH	H	>200	--	--	15
ATH8							
d4C	H	NH$_2$	H	0.30	30	100	17
d4T	CH$_3$	OH	H	4.1	110	27	17
5'-NH$_2$-d4C	H	NH$_2$	H	260	>500	>2	17
d4U	CH$_3$	OH	CN	>100	107	<1	17
C8166							
2'-F-d4T	CH$_3$	OH	H	100	>1000	>10	43
2'-F-d4C	H	NH$_2$	H	10	30	3	43
2'-F-d4U	H	OH	H	>100	--	--	43
CEM							
d4T	CH$_3$	OH	H	0.8	235	290	20
4-thio-d4T	CH$_3$	SH	H	39	>430	>11	20
H9							
3'-F-d4C	H	NH$_2$	F	6.2*	440*	71	23
2'-F-d4C	H	NH$_2$	H	3.1*	26*	>8	23
2'-F-d4T	CH$_3$	OH	H	NA	--	--	50
MT-4							
d4T	CH$_3$	OH	H	0.01	1.2	120	53
d4C	H	NH$_2$	H	0.13	7.9	61	53
3'-F-d4T	CH$_3$	OH	F	11	240	22	11
5-Cl-d4C	Cl	NH$_2$	H	22	185	8	52
5-Cl-d4U	Cl	OH	H	>20	44	2	54
d4U	H	OH	H	>125	27	<1	53
3'-CN-d4T	CH$_3$	OH	CN	>50	24	<1	9

Table 4. (continued)

PBL

d4C	H	NH$_2$	H	0.005	65	13,000	41
5-Me-d4C	CH$_3$	NH$_2$	H	0.009	80	>8800	41
d4T	CH$_3$	OH	H	0.009	70	>7700	42
d4U	CH$_3$	OH	H	68	>100	1	42
5-Et-d4U	C$_2$H$_5$	OH	H	76	>100	1	42

CONCLUSION

Pyrimidine dideoxynucleosides which target HIV reverse transcriptase have been the cornerstone of the ongoing effort to design and develop anti-HIV drugs. Since the discovery of the usefulness of AZT in the treatment of AIDS patients a variety of new pyrimidine nucleosides have been synthesized. As evidenced by this review, many successfully inhibit virus replication and lessen the pathogenic effects of HIV in vitro and some eventually may have clinical utility either as single agents or in combination with other antiviral agents. The emergence of AZT-resistant strains of HIV has spurred exploration of compounds with different modes of action than AZT. The discovery that potent antiviral activity could be exhibited by 4'-substituted-2'-deoxynucleosides which retain their 3'-OH functionality and the finding that substitution of the 5'-position by groups other than OH did not automatically preclude antiviral activity suggest that these compounds may inhibit virus replication by mechanisms other than established modes of inhibition of reverse transcriptase. The recent elucidation of the crystal structure of HIV reverse transcriptase and the discovery of non-nucleoside reverse transcriptase inhibitors may assist in a better understanding of the structure-activity correlations and modes of inhibition of these and other nucleoside inhibitors.

ACKNOWLEDGMENTS

We wish to thank Perlita M. Liwanag for her expert help in extracting data from our databases to prepare the tables contained herein and for her assistance in the preparation of this manuscript.

REFERENCES

1. H. Mitsuya, K.J. Weinhold, P.A. Furman, M.H. St. Clair, S. Nusinoff Lehrman, R.C. Gallo, D. Bolognesi, D.W. Barry, and S. Broder, 3'-Azido-3'-deoxythymidine (BW A509U): An antiviral agent that inhibits the infectivity and cytopathic effect of human T-lymphotropic virus type III/lymphadenopathy-associated virus in vitro, *Proc. Natl. Acad. Sci. USA* 82:7096 (1985).
2. P.A Furman, J.A. Fyfe, M. H. St. Clair, K. Weinhold, J.L. Rideout, G.A. Freeman, S. Nusinoff Lehrman, D.P. Bolognesi, S. Broder, H. Mitsuya, and D.W. Barry, Phosphorylation of 3'-azido-3'-deoxythymidine and selective interaction of the 5'-triphosphate with human immunodeficiency virus reverse transcriptase, *Proc. Natl. Acad. Sci. USA* 83:8333 (1986).
3. M.C. Stanes and Y.C. Cheng, Inhibition of human immunodeficiency virus reverse transcriptase by 2',3'-dideoxynucleoside triphosphates: Template dependence, and combination with phosphono-formate, *Virus Genes* 2:241 (1988).
4. K.J. Connolly and S.M. Hammer, Antiretroviral therapy: Strategies beyond single-Agent reverse transcriptase inhibition, *Antimicrob. Agents Chemother.* 36:509 (1992).
5. M. Nasr, C. Litterst, and J. McGowan, Computer-assisted structure-activity correlations of dideoxynucleoside analogs as potential anti-HIV drugs, *Antiviral Res.* 14:125 (1990).

6. M. Nasr, J. Cradock, and M.I. Johnston, Computer-assisted structure-activity correlations of halo-dideoxynucleoside analogs as potential anti-HIV drugs, AIDS Res. Human Retroviruses 8:135 (1992).

7. C. Génu-Dellac, G. Gosselin, F. Puech, J.-C. Henry, A.-M. Aubertin, G. Obert, A. Kirn, and J.-L. Imbach, Systematic synthesis and antiviral evaluation of α-L-arabinofuranosyl and 2'-deoxy-α-L-erythro-pentofuranosyl nucleosides of the five naturallly occurring nucleic acid bases, Nucleosides & Nucleotides 10:1345 (1991).

8. C. Génu-Dellac, G. Gosselin, A.-M. Aubertin, G. Obert, A. Kirn, and J.-L. Imbach, 3'-Substituted thymine α-L-nucleoside derivatives as potential antiviral agents: Synthesis and biological evaluation, Antiviral Chem. Chemother. 2:83 (1991).

9. M.-J. Camarasa, A. Diaz-Ortiz, A. Calvo-Mateo, F.G. De las Heras, J. Balzarini, and E. De Clercq, Synthesis and antiviral activity of 3'-C-cyano-3'-deoxynucleosides, J. Med. Chem. 32:1732 (1989).

10. T.-S. Lin, J.-Y. Guo, R.F. Schinazi, C.K. Chu, J.-N. Xiang, and W.H. Prusoff, Synthesis and antiviral activity of various 3'-azido analogues of pyrimidine deoxyribonucleosides against human immunodeficiency virus (HIV-1, HTLV-III/LAV), J. Med. Chem. 31:336 (1988).

11. A. Van Aerschot, P. Herdewijn, J. Balzarini, R. Pauwels, and E. De Clercq, 3'-Fluoro-2',3'-dideoxy-5-chlorouridine: Most selective anti-HIV-1 agent among a series of new 2'- and 3'-Fluorinated 2',3'-dideoxynucleoside analogues, J. Med. Chem. 32:1743 (1989).

12. H. Mitsuya and S. Broder, Toward the rational design of antiretroviral therapy for human immunodeficiency virus (HIV) infection, in: "The Human Retroviruses," R.C. Gallo and G. Jay, eds., Academic Press, San Diego (1991).

13. R.F. Schinazi, C.K. Chu, B.F. Eriksson, J.P. Sommadossi, K.J. Doshi, F.D. Boudinot, B. Oswald, and H.M. McClure, Antiretroviral activity, biochemistry, and pharmacokinetics of 3'-azido-2',3'-dideoxy-5-methylcytidine, Ann. NY Acad. Sci. 616:385 (1990).

14. H. Maag, R.M. Rydzewski, M.J. McRoberts, D. Crawford-Ruth, J.P.H. Verheyden, and E.J. Prisbe, Synthesis and anti-HIV activity of 4'-azido- and 4'-methoxynucleosides, J. Med. Chem. 35:1440 (1992).

15. C. O-Yang, W. Kurz, E.M. Eugui, M.J. McRoberts, J.P.H. Verheyden, L.J. Kurz, and K.A.M. Walker, 4'-Substituted nucleosides as inhibitors of HIV: An unusual oxetane derivative, Tetrahedron Lett. 33:41 (1992).

16. C. O-Yang, H.Y. Wu, E.B. Fraser-Smith, and K.A.M. Walker, Synthesis of 4'-cyanothymidine and analogs as potent inhibitors of HIV, Tetrahedron Lett. 33:37 (1992).

17. P. Herdewijn, J. Balzarini, E. Clercq, R. Pauwels, M. Baba, S. Broder, and H. Vanderhaege, 3'-Substituted 2',3'-Dideoxynucleoside analogues as potential anti-HIV (HTLV-III/LAV) agents, J. Med. Chem. 30:1270 (1987).

18. J. Balzarini, P. Herdewijn, R. Pauwels, S. Broder, and E. De Clercq, α,β- and β,γ-Methylene 5'-phosphonate derivatives of 3'-azido-2',3'-dideoxythymidine-5'-triphosphate, Biochem. Pharmacol. 37:2395 (1988).

19. T.R. Webb, H. Mitsuya, and S. Broder, 1-(2,3-Anhydro-ß-D-lyxofuranosyl)cytosine derivatives as potential inhibitors of the human immunodeficiency virus, J. Med. Chem. 31:1475 (1988).

20. E. Palomino, B.R. Meltsner, D. Kessel, and J.P. Horwitz, Synthesis and in vitro evaluation of some modified 4- thiopyrimidine nucleosides for prevention or reversal of AIDS-associated neurological disorders, J. Med. Chem. 33:258 (1990).

21. T.-S. Lin, Z.-Y. Shen, E.M. August, V. Brankovan, H. Yang, I. Ghazzouli, and W.H. Prusoff, Synthesis and antiviral activity of several 2,5'-anhydro analogues of 3'-azido-3'-deoxythymidine, 3'-azido-2',3'-dideoxyuridine, 3'-azido-2',3'-dideoxy-5-halouridines, and 3'-deoxythymidine against human immunodeficiency virus and Rauscher-murine leukemia virus, J. Med. Chem. 32:1891 (1989).

22. M.M. Mansuri, J.A. Wos, and J.C. Martin, A short synthesis of 3'-(methylsulfinyl)-3'-deoxythymidine and related analogues, Nucleosides & Nucleotides 8:1463 (1989).

23. R. Koshida, S. Cox, J. Harmenberg, G. Gilljam, and B. Wahren, Structure-activity relationships of fluorinated nucleoside analogs and their synergistic effect in combination with phosphonoformate against human immunodeficiency virus type 1, Antimicrob. Agents Chemother. 33:2083 (1989).

24. H. Bazin, J. Chattopadhyaya, R. Dateman, A.-C. Ericson, G. Gilliam, N.G. Johansson, J. Hansen, R. Koshida, K. Moelling, B. Oberg, G. Remaud, G. Stening, L. Vrang, B. Wahren, and J.C. Wu, An analysis of the inhibition of replication of HIV and MuLV by some 3'-blocked pyrimidine analogs, Biochem. Pharmacol. 38:109 (1989).

25. K.A. Watanabe, K. Harada, J. Zeidler, J. Matulic-Adamic, K. Takahashi, W.-Y. Ren, L.-C. Cheng, J.J. Fox, T.-C. Chou, Q.-Y. Zhu, B. Polsky, J.W.M. Gold, and D. Armstrong, Synthesis and anti-HIV-1 activity of 2'-"up"-fluoro analogues of active anti-AIDS nucleosides 3'-azido-3'-deoxythymidine (AZT) and 2',3'-dideoxycytidine (DDC), *J. Med. Chem.* 33:2145 (1990).

26. N.B. Tarussova, M.K. Kukhanova, A.A. Krayevsky, E.K. Karamov, V.V. Lukashov, G.B. Kornilayeva, M.A. Rodina, and G.A. Galegov, Inhibition of human immunodeficiency virus (HIV) production by 5'-hydrogenphosphonates of 3'-azido-2',3'-dideoxynucleosides, *Nucleosides & Nucleotides* 10:351 (1991).

27. P.F. Torrence, J.-E. Kenjo, K. Lesiak, J. Balzarini, and E. De Clercq, Aids dementia: Synthesis and properties of a derivative of 3'-azido-3'-deoxythymidine (AZT) that may become "locked" in the central nervous system, *FEBS Lett.* 234:135 (1988).

28. F. Puech, G. Gosselin, J. Balzarini, S.S. Good, L. Rideout, E. De Clercq, and J.-L. Imbach, Synthesis and biological evaluation of dinucleoside methylphosphonates of 3'-azido-3'-deoxythymidine and 2',3'-dideoxycytidine, *Antiviral Res.* 14:11 (1990).

29. J. Balzarini, M. Baba, R. Pauwels, P. Herdewijn, and E. De Clercq, Anti-retrovirus activity of 3'-fluoro- and 3'-azido-substituted pyrimidine 2',3'-dideoxynucleoside analogues, *Biochem. Pharmacol.* 37:2847 (1988).

30. L. Jie, A. Van Aerschot, J. Balzarini, G. Janssen, R. Busson, J. Hoogmartens, E. De Clercq, and P. Herdewijn, 5'-O-Phosphonomethyl-2',3'-dideoxynucleosides: Synthesis and anti-HIV activity, *J. Med. Chem.* 33:2481 (1990).

31. P. Wigerinck, A. Van Aerschot, G. Janssen, P. Claes, J. Balzarini, E. De Clercq, and P. Herdewijn, Synthesis and antiviral activity of 3'-heterocyclic substituted 3'-deoxythymidines, *J. Med. Chem.* 33:868 (1990).

32. J. Hiebl, E. Zbiral, J. Balzarini, and E. De Clercq, Synthesis, antiretrovirus effects, and phosphorylation kinetics of 3'-isocyano-3'-deoxythymidine and 3'-isocyano-2',3'-dideoxyuridine, *J. Med. Chem.* 33:845 (1990).

33. P. Herdewijn, J. Balzarini, M. Baba, R. Pauwels, A. Van Aerschot, G. Janssen, and E. De Clercq, Synthesis and anti-HIV activity of different sugar-modified pyrimidine and purine nucleosides, *J. Med. Chem.* 31:2040 (1988).

34. M.J. Bamford, P.L. Coe, and R.T. Walker, Synthesis and antiviral activity of 3'-deoxy-3'-C-hydroxymethyl nucleosides, *J. Med. Chem.* 33:2494 (1990).

35. A. Van Aerschot, J. Balzarini, R. Pauwels, L. Kerremans, E. De Clercq, and P. Herdewijn, Influence of fluorination of the sugar moiety on the anti-HIV-1 activity of 2',3'-dideoxynucleosides, *Nucleosides & Nucleotides* 8:1121 (1989).

36. D.E. Bergstrom, A.W. Mott, E. DeClercq, J. Balzarini, and D.J. Swartling. 3',3'-Difluoro-3'-deoxythymidine: Comparison of anti-HIV activity to 3'-fluoro-3'-deoxythymidine, *J. Med. Chem.* 35:3369 (1992).

37. G.A. Freeman, J.L. Rideout, M.H. St. Clair, G.B. Roberts, and P.A. Sherman, Antiviral activity of 5'-modified analogs of AZT against HIV-1 and 2, *Antiviral Res.* 15(Suppl. 1):54 (1991).

38. M.S. Motawia, E.B. Pedersen, and C.M. Nielsen, Synthesis of N-substituted 3'-amino-3'-deoxythymidines and their biological evaluation against HIV, *Arch. Pharm.* (Weinheim) 323: 971 (1990).

39. S.K. Aggarwal, S.R. Gogu, S.R.S. Rangan, and K.C. Agrawal, Synthesis and biological evaluation of prodrugs of zidovudine, *J. Med. Chem.* 33:1505 (1990).

40. R.F. Schinazi, J.-P. Sommadossi, V. Saalmann, D.L. Cannon, M.-Y. Xie, G.C. Hart, G.A. Smith, and E.F. Hahn, Activities of 3'-azido-3'-deoxythymidine nucleotide dimers in primary lyphocytes infected with human immunodeficiency virus type I, *Antimicrob. Agents Chemother.* 34:1061 (1990).

41. C.K. Chu, R.F. Schinazi, B.H. Arnold, D.L. Cannon, B. Doboszewski, V.B. Bhadti, and Z. Gu, Comparative activity of 2',3'-saturated and unsaturated pyrimidine and purine nucleosides against human immunodeficiency virus type I in peripheral blood mononuclear cells, *Biochem. Pharmacol.* 37:3543 (1988).

42. C.K. Chu, R.F. Schinazi, M.K. Ahn, G.V. Ullas, and Z.P. Gu, Structure-activity relationships of pyrimidine nucleosides as antiviral agents for human immunodeficiency virus type I in peripheral blood mononuclear cells, *J. Med. Chem.* 32:612 (1989).

43. J.A. Martin, D.J. Bushnell, I.B. Duncan, S.J. Dunsdon, M.J. Hall, P.J. Machin, J.H. Merrett, K.E.B. Parkes, N.A. Roberts, G.J. Thomas, S.A. Galpin, and D. Kinchington, Synthesis and antiviral activity of monofluoro and difluoro analogues of pyrimidine deoxyribonucleosides against human immunodeficiency virus (HIV-1), *J. Med. Chem.* 33:2137 (1990).

44. J.A. Warshaw, and K.A. Watanabe, 2'-Azido-2',3'-dideoxypyrimidine nucleosides. Synthesis and antiviral activity against human immunodeficiency virus, *J. Med. Chem.* 33:1663 (1990).

45. J. Hiebl, E. Zbiral, J. Balzarini, and E. De Clercq, Synthesis and antiretrovirus properties of 5'-isocyano-5'-deoxythmidine, 5'-isocyano-2',5'-dideoxyuridine, 3'-azido-5'-isocyano-3',5'-dideoxythymidine, and 3'-azido-5'-isocyano-2',3',5'-trideoxyuridine, *J. Med. Chem.* 34:1426 (1991).

46. E. Matthes, C. Lehmann, M. von Janta-Lipinski, and D. Scholz, Inhibition of HIV-replication by 3'-fluoro-modified nucleosides with low cytotoxicity, *Biochem. Biophys. Res. Commun.* 165:488 (1989).

47. A.E.-S. Abdel-Megied, E.B. Pedersen, and C.M. Nielsen, Synthesis of 2',3'-dideoxynucleosides from 5-alkoxymethyluracils, *Monatsh. Chem.* 122:59 (1991).

48. C.-H. Kim, V.E. Marquez, S. Broder, H. Mitsuya, and J.S. Driscoll, Potential anti-AIDS drugs. 2',3'-Dideoxycytidine analogues, *J. Med. Chem.* 30:862 (1987).

49. R.Z. Sterzycki, I. Ghazzouli, V. Brankovan, J.C. Martin, and M.M. Mansuri, Synthesis and anti-HIV activity of several 2'-fluoro-containing pyrimidine nucleosides, *J. Med. Chem.* 33:2150 (1990).

50. J.-T. Huang, L.-C. Chen, L. Wang, M.-H. Kim, J.A. Warshaw, D. Armstrong, Q.-Y. Zhu, T.-C. Chou, K.A. Watanabe, J. Matulic-Adamic, T.-L. Su, J.J. Fox, B. Polsky, P.A. Baron, J.W.M. Gold, W.D. Hardy, and E. Zuckerman, Fluorinated sugar analogues of potential anti-HIV-1 nucleosides, *J. Med. Chem.* 34:1640 (1991).

51. A. Van Aerschot, D. Everaert, J. Balzarini, K. Augustyns, L. Jie, G. Janssen, O. Peeters, N. Blaton, C. De Ranter, E. De Clercq, and P. Herdewijn, Synthesis and anti-HIV evaluation of 2',3'-dideoxyribo-5-chlororpyrimidine analogues: Reduced toxicity of 5-chlorinated 2',3'-dideoxynucleosides, *J. Med. Chem.* 33:1833 (1990).

52. J. Balzarini, A. Van Aerschot, P. Herdewijn, and E. De Clercq, 2',3'-Didehydro-2',3'-dideoxy-5-chlorocytidine is a selective anti-retrovirus agent, *Biochem. Biophys. Res. Commun.* 164:1190 (1989).

53. M. Baba, R. Pauwels, P. Herdewijn, E. De Clercq, J. Desmyter, and M. Vandeputte, Both 2',3'-dideoxythymidine and Its 2',3'-unsaturated derivative (2',3'-dideoxythymidinene) are potent and selective inhibitors of human immunodeficiency virus replication in vitro, *Biochem. Biophys. Res. Commun.* 142:128 (1987).

54. J. Balzarini, A. Van Aerschot, P. Herdewijn, and E. De Clercq, 5-Chloro-substituted derivatives of 2',3'-didehydro-2',3'-dideoxyuridine, 3'-fluoro-2',3'-dideoxyuridine and 3'-azido-2',3'-dideoxyuridine as anti-HIV agents, *Biochem. Pharmacol.* 38:869 (1989).

STEREOCONTROLLED ROUTES FOR THE SYNTHESIS OF ANTI-HIV AND ANTI-HBV NUCLEOSIDES

J. Warren Beach, Lak S. Jeong, Hea O. Kim,
S. Nampalli , K. Shanmuganathan and Chung K. Chu

Department of Medicinal Chemistry, College of Pharmacy
The University of Georgia, Athens, Georgia 30602

INTRODUCTION

The discovery of the clinical usefulness of 2'-deoxy- and 2',3'-dideoxy nucleoside analogues for the treatment of viral infections such as acquired immunodeficiency syndrome, hepatitis B virus, cytomegalovirus and herpes simplex, and for the treatment of cancer has lead to the development of synthetic methodologies for the total synthesis of these types of agents. The advantage of a total synthetic approach is the ability to synthesize nucleoside analogues in which the base or the sugar portion of the molecule is not of a natural type. It also circumvents the use of naturally occurring nucleosides as starting material, which, in some cases are not readily available in large quantities and /or are of high cost.

The disadvantage of this methodology is the production of varying amounts of the unwanted, in most cases, α anomers, necessitating the sometimes tedious chromatographic separation of this anomeric mixture.

In the synthesis of 2'-deoxynucleoside, some control of the α:β ratio can be gained by the use of 2-deoxy-3,5-di-O-(p-chlorobenzoyl or p-toluoyl)-α-**D**-erythro-pentofuranosyl chlorides and have found much utility in the synthesis of both pyrimidine and purine analogues. These α-halo sugar derivatives, which are obtained as crystalline materials give almost exclusively the β anomer due to a S_N2 type (inversion) reaction. Aoyama [1] has reported the condensation of 5-substituted uracil derivatives (H,CH_3,F,Cl,Br,I) with 2-deoxy-3,5-di-O-(p-chlorobenzoyl)-ribofuranosyl chloride (**1**) (Scheme 1).

It was found that condensation in the presence of Brönsted acids, such as p-nitrophenol, gave high yields (96-87%) of the β anomer **2**, with a trace to 10% α anomer **3**. However, in the presence of Brönsted bases such as triethylamine, pyridine and picoline, high yields of the α anomers **3** were obtained.

Nucleosides and Nucleotides as Antitumor and Antiviral Agents,
Edited by C.K. Chu and D.C. Baker, Plenum Press, New York, 1993

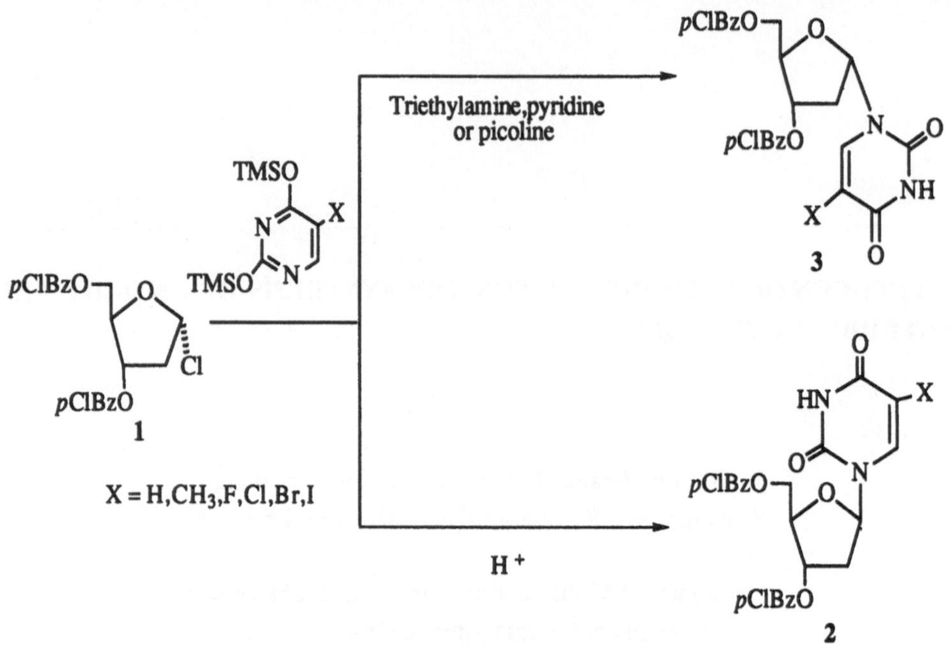

Scheme 1. Synthesis of α or β 2'-deoxynucleosides.

This was explained (Fig 1) on the basis of an initial attack of the organic base on the anomeric position from the β face followed by a secondary attack of the uracil base from the α face.

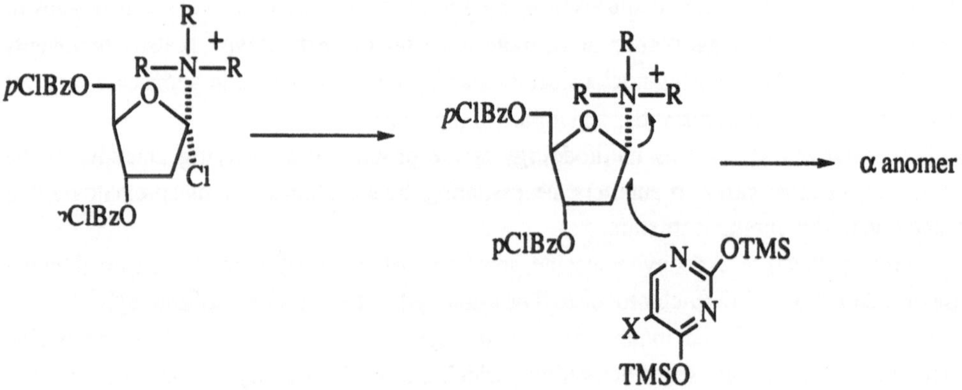

Figure 1. Proposed mechanism of α anomer formation.

In a similar vein, Hildebrand and Wright [2] have reported on the glycosylation of 2-deoxy-3,5-O-di-(toluly)-α-**D**-ribofuranosyl chloride (**4**) with the sodium salts of 6-halo and 2,6-dihalopurines (Scheme 2). It was found that the N^9 and N^7 β isomers (**5** and **7**) were the major products formed, however, in the case of 2,6-dibromo and 6-bromopurine, the N^9 α anomer **6** was formed to a slightly greater extent than in 2,6-dichloro and 6-chloropurine.

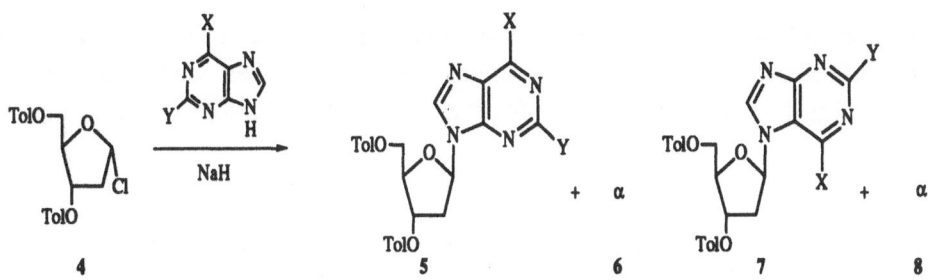

Scheme 2. Synthesis of 6-substituted and 2,6-disubstituted purine nucleoside using the sodium salt method.

This result was explained in terms of the rate of glycosylation being slower than the rate of anomerization of the glycosyl chloride.

Niedballa and Vorbrüggen[3] have shown that the use of this α-chloro sugar (**4**) in the condensation with 5-ethyluracil in the presence of stannic chloride produced approximately a 1:1 (α:β) anomeric mixture of **10** and **9** under a variety of reaction conditions (Scheme 3). This was apparently due to the use of stannic chloride which either causes anomerization of the chloro sugar or more likely formation of a carbocation that has an equal chance of α or β attack.

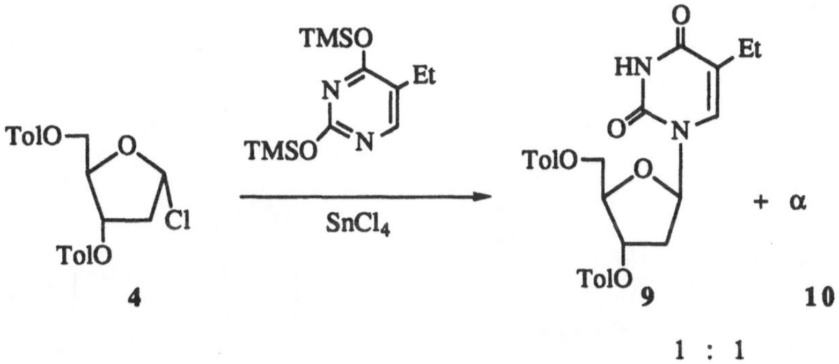

Scheme 3. Influence of a Lewis acid on the anomeric distribution in the condensation with an glycosyl chloride.

The main disadvantage of the glycosyl halides is their instability, making fresh preparation or crystallization necessary. The requirement of the α configuration at the anomeric position can be troublesome if this anomer is not crystalline or is not separable from the β anomer in some way. In some cases the production of a glycosyl halides is not possible or practical due to other substituent or protecting group, such is the case with most of the dideoxysugar. In most instances the syntheses reported have used an acetyl or methoxy group at the anomeric position.

In these cases it has been observed that the ratio of the α to β anomers varies in some instances with the length of reaction, the solvent use for condensation, the nucleoside base employed , the sugar portion to be introduced, and the catalyst used.

In studies directed at an efficient total synthetic route [4] (Schemes 4 and 5) to 3'-azido nucleosides such as AZT and AZDU, we condensed a silyl-protected 1-O-acetyl-3-azido-2,3-dideoxy-D-ribosefuranose (14) with various heterocyclic bases. Compound 14 was synthesized as shown in Scheme 4 by the Michael addition of LiN₃ or NaN₃ to the unsaturated lactone 11.

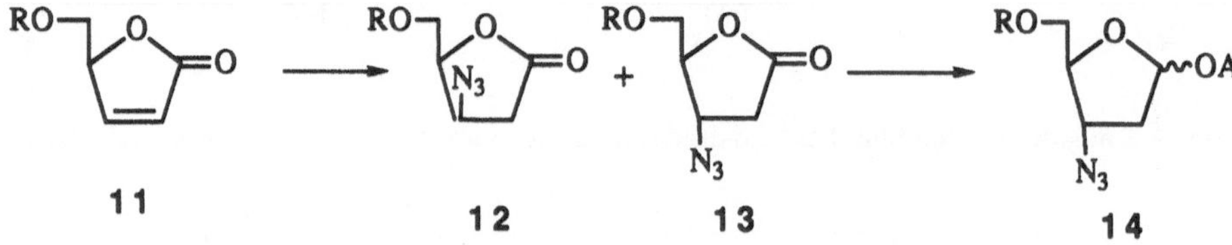

Scheme 4. Synthesis of 1-O-acetyl-3-azido-2,3-dideoxy-D-ribofuranose (14).

It was found that the amount of the undesired threo azide (12) to the desired erthyro azide (13) could be controlled by the use of a bulky protecting group at the 5 position. The t-butyldiphenylsilyl group was found to be the best in that no threo azide was formed. Reduction with Dibal-H and acetylation gave 14. On condensation of silylated uracil with 14 (Scheme 5), the α to β ratio (15a:16a) varied from 7:3 for SnCl₄/acetonitrile to 1:1 for TMSOTf/acetonitrile to 3:7 for TMSOTf/1,2-dichloroethane. However, for the condensation of silylated thymine with the same sugar the ratio of 15b to 16b was found to be 1:1 for all of the above reaction conditions.

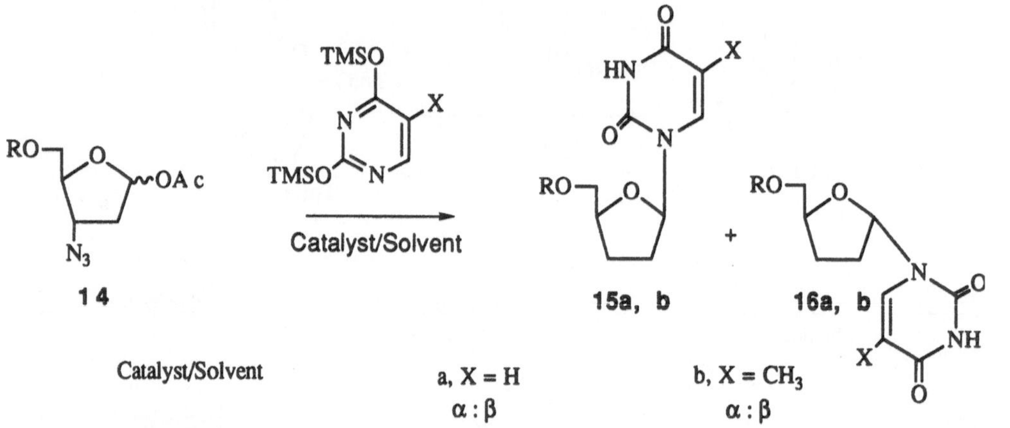

Catalyst/Solvent	a, X = H	b, X = CH₃
	α : β	α : β
SnCl₄/Acetonitrile	7 : 3	1 : 1
TMSOTf / Acetonitrile	1 : 1	1 : 1
TMSOTf /1,2 DCE	3 : 7	1 : 1

Scheme 5. Influence of solvent and catalyst of the anomeric distribution of AZDU and AZT.

Okabe et al. [5] (Scheme 6) found the α:β ratio also varied with the use of different Lewis acids and time of reaction for the condensation of silylated cytosine with 5-O-t-butyldimethylsilyl-2,3-dideoxy-D-glycero-pentofuranose (17). When stannic chloride in dichloro-methane was used a 2:1 α:β ratio (18:19) was produced. It was found that this ratio could be improved by longer eaction times, so called "aging the reaction". However, decomposition of both the anomers vas found to be faster than α to β isomerization.

A variety of other Lewis acids such as titanium tetrachloride, boron trifluoride etherate, TMSOTf and ethylaluminum dichloride, all in dichloromethane, were found to be effective :oupling reagents without causing decomposition of the products as with stannic chloride. Of he above Lewis acids, ethylaluminum dichloride was found to be the best, producing an α:β atio of 2:3 in 71% yield. Titanium tetrachloride gave a 1:1.2 α:β mixture in 61% yield.

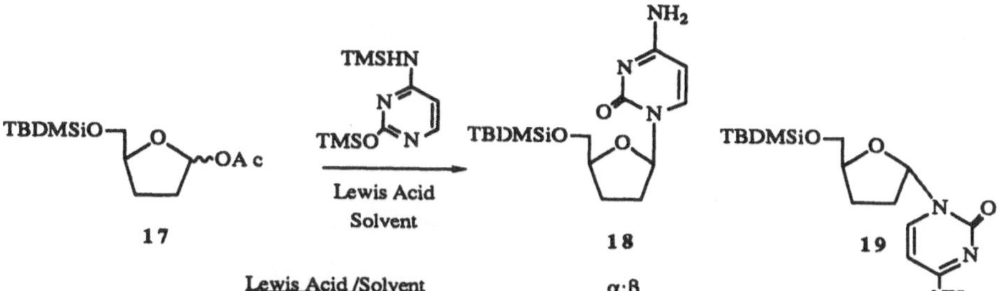

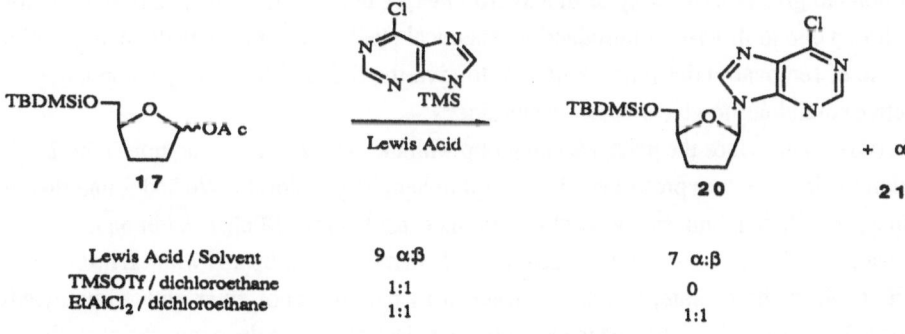

Lewis Acid /Solvent	α:β
SnCl4/ dichloromethane	2:1
TiCl4/ dichloromethane	1:1.2
EtAlCl2/ dichloromethane	2:3
TMSOTf/1,2dichloroethane	2:1

Scheme 6. Influence of catalyst on the anomeric distribution of ddC.

It has also been reported by us that the condensation of the same sugar with silylated cytosine in the presence of TMSOTf gave a 2:1 α:β mixture[6]. Futhermore condensation of **17** with 6-chloropurine(Scheme 7) also gave a 1:1 mixture of the N^9 anomers (**20** and **21**)[6,7]. When ethyl aluminum dichloride was employed in this system it still produced a 1:1 mixture of the N^9 anomers as well as a large quantity of the N^7 anomers.

In order to try to improve the α:β ratio of the these coupling reactions, we used a bulky group at the 3 position[6]. It was thought that this steric bulk on the α face of the sugar might deter the α face approach of the heterocyclic base and thus increase the amount of β anomer produced. A phenylthio group was chosen for substitution at the 3 position due to its ease of introduction as well as removal under reductive conditions to the dideoxy series of nucleoside or under oxidative condition to give the didehydro-dideoxy nucleoside analogues.

The thiophenyl group was introduced by Michael addition to the unsaturated lactone **11** in

Lewis Acid / Solvent	9 αβ	7 α:β
TMSOTf / dichloroethane	1:1	0
EtAlCl2 / dichloroethane	1:1	1:1

Scheme 7. Influence of Lewis acid on the product distribution of 2,3-dideoxy-6-chloropurine nucleoside.

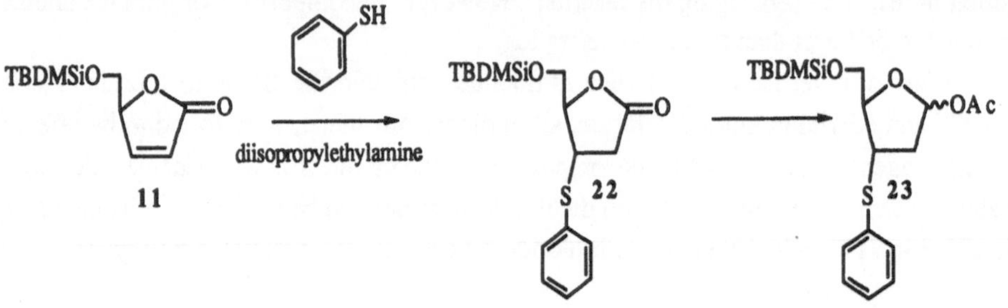

Scheme 8. Synthesis of 1-*O*-acetyl-5-*O*-(*t*-butyldimethylsilyl)-2deoxy-3*S*-phenyl-3-thio-D-ribofuranose (**23**).

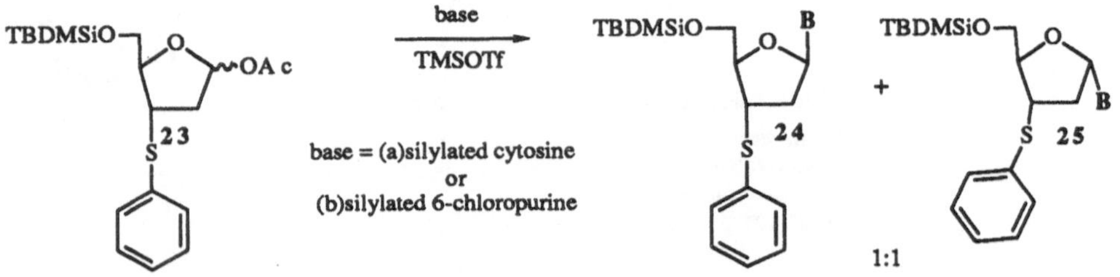

Scheme 9. Condensation of compound **23** with cytosine and 6-chloropurine

the presence of a trialkylamine (Scheme 8). This addition reaction gave, as in the case of azide addition, almost exclusively the α product. Reduction with Dibal-H, followed by acetylation, gave the acetate **23**.

Condensation was carried out with various bases such as cytosine or 6-chloropurine using TMSOTf (Scheme 9). However, the condensation produced a disappointing 1:1 anomeric mixture of **24** and **25**.

Having had little success at the 3-position, our attention turned to the possibility of a group at the 2-position of the sugar moiety. This group would have to be able to increase the β:α ratio either through a steric mechanism or a stabilizing transition state mechanism as in a normal ribose with acyl protection (Fig. 2). It would have to be easily removable under mild conditions to give both dideoxy or didehydro-dideoxy nucleosides. The phenylseleno group was chosen due to its ease of introduction, chemical stability during synthetic manipulations and ease of removal under mild oxidation, to give the didehydro-dideoxy compounds, and reductive conditions, to give dideoxy nucleosides.[8]

In order to introduce the phenylseleno group with the desired α configuration at the 2-position the lactone was protected with t-butyldiphenylsilyl chloride. We had found that this group gave sufficient bulk to give exclusively the α azide in the Michael addition of LiN$_3$ to the unsaturated lactone. The silylated lactone **27** (Scheme 10) was treated with LiHMDS at - 78°C to form the enolate, which was trapped as the silyl enol ether with trimethylsilyl chloride. This material was treated in situ with phenylselenue bromide to give the phenylseleno lactones **28** and **29** in a 2:1 α:β ratio, which could be readily separated by silica gel chromatography. It was also found that the undesired β isomer **29** could be equilibrated to a

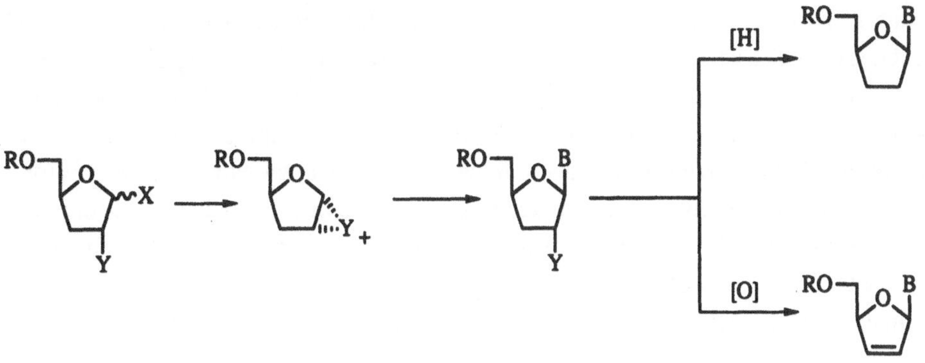

Figure 2. Theoretical 2-position directing group that could be used to obtain both dideoxy and didehydro-dideoxy nucleosides

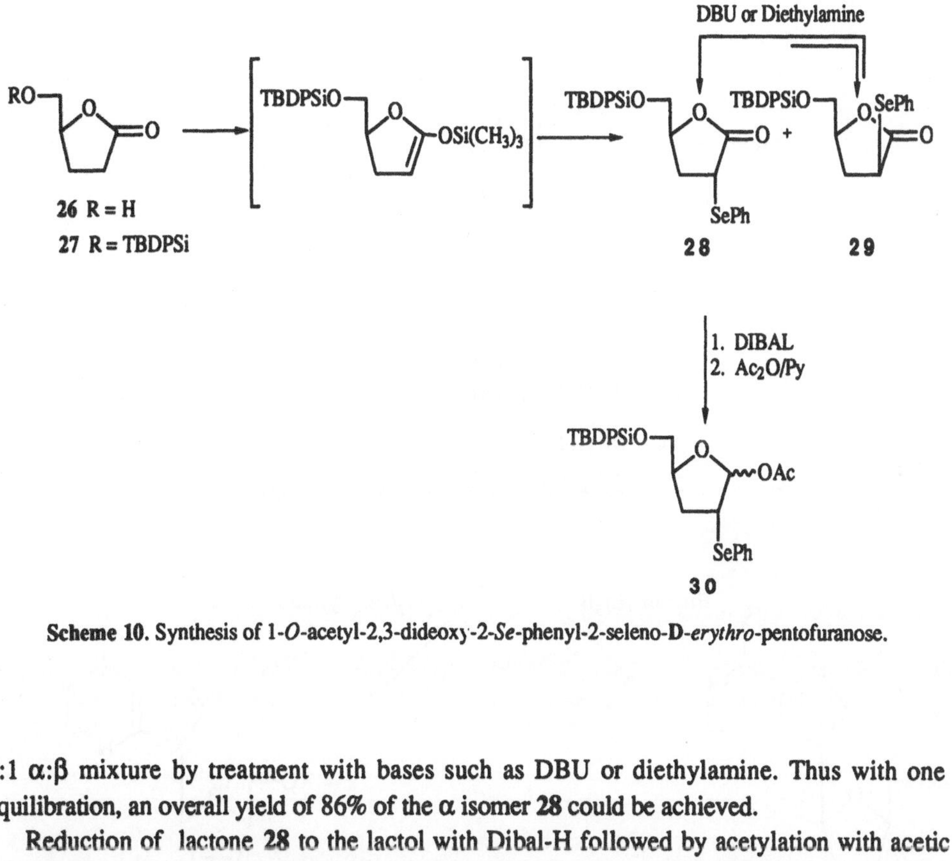

Scheme 10. Synthesis of 1-*O*-acetyl-2,3-dideoxy-2-*Se*-phenyl-2-seleno-D-*erythro*-pentofuranose.

2:1 α:β mixture by treatment with bases such as DBU or diethylamine. Thus with one equilibration, an overall yield of 86% of the α isomer **28** could be achieved.

Reduction of lactone **28** to the lactol with Dibal-H followed by acetylation with acetic anhydride/pyridine gave the desired key intermediate **30**. The lactol could not be purified due to isomerization of the α phenylseleno group to the β position, and the acetate was found to be hydrolytically unstable on silica gel .

The acetate **30** could be condensed (Schemes 11,12 and 13) with a variety of purines (6-chloro and 6-chloro-2-fluoropurine) and pyrimidines (thymine,uracil and cytosine) to give the corresponding 2-phenylseleno-β-nucleosides in good to excellent yields in 95 to 99% anomeric purity. The 6-chloropurine condensation product **31** (Scheme 11) is a versatile intermediate, which under the appropriate conditions can be converted to inosine (**32**), adenosine (**33**), *N*6-

methyl adenosine (**34**) derivatives. These phenylseleno derivatives may be further converted to the dideoxy nucleosides (**40,41,42**) under mild reductive conditions using tributyltin hydride and triethylborane at room temperature, followed by deprotection. The phenylseleno group can also be removed from the 6-chloropurine derivative **31** to yield dd6ClP **39** without removal of the chlorine. These compounds could also be converted to their unsaturated derivatives (**45** and **46**) by treatment with hydrogen peroxide in the presence of a catalytic amount of pyridine oxidizing the phenylseleno group to the selenium oxide, which undergoes spontaneous elimination, followed by desilylation.

In the 6-chloro-2-fluoropurine derivative **47** (Scheme 12), prepared by condensation of

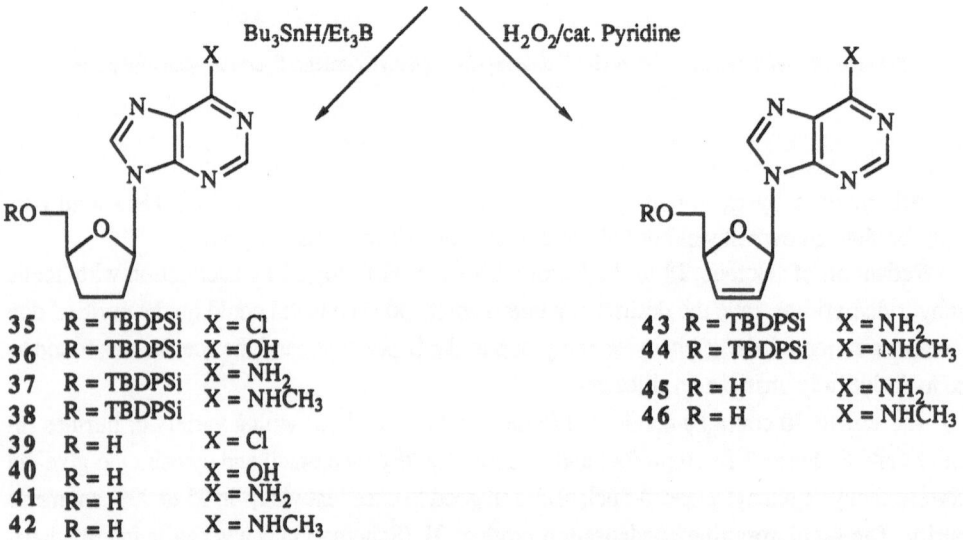

Scheme 11. Synthesis of 6-substituted purine dideoxy and didehydro-dideoxy nucleosides.

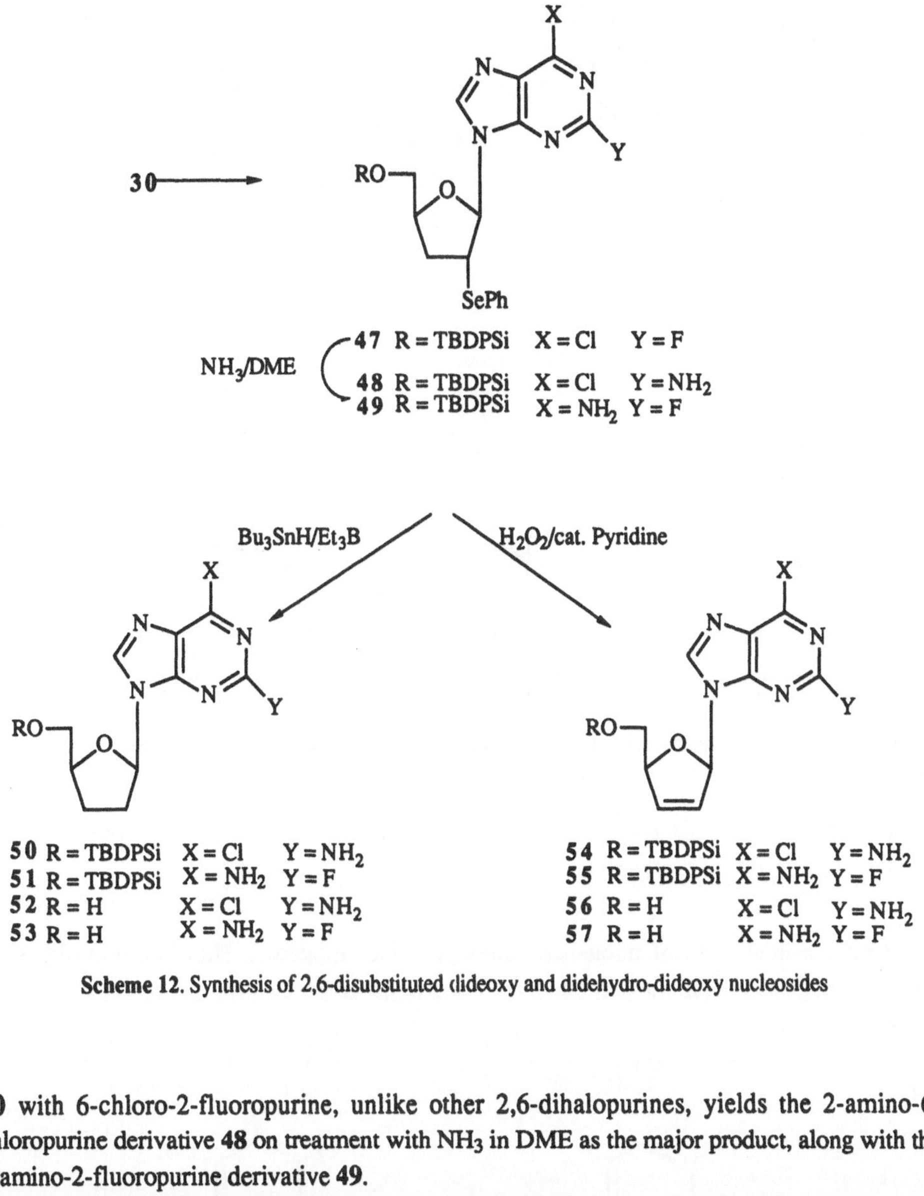

Scheme 12. Synthesis of 2,6-disubstituted dideoxy and didehydro-dideoxy nucleosides

30 with 6-chloro-2-fluoropurine, unlike other 2,6-dihalopurines, yields the 2-amino-6-chloropurine derivative **48** on treatment with NH_3 in DME as the major product, along with the 6-amino-2-fluoropurine derivative **49**.

This unusual reverse reactivity does not follow the usual course of 2,6-dihalopurine derivatives such as 2,6-dichloro and 2,6-dibromopurine, where the product of NH_3 treatment is the 6-amino-2-halopurine. The 2-amino-6-chloro (**48**) and 6-amino-2-fluoro (**49**) compounds could also be converted to their corresponding dideoxy (**52** and **53**) and unsaturated derivative (**56** and **57**) using tributyltin hydride/triethylborane and hydrogen peroxide/ cat. pyridine, respectively, followed by deprotection.

The condensation of thymine, uracil and cytosine with **30** went smoothly to give the corresponding 2'-phenylseleno nucleosides (**58**, **59** and **60**) (Scheme 13). For the cytidine derivative **60**, it was found that the use of N^4-acetyl cytosine gave the best results in regard to yield and ease of purification. These compounds underwent smooth conversion to the dideoxy analogues ddT (**63**), ddU (**64**) and ddC (**67**) on treatment with tributyltin hydride/triethylborane at room temperature, followed by deprotection, or to the unsaturated analogues d4T (**70**) and d4U (**71**) on treatment with hydrogen peroxide/cat. pyridine and desilylation.

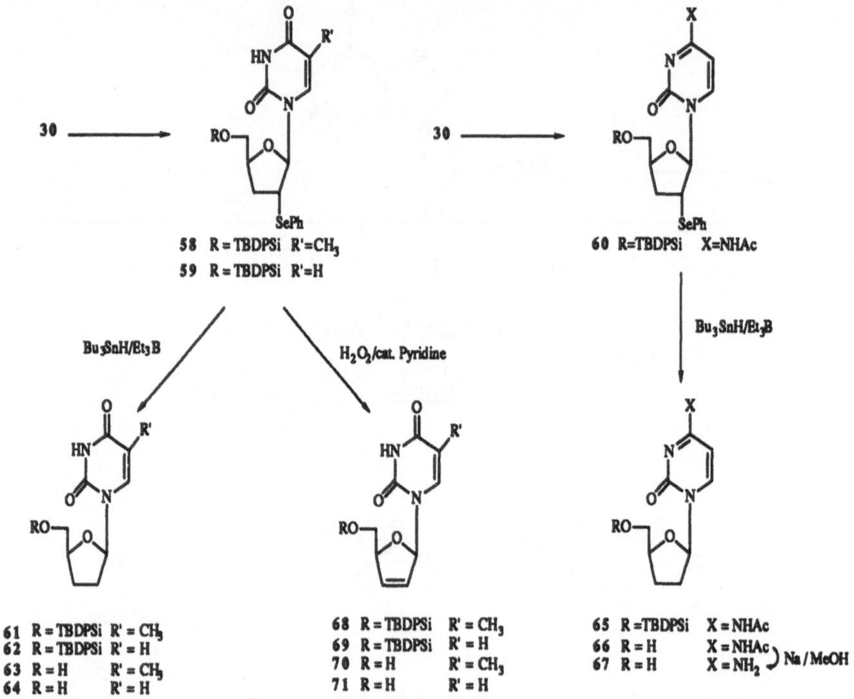

Scheme 13. Synthesis of pyrimidine dideoxy and didehydro dideoxy nucleosides

3'-HETERONUCLEOSIDES

Recently a new class of nucleoside analogues has appeared. These unique dideoxy nucleoside analogues have the 3' methylene group replaced by a sulfur[9] or oxygen[9,10] atom (Fig. 3).

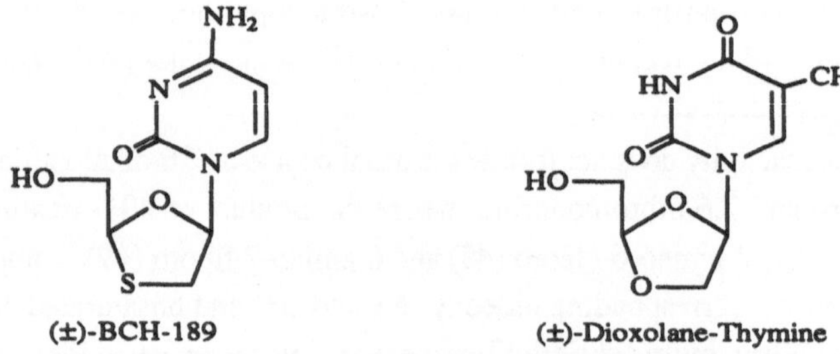

(±)-BCH-189   (±)-Dioxolane-Thymine

Figure 3. (±)-BCH-189 and (±) Dioxolane-T

These heteronucleosides were originally synthesized by Belleau *et al.*[9] (oxygen and sulfur) and Norbeck *et al.*[10] (oxygen) as racemic mixtures. In most biologically active compounds that are racemates, all or most of the biological activity lies with one of the enantiomers. In order to determine which one is active, the mixture must be separated by chromatographic means (chiral HPLC) or converted to diastereomers by reaction with a chiral reagent. These two methods have disadvantages and limitations in the amount that can be separated at any one time, and, if only one isomer is active, half of the mixture is wasted. The third alternative is a total synthetic approach using a chiral template as a starting material. This method gives flexibility in a structure-activity study, and one or the other isomer may be synthesized at will, for it is possible that one enatiomer will have one type of activity whereas the other may have activity in a different biological system (for example quinine and quinidine).

It was determined that the dioxolane ring system of a 1,6-anhydrohexose would serve well as a chiral template for the dioxolane nucleosides. Thus 1,6-anhydromannose[11,12] was chosen for its ease of large-scale preparation and the configuration of its hydroxyl groups at 2, 3 and 4 (2 and 3 cis, and 3 and 4 trans) which allows for selective protection prior to oxidative cleavage (Fig. 4).

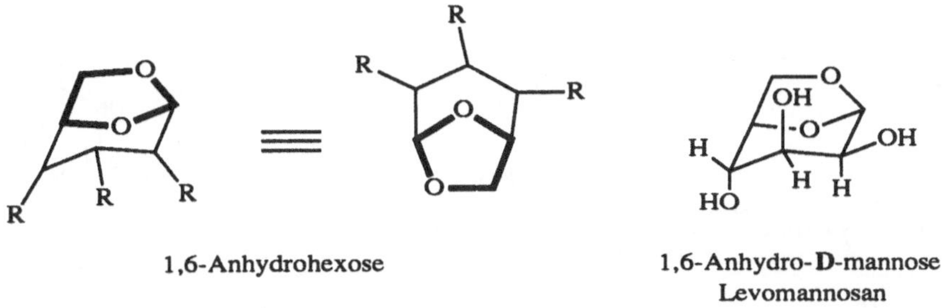

1,6-Anhydrohexose

1,6-Anhydro-**D**-mannose
Levomannosan

Figure 4. 1,6-Anhydro-**D**-mannose

1,6-Anhydro-**D**-mannose or levomannosan was originally synthesized by Hudson and co-workers[11] by the destructive distillation of ivory nut turnings and more recently on a preparative scale by Fraiser-Reid and co-workers[12] from **D**-mannose. In the later process the levomannosan is isolated as its 2,3-isopropylidene derivative (**72**). Benzoylation of the 4 position to give **73** (Scheme 14),[13] followed by removal of the isopropylidene group using dilute H_2SO_4 in aqueous dioxane at 75°C, gave the diol **74**. Treatment of the diol with $NaIO_4$ in aqueous methanol, followed by reduction with $NaBH_4$, gave the dioxolane derivative **75**.

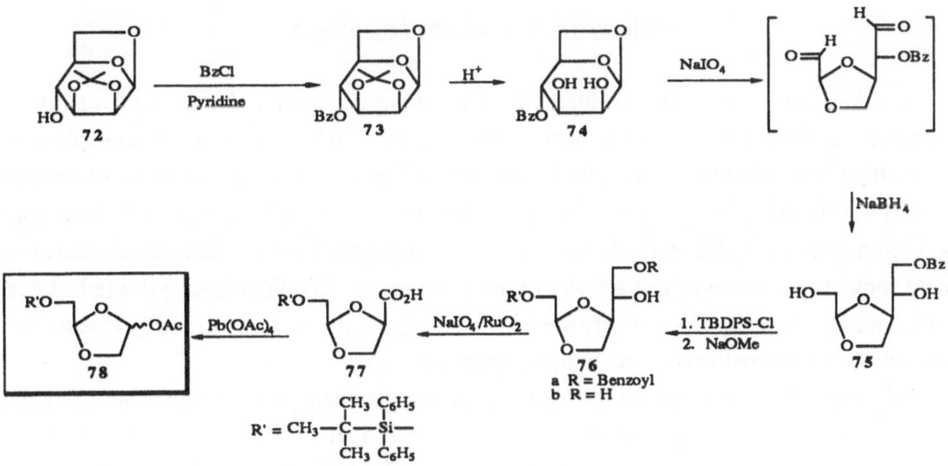

Scheme 14. Synthesis of dioxolanyl acetate

Note that the benzoate migrated from the secondary to the primary alcohol under the conditions of the reaction. This fortunate occurrence allowed the selective protection of the remaining primary alcohol with t-butyldiphenylsilyl chloride to give **76a**, which was deprotected using NaOMe in MeOH to give **76b**. The diol **76b** was converted to the acid by a process similar to that described by Nunez and Martin[14] using $NaIO_4$ and RuO_2 in a two-phase reaction system. The acid **77** was subjected to oxidative decarboxylation using $Pb(OAc)_4$/pyridine in ethyl acetate as described by Dhavale et al [15] gave the acetate **78**. It was later found that dry THF was a better solvent, giving faster reactions and better yields.

Condensation of the acetate **78** with silylated thymine (Scheme 15) using TMSOTf as Lewis acid gave a 1:1.5 α:β mixture of the protected nucleosides **79** and **80**, which were separated by column chromatography over silica gel. Deprotection with TBAF gave the desired free α (**81**) and β (**82**) dioxolane-T[13].

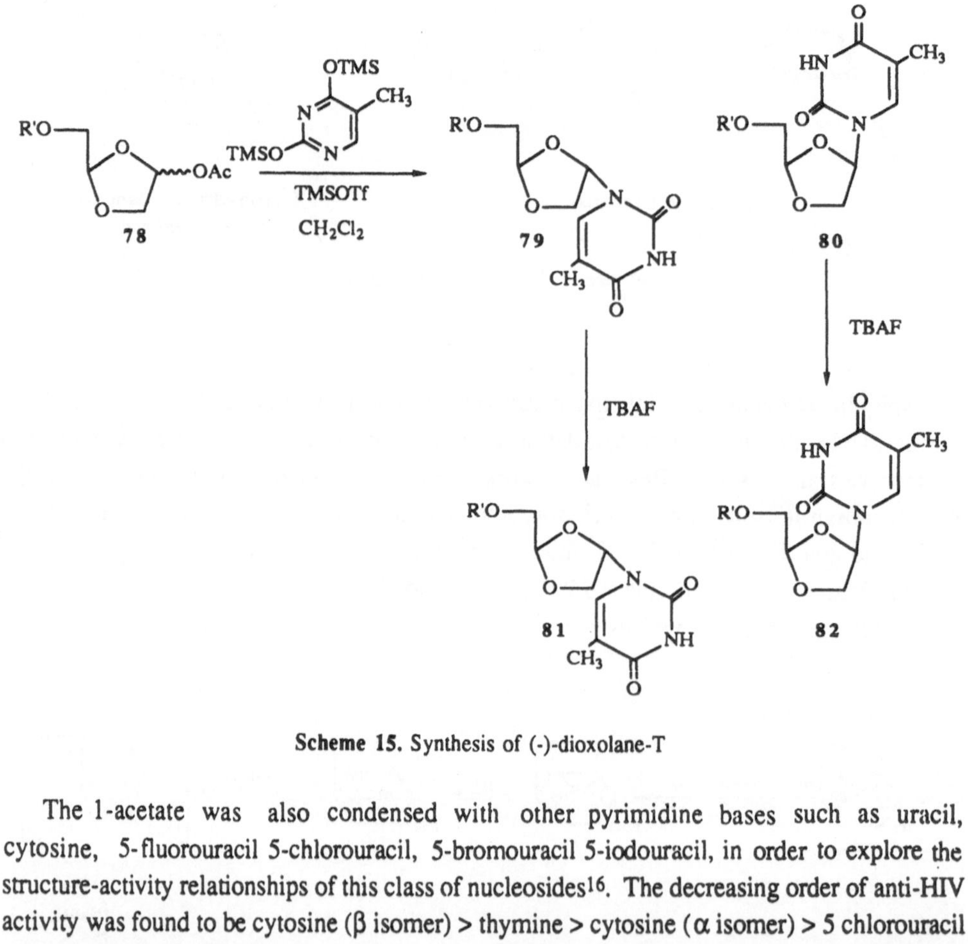

Scheme 15. Synthesis of (-)-dioxolane-T

The 1-acetate was also condensed with other pyrimidine bases such as uracil, cytosine, 5-fluorouracil 5-chlorouracil, 5-bromouracil 5-iodouracil, in order to explore the structure-activity relationships of this class of nucleosides[16]. The decreasing order of anti-HIV activity was found to be cytosine (β isomer) > thymine > cytosine (α isomer) > 5 chlorouracil > 5-bromouracil > 5-fluorouracil. The uracil, 5-methylcytosine and 5-iodouracil derivatives were found to be inactive. The activity of the α anomer of the dioxolane-cytosine had been previously reported[9] and was confirmed. The α and the β anomers were subjected to NOE studies in order to confirm the anomeric assignments.

Other important members of this class of nucleoside analogues are those that contain an oxathiolane ring system, in particular, the cytosine derivative, BCH-189. This compound showed potent anti-HIV activity. Again it was observed that the five-membered ring system of a thioanhydrohexose could be used to fix the stereochemistry of what would become the 4' position in the final nucleoside. Two such sugars were chosen as candidates for use as a chiral template due to the cis relationship of two adjacent hydroxyls and their low cost. These were the unknown 1,6-thioanhydro-D-mannose[17] and the known 1,6-thioanhydro-D-galactose[18] (Fig. 5). The synthesis of 1,6-thioanhydro-D-mannose was accomplished in four steps[17] (Scheme 16). Selective tosylation of D-mannose (82) at the 6 position, using a modification of the

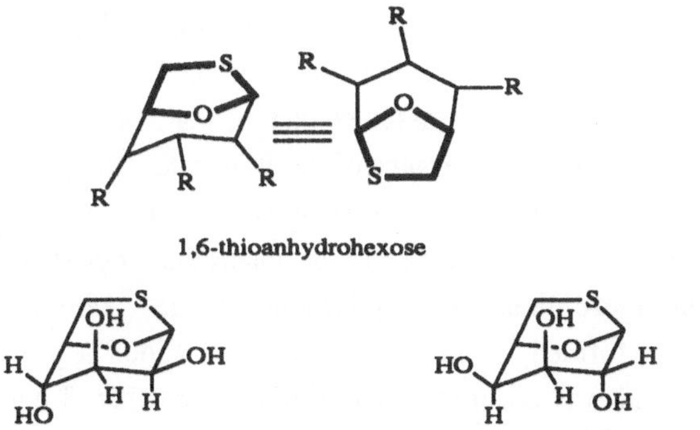

1,6-thioanhydrohexose

1,6-Thioanhydro-D-mannose   1,6-Thioanhydro-D-galactose

Figure 5. 1,6-Thioanhydrohexoses; 1,6-thioanhydro-D-mannose and 1,6-thioanhydro-D-galactose

procedure for 1,6-anhydro-D-mannose, followed by acetylation, gave 1,2,3,4,-tetra-O-acetyl-6-O-tosyl-D-mannose (83) as a foam, which, without further purification, was treated with HBr/Acetic acid using acetic acid as solvent, to give the bromo sugar 84. Reaction of 84 with 3 equiv of potassium ethyl xanthate in DMF at 0°C, followed by stirring overnight at room temperature, gave 2,3,4, tri-O-acetyl-1,6-thioanhydro-D-mannose (85), which was not isolated but subjected to deprotection using NH$_4$OH in MeOH. Concentration of the reaction mixture followed by chromatography over silica gel gave the desired product 86 in approx. 40% yield.

The classical method for thioanhydro formation[19] involves using 1.5 to 2 equiv of potassium ethyl xanthate in refluxing acetone, which produces the xanthate intermediate. This

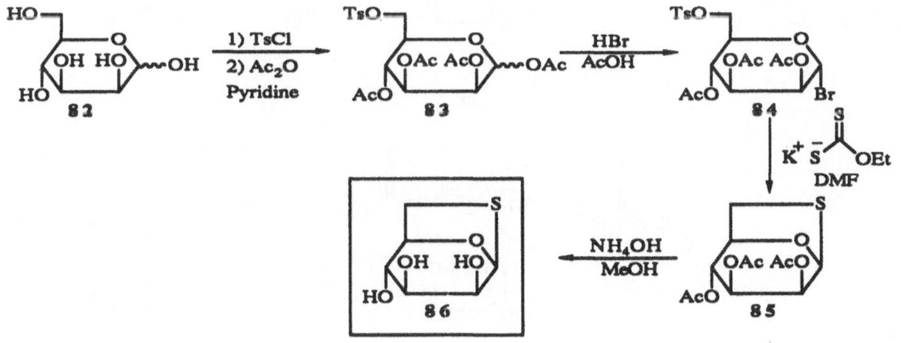

Scheme 16. Synthesis of 1,6-thioanhydro-D-mannose.

material is isolated, usually by crystallization, followed by treatment with NaOMe, which hydrolyzes the xanthate as well as the acetates, producing a sulfide anion that attacks the 6 position, displacing the tosyl group and forming the thioanhydro ring. The majority, if not all, previously known thioanhydro sugars have an α hydroxyl group and thus an α acetate at the 2 position. This assists in the formation of the β xanthate and the thioanhydro sugar. However, in the case of mannose, the 2 position is β, and thus little if any of the desired β xanthate could form when using the above conditions. This was observed in our initial experiment on the thioanhydro formation where we could only obtain approximately 5% of the desired product. Subsequently, we reasoned that by using a solvent that would, 1) dissolve potassium ethyl xanthate at less than room temperature and 2) be conducive to an S_N2 type reaction, would improve the yield. We assumed that the major anomer of the bromo sugar was α due to electronic effects and the presence of the β-acetate. DMF was chosen based on the yield it gave in intial experiments as well as ease of workup. The second major modification of the classical procedure was the use of 3 equiv of potassium ethyl xanthate. It was found by accident that when more than two equivalent of the xanthate were used, some thioanhydrosugar was formed, as more xanthate was added a proportional increase in thioanhydro compound was observed. At slightly more than 3 equiv, the intermediate xanthate had completely disappeared giving the thioanhydro as the major product, with a minor amount of dixanthate. A probable mechanism is shown in Fig. 6.

Isopropylidenation of the 2,3-cis diol of **86** (Scheme 17), followed by protection of 4 position with benzoyl chloride, gave **87,** which was deprotected using dilute H_2SO_4 in aqueous dioxane at 75-80°C, furnished the desired intermediate **88.** Treatment of **88** with $Pb(OAc)_4$ in ethyl acetate gave the dialdehyde, which, after partial purification to remove lead salts, was reduced with $NaBH_4$ in MeOH. As was seen in the dioxolane series earlier, the benzoyl group underwent a secondary-to-primary migration to give **89.** Selective protection of the remaining primary alcohol with t-butyldiphenylsilyl chloride and removal of the benzoate protection gave the diol **90.** Treatment of **90** with $Pb(OAc)_4$ in ethyl acetate gave the aldehyde, which was not isolated, but was further oxidized to the acid **91** using $NaClO_2$ in aqueous t-butanol.[20] Unfortunately, this also oxidized the sulfur atom of the ring to a

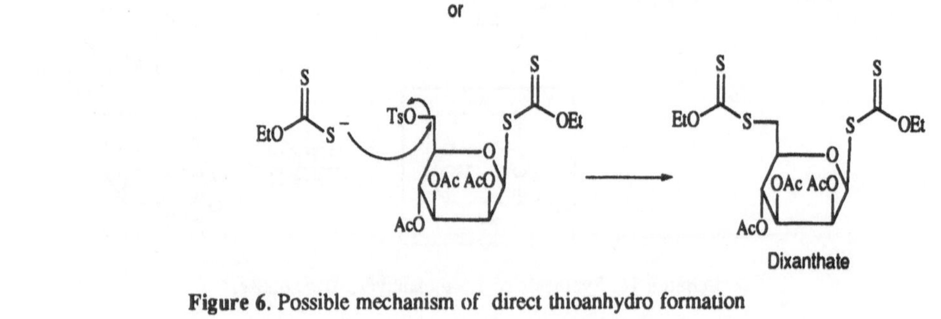

or

Figure 6. Possible mechanism of direct thioanhydro formation

mixtureof endo- and exo-sulfoxides. It was found that the sulfoxides would not undergo oxidative decarboxylation; therefore, the mixture of sulfoxides was esterified using dimethyl sulfate/K_2CO_3/acetone at room temperature to give **92**. Treatment of **92** with the dichloroborane-dimethylsufide complex[21] in dry THF at <0°C cleanly converted the mixture

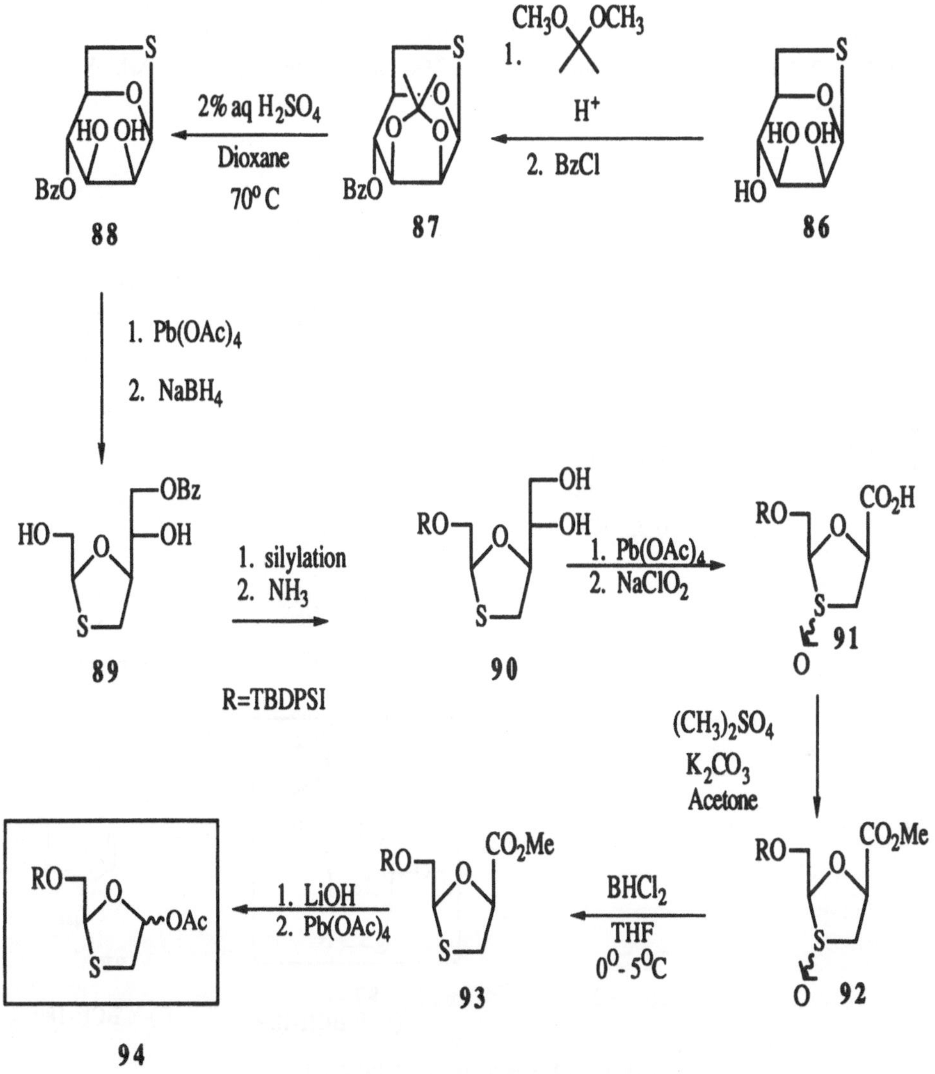

Scheme 17. Synthesis of oxathiolanyl acetate

to the sulfide ester **93**, which was hydrolyzed to the acid with LiOH in 4:1 THF/H_2O[10] and converted to the key intermediate acetate **94** by treatment with lead tetraacetate/pyridine[10,15] in ethyl acetate. Later THF was found to give faster reactions and homogeneous solutions. This may be due to the fact that oxygen tends to inhibit the reaction, as the THF was freshly distilled under nitrogen from Na/benzophenone. Condensation of the acetate with silylated N^4-acetyl cytosine gave a 2:1 β:α mixture of **95** and **96** (Scheme 18), which were separated by chromatography and subjected to deprotection with methanolic ammonia and TBAF to give the final free (+)-β **97** and (-)-α **98** BCH-189. These compounds were subjected to chiral HPLC analysis using a β-cyclodextrin acetate column to confirm their enantiomeric purity, and both were found to be homogeneous.

An alternative and more efficient synthesis[18] of the (+)-β isomer of BCH-189 **98** uses 1,6-thioanhydro-D-galactose as the starting material. 1,6-Thioanhydro-D-galactose was originally prepared by Whistler and Sieb[19b] (Fig. 7) and is easily scaled to large quantities.

The key to the efficiency of this process (Scheme 19) is the selective cleavage of the cis diol of **99** without protection followed by reduction with $NaBH_4$. The triol produced without isolation is then protected as its isopropylidene **100** and isolated. The diol, which is a side

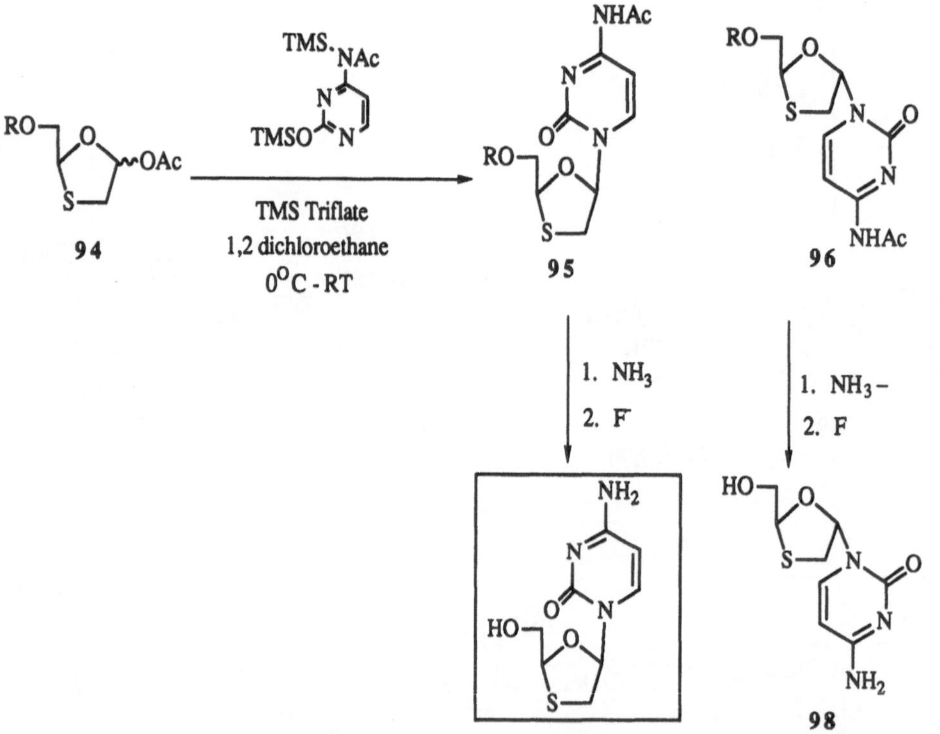

97
(+)-β-BCH-189

(-)-α-BCH-189

Scheme 18. Synthesis of (+)-BCH-189

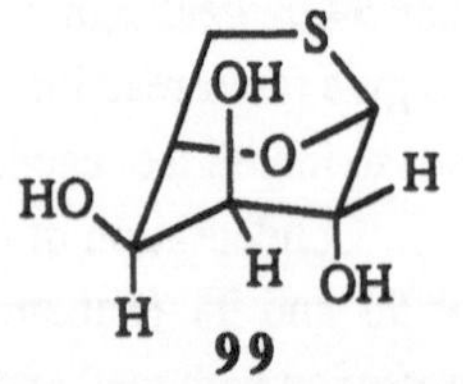

99

Figure 7. 1,6-Thioanhydro-D-galactose

product of the NaIO₄ reaction is washed out of the reaction mixture during work up. The remaining primary hydroxyl group of **100** is benzoylated, and the isopropylidene function is removed. The diol produced is cleaved and reduced to the alcohol **101**, which is protected with t-buyldiphenylsilyl chloride. This fully protected oxathiolane **103a** was then hydrolyzed with NaOMe to give the alcohol **103b**. Oxidation of **103b** with pyridium dichromate in DMF[22] gave the acid **104** without oxidation of the ring sulfur atom. Oxidative decarboxylation as before with lead tetraacetate/pyridine in THF gave the acetate **94**, which could be used for condensation as before.

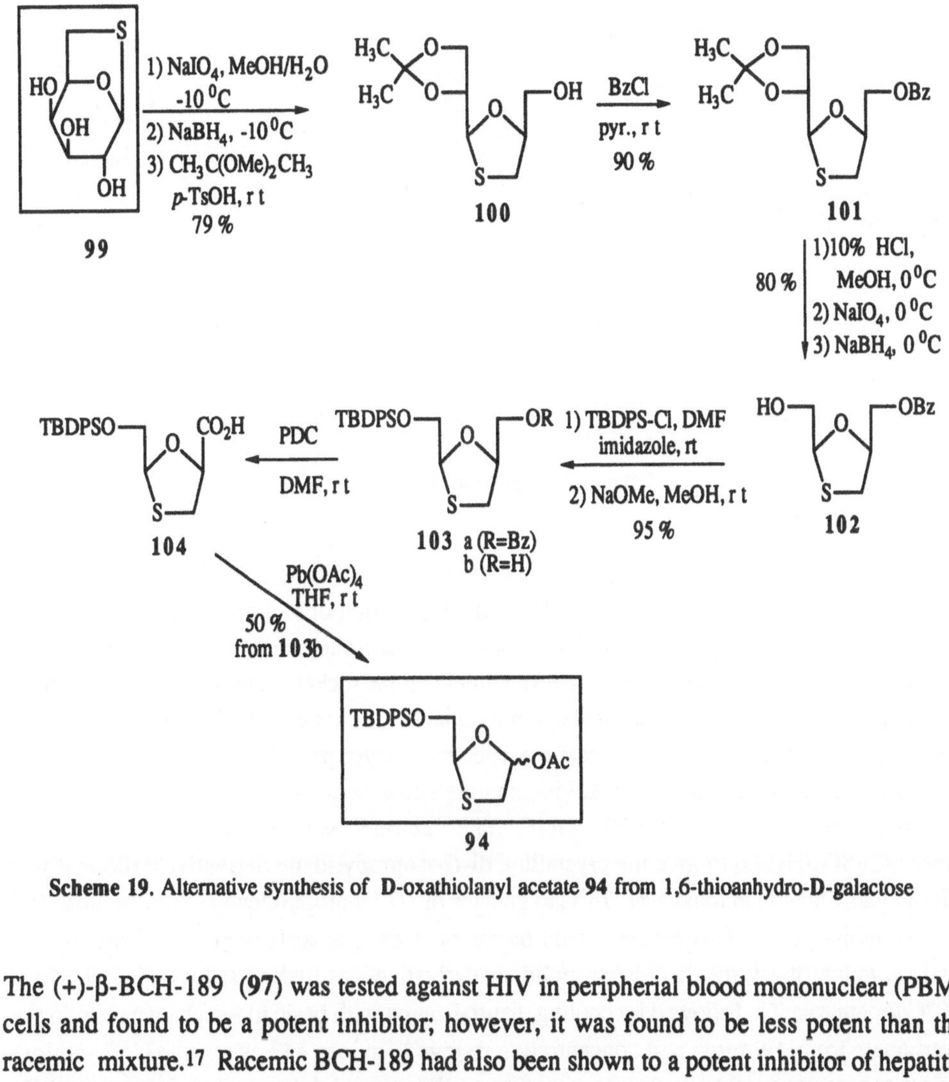

Scheme 19. Alternative synthesis of D-oxathiolanyl acetate **94** from 1,6-thioanhydro-D-galactose

The (+)-β-BCH-189 (**97**) was tested against HIV in peripherial blood mononuclear (PBM) cells and found to be a potent inhibitor; however, it was found to be less potent than the racemic mixture.[17] Racemic BCH-189 had also been shown to a potent inhibitor of hepatitis B virus (HBV) in 2.2.15 cells[23]. When the (+)-β-isomer **97** was tested in this system, it was also found to be less active than the racemate. In the case of (-)-Dioxolane-T, it as well was found to be less potent against HIV than the racemic mixture. Thus the L-like enantiomer must be the more potent species in both the oxathiolane as well as the dioxolane system.

In the search for an L-sugar that would serve as a chiral template for the synthesis of both the oxathiolane as well as dioxolane nucleosides, L-mannose was considered. L-mannose, however, suffers from several disadvantages in that it is not readily available or easily synthesized from readily available starting materials. It is expensive and would cause the same problem in the synthesis of the thioanhydrosugar namely, moderate yields.

The solution was found in the 5 "epimer" of D-mannose, namely L-gulose 110 (Scheme 20), which can be readily obtained from commercially available L-gulono-γ-lactone[24] 107.

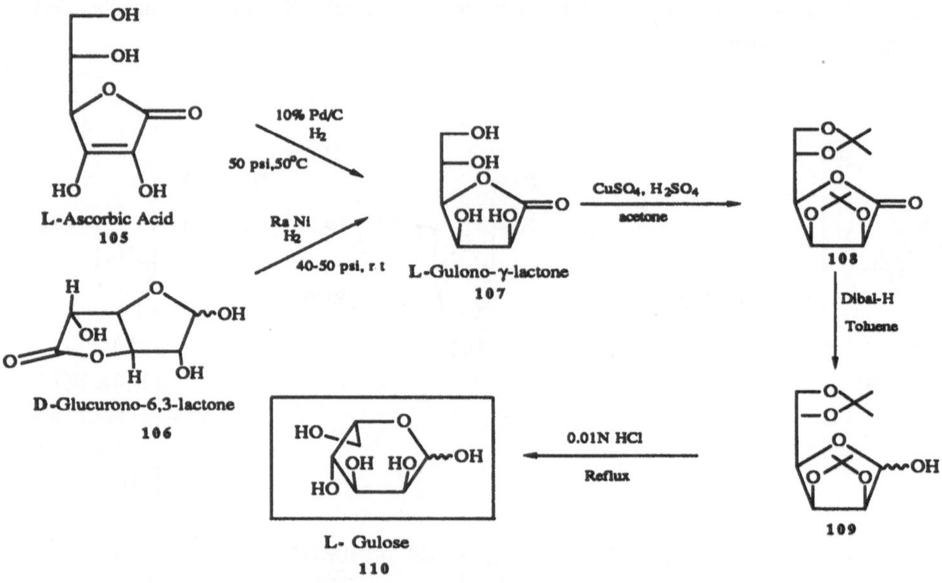

Scheme 20. Synthesis of L-gulose

The lactone may also be obtained by catalytic hydrogenation of L-ascorbic acid[25] 105 over 10% Pd/C at 50°C or D-glucurono-6,3-lactone[26]106 over Raney nickel at room temperature. The latter method was found to be more efficient in that the nickel catalyst could be reused for up to one kilogram of starting material without noticeable changes in yield or reaction length. The complete reaction mixture was decanted from the catalyst, passed through a strongly acidic ion-exchange column to remove nickel ion and the eluant concentrated to give crystalline L-gulono-γ-lactone 107 in 85-90% yield. The lactone was isopropylidenated using acetone/$CuSO_4$/H_2SO_4 to give the crystalline di-O-isopropylidene derivative[24]108 that was reduced with Dibal-H in toluene at -78°C to give the di--O-isopropylidene-L-gulofuranose 109 as a crystalline solid. This process could be run on a 250 gm scale with little difficulty. To obtain L-gulose itself, the di-O-isopropylidene derivative was hydrolysed with 0.1 N HCl at 100°C for approx. 2 h. followed by cooling, neutralization with basic ion-exchange resin in the bicarbonate form, filtration and concentration to give L-gulose 110 as a syrup[27]. L-Gulose has never been reported in the literature as a crystalline material; however, it is commercially available as a crystalline solid. A sample, kindly provided by Dr. J. Beadle, Biospherics Inc., Beltsville, MD was used as a seed to obtain crystalline L-gulose.

As with D-mannose, the L-gulose[28]110 was selectively to tosylated, followed by acetylation, to obtain 1,2,3,4-tetra-O-acetyl-6-O-tosyl-L-gulopyranose 111 (Scheme 21).

Treatment of this material with HBr/AcOH in acetic acid at room temperature gave the glycosyl bromide **112** as a foam, which was reacted with 3 equiv of potassium ethylxanthate in refluxing acetone to give the 2,3,4-tri-O-acetyl-1,6-thioanhydro-L-gulose, which was not isolated but subjected to hydrolysis using NH$_4$OH in MeOH to give, after column chromatography, 1,6-thioanhydro-L-gulose **113** in approx. 75% yield from L-gulose.

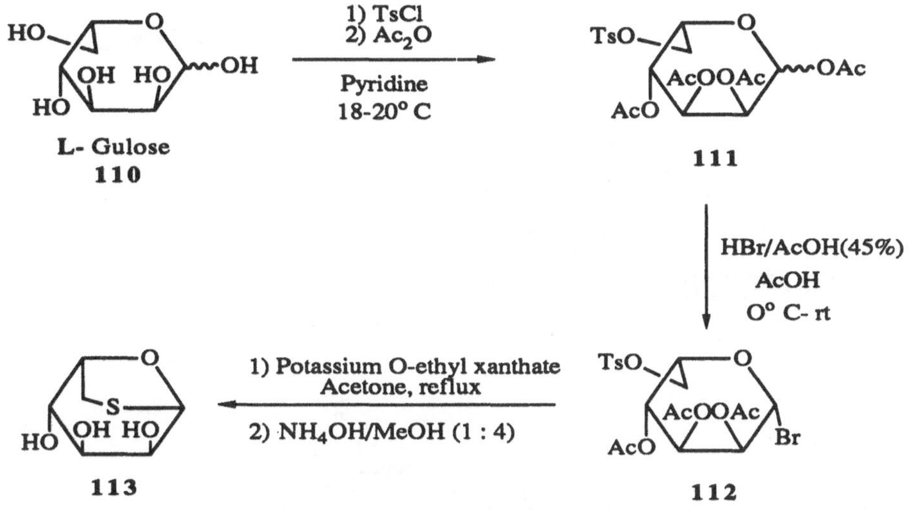

Scheme 21. Synthesis of 1,6-thioanhydro-L-gulose

The direct-cleavage methodology, developed for 1,6-thioanhydro-**D**-galactose was applied to 1,6-thioanhydro-L-gulose (Scheme 22). Treatment of the thioanhydro-L-gulose **113** with NaIO^4 followed by reduction of the resulting aldehyde with NaBH^4 gave the triol which was not isolated as such but converted to its isopropylidene **114** that was purified. Protection of the remaining alcohol with t-butyldiphenylsilyl chloride, followed by removal of the isopropylidene group, gave the protected oxathiolane diol **115**. Treatment of this material with lead tetraacetate to give the aldehyde, followed by further oxidation with pyridium dichromate in DMF, gave the acid **116**. Oxidative decarboxylation using lead tetraacetate/pyridine in THF gave the chiral acetate **117**. Condensation of **117** with silylated N^4-acetyl cytosine in the presence of TMSOTf/1,2-DCE gave, as before, a 1:2 α:β mixture of BCH-189 (**119** and **118**). The α,β mixture was separated and treated with methanolic ammonia and TBAF to give the individual (+)-α (**121**) and (-)-β (**120**) BCH-189. Chiral HPLC showed that again these were homogeneous and enantiomerically pure. The (+)-α **121** and (-)-β **120** anomers were evaluated against HIV-1 in PBM cell[29] in comparison to racemic and the (+)-β BCH-189 **97** (Table 1), which indicated the this L-like enantiomer was the most potent. The (-)-β BCH-189 was also found to be the most potent compound against HBV.[30] (Table 1)

It was found that both the (+)- and (-)-β enantiomers undergo phosphorylation in vitro; however, the (+) enantiomer was found to undergo deamination by cytidine deaminase, whereas the (-) enantiomer was not a substrate for the enzyme[30]. This may partially explain the difference in potency of the two enantiomers, as the uracil derivative is inactive.

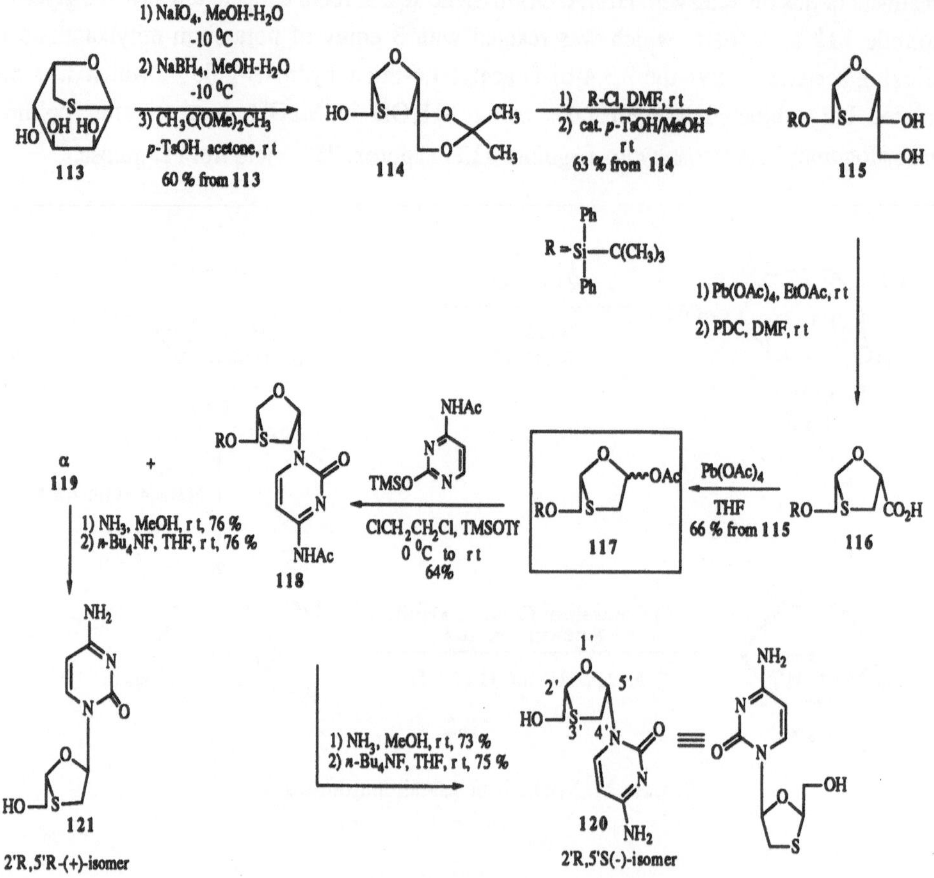

Scheme 22. Synthesis of (-)-BCH-189

Table 1. Biological activity of the isomers of BCH-189

Compounds		Optical Rotation $[a]_D^{25}$	Anti-HIV Activity in PBM Cells EC$_{50}$	Anti-HBV Activity[a] EC$_{50}$	Cytotoxicity in CEM cells IC$_{50}$
(±)-2'RS,5'RS		---	0.02-0.06 µM	0.05 µM	20 - 40 µM
(+)-2'S,5'R	97	+120.96° (c 1.04, MeOH)	0.21 µM	0.5 µM	2 µM
(-)-2'R,5'S (3TC)	120	-121.6° (c 1.1, MeOH)	0.0018 µM	0.01 µM	> 50 µM
(-)-2'S,5'S	98	-143.2° (c 0.62, MeOH)	> 100 µM	>5 µM	>100 µM
(+)-2'R,5'R	121	+146.6° (c 0.55, MeOH)	10.1 µM	>5 µM	ND

[a]Mean of triplicate assays; standard deviation <10%

For the preparation of the L-dioxolane nucleosides, L-gulose **110** was again used as the starting material; however, in this case the anhydrosugar may be obtained directly by heating 2,3,5,6-di-*O*-isopropylidene-L-gulose **109** in 0.5 N HCl[31] (Scheme 23). This process causes deprotection, followed by anhydro formation 1,6-Anhydro-L-gulose **122** may be obtained after chromatography in a crystalline form in approx. 45% yield.

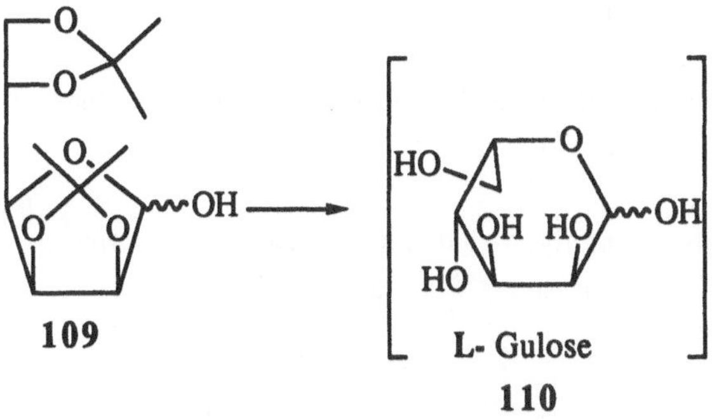

Scheme 23. Synthesis of 1,6 anhydro-L-gulose

As with the oxathiolane, direct cleavage of the cis diols of **122** with NaIO4, followed by NaBH4 reduction and reaction with TsOH in acetone, gave the isopropylidene derivative **123** (Scheme 24). Protection of the remaining hydroxyl as its benzoate, followed by removal of the isopropylidene group, gave the protected dioxolane diol **124**.

Treatment of this material with NaIO4 RuO2 gave the acid **125** that was converted to the acetate **126** as before with lead tetraacetate/pyridine in THF. This chiral acetate was used as before for condensation with various heterocyclic bases.

The dioxolane cytosine derivative was evaluated against HIV-1 in PBM cells. The β cytosine

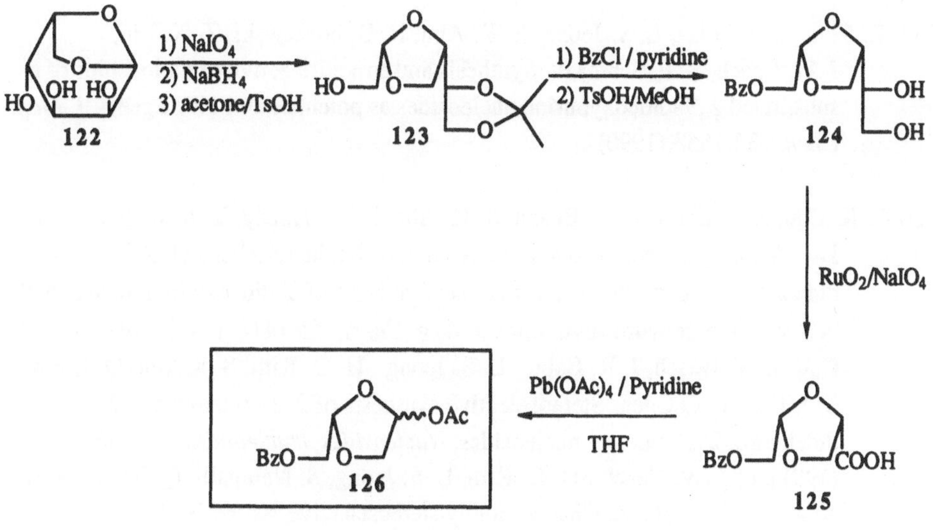

Scheme 24. Synthesis of L-dioxolanyl acetate

derivative was extremely potent; however, it was also very toxic to PBM and CEM cell lines. The α cytosine was also active as before.

The synthesis of other derivatives is in progress in order to determine the complete structure-activity relationships of this novel class of nucleoside analogues.

References

1. H. Aoyama. Stereoselective synthesis of anomers of 5-substituted 2'-deoxyuridines, *Bull.Chem. Soc. Jpn.* 60:2073 (1987).

2. C. Hildebrand, G. E. Wright. Sodium salt glycosylation in the synthesis of purine 2'-deoxyribonucleosides: Studies of isomer distribution. *J. Org. Chem.* 57:1808 (1992).

3. U. Niedballa, H. Vorbrüggen. A general synthesis of *N*-glycosides I. Synthesis of pyrimidine nucleosides, *J. Org. Chem.* 39:3654 (1974).

4. C. K. Chu, J. W. Beach, G. V. Ullas, Y. Kosugi. An efficient total synthesis of 3'-azido 3'-deoxythymidine (AZT) and 3'-azido 2', 3'-dideoxyuridine (AZDDU, CS-87) from **D**-mannitol. *Tetrahedron Lett.* 29:5349 (1988).

5. M. Okabe, R.-C. Sun, S. Y.-K Tam, L. J. Todaro, D. L. Coffen. Synthesis of the dideoxynucleosides ddC and CNT from glutamic acid, ribonolactone and pyrimidine bases. *J. Org. Chem.* 53:4780 (1988).

6. C. K. Chu, R. Raghavachari, J. W. Beach, Y. Kosugi, G. V. Ullas. A general synthetic method for 2',3'-dideoxynucleosides: Total synthetic approach. *Nucleosides Nucleotides* 8:903 (1989).

7. C. K. Chu, G. V. Ullas, L. S. Jeong, S. K. Ahn, B. Doboszewski, Z. X. Lin, J. W. Beach, R. F. Schinazi. Synthesis and structure-activity relationships of 6-substituted 2',3'-dideoxypurine nucleosides as potential anti-HIV agents. *J. Med. Chem.* 33:1553 (1990).

8.a) C. K. Chu, J. R. Babu, J. W. Beach, S. K. Ahn, H.-Q. Huang, L. S. Jeong, S. J. Lee. A highly stereoselective glycosylation of 2-(phenylselenenyl) 2',3 dideoxyribose derivative with thymine: Synthesis of 3'-dideoxythymidine and 3'-deoxy 2',3'-didehydro thymidine. *J. Org. Chem.* 55:1418 (1990). b) C. K. Chu, J. W. Beach, J. R. Babu , L. S. Jeong , H. O. Kim, S. K.Ahn, Q. Islam, S. J. Lee, Y.-Q Chen. Stereoselective synthesis of 2',3'-dideoxy and 2',3'-didehydro, 2',3'-dideoxy nucleosides, *Nucleosides Nucleotides* 10:423 (1991). c) J. W. Beach , H. O. Kim, L. S. Jeong, S. Nampalli, Q. Islam , S. K. Ahn, J. R. Babu, C. K. Chu. A highly stereoselective synthesis of anti-HIV

2',3'-dideoxy and 2',3'-didehydro 2',3'-dideoxy nucleosides, *J. Org. Chem.* 57:3887 (1992).

9. B. Belleau, D. Dixit, N. Nguyen-Ba, J.L. Kraus. Fifth International Conference on AIDS, Montreal, Canada, June 4-9, 1989, paper no. TCO1.

10. D. W. Norbeck, S. Spanton, S. Broder, H. Mitsuya. ((-1-[2β,4β)-2 hytroxymethyl-4-dioxolane thymine). A new 2'3'-dideoxy nucleoside prototype with *in vitro* activity against HIV. *Tetrahedron Lett.* 30: 6263 (1989).

11. A. E. Knauf, R. M. Hann, C. S. Hudson. D-Mannosan <1,5>β<1,6> or Levomannosan. *J. Am. Chem. Soc.* 63:1447 (1941).

12. M. A. Zottola, R. Alonso, G. D. Vile, B. Fraser-Reid. A practical, efficient, large-scale synthesis of 1,6-anhydrohexopyranoses. *J. Org. Chem.* 54: 6123 (1989).

13. C. K. Chu, S. K. Ahn, H. O. Kim, J. W. Beach, A. J. Alves , L. S. Jeong, Q. Islam, P. Van Roey, R. F. Schinazi. Asymmetric synthesis of enantiomerically pure (-) - (1'R, 4'R) - dioxolane thymine and its anti-HIV activity. *Tetrahedron Lett.* 32:3791 (1991).

14. M. T. Nunez, V. S. Martin. Efficient oxidation of phenyl group to carboxylic acids with rutherniun tetroxide. A simple synthesis of (R)-γ-caprolactam, the pheromone of *Trogoderma granarium*. *J. Org. Chem.* 55:1928 (1990).

15. D. Dhavale, E. Taghavini, C. Trombini, A. Umani-Romchi. A novel synthetic equivalent of differentially protected tartaric aldehyde. A simple route to useful C-4 chiral synthesis. *Tetrahedron Lett.* 29:6123 (1988).

16. H. O. Kim, S. K. Ahn, A. J. Alves , J. W. Beach, L. S. Jeong, B. G. Choi, P. Van Roey, R. F.Schinazi, C. K.Chu Asymmetric synthesis of 1,3-dioxolane pyrimidine nucleosides and their anti-HIV Activity. *J. Med. Chem.* 35:1987 (1992).

17. C. K. Chu, J. W. Beach, L. S. Jeong, B. G. Choi, F. I Comer, A. J. Alves , R. F. Schinazi. Enantiomeric Synthesis of (+) BCH-189. [(+)-2S,5R)-1-[2-(Hydroxymethyl)-1,3-oxathiolane-5-yl) cytosine] from D-mannose and its anti-HIV activity. *J. Org. Chem.* 56:6503 (1991).

18. L. S. Jeong, A. J. Alves, S. W. Carrigan, H. O. Kim, J. W. Beach, C. K. Chu. An Efficient synthesis of enantiomerically pure (+) (2S,5R)-1-[2-(hydroxymethyl)-1,3 oxathiolane-5-yl] cytosine [(+) BCH-189] from D-galactose. *Tetrahedron Lett.* 33: 595 (1992).

19.a) M. Akagi, S. Tejima, M. Haga. Biochemical studies in thio sugars IV. Synthesis of 1,6-anhydro-1,6-sulfide β-**D**-glucopyranose (thiolevoglucosan) and 6-deoxy-6-mercapto-1-thio-**D**-glucose. *Chem. Pharm. Bull.* 11: 58 (1963). b) R. L. Whistler, P. A. Seib. Alkaline degradation of 6-thio derivatives of **D**-glucose and **D**-galactose. *Carbohydr. Res.* 2:93 (1966).

20. G. A. Kraus, M. J. Taschner. Model studies for the synthesis of quassinoids. Construction of the BCE ring system. *J. Org. Chem.* 45:1175 (1980).

21. H. C. Brown, N. Ravindran. An exceptionally rapid and selective deoxygenation of aliphatic sulfoxides to sulfides under mild conditions with a new reducing agent, dichloroborane. *Synthesis* 42 (1973).

22. E. J. Corey, G. Schmidt. Procedures for the oxidation of alcohols involving pyridinium dichromate in aprotic media. *Tetrahedron Lett.* 379 (1979).

23. S.-L Doong, C.-H Tsai, R. F. Schinazi, D.C. Liotta, Y.-C. Cheng. Inhibition of the replication of hepatitis B Virus in vitro by 2',3' dideoxy-3'-thiacytidine and related analogues, *Proc. Natl. Acad. Sci., USA* 88:8495 (1991).

24. L.M. Lerner, B. D. Kohn, P. Kohn. Preparation of nucleosides *via* isopropylidene sugar derivatives III. *J. Org. Chem.* 33:1780 (1968).

25. G. Andrews, T.C. Crawford, B.E. Bacon. Stereoselective, catalytic reduction of L-ascorbic Acid. A convent synthesis of L-gulono-1,4-lactone. *J. Org. Chem.* 46:7976 (1981).

26. M. Ishedate, Y. I. Mai, Y. Hirasaka, K. Umemoto, A new action of ion-exchange resins on the lactone ring of some carbohydrates. *Chem. Pharm. Bull.* 13:173 (1965).

27. M. E. Evans, F. W. Parrish. A simple synthesis of L-gulose. *Carbohydr. Res.* 28: 359 (1973).

28. J. W. Beach, L. S. Jeong, A. J. Alves, D. Pohl, H. O. Kim, C.-N. Chang, S. L. Doong, R. F. Schinazi, Y.-C. Cheng, C. K. Chu. Synthesis of enantiomerically pure 2'R,5'S-(-)-1-[2-(Hydroxymethyl)-oxathiolane-5-yl] cytosine as a potent antiviral agent against hepatitis B virus (HBV) and human immunodeficiency virus (HIV). *J. Org. Chem.* 57:2217 (1992).

29. R. F. Schinazi, C. K. Chu, A. Peck, R. McMillan, R. Mathis, D. Cannon, L. S. Jeong, J. W. Beach, W. B. Choi, S. Yeola, D. C. Liotta. Activities of the four optical isomers of 2',3'-dideoxy 3' thiacytidine (BCH-189) against human immunodeficiency virus type 1 in human lymphocytes. *Antimicrob. Agents Chemother.* 36:672 (1992).

30. C. N. Chang, S.-L Doong, J. H. Zhou, J. W. Beach, L. S. Jeong, C. K. Chu, C. H. Tsa, Y.-C. Cheng. Deoxycytidine deaminase resistent stereoisomer is the active form of (±) 2',3'-dideoxy 3'-thiacytidine in the inhibition of hepatitis B virus replication. *J. Biol.Chem.* 267:13938 (1992).

31. L. C. Stewart, N. K. Richtmyer. Transformation of **D**-gulose to 1,6 anhydro β–**D**-gulopyranose in acid solution. *J. Am. Chem. Soc.* 77:1021 (1955).

32. H. O. Kim, K. Shanmuganathan, S. Nampalli, A. J. Alves, L. S. Jeong, J. W. Beach, R. F. Schinazi, C. K. Chu. Synthesis of 1,3-dioxolane-purine nucleosides and their anti-HIV Activity. *J. Med. Chem.*, in press, (1992).

5-SUBSTITUTED ARABINOFURANOSYLURACIL NUCLEOSIDES

AS ANTIHERPESVIRUS AGENTS: FROM araT TO BV-araU

Haruhiko Machida[1] and Shinji Sakata[2]

Biology[1] and Chemistry[2] Laboratories
R & D Division, Yamasa Corporation

10-1 Araoicho 2-chome, Choshi, Japan 288

INTRODUCTION

After the discovery of acyclovir, much effort was expanded to find new antiherpesviral agents. Before that time, however, application of thymidine analogues as antiherpesviral agents had been tried. The origin of this study is $2'$-deoxy-5-iodouridine (IDU). Although classical thymidine analogues such as IDU, trifluorothymidine, and $2'$-deoxy-5-ethyluridine (EDU), are used only topically due to their narrow selectivity, approach to application of thymidine analogues for selective antiviral agents was strongly emphasized by the findings by Jamieson et al.[1,2] that herpes virus encodes deoxypyrimidine nucleoside kinase (dPyK or viral thymidine kinase), and this enzyme differs far from cellular thymidine kinase (TK) in substrate specificity. Then, in a decade from 1975, a variety of thymidine analogues and acyclic guanosine derivatives were synthesized and tested for antiherpesviral activity to seek new antiviral agents for clinical use. The majority of compounds showing selective antiherpesviral activity are a variety of 5-substituted $2'$-deoxyuridine (dUrd) and arabinofuranosyluracil (araU) derivatives, acyclic guanosine analogues, and $2'$-fluoro-arabinosylpyrimidine nucleosides. The rationale for the selective antiherpesviral agents is based on the preferential phosphorylation by the viral dPyK having broad substrate specificity in the infected cells.

Based on the above knowledge and a finding on the selective antiherpesviral action of 1-β-D-arabinofuranosylthymine (araT) by Gentry and Aswell,[3] we started studying 5-substituted araU analogues for their synthesis and antiherpesviral activity. We found several novel antiherpesviral thymidine analogues such as 5-ethyl-araU, 5-propyl-araU, 5-vinyl-araU, and 5-propenyl-araU,[4,5] and reached to find a novel selective and potent antiherpesviral agent, 1-β-D-arabinofuranosyl-*E*-5-(2-bromovinyl)uracil (BV-araU; generic name: SORIVUDINE) in 1980.[6,7,8] BV-araU was strongly active against varicella-zoster virus (VZV)[9] and showed metabolic stability,[10] Then, we proposed that BV-araU was the most promising agent against VZV infections. Clinical trials conducted in Japan after preclinical toxicological and pharmacological studies indicated that oral administration of BV-araU was effective for treatment of both immunocompetent and immunocompromised patients with herpes zoster in reducing the time of vesicles, erythema, and pain.[11,12,13] In the present paper, we describe antiviral properties and mechanism of action of BV-araU and other 5-substituted arabinosyluracil analogues, and the structure–activity relationships of them.

Nucleosides and Nucleotides as Antitumor and Antiviral Agents,
Edited by C.K. Chu and D.C. Baker, Plenum Press, New York, 1993

5-ALKYL-araU

The selective antiherpesviral action of araT was first reported by Gentry and Aswell in 1975.[3] At the time, antiherpesviral activity of 5-substituted 2'-deoxyuridine derivatives,[14, 15] including 5-ethyl pyrimidine deoxynucleosides,[16] was extensively studied. The analysis of the mode of action of araT by Aswell et al.[17] indicated that araT inhibited DNA synthesis only in herpes simplex virus type 1 (HSV-1)-infected cells but not in uninfected cells, and that this compound was phosphorylated by extracts of HSV-1-infected cells but not by those of uninfected cells. They concluded that viral dPyK was responsible for the selective antiherpesviral activity of araT, which could not be a substrate for cellular TK. This conclusion was supported by the findings that araT was not inhibitory against a variety of dPyK deficient mutants of HSV, and that araT was also effective against VZV but not active against human cytomegalovirus (HCMV) shown later by Miller et al.[18]

The above search on the selective antiviral action of araT prompted us to study 5-alkyl-araU analogues as antiherpesviral agents. 5-Methyl, 5-ethyl, 5-propyl, and 5-butyl arabinosyluracil derivatives were synthesized via 2,2'-anhydro-1-β-D-arabinofuranosyl-5-alkyluracils. These intermediates were synthesized by coupling of 2,4-bis-trimethylsily-oxy-5-alkylpyrimidines and 1,2-di-O-acetyl-3-O-p-toluenesulfonyl-5-O-benzoyl-D-xylo-furanose or from 5-alkyluridines obtained by condensation of 5-alkyluracils with 1-O-acetyl-2,3,5-tri-O-benzoyl-β-D-ribofuranose followed by deprotection.[19] These arabinosyluracil nucleosides exhibited selective anti-HSV-1 activity when determined by the virus-induced cytopathic effect inhibition test on human embryonic lung (HEL) cells.[4] They were completely inactive against growth of HEL cells except that a weak antiproliferative effect was noted for araT. Antiviral potency of the 5-alkyl-araU analogues critically depended on the C-5 side-chain length. 5-Ethyl-araU, a novel thymidine analogue, had almost equivalent potency to that of IDU and EDU against HSV-1, but its activity against herpes simplex virus type 2 (HSV-2) was much less potent. However, as 5-ethyl-araU had no antiproliferative effect on HEL cells, it gave a greater in vitro chemotherapeutic index (ID_{50} for HEL cell growth divided by ED_{50} for HSV-1) than IDU, EDU, and araT. 5-Ethyl-araU was completely inactive against IDU-resistant HSV-1 mutant. Synthesis and the selective antiherpesviral activity of 5-ethyl-araU were also reported by Kulikowski et al.[20] at the same time.

Since araT showed the most potent anti-HSV activity among 5-alkyl-araU analogues in cell culture, we tested effect of araT against experimental encephalitis caused by intracerebral (i.c.) inoculation with HSV-1 in mice. The efficacy observed was not so significant, because a highly pathogenic virus was used as the challenge virus in the study.[21] AraT at a dose of 100 mg/kg twice a day for 9 times (4.5 days) reduced the mortality or increased the mean survival time of the infected mice, irrespective of the route of administration including oral treatments. This effect was equivalent to that of araA 5'-monophosphate at a dose of 50 mg/kg. AraT was well tolerated for mice as a large single dose or at any doses used for the therapy of the experimental infection. In our subsequent studies using a low dose of the same challenge virus, oral araT treatments significantly increased the survival rate of mice at a dose of 50 mg/kg in the same regimen.[22] However, it showed severe neurotoxicity in monkeys,[23] and it was dropped from clinical development. In vivo antiherpesviral activity of 5-ethyl-araU was marginal (Machida, unpublished).

To elucidate the antiherpesviral potencies of 5-substituted arabinosyluracil analogues and compare their antiviral activities with those of corresponding 2'-deoxyuridine analogues, we synthesized various 5-substituted araU and dUrd derivatives containing a new carbon chain at the C-5 position, and tested them for their antiherpesviral activities.[6, 8] Among the substituted nucleosides we newly synthesized, 5-acetonyl-dUrd (aceto-dUrd) exhibited significant anti-HSV-1 activity but did not have anti-HSV-2 effect, like 5-ethyl-araU. 5-Hydroxy-araU and 5-methoxycarbonylmethyl-dUrd (MCM-dUrd) showed moderate or modest antiviral effect, while arabinosyluracil congeners of aceto-dUrd and MCM-dUrd (aceto-araU and MCM-araU, respectively) were almost inactive (Table 1).

5-Alkenyl-araU

As described above, substitution for deoxyribose of certain 5-substituted 2′-deoxy-uridine derivatives by arabinose reduced or sometimes lost their antiproliferative effect, preserving their anti-HSV-1 activity. In this case, their anti-HSV-2 potency was also reduced or lost. Then, we expanded our research to seek the second series of 5-substituted arabinosyluracil analogues. We selected 5-alkenyl-araU analogues for the target. Since 5-vinyl-dUrd was reported to have potent antiherpesviral activity, but it was rather inhibitory to growth of WI-38 cells,[14] and 5-vinyluridine had been just synthesized (Ikeda, personal communication), we had expected particularly selective antiherpesviral activity of 5-vinyl-araU, at least with less toxic than 5-vinyl-dUrd. 5-Alkenyl-araU analogues were obtained by condensation of aldehydes with the 5-methylenetriphenylphosphorane derivative, which was obtained from 5-chloromethyl-araU (Ikeda, unpublished). As expected, 5-vinyl-araU showed marked antiviral effects on both HSV-1 and HSV-2 in the assay system described above.[5] The anti-HSV activities were almost comparable to those of araT and higher than those of IDU, EDU, and 5-ethyl-araU. 5-Vinyl-araU showed only marginal inhibitory effect, but a lesser extent than araT, against HEL cell growth. 5-Propenyl-araU and 5-butenyl-araU showed moderate and relatively weak activities against HSV-1, respectively.

Table 1. Antiherpesviral activity of 5-substituted arabinosyluracil and deoxy-uridine analogues assayed by the cytopathic effect inhibition and their anti-HEL cell growth activity

Compound	ED_{50} for HSV (μg/mL)		ID_{50} for HEL cells (μg/mL)
	HSV-1	HSV-2	
AraU	100	100	>1,000
AraT	1	1	100
5-Ethyl-araU	3.2	100	>1,000
5-Propyl-araU	32	320	>1,000
5-Butyl-araU	100	>1,000	>1,000
5-Hydroxy-araU	32	100	>1,000
Aceto-araU	320	>1,000	Not tested
MCM-araU	320	>1,000	Not tested
5-Vinyl-araU	0.32	1	650
5-Propenyl-araU	10	320	>1,000
5-Butenyl-araU	32	>1,000	>1,000
BV-araU	0.032	100	1,000
CV-araU	0.1	1,000	>1,000
5-Hydroxy-dUrd	32	10	Not tested
Aceto-dUrd	3.2	>1,000	>1,000
MCM-dUrd	32	32	400
IDU	3.2	3.2	8
EDU	3.2	10	40
BVDU	0.032	100	150
Acyclovir	0.1	0.1	260

Alkyl: R= CH$_3$,C$_2$H$_5$, C$_3$H$_7$,C$_4$H$_9$

Alkenyl: R= CH=CH$_2$, CH=CHCH$_3$, CH=CHCH$_2$CH$_3$

Alkynyl: R= C≡CH C≡CCH$_3$

Halogenovinyl: R= *E*-CH=CHBr, *E*-CH=CHCl *E*-CH=CHI

Data from Machida et al.[4,6] and Machida.[24]

But they were almost inactive against HSV-2. Thus the antiherpesviral activity of 5-alkenyl-araU analogues decreased with increasing the side-chain length at the C-5 position as seen in the activity of 5-alkyl-araU analogues. Table 1 summarizes the anti-HSV and anti-HEL cell growth activities of 5-substituted arabinosyluracil and 2′-deoxyuridine derivatives. Selective antiherpesviral activity of 5-propenyl-dUrd was also shown, which was preferentially phosphorylated by the viral dPyKs,[25] while Stening et al.[26] reported that 5-propenyl-dUrd showed potent anti-HSV activity, but 5-propenyl-araU was inactive when assayed on Vero cells.

The structure–activity relationship in the chain length of the 5-substituent, similar to that observed in 5-alkenyl-araU analogues, was reported concerning 5-alkenyl-dUrd analogues, in which 5-substituent was not longer than four carbon atoms in length.[27, 28] Moreover, it was shown that analogues in which the C=C double bond of the olefinic 5-substituent was conjugated to the pyrimidine ring were more potent antiviral drugs than the corresponding non-conjugated analogues and those with fully saturated hydrocarbon (alkyl) 5-substituent; E-5-(1-propenyl)-dUrd was more potent than 5-(2-propenyl)-dUrd (5-allyl-dUrd), and E-5-(1-butenyl)-dUrd was more potent than E-5-(2-butenyl)-dUrd and 5-butyl-dUrd.[28, 29] This suggested importance of conjugation of C=C unsaturated bond as a key molecular feature affecting antiherpesviral activity of 5-substituted deoxyuridines. Probably, this is true in 5-substituted arabinosyluracil analogues, because the structure–activity relationship with respect to the side-chain of the 5-substituent is completely the same between deoxyuridine and arabinosyluracil nucleosides as described here.

Since a large amount of 5-vinyl-araU could not be synthesized by the described method, testing 5-vinyl-araU for in vivo antiviral activity was not available for us. Reefschlaeger et al.[30] reported excellent in vivo activity of 5-vinyl-araU against experimental encephalitis caused by i.c. inoculation with HSV-1 in mice, which was as much potent as BV-araU and superior to acyclovir.

5-Alkynyl-araU

As antiherpesviral activity of 5-ethynyl-dUrd[31] and its growth inhibition for L5178Y cells were reported[32], 5-ethynyl-araU had been expected as a new selective antiherpesviral compound as a member of 5-substituted arabinosyluracil analogues. The new 5-alkynyl series of arabinosyluracils, however, were synthesized and reported by De Clercq et al.[33] as congeners of 5-alkynyl-2′-deoxyuridine after findings of potent antiherpesviral agents, E-5-(2-bromovinyl)-2′-deoxyuridine (BVDU)[31] and BV-araU[7, 8] described below. The 5-alkynyl series of arabinosyluracils were synthesized by cross-coupling reaction of 5-iodo-2′,3′,5′-tri-O-p-toluoyl-araU with terminal alkynes, followed by deprotection of the intermediates. 5-Ethynyl- and 5-propynyl-araU, however, were less active against HSV-1 and HSV-2 than expected. Their anti-HSV-1 activity was 8 to 70-fold less than that of the corresponding 5-alkynyl-dUrd and representative 5-alkenyl-araU analogues, while they exhibited weak antiproliferative activity resulting in having small in vitro chemotherapeutic index. In recent works, we also have found similar potencies of 5-ethynyl derivatives of dUrd and araU against HSV, but observed considerably potent effect of 5-ethynyl and 5-propenyl analogues of them on VZV (Table 2). The antiviral potency of the series of 5-alkynyl-araU, as well as that of 5-alkynyl-dUrd, also critically depended on the chain length of the C-5 side-chain as recorded for 5-alkyl-araU analogues mentioned above.

5-HALOGENOVINYL-ARABINOSYLURACILS AND THEIR ANTI-VZV POTENCY

As derivatives of 5-vinyl-dUrd, De Clercq et al.[31] found BVDU and 2′-deoxy-E-5-(2-iodovinyl)uridine (IVDU) with potent and selective antiherpesviral activity. Although these halogenovinyl compounds were very weakly active against HSV-2, BVDU was the most potent against HSV-1 among representative antiherpesviral nucleoside analogues, and the effectiveness of its topical application in a mouse model infection was demonstrated. In contrast, 2′-deoxy-(Z)-5-(2-bromovinyl)uridine (Z isomer of BVDU) was much less active than BVDU.[34] This is true concerning BV-araU; 1-β-D-arabinofuranosyl-Z-5-(2-bromovinyl)uracil (Z-BV-araU) is much less potent than BV-araU (Table 2). In structural–activity study of 5-substituted deoxyuridines, Goodchild et al.[27] pointed out for optimum inhibition of HSV-1 in cell culture that the unsaturated 5-substituent, conjugated with the pyrimidine ring, had E stereochemistry, and included a hydrophobic, electronegative function, but did not contain a branching point. BVDU fully contains such features. BVDU has been most extensively studied for its in vitro and in vivo antiviral property among antiherpesviral nucleosides except acyclovir. Its antiviral spectrum was reviewed.[35] The biochemical aspect of its selective antiherpesviral activity is almost the same to that described here with respect to selective thymidine analogues, araT, BV-araU and other thymidine analogues.[36]

For introducing the halogenovinyl group into the C-5 side-chain of 5-substituted arabinosyluracils, we first synthesized BV-araU and 1-β-D-arabinofuranosyl-E-5-(2-chlorovinyl)uracil (CV-araU) from 5-vinyl-araU by treating it with N-bromosuccinimide and N-chlorosuccinimide, respectively.[8] For a large-scale preparation of BV-araU, the original method was improved as follows: 1-β-D-arabinofuranosyl-E-5-(2-carboxyvinyl)uracil (CAV-araU) was obtained from 2′,3′,5′-tri-O-acetyl-1-β-D-arabinofurano-

Table 2. Comparison of anti-VZV and anti-HSV activities of 5-substituted arabinosyluracils and related compounds determined by the plaque reduction method.

Compound	ED_{50} (µg/mL) for[a];			Ratio of anti-HSV-1 to anti-VZV[b]
	HSV-1	HSV-2	VZV	
AraT	0.46	0.45	0.37	[0.10]
5-Ethyl-araU	2.03	65.5	>160	[<-1.9]
5-Vinyl-araU	0.10	4.45	1.10	[-1.04]
5-Ethynyl-araU	2.45	16.4	0.27	[0.96]
5-Propynyl-araU[c]	Not tested	Not tested	0.46	
BV-araU	0.021	48	0.0008	[1.42]
CV-araU	0.026	156	0.0017	[1.18]
IV-araU	0.047	38	0.0014	[1.53]
Z-BV-araU	14.6	>320	0.41	[1.55]
Z-CV-araU	6.43	>320	0.84	[0.89]
E-5-Trifluoropropenyl-araU	5.82	291	0.16	[1.55]
E-5-Phenylthiovinyl-araU	>320	>320	45.2	[>0.85]
5-Dibromovinyl-araU	300	>320	8.03	[1.57]
5-Ethyl-dUrd (EDU)	8.08	10	>160	[<-1.3]
5-Vinyl-dUrd	0.11	0.71	0.30	[-0.45]
5-Ethynyl-dUrd[c]	0.64	2.39	0.10	[0.81]
5-Propynyl-dUrd[c]	Not tested	Not tested	0.85	
BVDU	0.039	192	0.014	[0.44]
2′-Fluoro-5-methyl-araU (FMAU)	0.021	0.042	0.028	[-0.12]
2′-Fluoro-5-ethyl-araU (FEAU)	0.060	0.21	0.13	[-0.34]
2′-Fluoro-5-vinyl-araU (FVAU)	0.020	0.30	0.084	[-0.62]
2′-Fluoro-BV-araU (FBVAU)	0.049	8.35	0.0081	[0.78]

FEAU is a gift from Dr. Watanabe of Sloan-Kettering Institute.

[a] ED_{50} was determined for HSV-1 VR-3, HSV-2 MS, and VZV Oka.

[b] $Log_{10}(ED_{50}$ for HSV-1/ED_{50} for VZV).

[c] Data from Ashida et al. (unpublished).

syl-5-iodouracil by a palladium-catalyzed coupling reaction with methyl acrylate and subsequent hydrolysis. CAV-araU was converted into BV-araU with N-bromosuccinimide by the procedure used for synthesis of BVDU[37, 38] with minor modifications.[39] 1-β-D-Arabinofuranosyl-E-5-(2-iodovinyl)uracil (IV-araU) could be obtained by the method similar to the improved method (Sakata, unpublished). These 5-halogenovinyl-araU analogues were strongly active against HSV-1. They were slightly more active than 5-vinyl-araU, and almost as active as BVDU, but weakly active against HSV-2 as was BVDU, while they were less inhibitory to HEL cell growth than BVDU, giving a higher in vitro chemotherapeutic indexes than BVDU.[6, 7] IV-araU also showed antiherpesviral activities similar to those of BV-araU in vitro[22] and in vivo (Machida, unpublished).

Miller et al.[18] reported inhibition of VZV replication by araT, of which biochemical basis is in accordance with the fact that VZV encodes virus-specified dPyK.[40, 41] We also considered application of 5-substituted arabinosyluracil nucleosides to this area and tested them for anti-VZV activity. The bases for our approach were, i) VZV induces viral dPyK, ii) the morbidity rate of primary infection with VZV is much higher than that of primary infection with HSV, iii) in Japan, zoster is a more socially important disease than are HSV infections, iv) the chemotherapy for HSV infections seemed to be controlled by acyclovir, and v) acyclovir had a relatively weak activity against VZV compared with its anti-HSV potential. At that time, HSV-1, HSV-2, and VZV were known to induce viral dPyK among human herpesviruses, and it had been generally accepted that nucleoside analogues with anti-HSV-1 activity were also likely to be active against VZV. However, it was possible that there was a marked difference between HSV-1 and VZV in the susceptibility to certain nucleoside analogues, as seen in the difference between the susceptibilities of HSV-1 and HSV-2 to 5-ethyl-araU, aceto-dUrd, BVDU, and BV-araU as shown in Table 1.

In comparison of activities of nucleoside analogues against the viral dPyK-encoding herpesviruses, the most surprising phenomenon we found was that 5-ethyl-araU was almost inactive against VZV.[24] Acyclovir, EDU, aceto-dUrd, and 5-vinyl-araU were less active against VZV than against HSV-1. In contrast, BVDU, MCM-dUrd, and some arabinosyluracils showed higher potency against VZV than against HSV-1. Particularly, BV-araU was most potent against VZV among the compounds tested. Since the ED_{50} of BV-araU for VZV was 0.1–0.6 ng/mL and the ID_{50} for HEL cell growth was 800 μg/mL or more, in vitro chemotherapeutic index of BV-araU for VZV was calculated to be more than 1,000,000. Other E-5-(2-halogenovinyl)-araU derivatives (CV-araU and IV-araU) also showed marked anti-VZV effect almost equivalent to that of BV-araU (Table 2). The significant anti-VZV potency of BV-araU was confirmed by Baba et al.[42] by using recent clinical isolates of VZV. Interestingly, our subsequent study shows other 5-substituted arabinosyluracils such as 5-ethynyl-araU, 5-propynyl-araU, Z-BV-araU, 1-β-D-arabinofuranosyl-Z-5-(2-chlorovinyl)uracil (Z-CV-araU), and E-5-trifluoropropenyl-araU have antiviral activity selective for VZV. Their anti-HSV-1 activity is relatively weak, but they are 10-fold or more highly active against VZV than against HSV-1. Thus, a series of 5-substituted araU analogues show selective anti-VZV effect and the activity of E-5-(2-halogenovinyl)-araUs is particularly marked. It is noteworthy that E-5-(2-bromovinyl)-2$'$-deoxy-4$'$-thiouridine showed antiviral effect specific for VZV[43]. Also, Slusarchyk et al.,[44] have reported selective anti-VZV effect of 1-cyclobutyl-E-5-(2-halogenovinyl)uracil nucleosides. These carbocyclic nucleosides were moderately active against HSV-1, while their anti-VZV effect was 50 to 100 times more potent than anti-HSV-1 activity of them (ED_{50}s for several VZV strains were ranging from 0.02 to 0.2 μM). We have found that 2'-deoxy-2'-methylideneuridine analogues having the E-(2-halogenovinyl) group at the C-5 position also shows VZV-selective antiviral activity[39]. These recent findings suggest that irrespective of the kind of sugar moiety of thymidine analogues, E-5-(2-halogenovinyl)uracil can be easily recognized by VZV-dPyK rather than by HSV-1-dPyK resulting in highly potent anti-VZV activity of the corresponding nucleosides. This is also supported by our previous observation that 2$'$-fluoro-arabinosyl-E-5-(2-bromovinyl)uracil (FBVAU) shows the most potent anti-VZV activity among the 2$'$-fluoro-arabinofuranosyluracil analogues (Table 2).

2'-FLUORO-araU

This unique group of the 5-substituted arabinosyluracil analogues were synthesized by Watanabe et al.[45, 46] as a series of 2'-fluoro-arabinofuranosylpyrimidines, which were obtained by condensation of 3-O-acetyl-5-O-benzoyl-2-deoxy-2-fluoro-D-arabinofranosyl bromide with modified or unmodified pyrimidine bases. 2'-Fluoro-arabinosylpyrimidine nucleosides exhibited more potent antiherpesviral activity than did their 2'-chloro or 2'-bromo congeners.[47] The importance of the fluorine in the 2'-"up" (arabino) configuration for effective antiherpesviral activity was also demonstrated. Among the 2'-fluoro-arabinofuranosylpyrimidine nucleosides, 2'-fluoro-arabinosyl-5-iodocytosine (FIAC), 2'-fluoro-arabinosyl-5-iodouracil (FIAU), 2'-fluoro-arabinosyl-5-methyluracil (FMAU), 2'-fluoro-arabinosyl-5-ethyluracil (FEAU), and 2'-fluoro-arabinosyl-5-vinyluracil (FVAU) showed particularly potent antiherpesviral activity, while introducing an E-(2-bromovinyl) residue at the C-5 position did not improve the anti-HSV efficacy of the 2'-fluoro-substituted derivatives[46] (see also Table 2). FIAC was first studied in detail for the antiherpesviral property. This compound was active against both HSV-1 and HSV-2, but it was 8,000 times less active against dPyK-deficient mutant of HSV-1.[48] It was slightly more active against HSV-1 and slightly more inhibitory to normal cell growth than acyclovir. More interestingly, these 2'-fluoro-arabinofuranosyl pyrimidine nucleosides showed potent anti-HCMV activity, though FIAU and FMAU gave low selectivity due to their relatively high cell toxicity.[49] Efficacies of nontoxic intraperitoneal or oral treatments with FMAU, FIAC, and FIAU against i.c. infection of mice were demonstrated.[50, 51] It was also shown that FIAU and FMAU were effective in preventing the development of evidences of clinical infection of African green monkeys with simian varicella virus (SVV).[52]

ANTIVIRAL PROFILE AND MODE OF ACTION OF BV-araU

In Vitro Antiviral Spectrum

In the comparative study of in vitro antiviral activities of BV-araU and representative antiherpesviral agents by the plaque reduction method in a re-examination, BV-araU showed the most marked antiviral activity against all five strains of VZV tested (Table 3). Significant activity of BV-araU against all seven strains of HSV-1 tested was also demonstrated. Average ED_{50} of BV-araU for VZV and HSV-1 were 0.4 and 22 ng/mL, respectively. Based on the ED_{50} for VZV, BV-araU was over 2,000 times more active than acyclovir and 4,000 times more active than vidarabine, being consistent with our previous findings.[24] We achieved the significant potential of BV-araU against recent clinical isolates obtained from immunocompetent patients enrolled in the clinical trials of oral treatment with BV-araU for therapy of zoster; average ED_{50} of BV-araU for 101 isolates was 1.3 ng/mL.[53] This value was higher than the ED_{50} for the laboratory-stocked VZV strains described above, but these clinical isolates were also less susceptible to acyclovir than laboratory-stocked strains. BV-araU was 3,000-fold more potent than acyclovir for the clinical isolates based on the ED_{50} as was determined for laboratory-stocked strains. BV-araU showed marginal or no effect on HSV-2 replication and had little effect on HCMV.[54] Effectiveness of BV-araU for Epstein-Barr virus (EB virus) was also demonstrated.[55] Recently, Lin et al.[56] showed a structure–activity relationship between 5-vinyl-araU and BV-araU in inhibition of EB virus replication. The ED_{50} of 5-vinyl-araU and BV-araU for viral DNA replication were 0.005 and 0.3 μg/mL, respectively. 5-Vinyl-araU was the most potent among nucleoside analogues which had been tested, and inhibited the synthesis of some small viral polypeptides, but BV-araU had a more prolonged inhibitory effect than 5-vinyl-araU, indicating that the substitution of the 5-vinyl group for hydrogen by bromine resulted in reduction in anti-EB virus activity, while prolonging the drug effect and diminishing cytotoxicity.

Table 3. Antiherpesvirus profile of BV-araU and other representative antiviral agents determined by the plaque reduction method on HEL cells

Compound	ED_{50} of drug $(\mu g/mL)^a$						
	VZV (5)^b		HSV-1 (7)		HSV-2 (3)		HCMV (1)
	Mean	Range	Mean	Range	Mean	Range	
BV-araU	0.00040	0.00015 - 0.00079	0.022	0.012 - 0.042	56	23 - >320	569
BVDU	0.0051	0.0026 - 0.014	0.043	0.032 - 0.056	>320	>320	85
Acyclovir	0.98	0.50 - 1.78	0.10	0.088 - 0.13	0.21	0.18 - 0.29	11.3
IDU	0.72	0.35 - 1.69	0.59	0.28 - 0.95	2.0	1.7 - 2.7	25
Vidarabine	2.03	1.51 - 2.60	13.2	9.1 - 17.4	4.6	2.0 - 10	6.8

Data from Machida.[54]

^{a} Geometric mean of 2 to 4 separate experiments for each strain.

^{b} Number of strains tested.

Host Cell Dependency for Antiviral Activities of BV-araU

An interesting antiviral property of BV-araU is the host cell dependence of its anti-HSV-1 activity. The significant host cell dependence of anti-HSV-1 activity of BV-araU was first reported by Reefschlaeger et al.[57]; anti-HSV-1 potencies of BV-araU and 5-vinyl-araU in Vero cells were nearly one-100th of those seen in HEL cells or sometimes even completely absent, although anti-HSV activities of nucleoside analogues generally vary with the cell line used in antiviral activity assays.[58] We also confirmed this markedly different antiviral potencies of BV-araU against HSV-1 on human cell lines and Vero cells, compared various nucleoside analogues for their host cell dependence, and tested anti-HSV-1 activities of BV-araU in various cell lines.[59] All 5-halogenovinyl-araU analogues were inactive against laboratory-stocked strains of HSV-1 when tested in Vero cells but exerted moderate antiviral effect on three recent clinical isolates tested. Antiviral potencies of representative 2'-deoxyuridine, 2'-fluoro-arabinosylpyrimidine, and acyclic nucleoside analogues against the laboratory-stocked strains were not so markedly reduced in Vero cells compared with those in HEL cells. Generally, arabinosyluracil nucleosides were likely to exhibit inactivity or low activity against HSV-1 in Vero cells. The discrepancy in findings by us and by Stening[26] on the anti-HSV-1 activity of 5-propenyl-araU, mentioned previously, can be explained by this phenomenon. Further experiments showed that BV-araU had marked antiviral potencies against the laboratory-stocked strains in other human and monkey fibroblast cell lines including an immortalized cell line (Table 4), and that it showed moderate antiviral action on four of eleven clinical isolates. There were two types in HSV-1 strains based on the susceptibility to BV-araU in Vero cells. In the comparison of anti-HSV-1 activities of BV-araU and acyclovir on mouse 3T3 cells, BV-araU was less potent than acyclovir; the mean ED_{50}s of BV-araU and acyclovir for three other strains were 0.13 and 0.073 $\mu g/mL$, respectively, while BV-araU was more active in HEL cells.[60]

Antiviral activity of BV-araU against a laboratory strain and a clinical isolate of HSV-1 in Vero cells was significantly potentiated by the addition of 0.1 μM aminopterin or 2'-deoxy-5-fluorouridine (FUdR), an inhibitor of thymidylate synthase, to the level as marked as the activity in HEL cells. These results suggested that being highly active in thymidylate biosynthesis and a large thymidylate pool size in Vero cells might be responsible for the observed potency of BV-araU against HSV-1 in Vero cells. In contrast to the activity against HSV-1, BV-araU showed extremely marked antiviral activity against VZV even on Vero cells (Table 5). The reason why the host cell dependence of anti-HSV-1 and anti-VZV activities of these compounds varies is not yet known.

Table 4. Comparison of anti-HSV-1 activities of BV-araU and acyclovir on various cell lines

Cell lines (origin)	Drug	ED_{50} of drug (50% plaque reduction dose; μg/mL)			
		Laboratory-stocked strain (3)[a]		Clinical isolates (3)	
		Mean	Range	Mean	Range
HEL	BV-araU	0.013	0.016 - 0.027	0.012	0.004 - 0.030
(Human embryo lung)	Acyclovir	0.049	0.020 - 0.12	0.046	0.007 - 0.14
FS-4	BV-araU	0.025	0.019 - 0.035	0.017	0.009 - 0.027
(Human foreskin)	Acyclovir	0.062	0.022 - 0.145	0.115	0.085 - 0.19
Flow 4000	BV-araU	0.014	0.010 - 0.023	0.141	0.005 - 0.013
(Human whole embryo)	Acyclovir	0.009	0.12 - 0.18	0.097	0.066 - 0.13
KMS-6	BV-araU	0.032	0.021 - 0.030	0.082	0.006 - 0.015
(Human whole embryo)	Acyclovir	0.082	0.060 - 0.11	0.054	0.030 - 0.079
KMST-6	BV-araU	1.00*	0.72* - 1.38*	1.66	0.15 - 0.66
(Immortalized KMS-6)	Acyclovir	0.36	1.35 - 2.0	1.32	0.95 - 1.95
DBS-FCL2	BV-araU	0.13	0.057 - 0.30	0.077	0.032 - 0.058
(Monkey lung)	Acyclovir	0.053	0.064 - 0.89	0.20	0.055 - 0.11
Vero	BV-araU	>400	>400	2.4*	0.59 - 10
(Monkey kidney)	Acyclovir	0.60	0.20 - 1.8	0.24	0.21 - 0.30

[a]Number of strains tested.
* Complete inhibition of plaque formation was not achieved at concentrations up to 320 μg/mL.

Table 5. Anti-VZV activities of nucleoside analogues on HEL and Vero cells

Compound	Mean ED_{50} on (ng/mL);[a]		Ratio of mean ED_{50}[b]
	HEL	Vero	
BV-araU	0.37	0.57	1.6
IV-araU	0.33	0.45	1.3
CV-araU	0.45	0.78	1.7
BVDU	5.5	0.71	0.1
FMAU	4.9	6.4	1.3
FEAU	35	29	0.8
AraT	80	132	1.7
5-Vinyl-araU	186	1,235	6.6
Acyclovir	460	654	1.4

Data from Machida et al.[61]
[a]Geometric mean of ED_{50} for four strains of VZV.
[b] ED_{50} on Vero/ED_{50} on HEL.

Mode of Selective Action of BV-araU

Neither BV-araU nor acyclovir significantly affected uptake of ^{3}H-thymidine by HSV-1- and VZV-infected cells, but both compounds markedly inhibited incorporation of ^{3}H-thymidine into the DNA fraction of the infected cells.[62] The concentrations, at which inhibition of the DNA synthesis was observed, corresponded well with those inhibitory to plaque formation. The concentration of BV-araU inhibiting VZV DNA synthesis was about 3 ng/mL and more than 1,000-fold lower than that of acyclovir. The ratio of the inhibitory concentrations for VZV DNA synthesis was just the same as that in the inhibition of VZV plaque formation.

As molecular basis for the selective inhibition of herpesvirus replication by BV-araU, it was indicated that BV-araU could act as an alternative substrate of thymidine for HSV-1- and HSV-2-encoded dPyKs and was effectively phosphorylated by these enzymes, while cellular TK could not phosphorylate this compound.[63] BV-araU also showed high affinity for VZV dPyK (Table 6). The K_m of purified VZV dPyK for thymidine was 0.87 μM, while that for BV-araU was 0.19 μM. BV-araU acts as a competitive inhibitor in the phosphorylation of thymidine. The K_i of BV-araU for VZV dPyK was 0.19 μM, which was lower than that for HSV-1 dPyK. In contrast, the K_i of acyclovir for VZV dPyK was more than three orders of magnitude higher than that of BV-araU. The low affinity of VZV dPyK for acyclic guanosine analogues was reported to be related to the relatively low activity of acyclovir against VZV.[64] Like BVDU triphosphate (BVDU-TP), BV-araU triphosphate (BVAU-TP) strongly inhibited HSV polymerase.[65] The K_i of BVAU-TP for HSV-1 DNA polymerase was 20-fold lower than that for cellular DNA polymerase α. However, BVAU-TP could not be an alternative substrate of TTP. BVAU-TP was also strongly inhibited VZV DNA polymerase (Table 7), but it was not incorporated into viral DNA obtained from VZV-infected cells.[66] In contrast, IVDU triphosphate was well incorporated into viral DNA.[67, 68] BV-araU is several times more potent against VZV than BVDU; nevertheless VZV dPyK shows higher affinity for BVDU than for BV-araU. The slightly lower K_i of BVAU-TP for VZV DNA polymerase than that of BVDU-TP might reflect the more potent anti-VZV activity. Alternatively, BV-araU monophosphate (BVAU-MP) may be more effectively phosphorylated to its diphosphate (BVAU-DP) by VZV dPyK-associated thymidylate kinase than BVDU monophosphates (BVDU-MP).

Studies on metabolic fate of BV-araU indicated that most of radio-activity was recovered from the acid-soluble fraction of ^{14}C-labeled BV-araU-treated HSV-1- and VZV-infected cells, irrespective of the duration of incubation and the concentration of BV-araU.[66, 69, 70] No elevation of the uptake with time was shown in mock-infected and dPyK-deficient virus-infected cells. The uptake of BV-araU may be regulated by the amount of virus-induced dPyK activity. A significant amount of BVAU-MP and detectable amounts of BVAU-DP and BVAU-TP were found in the acid-soluble fraction of HSV-1- or VZV-infected cells, while only a minimum amount of BVAU-TP was accumulated in HSV-2-infected cells. By HPLC analysis, Ayisi et al.[71] observed that BVAU-DP and BVAU-TP were formed in HSV-1-infected cells, but not in HSV-2-infected cells. The prominently low ability of HSV-2 dPyK to phosphorylate BVAU-MP to BVAU-DP can be associated

Table 6. Interaction of BV-araU and other antiviral nucleosides with various thymidine kinases

Compound	K_m of VZV dPyK (μM)	K_i for thymidine kinases (μM)			
		VZV	HSV-1	HSV-2	Cytosol[a]
Thymidine	0.87				
BV-araU	0.19	0.19	8	100	274
BVDU		0.08	0.2	4.1	150
IVDU	0.07	0.06	0.06	0.05	
Acyclovir		820	47	180	1,000

Data from Yokota et al.[66]

[a] Prepared from HEL cells.

Table 7. Interaction of BVAU-TP and other nucleoside triphosphates with VZV DNA polymerase

Compound	K_m (μM)	K_i for VZV DNA polymerase (μM)
TTP	1.43	
BVAU-TP		0.34
BVDU-TP		0.55
Acyclovir-TP		0.29

Data from Yokota et al.[66]

with low activity of BV-araU against this virus. Similarly, Fyfe[72] demonstrated the ability of HSV-1 and VZV dPyKs to phosphorylate BVDU-MP to BVDU 5′-diphosphate and the inability of HSV-2 dPyK or cellular TK to effect their conversion. More interestingly, exhaustion of BVAU-MP and accumulation of a large amount of BVAU-TP was seen when thymidylate synthase activity was inhibited by treating HSV-1-infected cells with FUdR, and as a consequence, de novo synthesis of TMP/TTP was suppressed.[70] Phosphorylation of BVAU-MP to BVAU-DP seems to compete with that of thymidylate and to be the limiting step for activation of BV-araU, i.e. formation of BVAU-TP. There is no evidence that BV-araU is incorporated into viral DNA even at the 5′-termini by the analysis of the acid-insoluble materials obtained from ^{14}C-labeled BV-araU-treated HSV-1-infected cells (Ashida and Machida, unpublished). We propose that the potent antiherpesviral action of BV-araU is attributable to the direct inhibition of viral DNA synthesis by BVAU-TP in the infected cells.

In conclusion, the high affinity of BV-araU to VZV dPyK and the effective phosphorylation of BVAU-MP to BVAU-TP in VZV-infected cells correlates with the selective and potent anti-VZV action of BV-araU, although the direct phosphorylation of BVAU-MP to BVAU-DP by VZV-induced dPyK-associated thymidylate kinase has not yet been demonstrated.

ANTIHERPES ACTIVITY OF BV-araU IN EXPERIMENTAL ANIMALS

Effects on Mouse Model Infections with HSV-1

Effects of BV-araU on mouse model infections with HSV-1 were tested by using young male mice, which were infected either i.c. or i.p. with HSV-1 WT-51 strain. Over 90% and 80% of placebo-treated control mice bearing i.c. and i.p. infections, respectively, died. Significant effects of oral and i.p. treatments with BV-araU against experimental encephalitis caused by the i.c. infection were clearly demonstrated at a dose-dependent manner.[73] Treatments with BV-araU at a dose as low as 12.5 mg/kg twice a day for 7 days were effective in increasing both the survival rate and the mean survival time with statistical significance. Efficacies of BV-araU against the i.p. infection were demonstrated equally to those found against i.c. infection. The antiviral potency of BV-araU in mice was affected by the degree of virulence of the challenge virus strain used. We suggested that this was the reason for variation of the degree of efficacies of a certain nucleoside analogue in a mouse model infection with HSV reported by different researchers. Efficacy of i.p. administration of BV-araU was almost equivalent to that of oral treatments, irrespective of the route of virus inoculation. This is reasonable, because BV-araU was well absorbed through gastrointestinal tract after oral administration into mice and any identified metabolites were not found in the urine.[10] In contrast, when BVDU was administered into mice considerable amounts of *E*-(2-bromovinyl)uracil (BV-uracil) were found in plasma and the urine as a metabolite, and the blood concentration of BVDU rapidly decreased. It has been recently recognized, however, that when BV-araU is given orally or intravenously into rats BV-uracil is found in plasma, which is excreted into the urine, and that BV-araU is degraded by deglycosylating action of entero-bacteria, resulting in formation of BV-

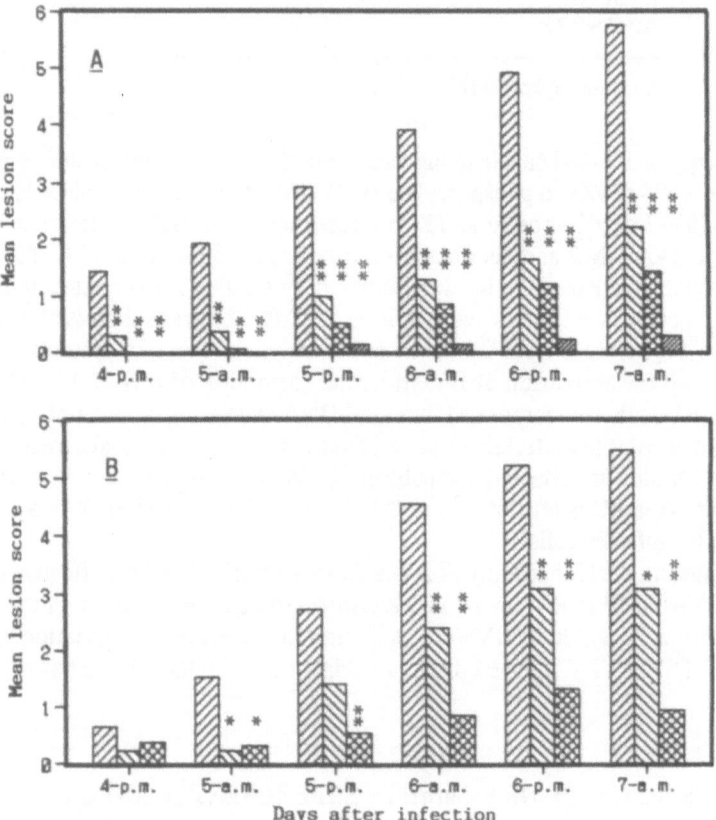

Fig. 2. Effect of oral treatment with BV-araU on cutaneous infections with HSV-1 F and WT-51 strains in shaved Balb/c mice. Shaved male Balb/c mice, in groups of 14 animals, were infected with HSV-1 F strain (A) or female mice were infected with WT-51 strain (B). Infected mice were treated orally with BV-araU twice daily for 7 days beginning one day after inoculation. ▨ control, ▨ 20 mg/kg, ▨ 50 mg/kg, and ▨ 100 mg/kg. *: $P<0.05$, **: $P<0.01$ by Mann-Whitney U-test.

uracil, but not by incubation with rat liver extract (Ijichi et al. unpublished).

For the evaluation of efficacies against cutaneous infection with HSV-1, we used 7 to 8 weeks old shaved Balb/c mice, which were cutaneously infected with HSV-1. Local skin symptoms were observed twice a day and scored according to the criteria shown by us.[60] Oral treatments with BV-araU at a dose of 20 mg/kg twice a day suppressed development of skin symptoms and markedly reduced the mortality (Fig. 2). Strong or almost complete suppression of development of symptoms on the skin was seen in groups of mice that received 50 and 100 mg/kg of BV-araU without infection death. The minimum dose effective in reducing the mortality was estimated to be 20 mg/kg twice a day. Topical treatments with 5% BV-araU cream 4 times a day beginning one day postinfection almost completely suppressed progression of cutaneous symptoms caused by the infection with a low virulent strain of HSV-1.[74] The 5% BV-araU cream was effective for reduction of the mean maximum lesion score (p<0.001), but not for decrease in the morbidity rate of the infected mice by delayed therapy (treatment began 2 days postinfection). Virus titers at the infected skin site were markedly decreased, and viral growth in the lower flank site (symptom-appearing site) was almost completely suppressed by the treatments with the BV-araU cream beginning one day postinfection. Topical application of BV-araU can be useful therapy for HSV-1 infections in humans.

Table 8. Comparison of efficacies of BV-araU and acyclovir on mouse model infections with HSV-1 WT-51 strain

Infection route	Administration route	Treatment[a] Drug	Dose (mg/kg)	Survivors per total	Mean survival time of mice that died (days ± S.E.)
i.c.	Oral	None (control)	0	2/21	5.42 ± 0.32
"	"	BV-araU	25	15/22 $P<0.001^b$ $(P<0.001)^d$	8.86 ± 1.22 $P<0.01^c$ $(P<0.01)^e$
"	"	BV-araU	100	19/22 $P<0.001^b$ $(P<0.001)^d$	12.33 ± 2.33 $P<0.001^c$ $(P<0.001)^e$
"	"	Acyclovir	25	1/21	6.05 ± 0.37
"	"	Acyclovir	100	1/22	7.00 ± 0.38 $P<0.01^c$
i.c.	i.p.	None (control)	0	0/21	5.19 ± 0.34
"	"	BV-araU	10	1/21	6.25 ± 0.32 $P<0.05^c$
"	"	BV-araU	50	16/20 $P<0.001^b$ $(P<0.001)^d$	9.00 ± 1.63 $P<0.01^c$
"	"	Acyclovir	10	1/20	5.74 ± 0.32
"	"	Acyclovir	50	1/20	6.58 ± 0.37 $P<0.05^c$
i.p.	Oral	None (control)	0	2/20	6.78 ± 0.31
"	"	BV-araU	5	8/20	8.71 ± 0.79 $P<0.05^c$
"	"	BV-araU	15	14/20 $P<0.001^b$	8.83 ± 1.65
"	"	BV-araU	50	19/20 $P<0.001^b$	12.00 ± 0.00
"	"	Acyclovir	5	9/20 $P<0.05^b$	8.14 ± 1.04
"	"	Acyclovir	15	19/20 $P<0.001^b$	10.00 ± 0.00
"	"	Acyclovir	50	20/20 $P<0.001^b$	
i.p.	i.p.	None (control)	0	3/19	6.38 ± 0.40
"	"	BV-araU	12.5	8/19	7.45 ± 0.37
"	"	BV-araU	50	11/19 $P<0.05^b$	7.75 ± 0.53 $(P<0.05)^e$
"	"	Acyclovir	12.5	13/19 $P<0.01^b$	7.50 ± 0.92
"	"	Acyclovir	50	16/19 $P<0.001^b$	11.67 ± 2.03 $P<0.001^c$

Data from Machida et al.[75]

[a] Four weeks old random-bred albino ICR mice were infected either i.c. with 19 LD_{50} or i.p. with 90 LD_{50} of HSV-1 WT-51, treated with BV-araU or acyclovir twice a day for 7 to 9 days beginning 4 h postinfection, and observed twice a day for death for 21 days.

[b] Significance in difference from the saline-treated control (χ^2 analysis with Yates' correction).

[c] Significance in difference from the saline-treated control (Student's t test).

[d] Significance in difference from acyclovir-treated group at the same dose (χ^2 analysis with Yates' correction).

[e] Significance in difference from acyclovir-treated group at the same dose (Student's t test).

Table 8 summarizes comparative antiviral efficacies of BV-araU and acyclovir against lethal infections with HSV-1 in mice. BV-araU was highly effective against i.c. infection, while only marginal effects, if any, of oral and i.p. treatments with acyclovir were observed. Acyclovir, however, was effective at lower doses than BV-araU in reducing the mortality of i.p. infected mice[75] and in inhibiting the progression of local symptoms of the cutaneously infected mice.[60] The higher activity of acyclovir against HSV-1 in mouse cells than BV-araU may reflect these observations in mouse model infections.[75] Reefschlaeger et al.[30] also found that BV-araU was highly effective against the experimental encephalitis in mice, which was superior to acyclovir. BV-araU was reported to be effective against experimental HSV-1 keratitis in rabbits when it was given topically as a 0.1% solution 5 times daily beginning 3 days post infection, though it was slightly less effective than 0.1% BVDU.[76]

Antiviral Effects in Immunosuppressed Mice

Since herpesvirus infections are often seen in immunocompromised patients, and these infections tend to become life-threatening, it is important that antiviral drugs are effective even in immunosuppressed animals. Effects of BV-araU in immunosuppressed mice were also examined by using a couple of model infections. For inducing immunosuppression, mice were treated i.p. with 200 mg/kg of cyclophosphamide 24 to 48 hours before virus inoculation. The immunosuppressed mice were inoculated with a relatively low level of challenge virus, but enough to cause a lethal infection. As shown in Table 9, oral treatment with BV-araU was effective even in immunosuppressed mice, although a higher dose of the drug was necessary for effective reduction of the mortality than that in cyclophosphamide-non-treated normal mice.[77] Treatments with BV-araU at a dose as high as 100 mg/kg was effective in increasing both the survival rate and the mean survival time with statistical significance in the immunosuppressed mice. BV-araU at a dose of 20 or 50 mg/kg led to extension of the mean survival time, but not reduction of the mortality. At these doses, treatments with BV-araU were effective in decreasing the mortality of normal mice. A similar effect of BV-araU treatments was achieved in the efficacy against cutaneous infection of immunosuppressed mice.[60] Treatments with BV-araU at a dose of 100 mg/kg were effective in both delaying skin lesion development and reducing the mortality, and the treatments at doses of 20 and 50 mg/kg were effective in causing a delay in skin lesion development but did not result in significant decrease in the final mortality.

Table 9. Effect of oral treatment with BV-araU on intraperitoneal infection with HSV-1 in immunosuppressed mice

Dose of BV-araU (mg/kg)	Survivors per total		Mean survival time of mice that died (days ± S.E.)	
0 (control)	3/20		5.65 ± 0.47	
10	0/20		6.43 ± 0.21	
20	3/20		7.26 ± 0.14	$P<0.01^a$
50	7/20		8.77 ± 0.32	$P<0.001$
100	15/20	$P<0.001^b$	11.30 ± 1.79	$P<0.001$

Data from Ijichi et al.[77]
Immunosuppressed mice were infected with 8 LD_{50} of HSV-1 WT-51 and orally given BV-araU twice a day for 7 days beginning 4 h post infection.
[a] Student's t-test.
[b] χ^2-analysis with Yates' correction.

Effect on Simian Varicella Virus Infection

The effectiveness of BV-araU on SVV infection of African green monkeys, which mimics VZV infections in humans, was demonstrated by Soike et al.[23] Intramuscular treatments with BV-araU at a dose of 20 mg/kg/day eliminated the development of rash and substantially suppressed viremia. Oral BV-araU was effective for the monkey model infection at a dose as low as 1 mg/kg/day (Soike, personal communication). In their recent study, it has been shown that viremia is essentially eliminated and rash is completely suppressed in animals orally treated with BV-araU at a dose as low as 0.4 mg/kg/day.[78] The SVV-infected monkeys treated with BV-araU at a dose of 0.4 mg/kg/day remained free of rash and developed little or no viremia, even if initiation of the treatment was delayed until 96 h after infection. BV-araU seems the most effective drug orally available against SVV infection. They also describe that BV-araU shows good oral bioavailability in monkeys and that BV-uracil is detected in plasma after oral dose of BV-araU, while only a minimum amount of BV-uracil was found after intravenous administration.

CONCLUSIONS

5-Substituted arabinosyluracil nucleosides have less potent antiproliferative effect than the corresponding 2′-deoxyuridine nucleosides, while the former is metabolically more stable. The structure–activity relationship between the two groups varies on species of herpes viruses and 5-substituent of the compounds; anti-HSV-1 activity of some arabinosyluracil nucleosides fully retains with the replacement for deoxyribose by arabinose in the deoxyuridines, while anti-HSV-2 potency is generally reduced in the arabinosyluracils. Particularly interesting is that certain arabinosyluracil nucleosides, such as BV-araU, CV-araU, and 5′-propynyl-araU, have the increased anti-VZV potency more or less compared with the corresponding 2′-deoxyuridines, and BV-araU is the most potent antiviral nucleoside against VZV. In addition, introducing the E-(2-halogenovinyl) group at the C-5 position results in having highly selective anti-VZV effect irrespective of the kind of sugar moiety but in marked reduction of anti-HSV-2 potency. BV-araU has two important structural units favorable to exhibit selective anti-VZV effect. It scarcely shows antiproliferative effect and is metabolically stable compared with the 2′-deoxyuridine congener.

Acknowledgments

Our studies described here have been carried out in collaboration with many chemists who supplied us with various arabinosyluracil analogues. We wish to gratefully thank them; Dr. K. Ikeda of Hokkaido University, and Dr. M. Saneyoshi, Dr. C. Nakayama, and Dr. Y. Mizuno who studied in Faculty of Pharmaceutical Sciences of Hokkaido University, and Dr. S. Shibuya, Dr. T. Ikeda, Mr. M. Kumagai, Mr. T. Yamaguchi, and Mr. F. Kano of our Chemistry Laboratory for their excellent work. Collaborations of Dr. T. Yokota of Department of Microbiology of Fukushima Medical College, Dr. T. Suzutani of Department of Microbiology of Asahikawa Medical College, and Mr. K. Ijichi, Mr. N. Ashida, and Mr. J. Takezawa of our Biology Laboratory, in the virological and biochemical studies on BV-araU are gratefully acknowledged. Continued support and interest from Dr. A. Kuninaka and Mr. Yoshino of Yamasa Corporation and excellent technical assistance of Miss Y. Watanabe, Miss M. Nishitani, Miss M. Miyazaki, and Miss M. Ebata for biological examination of the compounds is also acknowledged.

It was over fifteen years ago that I met first Prof. Tohru Ueda. He, a young professor, visited our company and gave us a lecture. He mentioned that EDU was notable as an antiviral agent. From that time, we had interest in the application of nucleoside analogues as antiviral agents. We wish to express our deepest sympathy for the loss of Prof. Ueda.

REFERENCES

1. A. T. Jamieson, G. A. Gentry, and J. H. SubaK-Sharpe. Induction of both thymidine and deoxycytidine kinase activity by herpes viruses. *J. Gen. Virol.* 24:465 (1974).

2. A. T. Jamieson, and J. H. Subak-Sharpe. Biochemical studies on the herpes simplex virus-induced deoxypyrimidine kinase activity. *J. Gen. Virol.* 24:481 (1974).

3. G. A. Gentry, and J. F. Aswell. Inhibition of herpes simplex virus replication by araT. *Virology* 65:294 (1975).

4. H. Machida, S. Sakata, A. Kuninaka, H. Yoshino, C. Nakayama, and M. Saneyoshi. In vitro antiherpesviral activity of 5-alkyl derivatives of 1-β-D-arabinofuranosyluracil. *Antimicrob. Agents Chemother.* 16:158 (1979).

5. H. Machida, A. Kuninaka, H. Yoshino, K. Ikeda, and Y. Mizuno. Antiherpersviral activity and inhibitory action on cell growth of 5-alkenyl derivatives of 1-β-D-arabinofuranosyluracil. *Antimicrob. Agents Chemother.* 17:1030 (1980).

6. H. Machida, S. Sakata, S. Shibuya, K. Ikeda, C. Nakayama, and M. Saneyoshi. Selective antiherpesviral activity of 5-substituted derivatives of 1-β-D-arabinofuranosyluracil. p. 207. *In* "Antiviral Chemotherapy: Design of Inhibitors of Viral Functions," K. K. Gauri, ed., Academic Press, Inc., New York (1981).

7. H. Machida, S. Sakata, A. Kuninaka, and H. Yoshino. Antiherpesviral and anticellular effects of 1-β-D-arabinofuranosyl-*E*-5-(2-halogenovinyl)uracils. *Antimicrob. Agents Chemother.* 20:47 (1981).

8. S. Sakata, S. Shibuya, H. Machida, H. Yoshino, K. Hirota, S. Senda, K. Ikeda, and Y. Mizuno. Synthesis and antiherpesviral activity of 5-C-substituted uracil nucleosides. *Nucleic Acids Res.* Symp. Ser. 8:s39 (1980).

9. H. Machida, A. Kuninaka, and H. Yoshino. Inhibitory effects of antiherpesviral thymidine analogs against varicella-zoster virus. *Antimicrob. Agents Chemother.* 21:358 (1982).

10. H. Machida, S. Sakata, and K. Saito. Bromovinyl-arabinosyluracil (BV-araU): antiviral activity and pharmacology. p.63 *In* "Herpes Viruses and Virus Chemotherapy," R. Kono and A. Nakajima, eds., Excerpta Medica, Amsterdam (1985).

11. A. Hiraoka, T. Masaoka, K. Nagai, A. Horiuchi, A. Kanamaru, M. Niimura, T. Hamada, and M. Takahashi. Clinical effect of BV-araU on varicella-zoster virus infection in immunocompromised patients with hematological malignancies. *J. Antimicrob. Chemother.* 27:361 (1991).

12. M. Niimura. A double-blind clinical study in patients with herpes zoster to establish YN-72 (brovavir) dose, p.267. *In* "Immunobiology and Prophylaxis of Human Herpes Virus Infections," C. Lopez et al., ed., Plenum Press, New York (1990).

13. M. Niimura, M. Takahashi, T. Nishikawa, H. Ogawa, Y. Asada, and J. Ishii, J. Multicenter double-blind study of YN-72 (BV-araU, brovavir) in patients with herpes zoster. *Jpn. J. Clin. Dermatol.* 44:447 (1990).

14. Y.-C. Cheng, B. A. Domin., R. A. Sharman, and M. Bobek. Antiviral action and cellular toxicity of four thymidine analogues: 5-ethyl, 5-vinyl-, 5-propenyl, and 5-allyl-2'-deoxyuridine. *Antimicrob. Agents Chemother.* 10:119 (1976).

15. E. De Clercq, J. Descamps, P. F. J. Torrence, E. Krajewska, and D. Shugar. Antiviral activity of novel deoxyuridine derivatives, p.352. *In* "Current Chemotherapy. Proceedings of 10th International Congress of Chemotherapy," vol. 1, W. Siegenthaler and Luethy ed., American Society for Microbiology, Washington, D.C. (1977).

16. E. De Clercq, and D. Shugar. Antiviral activity of 5-ethyl pyrimidine deoxynucleosides. *Biochem. Pharmacol.* 24:1073 (1975).

17. J. F. Aswell, G. P. Allen, A. T. Jamieson, D. E. Campbell, and G. A. Gentry. Antiviral activity of arabinosylthymine in herpesviral replication: mechanism of action in vivo and in vitro. *Antimicrob. Agents Chemother.* 12:243 (1977).

18. R. L. Miller, J. P. Iltis, and F. Rapp. Differential effect of arabinofuranosylthymine on the replication of human herpesviruses. *J. Virol.* 23:679 (1977).

19. C. Nakayama, H. Machida, and M. Saneyoshi. Synthetic nucleosides and nucleotides. XII. Synthesis and antiviral activities of several 1-β-D-arabinofuranosyl-5-alkyluracils and their 5'-monophosphates. *J. Carbohydrates.nucleosides.nucleotides* 6:295 (1979).

20. T. Kulikowski, Z. Zawadzki, D. Shugar, J. Descamps, and E. De Clercq. Synthesis and antiviral activities of arabinofuranosyl-5-ethylpyrimidine nucleosides. Selective antiherpes activity of 1-(β-D-arabinofuranosyl)-5-ethyluracil. *J. Med. Chem.* 22:647 (1979).

21. H. Machida, M., Ichikawa, A., Kuninaka, M., Saneyoshi, and H. Yoshino. Effect of treatment with 1-β-D-arabinofuranosylthymine of experimental encephalitis induced by herpes simplex virus in mice. *Antimicrob. Agents Chemother.* 17:109 (1990).

22. H. Machida, and S. Sakata. In vitro and in vivo antiviral activity of 1-β-D-arabinofuranosyl-*E*-5-(2-bromovinyl)uracil (BV-araU) and related compounds. *Antiviral Res.* 4:135 (1984).

23. K. F. Soike, G. Baskin, C. Cantrell, and P. Gerone. Investigation of antiviral activity of 1-β-D-arabinofuranosylthymine (ara-T) and 1-β-D-arabinofuranosyl-*E*-5-(2-bromovinyl)uracil (BV-ara-U) in monkeys infected with simian varicella virus. *Antiviral Res.* 4:245 (1984).

24. H. Machida. Comparison of susceptibilities of varicella-zoster virus and herpes simplex viruses to nucleoside analogs. *Antimicrob. Agents Chemother.* 29:524 (1986).

25. Y.-C. Cheng, S. Grill, J. Ruth, and D. E. Bergstrom. Anti-herpes simplex virus and anti-human cell growth activity of *E*-5-propenyl-2'-deoxyuridine and concept of selective protection in antivirus chemotherapy. *Antimicrob. Agents Chemother.* 18:957 (1980).

26. G. Stening, B. Gotthammar, A. Larsson, S. Alenius, N. G. Johansson, and B. Oeberg. Antiherpes activity of [E]-5-(1-propenyl)-2'-deoxyuridine and 5-(1-propenyl)-1-β-D-arabinofuranosyluracil. *Antiviral Res.* 1:213 (1981).

27. J. Goodchild, R. A. Porter, R. H. Raper, I. S. Sim, R. M., Upton, J. Viney, and H. J. Wadsworth. Structural requirement of olefinic 5-substituted deoxyuridines for antiherpes activity. *J. Med. Chem.* 26:1252 (1983).

28. I. S. Sim, and R. H. Raper. 5-Substituted deoxyuridines - structural requirements for antiviral activity against herpes simplex virus types 1 and 2 and possible biochemical basis for relative potency. *Antiviral Res.* 4:159 (1984).

29. I. S. Sim, J. Goodchild, D. M. Meredith, R. A. Porter, R. H., Raper, J. Viney, and H. J. Wadsworth. Possible molecular basis for antiviral activity of certain 5-substituted deoxyuridines. *Antimicrob. Agents Chemother.* 23:416 (1983).

30. J. Reefschlaeger, P. Wutzler, K.-D. Thiel, and G. Herrmann. Treatment of experimental herpes simplex virus type 1 encephalitis in mice with (*E*)-5-(2-bromovinyl)- and 5-vinyl-1-β-D-arabinofuranosyl: comparison with bromovinyl-deoxyuridine and acyclovir.*Antiviral Res.* 6:83 (1986).

31. E. De Clercq, J. Descamps, P. De Somer, P. J. Barr, A. S. Jones, and R. T. Walker. (*E*)-5-(2-Bromovinyl)-2'-deoxyuridine: a potent and selective anti-herpes agent. *Proc. Natl. Acad. Sci. U.S.A.* 76:2947 (1979).

32. E. De Clercq, J., Balzarini, P. F. Torrence, M. P. Mertes, C. L. Schmidt, D. Shugar, P. J., Barr., A. S. Jones, G. Verhelst, and R. T. Walker. Thymidylate synthetase as target enzyme for the inhibitory activity of 5-substituted 2'-deoxyuridines on mouse leukemia L1210 cell growth. *Mol. Pharmacol.* 19:321 (1981).

33. E. De Clercq, J. Descamps, J. Balzarini, J. Giziewicz, P. J. Barr, and M. J. Robins. Nucleic acid related compounds. 40. Synthesis and biological activities of 5-alkynyluracil nucleosides. *J. Med. Chem.* 26:661 (1983).

34. A. S. Jones, S. G. Rahim, R. T. Walker, and E. De Clercq. Synthesis and antiviral properties of (Z)-5-(2-bromovinyl)-2'-deoxyuridine. *J. Med. Chem.* 24:759 (1981).

35. E. De Clercq. The antiviral spectrum of (*E*)-5-(2-bromovinyl)-2'-deoxyuridine. *J. Antimicrob. Chemother.* 14:Suppl. A,85 (1984).

36. E. De Clercq. Biochemical aspects of the selective antiherpes activity of nucleoside analogues. *Biochem. Pharmacol.* 33:2159 (1984).

37. A. S. Jones, G. Verhelst, and R. T. Walker. The synthesis of the potent anti-herpes agent, *E*-5-(2-bromovinyl)-2'-deoxyuridine and related compounds. *Tetrahedron Lett.* 45:4415 (1979).

38. M. Ashwell, A. S. Jones, A. Kumar, J. R. Sayers, R. T. Walker, T. Sakuma, and E. De Clercq. Synthesis and antiviral properties of (*E*)-5-(2-bromovinyl)-2'-deoxyuridine-related compounds. *Tetrahedron* 43:4601 (1987).

39. H. Machida, S. Sakata, N. Ashida, K. Takenuki, and A. Matsuda. *In vitro* antiherpesvirus activities of 5-substituted 2'-deoxy-2'-methylidene pyrimidine nucleosides. *Antiviral Chem. Chemother.* in press.

40. T. Ogino, T. Otsuka, and M. Takahashi. 1977. Induction of deoxypyrimidine kinase activity in human embryonic lung cells infected with varicella-zoster virus. *J. Virol.* 21:1232.

41. Y.-C. Cheng, Y. T. Tsou, T. Hackstadt, and L. P. Mallavia. Induction of thymidine kinase and DNase in varicella-zoster virus-infected cells and kinetic properties of the induced thymidine kinase. *J. Virol.* 31:172 (1979).

42. M. Baba, K. Konno, S. Shigeta, and E. De Clercq. Inhibitory effects of selected antiviral compounds on newly isolated clinical varicella-zoster virus strains. *Tohoku J. Exp. Med.* 148:275 (1986).

43. M. R. Dyson, P. L. Coe, and R. T. Walker. The synthesis and antiviral activity of some 4'-thio-2'-deoxy nucleoside analogues. *J. Med. Chem.* 34:2782 (1991).

44. W. A. Slusarchyk, G. S. Bisacchi, A. K. Field, D. R. Hockstein, G. A. Jacobs, B. McGeever-Rubin, J. A. Tino, A. V. Tuomari, G. A. Yamanaka, M. G. Young, and R. Zahler. Synthesis and antiviral activity of 1-cyclobutyl-5-(2-bromovinyl)uracil nucleoside analogues and related compounds. *J. Med. Chem.* 35:1799 (1992).

45. K. A. Watanabe, U. Reichman, K. Hirota, C. Lopez, and J. J. Fox. Nucleosides. 110. Synthesis and antiherpes virus activity of some 2'-fluoro-2'-deoxyarabinofuranosylpyrimidine nucleosides. *J. Med. Chem.* 22:21 (1979).

46. K. A. Watanabe, T.-L. Su, U. Reichman, N. Greenberg, C. Lopez, and J. J. Fox. Nucleosides. 129. Synthesis and antiviral nucleosides: 5-alkenyl-1-(2-deoxy-2-fluoro-β-arabinofuranosyl)uracils. *J. Med. Chem.* 27:91 (1984).

47. K. A. Watanabe, T.-L. Su, R. S. Klein, C. K. Chu, A. Matsuda, M. W. Chun, C. Lopez, and J. J. Fox. Nucleosides. 123. Synthesis of antiviral nucleosides: 5-substituted 1-(2-deoxy-2-halogeno-β-arabinofuranosyl)cytosines and -uracils. Some structure-activity relationships. *J. Med. Chem.* 26:152 (1983).

48. C. Lopez, K. A. Watanabe, and J. J. Fox. 2'-Fluoro-5-iodo-aracytosine, a potent and selective antiherpesvirus agent. *Antimicrob. Agents Chemother.* 17:803 (1980).

49. J. M. Colacino, and C. Lopez. Efficacy and selectivity of some nucleoside analogs as anti-human cytomegalovirus agents. *Antimicrob. Agents Chemother.* 24:505 (1983).

50. R. F. Schinazi, J. Peters, M. K. Sokol, and A. J. Nahmias. 1983. Therapeutic activities of 1-(2-fluoro-2-deoxy-β-D-arabinofuranosyl)-5-iodocytosine and -thymine alone and in combination with acyclovir and vidarabine in mice infected intracerebrally with herpes simplex virus. *Antimicrob. Agents Chemother.* 24:95 (1983).

51. R. F. Schinazi, J. J. Fox, K. A. Watanabe, and A. J. Nahmias. Activities of 1-(2-deoxy-2-fluoro-β-D-arabinofuranosyl)-5-iodocytosine and its metabolites against herpes simplex virus types 1 and 2 in cell culture and in mice infected intracerebrally with herpes simplex virus type 2. *Antimicrob. Agents Chemother.* 29:77 (1986).

52. K. F. Soike, C. Cantrell, and P. J. Gerone. 1986. Activity of 1-(2'-deoxy-2'-fluoro-β-D-arabinofuranosyl)-5-iodouracil against simian varicella virus infections in African green monkeys. *Antimicrob. Agents Chemother.* 29:20 (1984).

53. H. Machida, and M. Nishitani. Drug susceptibilities of isolates of varicella-zoster virus in a clinical study of oral brovavir. *Microbiol. Immunol.* 34:407 (1990).

54. H. Machida. *In vitro* anti-herpes virus action of a novel antiviral agent, brovavir (BV-araU). *Chemotherapy* 38:256 (1990).

55. J.-C. Lin, and H. Machida. Comparison of two bromovinyl nucleoside analogs, 1-β-D-arabinofuranosyl-*E*-5-(2-bromovinyl)uracil and *E*-5-(2-bromovinyl)-2'-deoxyuridine, with acyclovir in inhibition of Epstein-Barr virus replication. *Antimicrob. Agents Chemother.* 32:1068 (1988).

56. J.-C. Lin, J. Reefschlaeger, G. Herrmann, and J. S. Pagano. Structure-activity relationship between (*E*)-5-(2-bromovinyl)- and 5-vinyl-1-β-D-arabinofuranosyluracil (BV-araU, V-araU) in inhibition of Epstein-Barr virus replication. *Antiviral Res.* 17:43 (1992).

57. J. Reefschlaeger, G. Herrmann, D. Baerwolff, B. Schwarz, D. Cech, and P. Langen. Antiherpes activity of (*E*)-5-(2-bromovinyl)- and 5-vinyl-1-β-D-arabinofuranosyluracil and some other 5-substituted uracil arabinosyl nucleosides in two different cell lines. *Antiviral Res.* 3:175 (1983).

58. E. De Clercq. Comparative efficacy of antiherpes drugs in different cell lines. *Antimicrob. Agents Chemother.* 21:661 (1982).

59. H. Machida, N. Nishitani, T. Suzutani, and K. Hayashi. Different antiviral potencies of BV-araU and related nucleoside analogues against herpes simplex virus type 1 in human cell lines and Vero cells. *Microbiol. Immunol.* 35:963 (1991).

60. H. Machida, K. Ijichi, and J. Takezawa. Efficacy of oral treatment with BV-araU against cutaneous infection with herpes simplex type 1 in shaved mice. *Antiviral Res.* 17:133 (1992).

61. H. Machida, K. Ijichi, A. Ohta, H. Honda, and M. Niimura. Antiviral potencies of BV-araU and related nucleoside analogues against varicella-zoster virus in different cell lines. *Microbiol. Immunol.* 34:959 (1990).

62. H. Machida, and Y. Watanabe. Inhibition of DNA synthesis in varicella-zoster virus-infected cells by BV-araU. *Microbiol. Immunol.* 35:139 (1991).

63. Y.-C. Cheng, G. Dutschman, J. J. Fox, K. A. Watanabe, and H. Machida. Differential activity of potential antiviral nucleoside analogs on herpes simplex virus-induced and human cellular thymidine kinases. *Antimicrob. Agents Chemother.* 20:420 (1981).

64. A. R. Karlstroem, C. F. R. Kallander, G. Abele, and A. Larsson. Acyclic guanosine analogs as substrates for varicella-zoster virus thymidine kinase. *Antimicrob. Agents Chemother.* 29:171 (1986).

65. J. L. Ruth, and Y.-C. Cheng. Nucleoside analogues with clinical potential in antivirus chemotherapy. The effect of several thymidine and 2'-deoxycytiluracidine analogue 5'-triphosphates on purified human (α, β) and herpes simplex virus (types 1, 2) DNA polymerases. *Mol. Pharmacol.* 20:415 (1981).

66. T. Yokota, K. Konno, S. Mori, S. Shigeta, M. Kumagai, Y. Watanabe, and H. Machida. Mechanism of selective inhibition of varicella zoster virus replication by 1-β-D-arabinofuranosyl-*E*-5-(2-bromovinyl)uracil. *Mol. Pharmacol.* 36:312 (1989).

67. H. S. Allaudeen, M. S. Chen, J. J. Lee, E. De Clercq, and W. H. Prusoff. Incorporation of *E*-5-(2-halovinyl)-2'-deoxyuridines into deoxyribonucleic acids of herpes simplex virus type 1-infected cells. *J. Biol. Chem.* 257:603 (1982).

68. T. Yokota, K. Konno, S. Shigeta, A. Verbruggen, and E. De Clercq. Incorporation of *E*-5-(2-iodovinyl)-2'-deoxyuridine into deoxyribonucleic acids of varicella-zoster virus (TK$^+$ and TK$^-$ strains)-infected cells. *Mol. Pharmacol.* 31:493 (1987).

69. H. Machida, Y. Watanabe, T. Suzutani, and M. Azuma. Metabolism of BV-araU in HSV-infected cells. 5th International conference on antiviral research, Abs. I-41. Williamsburg, WV, U.S.A. (1988).

70. T. Suzutani, H. Machida, T. Sakuma, and M. Azuma. Effects of various nucleosides on antiviral activity and metabolism of 1-β-D-arabinofuranosyl-*E*-5-(2-bromovinyl)uracil against herpes simplex virus types 1 and 2. *Antimicrob. Agents Chemother.* 32:1547 (1988).

71. N. K. Ayisi, R. A. Wall, R. J. Wanklin, H. Machida, E. De Clercq, and S. L. Sacks. Comparative metabolism of *E*-5-(2-bromovinyl)-2'-deoxyuridine and 1-β-D-arabinofuranosyl-*E*-5-(2-bromovinyl)uracil in herpes simplex virus-infected cells. *Mol. Pharmacol.* 31:422 (1987).

72. J. A. Fyfe. Differential phosphorylation of (*E*)-5-(2-bromovinyl)-2'-deoxyuridine monophosphate by thymidylate kinases from herpes simplex viruses types 1 and 2 and varicella zoster virus. *Mol. Pharmacol.* 21:432 (1982).

73. H. Machida, and J. Takezawa. Effect of 1-β-D-arabinofuranosyl-*E*-5-(2-bromovinyl)uracil (brovavir) on experimental infections in mice with herpes simplex virus type 1 strains of different degrees of virulence. *Antimicrob. Agents Chemother.* 34:691 (1990).

74. H. Machida, K. Ijichi, N. Ashida, and S. Varia. Efficacy of topical treatment with BV-araU cream against cutaneous infection with herpes simplex virus type 1 in shaved mice. 5th International conference on antiviral research, Abs. 138. Vancouver, Canada (1992).

75. H. Machida, T. Ikeda, and N. Ashida. Comparison of antiviral efficacies of 1-β-D-arabinofuranosyl-*E*-5-(2-bromovinyl)uracil (brovavir) and acyclovir against herpes simplex virus type 1 infections in mice. *Antiviral Res.* 14:99 (1990).

76. H. Toepke, M. Graef, P. Wutzler, G. Herrmann, and J. Reefschlaeger. Evaluation of (*E*)-5-(2-bromovinyl)- and 5-vinyl-1-β-D-arabinofuranosyluracil (BrVaraU, VaraU) in the treatment of experimental herpes simplex virus type 1 keratitis in rabbits: comparison with (*E*)-5-(2-bromovinyl)-2'-deoxyuridine. *Antiviral Res.* 9:273 (1988).

77. K. Ijichi, N. Ashida, and H. Machida. Effect of 1-β-D-arabinofuranosyl-*E*-5-(2-bromovinyl)uracil against herpes simplex virus type 1 infection in immunosuppressed mice. *Antimicrob. Agents Chemother.* 34:2431 (1990).

78. K. Soike, J.-L. Huang, J.-I. Tu, B. Stouffer, J. G. Mitroka, M. Swerdel, S. Olsen, D. P. Bonner, A. V. Toumari, and A. K. Field. Oral bioavailability and anti-simian varicella virus efficacy of 1-β-D-arabinofuranosyl-E-5-(2-bromovinyl)uracil (BV-ara-U) in monkeys. *J. Infect. Dis.* 165:732 (1992).

CONFORMATIONAL STUDIES AND ANTI-HIV ACTIVITY OF

MONO- AND DIFLUORODIDEOXY NUCLEOSIDES

Victor E. Marquez, Benjamin B. Lim, Joseph. J. Barchi, Jr. and Marc C. Nicklaus

Laboratory of Medicinal Chemistry, DTP, DCT, National Cancer Institute NIH, Bethesda, MD 20892

INTRODUCTION

The enormous disparity in anti-HIV activity that is evident for a large number of 2',3'-dideoxynucleoside analogues belies their apparent structural similarity. Biochemically, the differences observed could result mainly from differences in the ability of the individual dideoxynucleosides to generate the corresponding 5'-triphosphates, combined with differences in the ability of the resulting anabolites to inhibit the target enzyme, reverse transcriptase (RT). However, knowledge that irrespective of the nature of the aglycon, the 5'-triphosphates of the most common dideoxynucleosides (U, C, T, A, G) and AZT inhibit HIV-RT within a similar range of submicromolar concentrations,[1,2] helps to establish the identity of the most critical step. Clearly, if the ability of these 5'-triphosphate metabolites to block RT is very similar, the crucial parameter has to be the efficiency with which cellular enzymes are capable of generating adequate amounts of these 5'-triphosphates. Indeed, for the same group of dideoxynucleosides, the great disparity in measured levels of 5'-triphosphates agrees well with the more than 1000-fold range in potency observed for these compounds as anti-HIV agents.[1] Achieving an effective intracellular concentration of the triphosphate metabolite depends, in turn, on the ease of transport of the drug as well as its particular affinity for the appropriate cellular nucleoside and nucleotide kinases. Of the three kinases that are necessary for full activation of dideoxynucleosides, the first kinase appears to be the most selective, and it is this particular phosphorylation step that is recognized as the most critical.[3] This first kinase, however, is not a universal enzyme. In fact, several different enzymes perform this first phosphorylation step depending on the nature of the aglycon. AZT, for example, is converted to the 5'-monophosphate by thymidine kinase,[4] ddC is phosphorylated by deoxycytidine kinase,[5] which can also phosphorylate ddA,[6] ddA is phosphorylated by adenosine kinase,[6] and ddI, ddG, and carbovir are phosphorylated by yet another enzyme, a phosphoribosyltransferase/5'-nucleotidase that utilizes IMP as a phosphate donor.[7] Based on the existence of this array of phosphorylating enzymes, it is important that structure-activity comparisons (SAR) be performed on groups of compounds expected to be phosphorylated by the same enzyme.

In this work we performed a limited SAR study on a few fluorine-substituted

Nucleosides and Nucleotides as Antitumor and Antiviral Agents,
Edited by C.K. Chu and D.C. Baker, Plenum Press, New York, 1993

Table 1. Monofluorodideoxy nucleosides

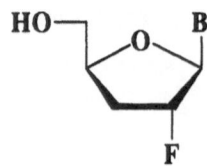

B	ED$_{50}$ (µM)	Cell	Ref.
T	0.001	MT4	12
T	1.4	ATH8	14
U	0.04	MT4	12
C	16	MT4	12
C	8	ATH8	14
A	50	MT4	13

B	ED$_{50}$ (µM)	Cell	Ref.
A	>625	MT4	13
G	inactive	MT4	15

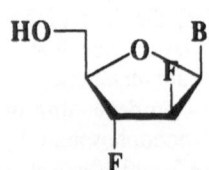

B	ED$_{50}$ (µM)	Cell	Ref.
T	53	MT4	16
T	inactive	H9	18
A	inactive	MT4	13
A	inactive	ATH8	21
C	inactive	C8166	17

B	ED$_{50}$ (µM)	Cell	Ref.
C	0.70	C8166	17
C	2.30	CEM	19
C	2.54	MT4	20
C	9.80	MT4	16
C	1.5	ATH8	22
A	35	MT4	13
A	5	ATH8	21
T	>25	CEM	19
U	>10	CEM	19
U	inactive	C8166	17
G	18	CEM	23

Table 2. Difluorodideoxy nucleosides

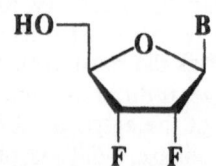

B	ED$_{50}$ (µM)	Cell	Ref.
C	10	C8166	17
U	inactive	C8166	17
T	inactive	H9	18

B	ED$_{50}$ (µM)	Cell	Ref.
T	inactive	H9	18
U	inactive	MT4	24
C	inactive	MT4	24

cytosine dideoxynucleosides for which we examined the relationship between their preferred form of ring-puckering in solution and anti-HIV activity. According to the above premise, these results might provide an indirect measure of the compounds' ability to function as adequate substrates for the first phosphorylation step.

The unsubstituted dideoxyribose ring is very flexible, and its conformation in solution differs from that found in the solid state.[8] For that reason, it was felt that a more reliable SAR study could emerge if a certain degree of rigidity were introduced into the system in order to favor specific forms of puckering. Subsequently, these specific and more rigid conformations could be examined for correlation with anti-HIV activity. Two preliminary NMR studies on the conformation of 3'- fluorothymidine (FddT) provided the initial thrust for this work.[9,10] These authors showed that for this very potent anti-HIV agent, a considerable amount of rigidity that favored a specific conformer was the direct result of fluorine substitution. For our part, we were interested in knowing the degree of rigidity and the identity of the preferred conformer that would result from changing the regiochemistry and stereochemistry of the fluorine substituent. The origin of the fluorine-induced rigidity appears to arise from the so-called *gauche* effect that results from the interaction between the ribose oxygen and the very electronegative fluorine atom.[11]

A review of the published literature on monofluoro dideoxynucleosides and their anti-HIV activity provided an additional incentive to pursue this study. For various aglycon moieties, a fluorine atom at positions 3'-"down" or 2'-"up" correlates with anti-HIV activity (Table 1). On the other hand, fluorine atoms at the same positions, but with inverted stereochemistry, consistently produced inactive compounds (Table 1). The variability in potency observed in Table 1 for the active compounds appears to be a function of the different nature of the aglycon and the different biochemical characteristics of the type of cells used in the bioassay.

Using a similar reasoning, one can expect that *gauche* effect interactions in difluoro-substituted dideoxyribosides could either enhance or diminish the rigidity of the molecule depending on the relative disposition of the fluorines. The number of difluorodideoxy compounds reported in the literature, however, is limited (Table 2), and no generalizations can be made *a priori* regarding their rigidity, preferred form of puckering, and biological activity. The dideoxydifluoroarabinofuranosyl template, for example, which was very attractive because it combined the optimal monofluoro stereochemistries at C-2' and C-3' in one molecule (Table 2), was synthesized and shown to have a significantly reduced level of potency relative to the 2'-"up" monofluoro in the cytidine series.[17] Compounds with two fluorines in a ribo configuration were also prepared and were found to be inactive as anti-HIV agents.[18,25,26]

An important generalization that attempts to correlate a specific solid-state conformation with anti-HIV activity in dideoxynucleosides has been suggested.[27] The proposed hypothesis appears to hold true for a number of different compounds with various aglycons when viewed in their crystalline state.[27] However, recent solution studies with NMR and Raman spectroscopy appear to indicate that the observed crystalline conformations may not necessarily coincide with the preferred conformation of the nucleosides in solution.[8,28]

In order to correlate a preferred conformation in solution with anti-HIV activity, we present here the results of a conformational analysis using NMR spectroscopy on a limited number of monofluoro- and difluoro-substituted pyrimidine dideoxynucleosides that exhibit a fair degree of rigidity and show distinct preferences for various forms of puckering that seem to correlated well with their level of anti-HIV activity.

PSEUDOROTATION AND CONFORMATION OF NUCLEOSIDES

There are three conformational degrees of freedom that fully describe ribofuran-

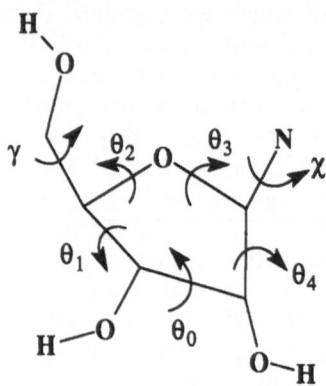

Figure 1. **Important torsion angles in a nucleoside**

osides either in solution or in the solid state. These are: a) the glycosyl torsion angle χ which determines the *syn* or *anti* disposition of the base relative to the sugar moiety, b) the C-4'—C-5' torsion angle γ which determines the orientation of the 5'-OH relative to the sugar ring, and c) the pseudorotational phase angle, P, which together with the median puckering amplitude, θ_m, determine the shape and extent of deviation from planarity of the sugar ring itself. In this work we have chosen to restrict ourselves to the study of the last two properties, P and θ_m, because of the well-documented correlation that exists between a specific value of P and both torsion angles χ and γ.[29-31] Furthermore, all of the compounds presented in this discussion are pyrimidine nucleosides which tend to exist exclusively in the *anti* form in which the C-2 carbonyl is opposite to the sugar ring.[32]

The puckering of the ribose ring expressed in terms of P (for a given θ_m) represents a conformation which defines a specific relationship between the five torsion angles (θ_0-θ_4) that characterize the non-planar five-membered ring (Figure 1). A value of P, therefore, corresponds to a specific point in the pseudorotation cycle (Figure 2). This is why for a given crystal structure there is a precise value of P. The value of θ_m ranges from 35° to 45°, and there appears to be no correlation between its magnitude and P.[31] A smaller value of θ_m means that the ribose ring is flattened slightly relative to a ring with a higher θ_m. From a large number of crystal structures of purine and pyrimidine nucleosides, two preferred pseudorotational domains in each hemisphere have been identified. For the northern hemisphere (N, P = 0° ± 90°) the preferred domain is in the 0°-36° range, whereas for the southern hemisphere (S, P = 180° ±90°) the preferred orientation is in the 144°-180° range. These two ranges are separated by two pseudorotational barriers that occur approximately in the eastern (P = 90°) and western (P =270°) regions (Fig. 2).[30]

For a nucleoside in solution, despite the fact that the entire pseudorotation cycle would appear to be accessible, experimental evidence seems to indicate that both preferred regions identified in the solid structures are accessible in solution and that both coexist in a dynamic N / S equilibrium. This equilibrium is defined by the relative amounts of these two populated domains in which each N and S conformation is, respectively, a blend of near neighbors of a P = 0° conformer and a P = 180° conformer. In general terms, the N and S conformations are described as C-3'-*endo* and C-2'-*endo*, respectively. *Endo* (up) and *exo* (down) indicate the direction of displacement (puckering) of a specific atom in relation to the plane of the other atoms in the ribose ring. For the various twist (T) and envelope (E) forms this is described with numbers in the form of superscripts (*endo*) and subscripts (*exo*) that indicate the atom being displaced above or below the plane (Figure 2). Finally, for the C-4'—C-5' torsion angle, there are three possible orientations (ap, +sc, and -sc, Figure 3a) of which the ap and +sc are the most populated.[32] In solution, there is also a strong correlation between the value of P and the preferred orientation of the 5'-OH.[29]

The direction of the N / S equilibrium in solution can be affected by various factors including steric and anomeric effects, as well as the already mentioned *gauche* effect. The *gauche* effect induces a strong tendency on vicinal electronegative atoms to adopt a *gauche* rotational arrangement (Figure 3c) instead of the expected trans-diaxial orientation (Figure 3b). Thus, for a tetrahydrofuran ring existing either as a pure N or

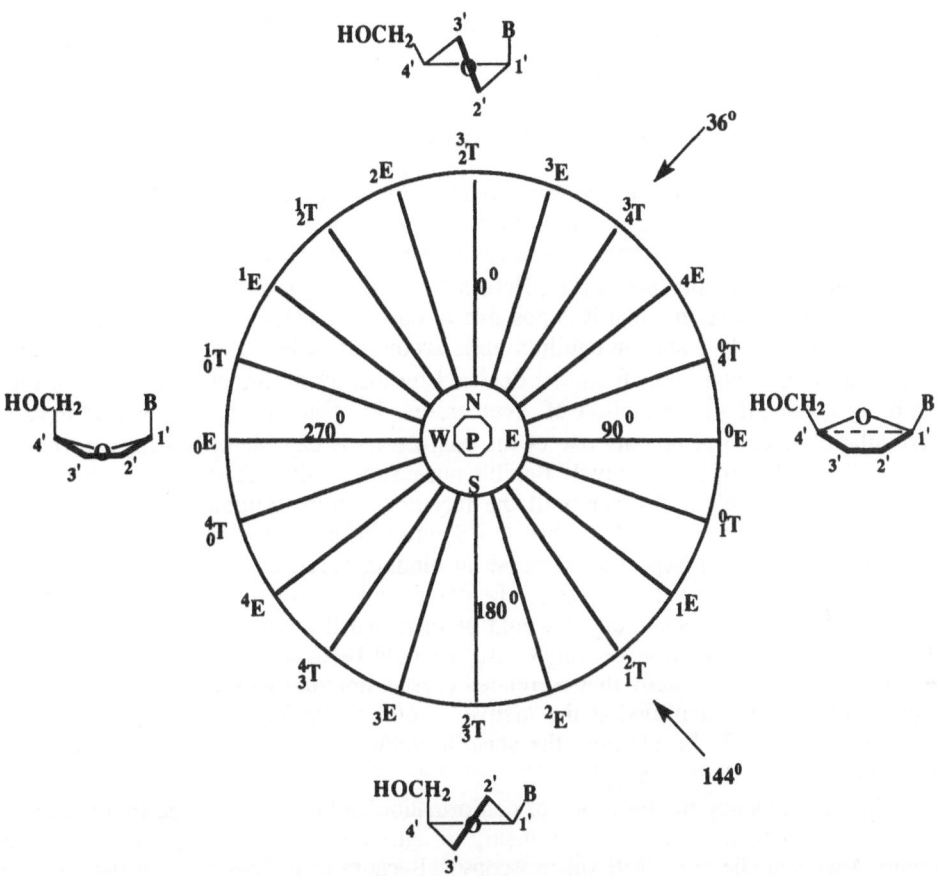

Figure 2. Pseudorotation cycle

pure S conformer (Figure 2) any pseudoaxial substituent at C-2' or C-3' is approximately in a *gauche* disposition relative to the furan oxygen. This effect appears to explain why 2'-deoxynucleosides prefer an S conformation and 3'-deoxynucleosides, on the contrary, prefer an N conformation.[33] The *gauche* effect increases in magnitude when in these molecules either hydroxyl group is replaced by the more electronegative fluorine atom.[11] Therefore, the orientation of a fluorine atom in a ribose ring should be a determining factor in inducing a specific conformational "stiffness" and puckering of the sugar ring.

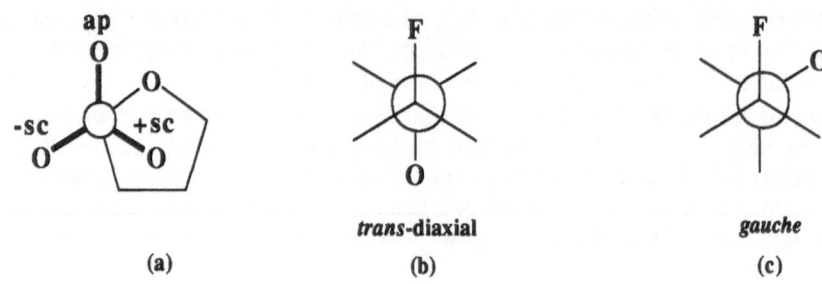

trans-diaxial   gauche

(a)   (b)   (c)

Figure 3. Newman projection of the C-4'--C-5' bond showing the different orientations of the 5'-O as determined by the value of γ (a). Newman projection of a fluorine and oxygen atoms in a *trans*-diaxial disposition (b) and in a *gauche* arrangement (c).

FLEXIBILITY OF THE FURANOSE RING

Conformational studies of nucleosides in solution indicate that the N/S interconversion rate is very rapid on the NMR time scale and that only small preferences towards either hemisphere are detectable.[34] However, incorporation of a very electronegative substituent (i.e., fluorine) into the sugar ring shifts the position of equilibrium to the extreme that it is possible to observe mainly one preferred conformer in solution.[9-11,35] This shift in equilibrium is accompanied by an increase in the rigidity of the ribose backbone which can be detected by measuring the changes in the proton coupling constants as a function of temperature.[10] One can speculate that if for a particular class of enzyme only one of the conformers is capable of binding effectively, the N/S equilibrium of a relatively flexible nucleosides will shift accordingly to provide more of the "active" conformer until the receptor is fully saturated. However, if by some means one were to shift this N/S equilibrium more towards the "active" conformer, one would expect an increase in binding based on entropy considerations, with optimum affinity occurring when the compound exists exclusively in the "active" conformation. The opposite effect would be observed if a compound were "frozen" in the "inactive" conformation in which case it would be expected to bind poorly to the enzyme. In a previous study that correlated crystal structure with anti-HIV activity, the S conformation was identified as the "active" conformation for a selected group of anti-HIV nucleosides.[27] In addition, the specific value of P was found to be critical for compounds within the S hemisphere.[27]

In order to study the influence of conformation and rigidity induced by fluorine on the dideoxyribose ring and its relationship to anti-HIV activity, compounds shown in Figure 4 were studied by NMR spectroscopy. Because of its high electronegativity and strong *gauche* effect interaction, a single fluorine atom was expected to influence the conformation and degree of rigidity of the molecule commensurate with its location and orientation on the dideoxyribose ring. Furthermore, we expected that an additional fluorine atom would also influence the conformation and rigidity of the molecule depending on its relative disposition and the strength of the interfluorine *gauche* effect.

As mentioned before, compounds with a C-3'-F "down" (i.e., compound **1**) are anti-HIV effective. This is particularly true for the case in which the base equals thymine or uracil (Table 1). The C-2'-F "up" substituent found in **2** is equally consistent with anti-HIV activity. However, inversion at either one of these two centers results in the complete disappearance of antiviral activity. One can invoke the aforementioned *gauche* effect between fluorine and the furanose oxygen to explain why in both active templates of **1** and **2** the S conformation is favored in solution. The inactivity of the inverted conformers, on the other hand, could be explained

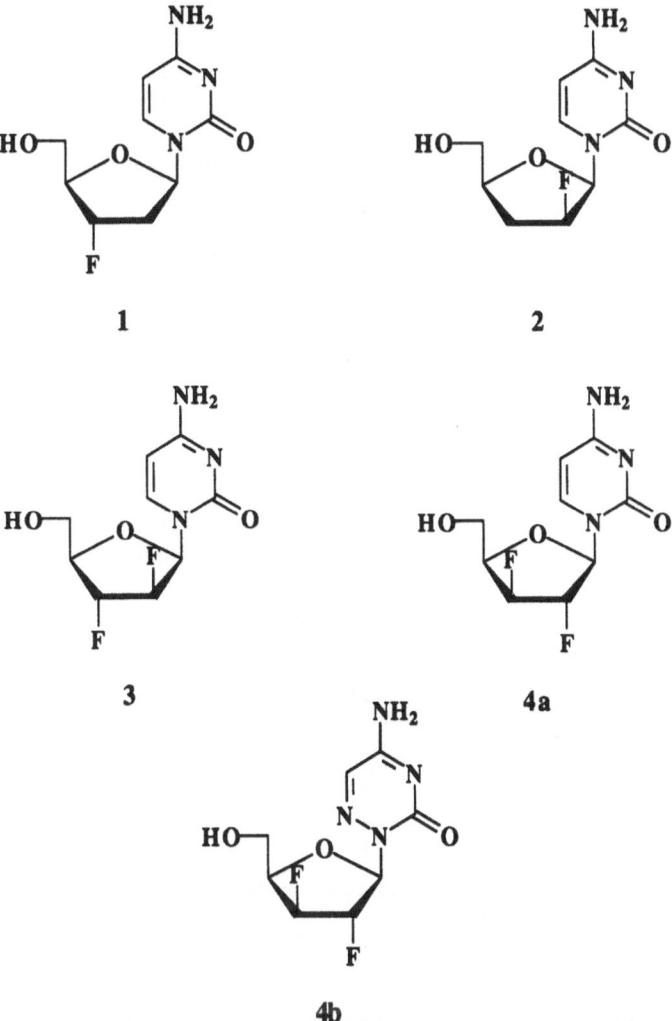

Figure 4. Selected mono- and difluorodideoxy cytosine nucleosides

by invoking that the N/S equilibrium has been severely shifted towards the unfavorable N hemisphere by the same *gauche* effect working in the opposite direction. An interesting example of clever drug design was provided by the analogue **3** in which the two S-inducing fluorine stereochemistries found in compounds **1** and **2** were combined in a single molecule.[17] Surprisingly, compound **3** was 14 times less potent than the corresponding monofluoro analogue **2**.[17] Even before this compound was made, it was suggested by Taylor et al.,[36] that in this molecule (**3**) the more powerful *gauche* effect between the two fluorine atoms would oppose their tendency to become individually *gauche* to the ring oxygen, thus, shifting the equilibrium towards the less active N conformation. As we shall see later, the magnitude of this predicted change in P was indeed measurable by NMR spectroscopy in solution. In order to test the same phenomenon and hopefully shift the equilibrium in the opposite direction, we prepared compound **4a**. In this compound, the disposition of the fluorines is such that, individually, they would favor the N conformation. However, the anticipated F—F *gauche* effect would be expected to force the molecule towards the opposite, "active" S conformation. Unfortunately, this was not realized and the compound exists exclusively

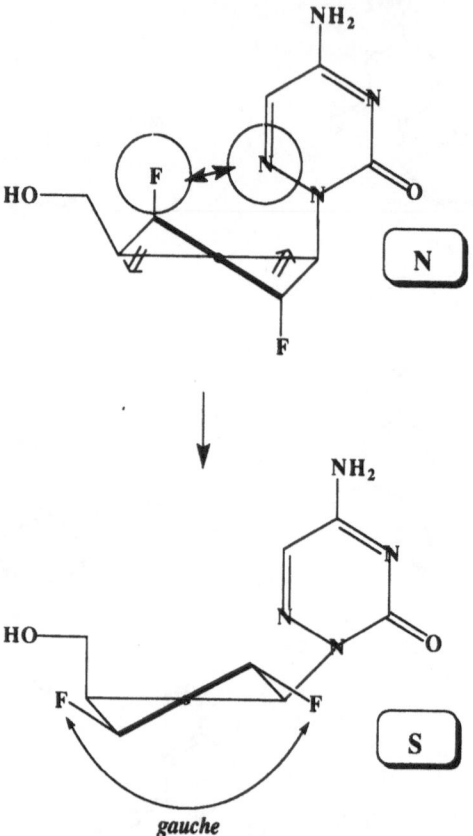

Figure 5. Expected direction of change in compound **4b** by the repulsive effect between 3'-F" up" and N-6.

in the N hemisphere and, because of this, it is totally inactive.[37] We were able to detect that in compound **4a** there is another important force besides the *gauche* effect acting on the molecule. This new force appears to have a strong influence on the conformation of this compound and other pyrimidine nucleosides with a C-3'-F "up"-substituent . For lack of a better term, and in view of the fact that we do not yet have a crystal structure of a C-3'-F "up" pyrimidine nucleoside to confirm it, we shall simply describe this force as a strong H-6—C-3'-F attraction that brings these two atoms close to each other. Such H-6—C-3'-F closeness which was described by Bergstrom et al.[38] in theoretical terms, appears to be supported by our NMR data (*vide infra*). As we shall see, this molecule not only exists exclusively in an N configuration, but in addition, it is extremely rigid, as variable temperature NMR measurements would indicate. In order to eliminate this strong H-6—C-3-'F interaction, the corresponding 6-azapyrimidine nucleoside **4b** was synthesized (Figure 4). The NMR study confirmed that indeed **4b** was less rigid than **4a** and behaved in terms of a conventional N/S equilibrium (*vide infra*). However, despite the fact that the equilibrium was shifted slightly towards the "active" S hemisphere, as predicted by the *gauche* effect, the molecule still preferred the N conformation and, accordingly, it was devoid of anti-HIV activity.[37] The same analysis holds true for the corresponding uridine analogues that were also made. The reason for choosing the 6-azapyrimidine moiety was based on the following considerations: a repulsive force between the C-3'-F "up" and the N-6 of the pyrimidine would replace the H-6—C-3'-F attractive force. As a result of this repulsion, the C-3'-*endo* carbon is expected to move in an *exo* direction (Figure 5).

Concomitanlty, the C-2' carbon should move in the opposite *endo* direction, causing the two fluorine atoms to be disposed in a *gauche* relationship to each other (Figure 5). Although, according to the NMR data these motions appeared to have taken place, the magnitude of the displacement caused by the interfluorine *gauche* effect was not sufficient to displace the ring pucker into the *S* hemisphere. This may indicate that a single and perhaps stronger *gauche* F—F interaction is not capable of overcoming two possibly weaker F—O *gauche* interactions.

SYNTHESIS

The new target difluoro-substituted compounds (**4a** and **4b**) that were chosen for this study have the two fluorine atoms in a *xylo*-configuration. Uridine (**6a**) and 6-azauridine (**6b**) were used as the starting materials (Scheme). After protection of the 5'-hydroxyl group as the trityl ether **7**, formation of the anhydro compound **8** was

Scheme

a. TrCl, pyr. 50 °C. *b*. Im₂C=S, Toluene, Δ. *c*. DAST, pyr. CH₂Cl₂, Δ. *d*. 1 N NaOH/CH₃CN. *e*. DAST, pyr. CH₂Cl₂, Δ. *f*. 1) POCl₃, 1,2,4-triazole 2) NH₄OH, Dioxane. *g*. 80% AcOH

induced via the transient 2',3'-cyclic thionocarbonate intermediate in refluxing toluene with bis(imidazol-1-yl)thione. DAST fluorination in refluxing CH_2Cl_2 in the presence of pyridine afforded, in both instances, a ca 55% yield of the desired monofluoro-substituted compounds **9** with the 3'-"up" fluoro configuration. Although an attractive single hydrofluorination of anhydrothymidine via soluble aluminum derivatives has been reported to proceed in useful yields,[39] a two step conversion was preferred. First, the anhydro ring was quantitatively opened at room temperature with NaOH, and the resulting hydroxyl moiety in **10** was then reacted with DAST, as before, to give the desired 2',3'-dideoxy-2',3'-difluoro-D-*xylo*-pentofuranosyl template (**11**). Removal of the trityl group from **11** in 80% AcOH afforded the uridine analogues (**12a** and **12b**), while formation of the 1,2,4-triazol-1-yl intermediates, followed by reaction with NH_4OH, produced the corresponding cytidines **13a** and **13b**. Removal of the trityl group provided the two target compounds **4a** and **4b**.

NMR SPECTRAL STUDIES

NMR spectroscopy was used to study the conformation and degree of rigidity of active and inactive fluorine-containing cytosine dideoxynucleosides (**1** - **4**, Figure 4). Changes in the proton coupling constants as a function of temperature were used to ascertain the degree of flexibility of these molecules in the range of 285 K (12 °C) to 350 K (77 °C). Since we did not have an authentic sample of compound **1**, we chose to use the values reported for 5-methyl-1-(3'-F-2',3'-dideoxy-*erythro*-ß-D-ribofuranosyl)uracil (FddT) by Chattopadhaya and co-workers[10] who recorded the changes in the NMR spectrum of this compound between 298 K and 353 K. The coupling constants are given in Table 3. The use of uridine values in a cytidine study is justified by our own observation that there are no differences in the coupling constants of the sugar ring protons between corresponding uracil and cytosine nucleosides possessing an identical fluorinated sugar. For compound **2**, the NMR studies were performed with a sample prepared earlier in this laboratory,[40] and the corresponding measured coupling constants appear in Table 4. A sample of the uridine analogue of **3**[17] was kindly provided by Dr. Joseph A. Martin (Roche Products, Ltd., UK), but the authentic cytosine derivative **3** was not available. However, as mentioned above, and later demonstrated for the difluoro analogues **4a** and **4b**, the coupling constants for **3** and its uridine precursor can be assumed to be identical. Therefore, the data for compound **3** (Table 5) was measured from the corresponding uridine analogue. The coupling constants for the other difluoro analogues (**4a** and **4b**) appear in Tables 6 and 7, respectively.

High-field ^{1}H NMR spectra (500 MHz) were recorded in D_2O at the temperatures indicated. The majority of the resonances were first-order, and hence coupling constants were measured directly. In the cases where second-order effects and spectral overlap obscured the desired J value, a combination of spin-spin decoupling with difference spectroscopy and spectral simulation using a LAOCOON type program called "PANIC" (supplied with standard Bruker DISNMR software) was employed. These values represent the time-average coupling constants for the N and S conformers in equilibrium.

PSEUROT

PSEUROT is an iterative least-squares computer program that was devised to obtain the best fit between experimental vicinal coupling constants and the pseudorotational parameters, P and θ_m, for the participating conformers in a N / S equilibrium.[41] In addition, it calculates the mole fraction X_N and X_S of each conformer. This program is available from Indiana University (QCPE #463) and runs on a PC with a Fortran compiler. After a number of iterations, PSEUROT converges to a set of ideal values for P_N, P_S, θ_N, θ_S and X_N, based on the experimentally determined coupling constants. This program uses a generalized Karplus equation (1) in which $J_{H,H}$ is expressed in terms of the dihedral angle θ plus a series of terms that take into account the electronegativity ($\Delta\chi_i$) of the substituents as well as their relative orientation (ξ_i) to the coupled protons.[41,42]

$$J_{H,H} = P_1\cos^2\theta + P_2\cos\theta + P_3 + \Sigma\Delta\chi_i\ [P_4 + P_5\cos^2(\xi_i\theta + P_6\ |\ \Delta\chi_i|)] \tag{1}$$

The influence of ß-substituents in modulating the electronegativity of the α-substituent is also taken into account (equation 2). Optimum values for the coefficients P_1-P_7 were empirically determined.[41,42]

$$\Delta\chi_i\text{group} = \Delta\chi_i\alpha\text{-substituent} - P_7 \Sigma\ \Delta\chi_j\beta\text{-substituent} \tag{2}$$

The program assumes a rapidly achieved N / S equilibrium and calculates the relative population of each conformers based on a linear relationship in which the time averaged coupling constants (measured values) are related to X_N in equation (3). J_N and J_S represent the coupling constants for a particular set of protons in each N and S conformer.

$$J_{exp} = X_N J_N + (1-X_N) J_S \tag{3}$$

In addition, the program incorporates the concept of pseudorotation developed by Altona and Sundaralingam[31] in which the five torsion angles of the ribose ring (θ_0-θ_4, Figure 1) are mathematically related to the phase angle of pseudorotation P. In the most elementary form, the relationship between the five endocyclic torsion angles (θ_i, i = 0 - 4) can be expressed as a simple cosine function.

$$\theta_i = \theta_m \cos (P + i\ \delta) \tag{4}$$

where i = 0 - 4 , $\delta = 144°$ and θ_m is the mean puckering amplitude.

For i = 0, the equation reduces to:

$$\theta_0 = \theta_m \cos P. \tag{5}$$

The value of $\delta = 144°$ represents the phase shift between two adjacent torsion angles.

Equations (4) and (5) have been empirically refined from a statistical survey of numerous crystal structures to account principally for the behavior of ribosides and 2'-deoxyribosides.[41] However, similar modified equations have also been developed for sugars having other patterns of substituents (*ara, xylo,* and *lyxo*).[42] The parameters characteristic for the ara-, xylo- and lyxo-configurations are phase-shifted by 120° with respect to either C-2' and C-3' of ribose. This is very important, given the fact that the experimental coupling constants have to be identified by PSEUROT as either cis or trans. For compounds 1—4 the modified equations published in ref. 42 were used.

By convention, a conformation in the pseudorotation itinerary for which $P = 0°$, corresponds to the North point of the cycle (Figure 2). In such reference conformation, C-2' and C-3' are maximally displaced from each other (3'-endo and 3'-exo) and the C2'—C-3' torsion angle (θ_0 in Figure 1) is represented by an equation of type (5). Equations representing torsion angles for C-1'—C2' or C-3'—C-4' will be shifted -144° or +144°, respectively, and they will correspond to equations of type (4).[42]

Given the values of the experimental coupling constants, PSEUROT computes the best values for five variables, P_N, θ_N, P_S, θ_S, and X_N. However, in instances where the number of coupling constants available is less than five, the system is underdetermined (i.e., for compounds such as **3** and **4a,b**, only three coupling constants can be measured). This situation is remedied by decreasing the number of parameters to be calculated and constraining one or several of them to reasonable values.[41] This situation works quite well specially in cases where the equilibrium is extremely biased to one specific conformer. The imposed restrictions used by us are indicated in Tables 3—7 as "fixed parameters".

RESULTS AND DISCUSSION

The calculations for compound **1** (Table 3) indicate that the solution equilibrium significantly favors a fairly stiff S-conformer which shows very little tendency to change with temperature ($\%S = 87.9 \rightarrow \%S = 84.2$). The calculated conformational parameters for the preferred S-conformer ($P = 160.5°$, $\theta_m = 36.9°$) agree very well with the X-ray data for one of the crystal forms ($P = 164°$, $\theta_m = 36.0°$).[43] The structure of the N-conformer, however, is intrinsically less accurate. In general, as the equilibrium is shifted more towards one hemisphere, the pseudorotational parameters of the less abundant conformer are less well defined by PSEUROT. When we constrained the values of P and θ_m for the less abundant N-conformer to 9° and 36.0°, respectively, the program gave the best convergence. The quality of convergence is measured by the difference between calculated and measured coupling constants. This is expressed in terms of a "root mean square" (rms) of the absolute value of these differences for each

Table 3. Coupling constants and pseudorotational parameters for 1 based on data from FddT (ref. 10)

1   FddT

Temp	J(Hz)*					%S
	1'-2'(t)	1'-2"(c)	2'-3'(c)	2"-3'(t)	3'-4'(t)	
298	9.1	5.7	5.3	1.5	1.5	87.9
303	9.1	5.7	5.3	1.4	1.5	88.3
333	8.9	5.8	5.5	1.5	1.6	86.4
353	8.7	5.9	5.6	1.7	1.7	84.2

N Conformer $P = 9.0°$, $\theta_m = 36.0°$ (fixed parameters)
S Conformer $P = 160.5°$, $\theta_m = 36.9°$ (calculated); rms = 0.124

* Coupling constants for cis and trans protons are indicated by (c) and (t).

temperature. The lower the value the better the convergence. Normally, an rms value of ca. ≤ 0.4 Hz is borderline.[41] Results are considered reliable if rms values are ≤ 0.1 Hz.[41] For compound **3** the rms value is 0.124 Hz.

Compound **2** also appears to exist preferentially as an *S*-conformer (Table 4). In this case, however, the rms value is not as good (rms = 0.447). We think that in this structure, as well as others with a fluorine in an "up" configuration (2' or 3'), there is a strong attraction between H-6 of the base and the "up" fluorine (H-6...F) that perturbs the puckering and is responsible for PSEUROT's poorer performance. Nevertheless, the calculations for this "active" compound indicate that the molecule remains preferentially, and fairly rigidly ($\Delta\%S = 5.3$), in the *S*-hemisphere with a P value of 150.1°

Table 4. Coupling constants and pseudorotational parameters for 2

2

			J(Hz)			%S
Temp	1'-2'(c)	2'-3"(t)	2'-3'(c)	3'-4'(c)	3"-4'(t)	
285	3.2	1.5	5.5	8.6	5.0	74.9
294	3.1	1.7	5.5	8.5	5.0	74.1
304	3.2	1.8	5.6	8.6	5.1	73.3
315	3.2	1.9	5.6	8.5	5.1	72.6
333	3.4	2.1	5.7	8.5	5.4	70.1
350	3.4	2.1	6.1	8.5	5.4	69.6

N Conformer $P = 36.0°$, $\theta_m = 50.0°$ (fixed parameters)
S Conformer $P = 150.1°$, $\theta_m = 33.0°$ (calculated); rms = 0.447

An explanation for the reduced potency for the dideoxydifluoroarabinofuranosyl cytosine (**3**) was advanced by Taylor et al.,[36] based on the argument that the expected *gauche* effect between the two very electronegative fluorines would force the molecule in a more northerly direction. Such an interfluorine *gauche* effect would oppose the F—C-3'—C-4'—O-4' and F—C-2'—C-1'—O-4' *gauche* effects. The calculations performed with this molecule (Table 5) gave a good rms value of 0.101 Hz despite the presence of a fluorine in an "up" configuration. It is possible that in compound **3** the motion caused by the F—F *gauche* interaction effectively pushes H-6 and 2'-F away from each other. This will diminish the H-6—C-2'-F attraction that was invoked earlier for compound **2**, and provides a system that is better predicted by PSEUROT. The results indicate that **3** still prefers the active *S* conformation in solution, but with a P value that is closer to the *N* hemisphere (P = 143°) than the P values for **1** and **2**. A rather convenient synoptic representation of the conformational equilibrium of compound **3** can be seen in a plot showing the theoretical dependence of the transoidal proton couplings ($J_{3',4'}$) and the cisoidal couplings ($J_{1',2'}$) for the entire pseudorotation cycle (P = 0° → 360°). This is

Table 5. Coupling constants and pseudorotational parameters for 3 based on the uridine analogue

3

Temp	J(Hz)			%S
	1'-2'(c)	2'-3'(t)	3'-4'(t)	
285	2.7	1.5	2.9	89.9
294	3.4	1.4	3.2	85.7
304	3.4	1.6	3.2	84.7
315	3.5	1.6	3.4	83.2
333	3.7	1.8	3.4	81.3
350	3.7	1.7	3.5	81.3

N Conformer P = 9.0° , θ_m = 36.0° (fixed parameters)
S Conformer P = 143.1°, θ_m = 35.7° (calculated); rms = 0.101

shown in Figure 6 for three curves with different puckering amplitudes (θ_m = 30°, 33°, and 36°). These curves are generated using the same PSEUROT equations but the results are more qualitative because only two sets of coupling constants are used. However, the position of the experimental data points is consistent with the calculated P value of 143° and indicates that the molecule is strongly biased to the S form. In this plot it is evident that the transoidal coupling $J_{3',4'}$ shows a greater variation than the cisoidal coupling $J_{1',2'}$ as a function of P. Similar plots like this can be constructed for other compounds.

Since compound 3 showed less anti-HIV potency than 2[17] (Table 1), this supports the hypothesis that lower values of P are associated with reduced potency. A tentative trend that becomes apparent for the three active compounds (1—3) is that, although they all exist preferentially as S-conformers in solution, a decreasing value of P correlates with lower potency. The difference in P values between 1 and 3 also represents an experimental measure of the extent to which the interfluorine *gauche* effect is capable of affecting the puckering in a dideoxyribose template.

The alternate difluoro arrangement in dideoxydifluoroxylofuranosyl cytosine (4a) results in an extremely rigid compound that, while not calculated well by PSEUROT, appears to exist exclusively as an N-conformer, there being no coupling detected between H-1' and H-2', and H-2' and H-3' (Table 6). These values remained nearly unchanged through most of the entire temperature range. In this case, we believe that the H-6—C-3'-F attraction appears to be stronger and overpowers the interfluorine *gauche* effect. This is probably the reason why PSEUROT fails (rms = 0.862) since such H-6—C-3'-F attraction is not factored into the equations used by the program. Thus, the value of P calculated by PSEUROT (P = -9.0°) can be considered only as an approximation.

The idea that the interference caused by the H-6—C-3'-F attraction might be eliminated proved to be quite informative. Indeed, the 6-azapyrimidine analogue 4b was

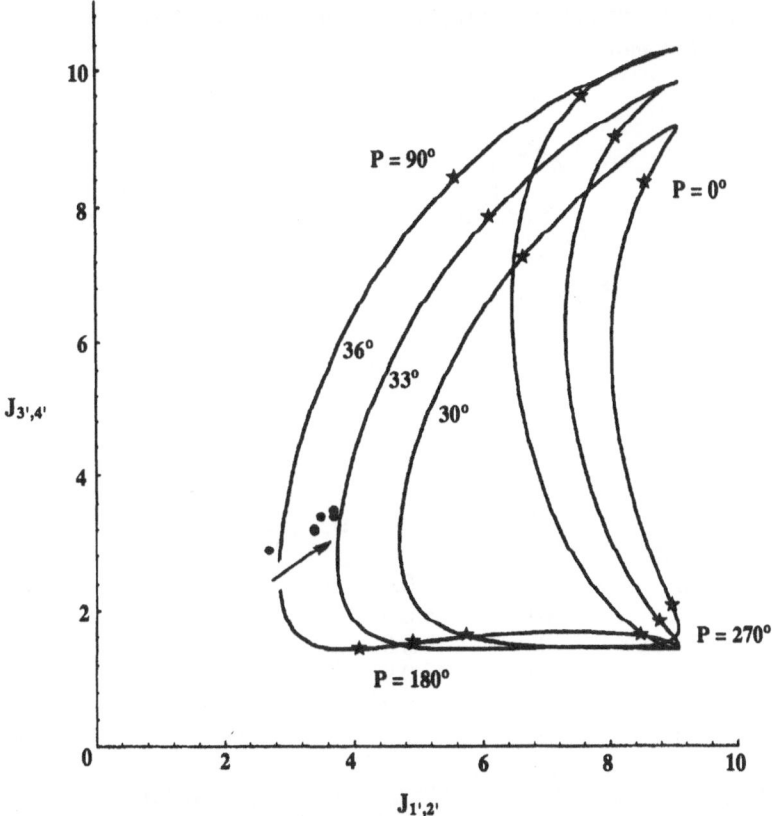

Figure 6. Theoretical dependence of the coupling constants on P for compound 3. The experimental values appear clustered in the S-hemisphere and the arrow indicates the direction of change with increasing temperature. The values of P for the four cardinal points are indicated with a star (*) in each of the curves with puckering amplitudes of 30°, 33°, and 36°.

handled quite well by PSEUROT (Table 7) and the results are very similar to the case of the difluoroarabinofuranosyl template **3**, but in the opposite direction. Although, the compound remained in the N-hemisphere, a P value of 39.0° indicated movement towards the S hemisphere (relative to **4a**) as a result of the interfluorine *gauche* effect. In this molecule, as well as **3**, the change in the direction of puckering caused by the interfluorine *gauche* effect appears to be predictable and corroborated by our NMR measurements. From our data, it would appear that the interfluorine *gauche* effect is unable to overcome the sum of two *gauche* effects that result from the interaction of two pseudo-axial fluorines with the furanose oxygen. For that reason, the inactive N-conformation of **4b** was unable to flip over to an active S-conformation exclusively on the strength of the interfluorine *gauche* effect. Nevertheless, the results are encouraging and based on the known isosteric relationship between an aromatic nitrogen and a C-F bond,[44] the synthesis of a compound with a 6-fluoropyrimidine aglycon is suggested. In this instance the magnitude of the C-3'-F...C-6-F repulsion should be greater than what was seen in **4b** and perhaps the molecule will shift towards the S hemisphere.

CONCLUSIONS

From this limited data we can conclude that anti-HIV activity in 2'- and 3'-monofluoro-, and 2',3'-difluoro-substituted cytidine nucleosides appears to be

Table 6. Coupling constants and pseudorotational parameters for 4a based on data for the uridine analogue

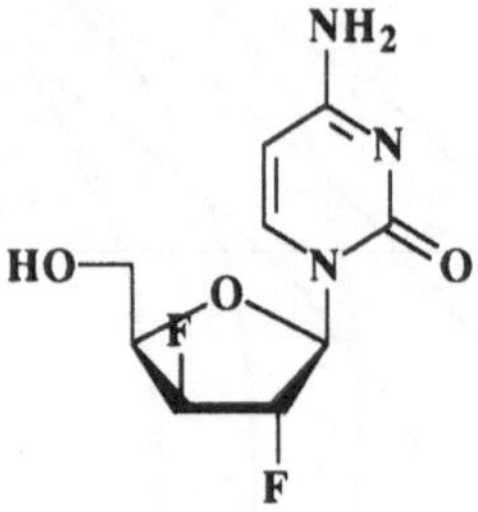

4a 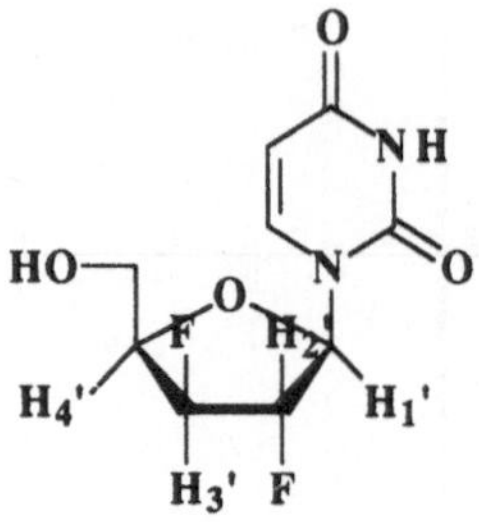   12a

	J (Hz)			%N
	1'-2'(t)	2'-3'(t)	3'-4'(c)	
Temp				
285	0.0	0.0	2.5	100
294	0.0	0.0	2.1	100
304	0.0	0.0	2.2	100
315	0.0	0.9	2.8	100
333	0.0	0.9	2.8	100
350	0.0	1.0	2.9	100

N Conformer P = -9.0° , θ_m = 32.7° (calculated); rms = 0.861
S Conformer P = 169.0°, θ_m = 36.0° (fixed parameters)

Table 7. Coupling constants and pseudorotational parameters for 4b

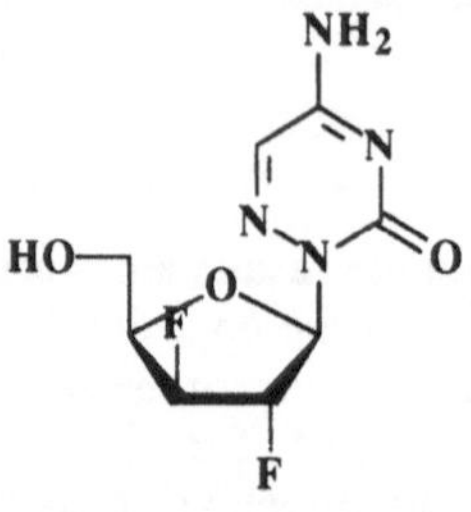

4b 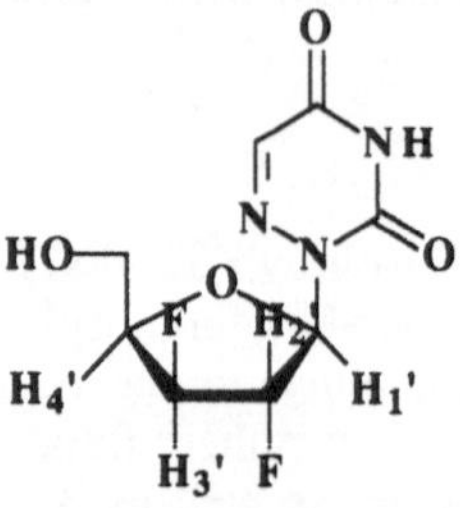   12b

	J (Hz)			%N
	1'-2'(t)	2'-3'(t)	3'-4'(c)	
Temp				
285	2.8	0.0	3.7	92.7
294	2.8	1.1	3.8	86.6
304	2.8	0.7	3.8	88.6
315	2.8	1.2	3.8	86.1
333	2.7	1.1	3.8	87.0
350	3.0	1.3	4.1	83.2

N Conformer P = 39.0° , θ_m = 30.1° (calculated); rms = 0.157
S Conformer P = 169.0°, θ_m = 36.0° (fixed parameters)

associated with compounds that prefer the *S*-conformation in solution. Moreover, for molecules in this hemisphere, it is suggested that potency is dependent on the value of P. Values of P that approach the lower end of the most commonly populated S-range (144°-180°) would appear to have reduced potency. It is clear that in these compounds fluorine substitution can induce a significant degree of stiffness to the dideoxyribose backbone that is related to the number of fluorine atoms and their preferred pseudoaxial orientation. An exceptional degree of rigidity was observed for a compound having a 3'-"up" fluorine. Such an unusual rigidity appears to be the result of a significant attraction between the pyrimidine H-6 and the 3'-"up" fluorine. When this interaction is disrupted by the use of 6-azapyrimidine bases, the molecule regains a certain degree of flexibility. It is concluded that the interfluorine *gauche* effect is not strong enough to overcome this unusual H-6...C-3'-F attraction. Finally, it appears that the interfluorine *gauche* effect in difluoro-substituted compounds is not capable of overpowering two O-4'—F *gauche* interactions in the same molecule.

ACKNOWLEDGMENTS

The authors wish to thank Drs. J. S. Driscoll and G. W. A. Milne for their valuable suggestions.

REFERENCES

1. Z. Hao, D.A. Cooney, N. R. Hartman, C. F. Perno, A. Fridland, A. L. DeVico, M. G. Sarngadharan, S. Broder, D. G. Johns. Factors Determining the Activity of 2',3'-Dideoxynucleosides in Suppressing Human Immunodeficiency Virus In Vitro, *Mol. Pharmacol.* **34**:431 (1988).

2. Z. Hao, D. A. Cooney, D. Farquhar, C. F. Perno, K. Zhang, P. Masood, Y. Wilson, N. R. Hartman, J. Balzarini, D.G. Johns. Potent DNA Chain Termination Activity and Selective Inhibition of Human Immunodeficiency Virus Reverse Transcriptase by 2',3'-Dideoxyuridine-5'-triphosphate, *Mol. Pharmacol.* **37**:157 (1990).

3. P. Van Roey, E.W. Taylor, C.K. Chu, R.F. Schinazi. Correlation of Molecular Conformation and Activity of Reverse Transcriptase Inhibitors, *Ann. N. Y. Acad. Sci.* **616**:29 (1990).

4. P.A. Furman, J.A. Fyfe, M. H. St. Clair, K. Weinhold, J.L. Rideout, A.G. Freeman, S. Nusinoff-Lehrman, D.P. Bolognesi, S. Broder, H. Mitsuya, D.W. Barry. Phosphorylation of 3'-Azido-3'-deoxythymidine and Selective Interaction of the 5'-Triphosphate with Human Immunodeficiency Virus Reverse Transcriptase, *Proc. Natl. Acad. Sci. U.S.A.* **83**:8333 (1986).

5. M.C. Starnes, Y.-C. Cheng. Cellular Metabolism of 2',3'-Dideoxycytidine, a Compound Active Against Human Immunodeficiency Virus In Vitro, *J. Biol. Chem.* **262**:988 (1987).

6. D.A. Cooney, G. Ahluwalia, H. Mitsuya, A. Fridland, M. Johnson, Z. Hao, M. Dalal, J. Balzarini, S. Broder, D.G. Johns. Initial Studies on the Cellular Pharmacology of 2',3'-Dideoxyadenosine, an Inhibitor of HIV Infectivity, *Biochem. Pharmacol.* **36**:1765 (1987).

7. A. Fridland, M.A. Johnson, D.A. Cooney, G. Ahluwalia, V.E. Marquez, J.S. Driscoll, D.G. Johns. Metabolism in Human Leukocytes of Anti-HIV Dideoxypurine Nucleosides, *Ann. N. Y. Acad. Sci.* **616**:205 (1990).

8. B. Jagannadh, D.V. Reddy, A.C. Kunwar. 1H NMR Study of the Sugar Pucker of 2',3'-Dideoxynucleosides with Anti-Human Immunodeficiency Virus (HIV) Activity, *Biochem. Biophys. Res. Commun.* **179**:386 (1991).

9. N. Hicks, O.W. Howarth, D.W. Hutchinson. NMR Studies of the Fexibility of the Glycosyl Ring of Thymidine and Uridine Nucleosides, *Carbohydr. Res.* **216**:1 (1991).

10. J. Plavec, L.H. Koole, A. Sandström, J. Chattopadhyaya. Structural Studies of Anti HIV 3'-α-Fluorothymidine and 3'-α-Azidothymidine by 500 MHz ^{1}H NMR Spectroscopy and Molecular Mechanics (MM2) Calculations, *Tetrahedron* **47**:7363 (1991).

11. W. Olson. How Flexible is the Furanose Ring? 2. An Updated Potential Energy Estimate, *J. Am. Chem. Soc.* **104**:278 (1982).

12. J. Balzarini, M. Baba, R. Pauwels, P. Herdewijn, E. De Clercq. Anti-Retrovirus Activity of 3'-Fluoro- and 3'-Azido-Substituted Pyrimidine 2',3'-Dideoxynucleoside Analogues, *Biochem. Pharmacol.* **37**:2847 (1988).

13. P. Herdewijn, R. Pauwels, M. Baba, J. Balzarini, E. De Clercq. Synthesis and Anti-HIV Activity of Various 2'- and 3'-Substituted 2',3'-Dideoxyadenosines: A Structure-Activity Analysis, *J. Med. Chem.* **30**:2132 (1987).

14. P. Herdewijn, J. Balzarini, E. De Clercq, R. Pauwels, M. Baba, S. Broder, H. Vanderhaeghe. 3'-Substituted 2',3'-Dideoxynucleoside Analogues as Potential Anti-HIV (HTLV-III/LAV) Agents, *J. Med. Chem.* **30**:1270 (1987).

15. F. Puech, G. Gosselin, J.-L. Imbach. Synthesis of 9-(3-Deoxy- and 2,3-Dideoxy-3-fluoro-ß-D-xylofuranosyl)guanines as Potential Antiviral Agents, *Tetrahedron Lett.* **30**:3171 (1989).

16. A. Van Aerschot, P. Herdewijn, J. Balzarini, R. Pauwels, E. De Clercq. 3'-Fluoro-2',3'-dideoxy-5-chlorouridine: Most Selective Anti-HIV-1 Agent among a Series of New 2'- and 3'-Fluorinated 2',3'-Dideoxynucleoside Analogues, *J. Med. Chem.* **32**:1743 (1989).

17. J.A. Martin, D.J. Bushnell, I.B. Duncan, S.J. Dunsdon, M.J. Hall, P.J. Machin, J. H. Merrett, K.E.B. Parkes, N.A. Roberts, G.J. Thomas, S.A. Galpin, D. Kinchington. Synthesis and Antiviral Activity of Monofluoro and Difluoro Analogues of Pyrimidine Deoxyribonucleosides Against Human Immunodeciency Virus (HIV-1), *J. Med. Chem.* **33**:2137 (1990).

18. J.-T. Huang, L.-C. Chen, L. Wang, M.-H. Kim, J.A. Warshaw, D. Armstrong, Q.-Y. Zhu, T.-C. Chou, K.A. Watanabe, J. Matulic-Adamic, T.-L. Su, J.J. Fox, B. Polsky, P.A. Baron, J.W.M. Gold, W.D. Hardy, E. Zuckerman. Fluorinated Sugar Analogues of Potential Anti-HIV Nucleosides, *J. Med. Chem.* **34**:1640 (1991).

19. R.Z. Sterzycki, I. Ghazzouli, V. Brankovan, J.C. Martin, M.M. Mansuri. Synthesis and Anti-HIV Activity of Several 2'-Fluoro-Containing Pyrimidine Nucleosides, *J. Med. Chem.* **33**:2150 (1990).

20. K.A. Watanabe, K. Harada, J. Zeidler, J. Matulic-Adamic, K. Takahashi, W.-M. Ren, L.-C. Cheng, J.J. Fox, T.C. Chou, Q.-Y. Zhu, B. Polsky, J.W.M. Gold, D. Armstrong. Synthesis and Anti-HIV-1 Activity of 2'-"Up"-Fluoro Analogues of Active Anti-AIDS Nucleosides 3'-Azido-3'-deoxythymidine (AZT) and 2',3'-Dideoxycytidine (DDC), *J. Med. Chem.* **33**:2145 (1990).

21. V.E. Marquez, C.K.-H. Tseng, S. Mitsuya, S. Aoki, J.A. Kelley, H. Ford, Jr., J.S. Roth, S. Broder, D.G. Johns, J.S. Driscoll. Acid Stable 2'-Fluoro Purine Dideoxynucleosides as Active Agents Against HIV, *J. Med. Chem.* **33**:978 (1990).

22. M.A. Siddiqui, J.S. Driscoll, V.E. Marquez, J.S. Roth, T. Shirasaka, H. Mitsuya, J.J. Barchi, Jr., J.A. Kelley. Chemistry and Anti-HIV Properties of 2'-Fluoro-2',3'-Dideoxyarabinofuranosyl Pyrimidines, *J. Med. Chem.* **35**:2195 (1992).

23. M.A. Siddiqui, J.S. Driscoll, V.E. Marquez. Unpublished results.

24. A. Van Aerschot, J. Balzarini, R. Pauwels, L. Kerremans, E. De Clercq, P. Herdewijn. Influence of Fluorination of the Sugar Moiety on the Anti-HIV-1 Activity of 2',3'-Dideoxynucleosides, *Nucleosides Nucleotides* **8**:1121 (1989).

25. A. Van Aerschot, P. Herdewijn, P. 2',3'-Difluoro- and 3'-Azido-2'-fluoro Substituted Dideoxypyrimidines as Potential Anti-HIV Agents, *Bull. Soc. Chim. Belg.* **98**:937 (1989).

26. P. Herdewijn, A. Van Aerschot. Synthesis of 2',3'-Disubstituted 3'-Deoxythymidine Derivatives, *Bull. Soc. Chim. Belg.* **98**:943 (1989).

27. P. Van Roey, J. M. Salerno, C. K. Chu, R. F. Schinazi. Correlation Between Preferred Sugar Ring Conformation and Activity of Nucleoside Analogues Against Human Immunodeficiency Virus, *Proc. Natl. Acad. Sci. U.S.A.* **86**:3929 (1989).

28. S. Dijkstra, J.M. Benevides, G. . Thomas, Jr. Solution Conformation of Nucleoside Analogues Exhibiting Antiviral Activity Against Human Immunodeficiency Virus, *J. Mol. Struct.* **242**:283 (1991).

29. F.E. Hruska, A.A. Smith, J.G. Dalton. A Correlation of Some Structural Parameters of Pyrimidine Nucleosides. A Nuclear Magnetic Resonance Study, *J. Am. Chem. Soc.* **93**:4334 (1971).

30. D.A. Pearlman, S.-H. Kim. Conformational Studies of Nucleic Acids II. The Conformational Energetics of Commonly Occurring Nucleosides, *J. Biomol. Struct. Dyn.* **3**:99 (1985).

31. C. Altona, M. Sundaralingam. Conformational Analysis of the Sugar Ring in Nucleosides and Nucleotides. A New Description Using the Concept of Pseudorotation, *J. Am. Chem. Soc.* **94**:8205 (1972).

32. W. Saenger. Principles in Nucleic Acid Structure, Springer-Verlag, New York (1984).

33. L.H. Koole, H.M. Buck, A. Nyilas, J. Chattopadhyaya. Structural Properties of Modified Deoxyadenosine Structures in Solution. Impact of the *Gauche* and Anomeric Effects on the Furanose Conformation, *Can. J. Chem.* **65**:2089 (1987).

34. C. Altona, M. Sundaralingam. Conformational Analysis of the Sugar Ring in Nucleosides and Nucleotides. Improved Method for the Interpretation of Proton Magnetic Resonance Coupling Constants, *J. Am. Chem. Soc.* **95**:2333 (1973).

35. W. Guschlbauer, K. Jankowski. Nucleoside Conformation is Determined by the Electronegativity of the Sugar Substituent, *Nucleic. Acids Res.* **8**:1421 (1980).

36. E.W. Taylor, P. Van Roey, R.F. Schinazi, C.K. Chu. A Stereochemical Rationale for the Activity of Anti-HIV Nucleosides, *Antivir. Chem. Chemother.* **1**:163 (1990).

37. V.E. Marquez et al. All new compounds reported in this article were evaluated as anti-HIV agents in ATH8 and CEM cells (results unpublished).

38. D. Bergstrom, D.J. Swartling, A. Wisor, M.R. Hoffmann. Evaluation of Thymidine, Dideoxythymidine, and Fluorine Substituted Deoxyribonucleoside Geometry by the MINDO/3 Technique: The Effect of Fluorine Substitution on Nucleoside Geometry and Biological Activity, *Nucleosides Nucleotides* **10**:693 (1991).

39. K. Green, D. M. Blum. Hydrofluorination of Anhydrothymidine via Soluble Aluminum Derivatives, *Tetrahedron Lett.* **32**:2091 (1991).

40. R.J. Wysocki Jr., M.A. Siddiqui, J.J. Barchi Jr., J.S. Driscoll, V.E. Marquez. A More Expedient Synthesis of Anti-HIV-Active 2,3-Dideoxy-2-fluoro-ß-D-*threo*-pentofuranosyl Nucleosides, *Synthesis,* 1005 (1991).

41. C.A.G. Haasnoot, F.A.A.M. de Leeuw, H.P.M. de Leeuw, C. Altona. The Relationship Between Proton-Proton NMR Coupling Constants and Substituent Electronegativities. II. Conformational Analysis of the Sugar Ring in Nucleosides and Nucleotides in Solution Using a Generalized Karplus Equation, *Orgn. Magn. Resns.* **15**:43 (1981).

42. F.A.A.M. de Leeuw, C. Altona. Conformational Analysis of ß-D-Ribo, ß-D-Deoxyribo-, ß-D-Arabino-, ß-D-Xylo, and ß-D-Lyxo-nucleosides from Proton—Proton Coupling Constants, *J. Chem. Soc., Perkin Trans. 2*, 375 (1982).

43. N. Camerman, D. Mastropaolo, A. Camerman. Structure of the Anti-Human Immunodeficiency Virus Agent 3'-Fluoro-3'-deoxythymidine and Electronic Charge Calculations for 3'-Deoxythymidines, *Proc. Natl. Acad. Sci. U.S.A.* 87:3534 (1990).

44. D.J. McNamara, P.D. Cook. Synthesis and Antitumor Activity of Fluorine-Substituted 4-Amino-2(1*H*)-pyridinones and their Nucleosides. 3-Deazacytosines, *J. Med. Chem.* 30:340 (1987).

CRYSTAL STRUCTURES AND MOLECULAR CONFORMATIONS OF ANTI-HIV NUCLEOSIDES

Patrick Van Roey[1*] and Chung K. Chu[2]

[1]Medical Foundation of Buffalo, Buffalo, NY 14203

[2]Department of Medicinal Chemistry, College of Pharmacy
University of Georgia, Athens, GA 30602

INTRODUCTION

Since the discovery of 3'-azido-3'-deoxythymidine (AZT) as the first drug for the treatment of AIDS, considerable efforts have been made to develop new nucleoside analogues that would be more active, less toxic inhibitors of the HIV-1 reverse transcriptase.[1-3] Many novel compounds have been synthesized and tested,[3-10] with only one general criterion for selection: the absence of a 3'-hydroxyl group so that the triphosphate nucleotide can inhibit HIV-1 reverse transcriptase by acting as a chain terminator. The molecular structures and conformations of many of these compounds have been examined in efforts to determine if structural or conformational features of the molecules can be correlated with activity.[11-33] However, the diversity of compounds that show at least some activity and the fact that apparently very similar compounds can have extremely different activities indicate that identification of a single or a few structural parameters that would be required for activity is unlikely. The variations in activity are at

---

* Present address: Wadsworth Center for Laboratories and Research, New York State Department of Health, Empire State Plaza, P.O. Box 509, Albany, NY 12201-0509.

Nucleosides and Nucleotides as Antitumor and Antiviral Agents,
Edited by C.K. Chu and D.C. Baker, Plenum Press, New York, 1993

least partially caused by the presence of several alternative metabolic pathways for the phosphorylation of the nucleoside to generate the active triphosphate nucleotide.[2] This limits the utility of the study of the nucleoside conformation in analyzing the interaction of the nucleotide with reverse transcriptase. However, it may be possible to determine conditions required for efficient phosphorylation of the nucleoside. An additional factor complicating the development of structure-activity relationships is the inherent conformational flexibility of the nucleoside.[34] For most nucleosides, several low-energy conformations can be identified and are accessible over low barriers to rotation. This precludes conclusions from the crystal structure determination of a single nucleoside but requires analysis of structural and conformational properties of a large group of related compounds with different activities. Most studies have focused on the modification of the deoxyribose unit rather than comparison of compounds with different bases for three reasons: (1) certain modification of the deoxyribose ring has been associated with high activity; (2) the conformation of the furanose ring determines the accessibility of the O5'-hydroxyl group; and, (3) nucleosides with different bases are frequently phosphorylated by different kinases.

In this chapter we describe and analyze the molecular conformations of several nucleosides, both active and inactive against HIV-1, that belong to four classes each with different modifications to the furanose ring. This chapter is not intended to be a general overview of the literature on nucleoside geometry or a review of all studies aimed at determining correlations with activity against HIV. The analysis is limited to those groups of compounds for which the crystal structures of several related compounds have been determined. Groups of compounds not included for which conformational data is available include 2',3'-fused analogues[35] and 2'-ara compounds.[36]

DESCRIPTION OF NUCLEOSIDE CONFORMATION

The conformational properties of nucleosides have been reviewed in great detail by Sundaralingam,[37] by Saenger,[34] by Pearlman and Kim[38] and by Birnbaum and Shugar.[39] The available data are the result of more than two hundred X-ray crystallographic studies,

NMR studies and conformational analysis by computational methods at various levels of complexity. Three parameters describe the primary features of the conformation of a nucleoside: the glycosylic bond torsion angle (χ) (torsion angle C2-N1-C1'-O4' for pyrimidines or C4-N9-C1'-O4' for purines) which describes the orientation of the base relative to the furanose ring; the C4'-C5' torsion angle (γ), describing the orientation of the 5'-hydroxyl group relative to the furanose ring (angle C3'-C4'-C5'-O5'); and, the furanose ring puckering which is fully described by the pseudorotational angle (P) and the maximum torsion angle (v_{max}). The puckering or pseudorotation is a distortion from planarity of the five-membered ring that consists of a displacement of one (envelope) or two (twist) atoms out of the plane of the ring. The pseudorotation angle is cyclic function, based on all the endocyclic torsion angles, that describes which atom or atoms are displaced out of the plane of the ring and in which direction. The maximum torsion angle indicates the overall degree or magnitude of the displacement. Several codes have been developed for the description of the furanose conformation. For the sake of simplicity we will limit ourselves to the use of the terms *endo* and *exo*, which refer to the displacement of an atom above or below the plane of the ring, respectively. Envelope conformations are indicated by a single displacement assignment, e.g., C3'-*exo*, and twist conformations, in which two atoms are significantly displaced, are indicated by a double assignment, e.g., C3'-*exo*/C2'-*endo*. Figure 1 illustrates the definitions of the three conformational parameters.

Nucleosides inherently are very flexible molecules and can adopt many conformations that are energetically equal or nearly equal and accessible over low barriers to rotation. Analysis of the preferred conformations, reviewed in detail by Saenger,[34] can be summarized as follows.

1.   Glycosylic bond: The *anti* region (χ negative) is preferred over the *syn* region (χ positive) because of potential steric contacts. Pyrimidine nucleosides are predominantly observed in an *anti* conformation with $-180 < \chi < -115°$. High-*anti* ($-60 < \chi < -100°$) is often observed for purine analogues and is in this case accessible because of the reduced bulk of the 5-membered ring of the base. Three distinct conformations are significant: $\chi \sim -160°$ which places C6 of the pyrimidine base eclipsed with O4'; $\chi \sim -120°$ which places C6 above the center of the ring and $\chi \sim -100°$ in which C6 would be eclipsed with C2'.

2. Furanose ring: two preferred areas of pseudorotational space have been commonly observed: those with $0 < P < 36°$, where the C3'-*endo* is the main displacement and $144 < P < 180°$ where C2'-*endo* is the main displacement. In 2'-deoxynucleosides, the C2'-*endo* conformation is much more likely to be observed than the C3'-*endo* conformation. The area around $P \sim 270°$ is much higher in energy than the others because in this conformation both bulky substituents, C5' and the base, are maximally axial and substituents on C2' and C3' are eclipsed. Transitions from the C2'-*endo* to the C3'-*endo* conformations are expected to happen over the $P \sim 90°$ conformation.

3. 5'-Hydroxyl group orientation: Three conformations have been observed: +*sc*, *ap* and -*sc*, corresponding to γ angles of 60, 180 and -60° respectively. Pyrimidine nucleosides have a strong preference for the +*sc* conformation while purine nucleosides are about equally common in the +*sc* and the *ap* conformation. The +*sc* conformation places O5' above the center of the furanose ring and the *ap* conformation places it outside the ring but on the O4' side. The O5'-hydroxyl orientation in crystal structures cannot be evaluated very reliably because the O5'-hydroxyl acts as a hydrogen bond donor and as such plays a major role in establishing the intermolecular contacts that are involved in crystallization.

Strong correlations exist between the three conformational parameters. The most obvious one is between the glycosylic bond and the ring puckering: C3'-*endo* with $-180 < \chi < -138°$ and C2'-*endo* with $-144 < \chi < -115°$.

CRYSTALLOGRAPHIC STUDIES

The crystal structures of representative compounds from the following four classes of nucleoside analogues have been determined and will be analyzed here:
(1) 2',3'-dideoxynucleosides (d2N), (2) 2',3'-didehydro-2',3'-didoxynucleosides (d4N), (3) 3'-substituted-2',3'-dideoxynucleosides (Sd2N) and (4) 1,3-dioxolane-pyrimidine nucleosides (dON). The majority of these compounds crystallize with two or four molecules in the crystallographic asymmetric unit giving two or four independent observations that are indicated with the capital letters A, B, C and D in the tables and discussion section. Figures 2 through 7 show stereo diagrams of molecules that represent

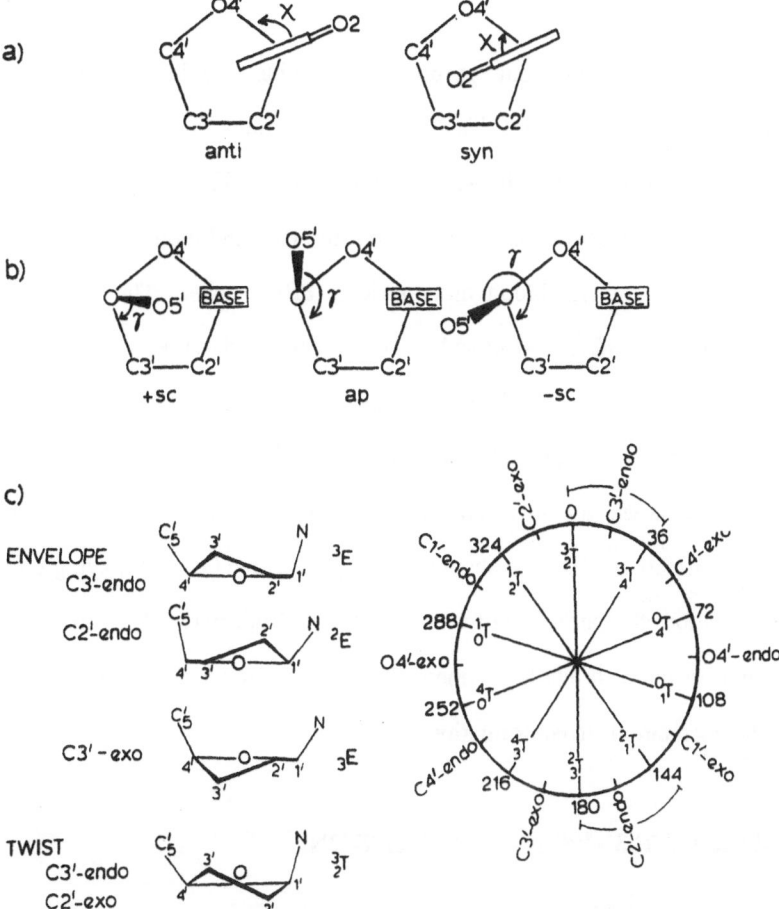

Figure 1. (a) glycosylic bond geometries for pyrimidine nucleosides. In the *anti* conformation, C6 is above the furanose ring. Observed conformations run from N1-C6 eclipsed with C1'-O4', $\chi \sim 180°$ to N1-C6 eclipsed with C1'-C2', $\chi \sim -90°$. In the *syn* conformation the more bulky C2-O2 side of the base would be above the ring. This conformation is therefore seldom observed. (b) orientation of the C5'-O5' bond relative to the furanose ring. The three possible conformations shown are possible but the *+sc* and *ap* conformations more commonly observed. (c) The diagram showing the relationship between the pseudorotational angle P and the puckering of the furanose ring. Some of the more commonly conformations are shown on the left.

the most important different conformations. These figures were produced using the program PLUTON.[40] Oxygen atoms are black and nitrogen atoms are cross hatched.

1. 2',3'-Dideoxynucleosides (d2N)

One purine, adenosine, and four pyrimidine, cytidine, uridine, thymidine and 5-ethyluridine (eU), were studied. Table 1 lists the most important conformational parameters of these nucleosides with their respective activity. The anti-HIV activity (EC_{50}) shown are the effective micromolar concentration for 50% inhibition of virus proliferation in HIV-1 infected peripheral blood mononuclear (PBM) cells. The glycosylic bond conformations are standard: $\chi \sim$ -130 and $\sim$ -160° for the pyrimidines and $\sim$ -100° for the purine analogue. Three furanose rings are in the C3'-*endo* conformation, two are in the C2'-*endo* conformation and two, d2C and d2A, which are the most potent anti-HIV compounds of the series, are in a more extreme C4'-*endo*/C3'-*exo* conformation (P > 190°). The conformation of D2A is shown in Figure 2. Three O5'-hydroxyl, groups are in the *ap* conformation: the purine and two pyrimidines. Both pyrimidines have $\chi \sim$ -160° but have opposite furanose conformations. The O5'-hydroxyl group of molecule B of d2eU is in the uncommon -*sc* conformation.

2. 3'-substituted-2',3'-dideoxynucleosides (Sd2N)

Conformational data on eigth pyrimidine analogues are listed in Table 2. The compounds include the 3'-azido-2',3'-dideoxy analogues of uridine, thymidine (AZT), 5-ethyluridine, cytidine and 5-methylcytidine (mC). Other compounds are the 3'-fluoro, 3'-amino-3'-deoxythymidine and 3'-(2-propenyl)-2',3'-dideoxyuridine (ped2U) analogue. Again two glycosylic bond conformations are observed $\chi \sim$ -120 and $\sim$ -165°. The furanose ring conformations of the uridine, thymidine and 5-ethyluridine are in the C2'-*endo* through C3'-*exo* region, except for the totally inactive 3'-amino and 3'-propenyl analogues, which are C3'-*endo*. The 3'-azido group appears to promote an extreme C3'-*exo*/C4'-*endo* formation with P > 200°, at least for the uridine analogues. Four of the five

Table 1.   2',3'-dideoxynucleosides.

		$EC_{50}(10^{-6}M)$	χ (°)	γ (°)	P (°)	furanose conf.	Ref.
d2C		0.011	-156.9	164.5	208.2	C3'-*exo*/C4'-*endo*	11
d2U		96.8	-163.5	177.9	7.2	C3'-*endo*/C2'-*exo*	12
d2T	A	0.17	-170.2	63.5	12.8	C3'-*endo*/C2'-*exo*	12
	B		-128.9	66.0	166.0	C2'-*endo*	
d2eU	A	4.9	-129.7	-78.2	186.8	C3'-*exo*/C2'-*endo*	13
	B		-167.4	57.8	12.5	C3'-*endo*/C2'-*exo*	
d2A		0.91	-96.1	180.0	193.4	C3'-*exo*/C2'-*endo*	14

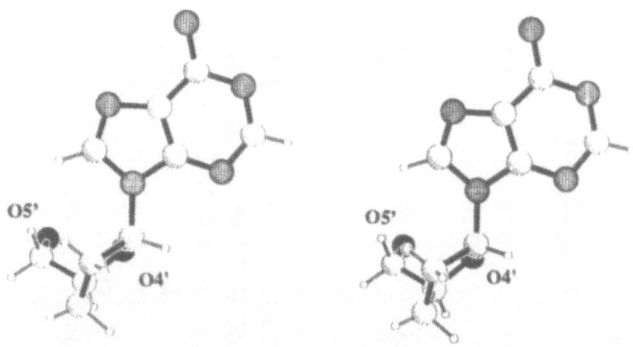

Figure 2. Molecular conformation of d2A. Significant features of the conformation are the extreme C3'-*exo* conformation of the furanose ring and the $\chi \sim -100°$ glycosylic torsion angle which places C8 above the middle furanose ring.

Table 2. 3'-substituted-2',3'-dideoxynucleosides.

	$EC_{50}(10^{-6}$ M)	χ (°)	γ (°)	P (°)	furanose conf.	Ref.
N_3d2U	0.18	-159.8	57.3	176.3	C2'-endo/C3'-exo	15
N_3d2T A	0.002	-124.4	50.9	174.9	C2'-endo/C3'-exo	15-20
B		-173.6	173.4	215.3	C3'-exo/C4'-endo	
N_3d2eU A	0.56	-129.9	54.0	169.3	C2'-endo	15
B		-170.8	175.1	203.3	C3'-exo/C4'-endo	
C		-108.8	48.6	168.5	C2'-endo/C3'-exo	
D		-170.6	49.6	209.6	C3'-exo/C4'-endo	
N_3d2C.HCl	0.66	-130.	48.	135.	C1'-exo/C2'-endo	21
N_3d2mC.HCl	0.081	-131.	49.	116.	C1'-exo/O4'-endo	13
Fd2T A	0.0086	-153.2	51.9	175.2	C2'-endo/C3'-exo	22
B		-149.4	50.6	176.3	C2'-endo/C3'-exo	
C		-137.6	48.3	173.5	C2'-endo/C3'-exo	
D		-129.5	45.8	171.1	C2'-endo/C3'-exo	
NH_2d2T	>100.	-176.4	57.0	3.8	C3'-endo/C2'-exo	23
Ped2U	100.	-161.6	179.4	15.8	C3'-endo/C2'-exo	24

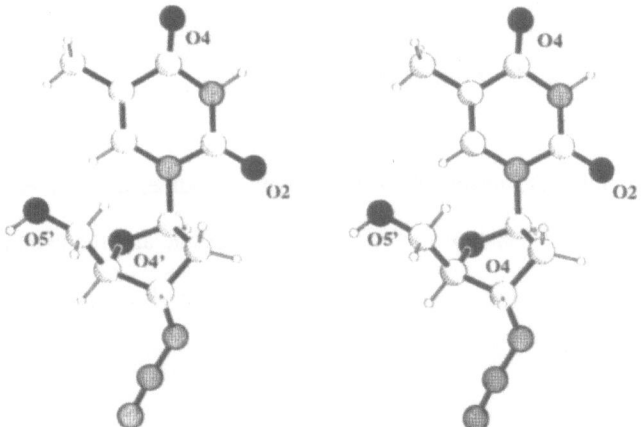

Figure 3. Molecular conformation of AZd2T, molecule A. The hydrogen atom of C6 is above the center of the furanose ring which is in a slight C3'-*exo* conformation.

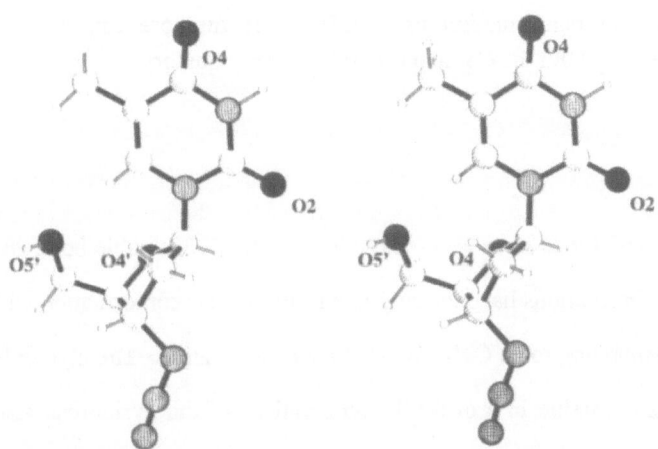

Figure 4. Molecular conformation of AZd2T, molecule B. The $\chi \sim -170°$ glycosylic torsion angle eclipses N1-C6 with C1'-O4'. The furanose ring is in an extreme C3'-*exo* conformation which makes all three substituent axial.

occurances of this extreme ring conformation, including d2C and d2A, are coupled with the *ap* O5'-orientation and all pyrimidine compounds in the group have $\chi \sim -160°$. Both cytidine analogues are closer to the C1'-*exo* region, which is also rather unusual. Figures 3, 4 and 5 show the conformations of both molecules of AZd2T and that of Ped2U.

3. 2',3'-Didehydro-2',3'-dideoxynucleosides (d4N)

Crystal structures are available for the cytidine, uridine, thymidine, 5-ethyluridine, adenosine and guanosine analogues. For both the thymidine and the cytidine compounds

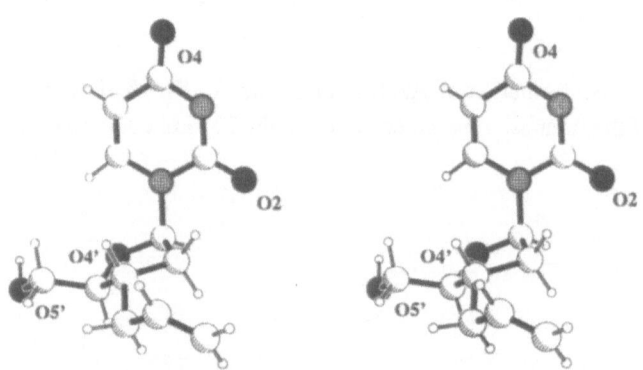

Figure 5. Molecular conformation of Ped2U. The furanose ring is in a C3'-*endo* conformation, making both the C3' and C4' substituents equatorial.

two different crystal forms have been determined. The 2',3'-double bond eliminates ring puckering. All observations have the ring in a nearly planar conformation with the largest deviations corresponding to an O4'-*exo* envelope conformation. The glycosylic bond is in the high-anti conformation in 4 of the 12 observations of the pyrimidine analogues. All three observations of the purine nucleosides are also in the high-*anti* conformation. The high-*anti* conformation is common for purine nucleosides but uncommon for pyrimidine analogues.

Table 3.   2',3'-didehydro-2',3'-dideoxynucleosides.

Compound		$EC_{50}(10^{-6}M)$	χ (°)	γ (°)	Ref.
d4U	A	73.8	-96.1	53.5	25
	B		-178.0	54.2	
d4T	A	0.009	-174.0	58.	25
	B		-81.4	56.	
d4T	C		-118.0	60.6	26
	D		-174.0	53.8	
d4eU	A	75.7	-98.6	48.	25
	B		-168.9	57.	
d4C	A	0.005	-82.7	55.1	25
			-87.8	47.3	
d4C	C		-118.7	49.8	27
	D		-160.2	165.0	
d4A		0.76	-100.2	179.8	14
d4G	A	>100.	-102.3	54.0	28
	B		-94.2	47.6	

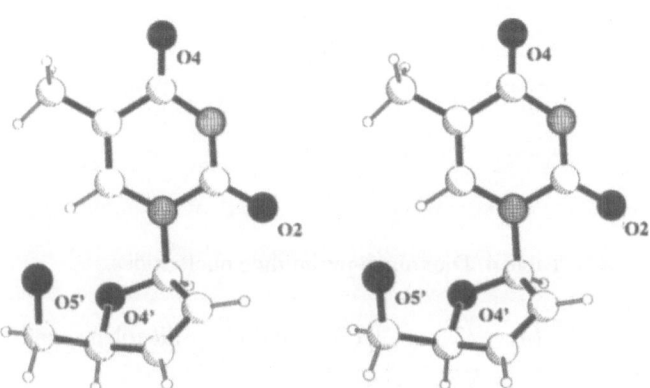

Figure 6. Molecular conformation of d4T, molecule A. N1-C6 is eclipsed with C1'-O4'
The conformation resembles that of AZd2T, molecule B (Figure 4), except that the planar
ring does not allow the C5' to become axial.

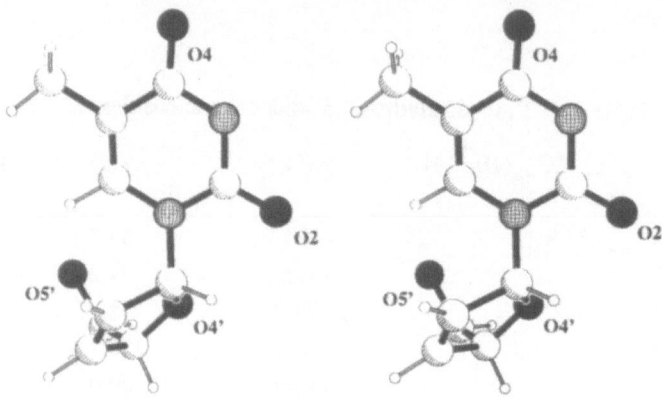

Figure 7. Molecular conformation of d4T, molecule B. The N1-C6 bond is eclipsed with the C1'-C2' bond, a conformation not observed in saturated pyrimidine nucleosides. Overall the conformation is very similar to that of d2A (figure 2).

4. 1,3-Dioxolane-pyrimidine Nucleosides (dON)

Three crystals structures have been determined: racemic thymidine analogue and enantiomerically pure (1'R,4'R)-thymidine and cytidine analogues. Neither the glycosylic bond nor the 5'-hydroxyl conformations are exceptional. All dioxolane rings are in the C3'-*endo* conformation.

Table 4. Dioxolane-pyrimidine nucleosides.

		$EC_{50}(10^{-6}M)$	χ (°)	γ (°)	P (°)	dioxolane conf.	Ref.
dOC	A	0.016	-163.	61.	18.	C3'-*endo*	29
	B		-157.	53.	13.	C3'-*endo*/C2'-*exo*	
dOT	A	0.39	-175.2	48.9	9.0	C3'-*endo*/C2'-*exo*	30
	B		-133.4	64.1	20.0	C3'-*endo*	
(±)dOT		0.09	-122	66	32	C3'-*endo*/C4'-*exo*	31

DISCUSSION

The basic assumption of our analysis is that the nucleosides that are the most efficiently phosphorylated by a specific kinase are those that are best suited to adopt the "active site" conformation. Because phosphorylation is base specific, one must assume that binding occurs at the base and that the furanose ring orientation, the furanose ring conformation and the C4'-C5' bond conformation are adjusted to place the O5' in the active site of the nucleoside kinase. These three conformational variables are described by the conformational parameters χ, P and γ, respectively. It is not to be assumed that the active nucleosides would always be observed in a specific "different" conformation but frequent occurence of conformations that are rarely observed for the inactive compounds would indicate that conformational effects may be present.

General trends for activity of compounds with the same base is as follows: AZd2N compounds are somewhat more active than the corresponding d4N analogues but both groups are much more active than the d2N compounds. The 3'-fluoro analogue of d2N is also very active but the other halo analogues are less active than the d2N compounds. Compounds with electropositive or alkyl 3'-substituents have no activity.

Therefore, the AZd2N and d4N compounds should at least occasionally be observed in conformations that are rarely observed for the less active d2N compounds and not for the inactive 3'-substituted compounds. The active 3'-substituted compounds as well as the very active unsubstituted compounds are frequently are observed in the extreme C4'-*endo*/C3'-*exo* conformation, while the inactive compounds are usually observed in a C3'-*endo* conformation. Taylor *et al.*[32] have demonstrated that this preference for the C4'-*endo*/C3'-*exo* conformation can be explained in terms of the *gauche* effect[41] involving interactions of O4', the lone pair electrons on O4' and the substituents on C3'. The main difference between these two conformations is the position of the C5' which is axial in the C4'-*endo*/C3'-*exo* conformation and equatorial in the C3'-*endo* conformation, as shown in Figure 8. This has the following effects: the distance between C5' and the base is shortened, O5' is brought higher above the plane of the ribose ring and the +*sc* conformation of O5' is stabilized in favor of the *ap* conformation. Figure 9 shows the distances between C5' and N1 in C3'-*exo* and C3'-*endo* conformations.

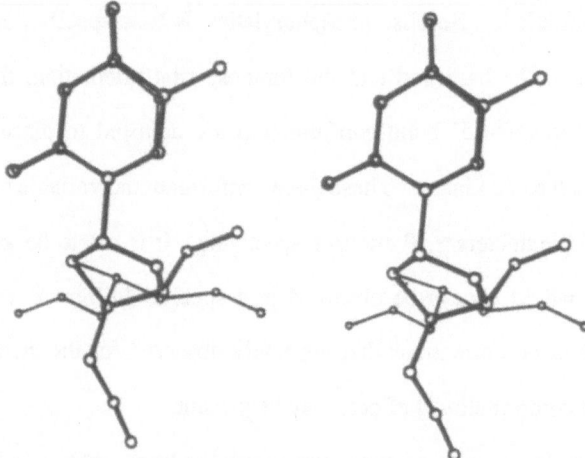

Figure 8. ORTEP[42] stereo diagram showing the overlap of the conformations of AZd2T, molecule B with Ped2U produced by least-squares fitting the atoms N1, C1', C2' and O4'. The change in the furanose ring conformations moves the C5' substituent from an axial to an equatorial position.

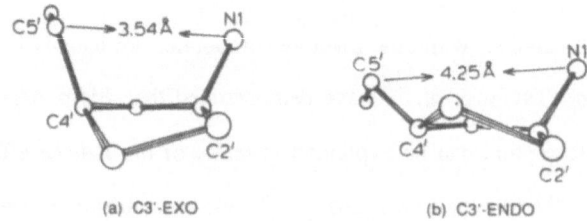

(a) C3'-EXO   (b) C3'-ENDO

Figure 9. Illustration of the effect of the change from an axial (a) to an equatorial position for the C5'-substituent.

The d4N pyrimidine nucleosides occur frequently (5 out of 12 observations) in the unusual high-anti glycosylic bond conformation. In this case, molecular mechanics calculations have shown that a much greater area of conformational space is available for the glycosylic bond conformation, primarily because of the absence of the "2'-up" hydrogen atom.[25] Also, although the crystallographic studies would not indicate so, the *ap* O5'-hydroxyl group conformation is more accessible and equal in energy to the *+sc* conformation. As mentioned before, the orientation of the O5' bond in the crystal structure is highly affected by the ability of O5' to act as a hydrogen bond donor, representing the main energy factor in forming intermolecular contacts in the crystal.

The dioxolane analogues form a separate class of nucleosides. The conformational parameters are given above for the (1'*R*,4'*R*)-enantiomer. It has now been demonstrated that the opposite enantiomer of the related 3'-thioxolane analogues is the more active one.[43] This would seem to be also the case for the dioxolane compounds because the racemic dioxolane compound is more active than the (1'*R*,4'*R*)-enantiomer. Therefore, the C3'-*endo* conformation of the dioxolane ring is not inconsistent with our conclusions regarding the preference for the C3'-*exo* conformation.

CONCLUSION

Using crystallographic studies we have been able to demonstrate that active anti-HIV compounds are capable of adopting unusual conformations. This shows that unusual conformational flexibility is an asset for activity and that the "active site" conformation of a nucleoside at the kinase is a special conformation.

ACKNOWLEDGEMENTS

Research supported in part by the Father Vincent Grant for AIDS Research

REFERENCES

1.   H. Mitsuya, R. Yarchoan, R. Kageyama, and S. Broder, Targeted therapy of human immunodeficiency virus-related disease, *FASEB J.*, 5:2369 (1991).

2.   R.F. Schinazi, J.R. Mead, and P.M. Feorino, Insights into HIV chemotherapy, *AIDS Res. Human Retroviruses*, 8:963 (1992).

3.   M. Nasr, C. Litterst and J. McGowan, Computer-assisted structure-activity correlations of dideoxynucleoside analogues as potential anti-HIV drugs, *J. Antiviral Res.*, 14:125 (1990).

4.   H. Mitsuya, R. Yarchoan and S. Broder, Molecular targets for AIDS therapy, *Science*, 249:1533 (1990).

5.   C.K. Chu, R.F. Schinazi, B.H. Arnold, D.L. Cannon, B. Doboszewski, V.B. Bhadti, and Z.P. Gu, Comparative activity of 2',3'-saturated and unsaturated pyrimidine and purine nucleosides against human immunodeficiency virus type 1 in peripheral blood mononuclear cells, *Biochem. Pharmacol.*, 37:3543 (1988).

6.   C.K. Chu, R.F. Schinazi, M.K. Ahn, G.V. Ullas and Z.P. Gu, Structure-activity relationships of pyrimidine nucleosides as antiviral agents for human immunodeficiency virus type 1 in peripheral blood mononuclear cells, *J. Med. Chem.*, 32:612 (1989).

7.   J. Balzarini, G.-J. Kang, M. Dalal, P. Herdewijn, E. De Clercq, S. Broder and D.G. Johns, The anti-HTLV-III (anti-HIV) and cytotoxic activity of 2',3'-didehydro-2',3'-dideoxyribonucleosides: a comparison with their parental 2',3'-dideoxyribonucleosides, *Mol. Pharmacol.*, 32:162 (1987).

8.   T.-S. Lin, R.F. Schinazi and W.H. Prusoff, Potent and selective *in vitro* activity of 3'-deoxythymdine-2'-ene (3'-deoxy-2',3'-didehydrothymidine) against human immunodeficiency virus, *Biochem. Pharmacol.*, 36:2713 (1987).

9.   M.M. Mansuri, J.E. Starrett Jr., I. Ghazzouli, M.J.M. Hitchcock, R.Z. Sterzycki, V. Brankovan, T.-S. Lin, E.M. August, W.H. Prusoff, J.-P. Sommadossi and J.C. Martin, 1-(2,3-dideoxy-β-D-glyceropent-2-enofuranosyl)thymine. A highly potent and selective anti-HIV agent, *J. Med. Chem.*, 32:461 (1989).

10.   J. Balzarini, R. Pauwels, P. Herdewijn, E. De Clercq, D.A. Cooney, G.J. Kang, M. Dalal, D.G. Johns and S. Broder, Potent and selective anti-HTLV-III/LAV activity of 2',3'-dideoxycytidine, the 2',3'-unsaturated derivative of 2',3'-dideoxycytidine, *Biochem. Biophys. Res. Commun.*, 140:735 (1986).

11.   G.I. Birnbaum, T.-S. Lin and W.H. Prusoff, Unusual structural features of 2',3'-dideoxycytidine, an inhibitor of the HIV (AIDS) virus, *Biochem. Biophys. Res. Commun.*, 151:608 (1988).

12.   P. Van Roey, J.M. Salerno, C.K. Chu and R.F. Schinazi, Correlation between preferred sugar ring conformation and activity of nucleoside analogues against human immunodeficiency virus, *Proc. Natl. Acad. Sci. USA*, 86:3929 (1989).

13.   P. Van Roey and C.K. Chu, unpublished.

14.   C.K. Chu, V.S. Bhadti, B. Doboszewski, Z.P. Gu, Y. Kosugi, K.C. Pullaiah and P. Van Roey, General synthesis of 2',3'-dideoxynucleosides and 2',3'-didehydro-2',3'-dideoxynucleosides, *J. Org. Chem.*, 54:2217 (1989).

15.   P. Van Roey, J.M. Salerno, W.L. Duax, C.K. Chu M.K. Ahn and R.F. Schinazi, Solid-state conformation of anti-human immunodeficinecy virus type-1 agents: crystal structures of three 3'-azido-3'-deoxythymidine analogues, *J. Am. Chem. Soc.*, 110:2277 (1988).

16.   G.I. Birnbaum, J. Giziewicz, E.J. Gabe, T.-S. Lin and W.H. Prusoff, Structure and conformation of 3'-azido-3'-deoxythymidine (AZT), an inhibitor of the HIV (AIDS) virus, *Can. J. Chem.*, 65:2135 (1988).

17. A. Camerman, D. Mastropaolo, D. and N. Camerman, Azidothymidine: crystal structure and possible functional role of the azido group, *Proc. Natl. Acad. Sci. USA*, 84:8239 (1987).

18. R. Parthasarathy and H. Kim, Conformation and sandwiching of bases by azido groups in the crystal structure of 3'-azido-3'-deoxythymidine (AZT), an antiviral agent that inhibits HIV reverse transcriptase, *Biochem. Biophys. Res. Commun.*, 152:351 (1988).

19. I. Dyer, J.N. Low, P. Tollin, H.R. Wilson and R.A. Howie, Structure of 3'-azido-3'-deoxythymidine, AZT, *Acta Cryst. Ser. C*, 44:767 (1988).

20. G.V. Gurskaya, E.N. Tsapkina, N.V. Skaptsova, A.A. Kracvskii, S.V. Lindeman and Y.T. Struchkov, *Dokl. Akad. Nauk SSSR*, 291:854 (1986).

21. J.N. Low and R.A. Howie, Structure of 3-(3-azido-2,3-dideoxy-β-D-*erythro*-pentofuranosyl)cytosine hydrogen choloride, *Acta Cryst. Ser. C*, 46:84 (1990).

22. P. Van Roey and R.F. Schinazi, The crystal and molecular structure of 3'-fluoro-3'-deoxythymidine, a potent anti-HIV-1 nucleoside, *Antiviral Chem. Chemother.*, 1:93 (1990).

23. G.V. Gurskaya and E.N. Tsapkina, *Dokl. Akad. Nauk. SSSR*, 303:1378 (1988).

24. C.K. Chu, B. Doboszewski, W. Schmidt, G.V. Ullas and P. Van Roey, Synthesis of pyrimidine 3'-allyl-2',3'-dideoxyribonucleosides by free-radical coupling, *J. Org. Chem.*, 54:2767 (1989).

25. P. Van Roey, E.W. Taylor, C.K. Chu, and R.F. Schinazi, Conformational analysis of 2',3'-didehydro-2',3'-dideoxypyrimidine nucleosides, *J. Am. Chem. Soc.*, in press.

26. W.E. Harte Jr., J.E. Starrett Jr., J.C. Martin and M.M. Mansuri, Structural studies of the anti-HIV agent 2',3'-didehydro-2',3'-dideoxythymidine (D4T), *Biochem. Biophys. Res. Comm.*, 175:298 (1991).

27. G.I. Birnbaum, J. Giziewicz, T.-S. Lin, and W.H. Prusoff, Structural features of 2',3'-dideoxy-2',3'-didehydrocytidine, a potent inhibitor of the HIV (AIDS) virus, *Nucleosides Nucleotides*, 8:1259 (1989).

28. P. Van Roey and C.K. Chu, The crystal and molecular structure of the complex of 2',3'-didehydro-2',3'-dideoxyguanosine with pyridine, *Nucleosides Nucleotides*, 11:1229 (1992).

29. H.O. Kim, S.K. Ahn, A.J. Alves, J.W. Beach, L.S. Jeong, B.G. Choi, P. Van Roey, R.F. Schinazi and C.K. Chu, Asymmetric synthesis of 1,3-dioxolane-pyrimidine nucleosides and their anti-HIV activity, *J. Med. Chem.*, 35:1987 (1992).

30. C.K. Chu, S.K. Ahn, H.O. Kim, J.W. Beach, A.J. Alves, L.S. Jeong, Q. Islam, P. Van Roey and R.F. Schinazi, Asymmetric synthesis of enantiomerically pure (-)-(1'R,4'R)-dioxolane-thymidine and its anti-HIV activity, *Tetrahedron Lett.*, 32:3791 (1991).

31. D.W. Norbeck, S. Spanton, S. Broder and H. Mitsuya, A new 2',3'-dideoxynucleoside prototype with *in vitro* activity against HIV, *Tetrahedron Lett.*, 30:6263 (1989).

32. E.W. Taylor, P. Van Roey, R.F. Schinazi and C.K. Chu, A stereochemical rationale for the activity of anti-HIV nucleosides, *Antiviral Chem. Chemother.*, 1:163 (1990).

33. P. Van Roey, E.W. Taylor, C.K. Chu and R.F. Schinazi, Correlation of molecular conformation and activity of reverse transcriptase inhibitors, *Ann. N. Y. Acad. Sci.*, 616:29 (1990).

34. W. Saenger, "Principles of Nucleic Acid Structure", Springer-Verlag, New York, (1984).

35. L.H. Koole, S. Neidle, M.D. Crawford, A.A. Krayevski, G.V. Gurskaya, A. Sandstroem, J.-C. Wu, W. Tong and J. Chattopadhyaya, Comparative structural studies of [3.1.0]-fused 2',3'-modified β-D-nucleosides by X-ray crystallography, NMR spectroscopy, and molecular mechanics calculations, *J. Org. Chem.*, 56:6884 (1991).

36. V.E. Marquez, C.K.-H. Tseng, J.A. Kelley, H. Mitsuya, S. Broder, J.S. Roth and J.S. Driscoll, 2',3'-dideoxy-2'-fluoro-ara-A. An acid-stable purine nucleoside active against human immunodeficiency virus (HIV), *Biochem. Pharmacol.*, 36:2719 (1987).

37. M. Sundaralingam, *in*: "Structure and Conformation of Nucleic Acid and Protein-Nucleic Acid interactions" M. Sundaralingam and S.T. Rao, eds. 487-536, University Park Press, Baltimore, MD (1975).

38. D. Pearlman and S.-H. Kim, Conformational studies of nucleic acids. II. The conformational energetics of commonly occuring nucleosides, *J. Biomol. Ster. Dynam.*, 3:99 (1985).

39. G.I. Birnbaum and D. Shugar, Biologically active nucleosides and nucleotides: conformational features and interactions with enzymes, *in*: " Nucleic Acid Structure", S. Neidle, ed., Part 3, 1, VCH, New York (1987).

40. A.L. Spek, 1992 American Crystallographic Association Meeting, Pittsburgh, PA, Abstract J05 (1992).

41. W.K. Olson, How flexible is the furanose ring? II. An updated potential energy estimate, *J. Am. Chem. Soc.*, 104:278 (1982).

42. C.K. Johnson, ORTEP-II, Report ORNL-5138 (Oak Ridge Natl. Lab., Oak Ridge, TN) (1976).

43. J.W. Beach, L.S. Jeong, A.J. Alves, D. Pohl, H.O. Kim, C.-N. Chang, S.-L. Doong, R.F. Schinazi, Y.-C. Cheng and C.K. Chu, Synthesis of enantiomerically pure (2'R,5'S)-(-)-1-[2-(hydroxymethyl)oxothiolan-5-yl]cytosine as a potent antiviral agent against hepatitis B virus (HBV) and human immunodeficiency virus (HIV) *J. Org. Chem.*, 57:2217 (1992).

MODIFIED OLIGORIBONUCLEOTIDES:

SYNTHESIS AND PRELIMINARY EVALUATION OF α-RNA

Françoise Debart, Carole Chaix,
Bernard Rayner and Jean-Louis Imbach

Laboratoire de Chimie Bio-Organique
U.R.A. 488 au C.N.R.S.
Université Montpellier II
Sciences et Techniques du Languedoc
Place Eugène Bataillon
34095 Montpellier Cédex 5 (France)

Modified oligodeoxyribonucleotides have been extensively studied for a few years as potential antisense compounds.[1,2] However, only a few modified oligoribonucleotides have been described. One of the reasons for this situation may be related to the fact that RNA synthesis is more sophisticated than that of DNA , due to the presence of the 2'-OH function, which necessitates the use of an added protection that is compatible with all the other protective groups.

Considering the RNA wild-type structure, one can envisage the introduction of modifications either on the phosphate backbone or on the sugar moiety (Figure 1) as has been extensively done for the DNA series.[2]

Phosphate backbone modifications have previously led us to the corresponding phosphorothiate series *i.e.*, S-RNA.[3] In this case, the corresponding monomeric units are commercially available. However, if one considers the sugar-modified RNA series, the situation is much more tedious, as we must design first the modified ribonucleosides (which are generally not commercially available), transform them into the corresponding suitably protected phosphoramidite synthons, and then proceed to the oligonucleotide synthesis.

Apart from some L-oligoribonucleotides that have been described by Van Boom and co-workers[4] in relation with prebiotic studies, only the 2'-OMe-ribo series have been

Nucleosides and Nucleotides as Antitumor and Antiviral Agents,
Edited by C.K. Chu and D.C. Baker, Plenum Press, New York, 1993

303

extensively studied by Ohtsuka's group[5] and more recently by B. Sproat.[6] In our opinion this latter series may be somehow considered as a modified DNA (*i.e.*, 2'-OMe substitution is very stable, absence of potential ribozymic activity ...), but in some cases they behave like RNA (strong stability of the corresponding RNA/RNA duplexes).

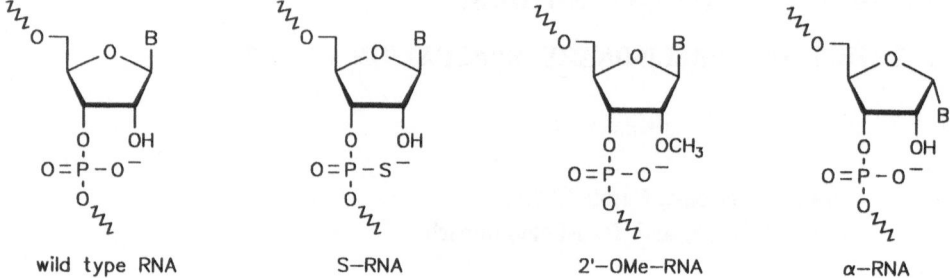

Figure 1. Different modified RNA series.

Due to the particular properties that we previously discovered in the α-DNA series,[7] we have envisaged exploration into the α-RNA series. We would like to report and comment herewith on the data that we recently obtained in this particular series.

The α-RNA synthesis was planned using the solid-phase technique in combination with the phosphoramidite methodology. Such an approach requires the preparation of the corresponding synthons, and, hence, preparation of the α-ribonucleosides. The synthesis of the necessary α-ribonucleosides in the pyrimidine series has been previously described,[8] and their syntheses are straightforward. As shown on Figure 2, α-rU is obtained in an overall yield of 35%, and the corresponding α-rC can be easily reached from the uridine derivative.[9]

The necessary purine derivatives namely α-rA and α-rG were also previously described.[10,11] However, due to the low yields obtained for the adenine derivative and the impossibility to reproduce Furukowa and co-workers experiments,[11] we used other approaches (Figure 2) which allowed us to isolate and characterize the expected N-protected nucleosides in sufficient quantities.

After DMTr protection of the 5'-OH nucleoside function, the TBDMS group was introduced at the 2'-OH position as previously described,[3,12] and the expected phosphoramidite synthons were obtained using standard procedures.[9,12] Preparation of the derivatized solid supports (long-chain alkylamine on controlled-pore glass beads (LCA-CPG)) was achieved as usual,[9,12] with the loaded yields ranging from 18 to 26 μmol g[-1].

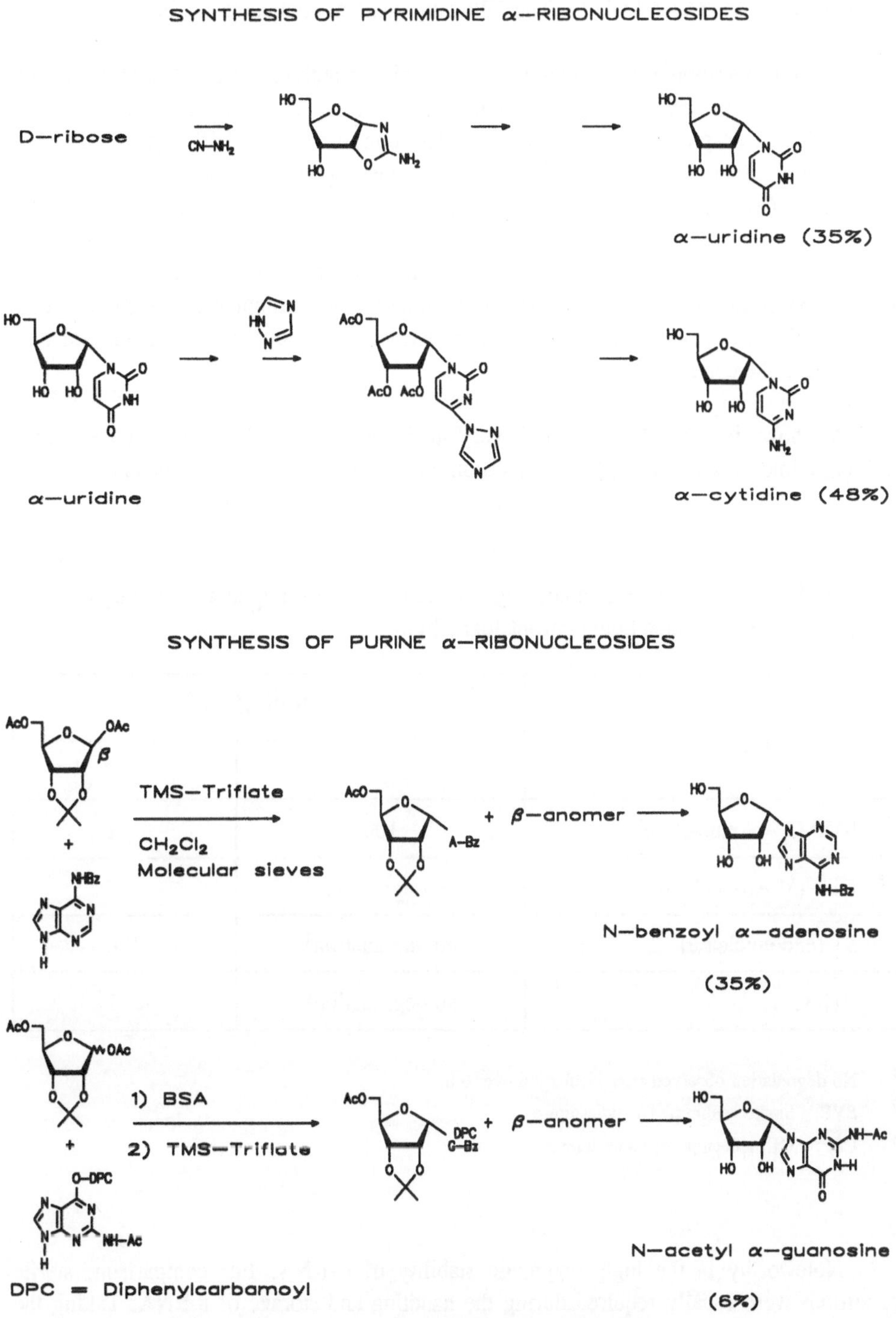

SYNTHESIS OF PYRIMIDINE α−RIBONUCLEOSIDES

D−ribose

α−uridine (35%)

α−uridine

α−cytidine (48%)

SYNTHESIS OF PURINE α−RIBONUCLEOSIDES

TMS−Triflate

CH₂Cl₂
Molecular sieves

+ β−anomer →

N−benzoyl α−adenosine

(35%)

1) BSA

2) TMS−Triflate

+ β−anomer →

N−acetyl α−guanosine

DPC = Diphenylcarbamoyl

(6%)

Figure 2. Principle of α-ribonucleoside syntheses.

Various α-RNA oligomers were first synthesized in order to assess the efficiency of the method (*i.e.*, α-rU$_6$ 1, α-rU$_{12}$ 2, α-rUCUUAACCCACA 3, α-rACUAAAUUCCCC 4, α-rUGUGAACGAAAACU 5). The following general comments can be made.

The condensation time required to obtain high coupling yields (> 97%) was 15 min using a 18-fold excess of phosphoramidite and tetrazole as activator. Deprotection was conducted as previously described.[9,12] The oligomers were HPLC purified, their homogeneity ascertained by gel and capillary electrophoresis, and their nucleoside composition (for 2 and 3) confirmed after degradation and HPLC analysis.

Using some of the previous sequences, we then determined their resistance to degradation by nucleases and their capacity to anneal with a complementary strand. As a first step the α-rU$_6$ was evaluated for its resistance to various purified nucleases and compared to the corresponding β-rU$_6$.

As shown in Table 1 the α-rU$_6$ is very stable and was only slightly degraded under such experimental conditions, by snake venom phosphodiesterase (a 3'-exonuclease).

Table 1. Comparative enzymatic degradation of α- and β-rU$_6$ as analysed by high-performance liquid chromatography.

Enzymes	Half-life (min)	
	α-rU$_6$[1]	β-rU$_6$[1]
SVP (3'-exonuclease)[2]	100	1
CSP (5'-exonuclease)[3]	no degradation[1]	17
S1 (endonuclease)	no degradation[1]	120
RNase A	no degradation[1]	< 1

[1] No degradation observed after incubation over 6 h.
[2] SVP: Snake venom phosphodiesterase
[3] CSP: Calf spleen phosphodiesterase

Noteworthy is the high enzymatic stability of α-RNA. For comparison, sterile conditions were usually required during the handling and storage of β-RNA. Taking the α-RNA mixed sequence 3 as an example, no sterile conditions were used, and this oligomer was found intact after being stored for several months at - 20°C.

In relation to its stability, the previous α-RNA sequence has been easily functionalized with 9-aminoellipticine (Figure 3) using the AP site methodology that we previously described.[13]

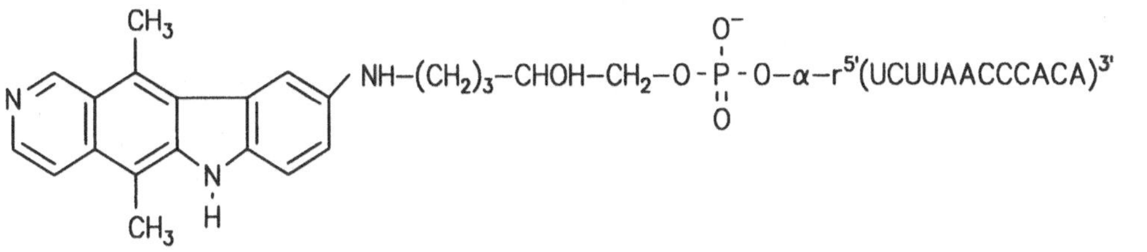

Figure 3. α-RNA functionalisation (Sequence 6)

In addition, we would like to mention that the oligomer $\underline{3}$ was designed to be complementary to the splice acceptor junction of the *tat* HIV gene (*vide infra*) in order to evaluate its inhibitory capacity on HIV-infected MT4 cells. As such experiments are done in culture medium (5 days at 37°C in RPMI 1640 containing 10% of heat-inactivated calf serum), we have been able to determine its stability under such stringent conditions using the ISRP on-line HPLC approach that we recently introduced.[14]

Under such conditions, the half-life of $\underline{3}$ is only 14 min ($k_1 = 5.2 \times 10^{-2}$ min^{-1}), but contrast to all the ß-oligos, which are processively degraded from their 3' end (n --> n-1 --> n-2 ...), α-RNA, like the others in α series, namely α-DNA and α-S-DNA, is degraded to only a shorter species (n-1 or n-2 ?), which is then very stable ($t_{1/2} = 45$ h for α-RNA). This led us to the general conclusion that extrapolation from purified enzyme stability data to serum behavior must be taken with care, whatever the oligo structure is.[15]

The second important point, we had to determine the α-RNA $\underline{3}$ binding capacity and, above all, its annealing polarity. For this purpose we evaluated its affinity by spectrophotometric analysis for complementary sense and antisense DNA sequence (T_m determination).

Using 1M NaCl no significant absorbance variation, versus temperature, was observed for an equimolecular mixture of $\underline{3}$ with the complementary ß-DNA antiparallel strand. However, a clear transition ($T_m = 25.5$°C) was observed upon annealing (80° to 0°C) in the case of an equimolecular mixture of $\underline{3}$ and the DNA parallel strand [ß-d(AGAATTGGGTGT)]. Note that, like α-DNA, the α-RNA binds to a complementary DNA strand with a parallel polarity. In addition, upon melting (from 0° to 80°C), a second transition was observed ($T_m = 44.4$°C). Such behavior may suggest triplex formation,[16] and this was ascertained with a mixing curves titration at 5°C from which a 2-α-RNA/1-ß-DNA stoichiometry was observed.[9]

However, near physiological NaCl concentration (NaCl 0.1 M), only one transition was observed ($T_m = 14.3$°C), and it was attributed to α-RNA/ß-DNA duplex formation on the basis of the stoichiometry experiment.

In order to ascertain if this behavior was not only correlated to this particular sequence $\underline{3}$, we looked at the fusion curves of $\underline{4}$ and $\underline{6}$ with their complementary parallel DNA sequence. Again two transitions were observed in the fusion experiments (see Figure 4 for $\underline{4}$).

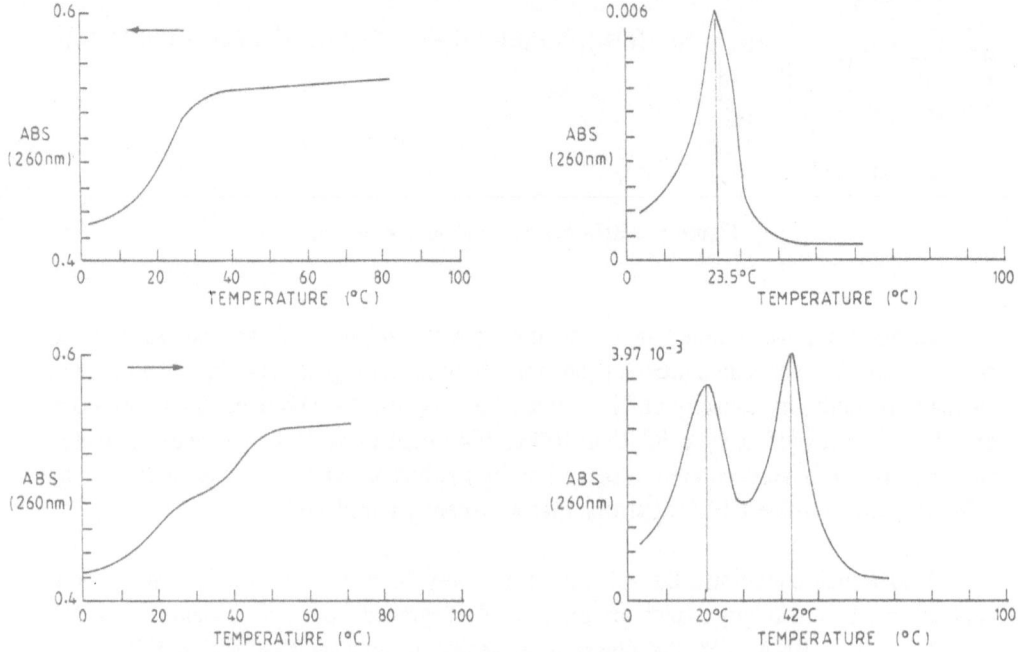

Figure 4. UV fusion and hybridization profiles and derivatives versus temperature of an equimolar mixture of α-RNA <u>4</u> and its complementary ß-DNA. Each strand concentration was 10 μM in 1 M NaCl and 0.01 M sodium cacodylate (pH 7).

This seems to indicate that triplex formation at high salt concentration is a general behavior of α-RNA sequences. However, it is noteworthy that mixed-base α-RNAs were used, and that, in the ß series, triplex formation has mainly been detected using homopyrimidine and homopurine sequences involving TAT or CGC$^+$ triads.

As triplex formation is very sensitive to a decrease in pH , since the cytosine must be protonated to form Hoogsteen hydrogen bonds, we determined the fusion temperature of the same sequences at pH 5.9. Surprisingly no stabilization effect was observed. This may indicate that CGC$^+$ triads are not involved in such triplex formation, and its structure remains unknown. Work is in progress along these lines.

Finally, we determined the ability of the antisense sequence <u>3</u> to block the *de novo* HIV-1 infection in MT4 cells. In this experiment the α-RNA random sequence <u>4</u> was used as a control, and the corresponding α-DNA, α-S-DNA and ß-DNA sequences were also comparatively evaluated.

As shown in Table 2, α-DNA, α-S-DNA and α-RNA sequences inhibit HIV-1 infection, whereas the ß-DNA antisense was inefficient at concentrations up to 100 μM. This lack of activity may be related to the extensive ß-DNA degradation in culture medium.[14]

Table 2. Comparative biological data of modified oligonucleotides complementary of the splice acceptor site for the *tat* Gene of HIV-1.

5'-----[5357]A G A A U U G G G U G U[5368]-----3' Target		
	ED$_{50}$ (μM)[1] /MT$_4$ cells	
	antisense	random
ß-DNA	> 100	nd
α-RNA	10 < < 20	10 < < 20
α-DNA	60	70 $\pm$ 5
α-S-DNA	0.25 $\pm$ 0	0.6 $\pm$ 0

[1] MT$_4$ cells + HIV were incubated at 4°C for 30 min (fixation of the virus on cells membrane), followed by a wash to eliminate the virus. Incubation at 37°C with addition of the oligo was carried out for 5 days. Syncytia formation and RT were measured.

However, it is evident that the observed inhibitory effects are non-sequence dependent, and this may be tentatively explained by unspecific interactions with membrane proteins, thus interfering with CD4/Gp120 interactions.

In conclusion, we have synthesized for the first time α-anomeric oligoribonucleotides (α-RNA). This series demonstrates enzymatic stability as compared to ß-RNA, and these anneal in parallel orientation with complementary DNA or RNA strands. However, a 2-α-RNA/1-ß-DNA stoichiometry was surprisingly observed at high salt concentration. In addition, when directed on the *tat* splice acceptor site of HIV, an α-RNA antisense compound unspecifically inhibits *de novo* HIV infection in MT4 cells.

REFERENCES

1. J.S. Cohen, "Oligodeoxynucleotides: Antisense Inhibitors of Gene Expression", The Macmillan Press Ltd, London, (1989).
2. E. Uhlman and A. Peyman, Antisense oligonucleotides: A new therapeutic principle, *Chem. Rev.* 90:543 (1990).
3. F. Morvan, B. Rayner and J.L. Imbach, Modified oligonucleotides: IV Solid phase synthesis and preliminary evaluation of phosphorothioate RNA potential antisense agents, *Tetrahedron Lett.* 31:7149 (1990).

4. G.M. Visser, J. Van Westrenen, C.A.A. Van Boeckel and J.H. Van Boom, Synthesis of the mirror image of the RNA fragment D-CAAGG: A model compound to study interactions between oligonucleotide of opposite handedness, *Recl. Trav. Chim. Pays-Bas* 105:528 (1986).

5. H. Inoue, Y. Hayase, A. Imura, S. Iwai, K. Mura and E. Ohtsuka, Synthesis and hybridization studies on two complementary nona (2'-O-methyl)ribonucleotides, *Nucl. Acids Res.* 15:6131 (1987).

6. B.S. Sproat, B. Beijer and A. Iribarren, New Synthetic routes to protected purine 2'-O-methyl riboside-3'-O-phosphoramidites using a novel alkylation procedure, *Nucl. Acids Res.* 18:41 (1990).

7. F. Morvan, B. Rayner and J.L. Imbach, α-Oligonucleotides: a unique class of modified chimeric nucleic Acids, *Anti-Cancer Drug Design* 6:521 (1991).

8. D.H. Shannahoff and R.A. Sanchez, 2,2'-Anhydropyrimidine nucleosides. Novel Syntheses and Reactions, *J. Org. Chem.* 38:593 (1973).

9. F. Debart; B. Rayner, G. Degols and J.L. Imbach, Synthesis and base-pairing properties of the nuclease-resistant α-anomeric dodecaribonucleotide α-[r(UCUUAACCCACA)], *Nucl. Acids Res.* 20:1193 (1992).

10. L. Pichat, P. Dufay and Y. Lamorre, Synthèse de l'adénosine [14]C-8. Formation simultanée des deux nucléosides anomères α et ß par la méthode de fusion, *C.R. Acad. Sc. Paris* 2453 (1964).

11. Y. Furukawa, K.I. Imai and M. Honjo, A novel method for the synthesis of purine α-Ribonucleosides, *Tetrahedron Lett.* 4655 (1968).

12. F. Debart, B. Rayner and J.L. Imbach, Sugar-modified oligonucleotides: II. Solid-phase synthesis of nuclease resistant α- anomeric uridylates as potential antisense agents, *Tetrahedron Lett.* 31:3537 (1990).

13. J.J. Vasseur, D. Péoc'h, B. Rayner and J.L. Imbach, Derivatization of oligonucleotides through abasic site formation, *Nucleosides Nucleotides* 10:107 (1991).

14. A. Pompon, I. Lefebvre and J.L. Imbach, "On-line internal surface reversed-phase cleaning": The direct HPLC analysis of crude biological samples, *Biochem. Pharmacol.* 43:1769 (1992).

15. F. Morvan, H. Porumb, G. Degols, I. Lefebvre, A. Pompon, B.S. Sproat, B. Rayner, C. Malvy, B. Lebleu and J.L. Imbach, Comparative evaluation of 8 oligonucleotide series as potential antisense agents, (accepted for publication in *J. Med. Chem.*).

16. J.G. Wetmur, DNA Probes: Applications of the principles of nucleic acid hybridization, *Crit. Rev. Biochem. and Mol. Biol.* 26:227 (1991).

TOWARDS SECOND-GENERATION SYNTHETIC BACKBONES

FOR ANTISENSE OLIGONUCLEOSIDES[§†]

Yogesh S. Sanghvi and P. Dan Cook

ISIS Pharmaceuticals, Department of Medicinal Chemistry

2280 Faraday Avenue, Carlsbad, California 92008, USA

INTRODUCTION

The artificial control of gene expression by modified oligonucleotides (ONs) provides an exciting opportunity for the rational design of antisense ONs and, in particular, for the design of antiviral and antitumor ONs that can be used for therapeutic purposes.[1] The effectiveness of antisense ONs *in vitro* and *in vivo* has been demonstrated by a first generation of phosphate-modified analogs such as phosphorothioates (PT)[2] and methylphosphonates (MP).[3] The initial choice of phosphate-modified ONs arose from their ease of synthesis and resistance to nuclease degradation. However, both the PT and MP modifications lead to the introduction of diastereoisomers, which results in a $2^{(n-1)}$ mixture of diastereomers in the modified ONs. The presence of multiple diastereomers considerably weakens the capability of the ONs to hybridize with the target sequences. Duplexes formed between ON diesters and MP ONs of defined stereochemistry show an enhanced T_m relative to their racemic counterparts.[1g] Also, stereoselective automated synthesis for PT or MP is not available. Despite these stereochemical problems, both MP and PT ONs are clinical candidates at the present time.[4] This is creating a need for cost-effective scale-up processes for producing MP and PT ONs. The demand is currently being met by a solid-support automated synthesis, which may not be suitable for producing drugs on a large-scale for human applications.[2]

Some researchers have tried to circumvent the chirality and cost problems associated with MP and PT by replacing the four-atom, anionic phosphate backbone with a neutral and achiral non-phosphate backbone[1e] (Figure 1). Neutral ONs such as MP, however, may not have sufficient water solubility for effective therapeutic applications. Resistance to degradation by nucleases is another desirable trait of an antisense ON. Because cellular nucleases recognize and cleave the sugar–phosphate linkage, it is believed that non-phosphate backbone modifications will not serve as substrates for nucleases. ONs with a combination of neutral (non-phosphate) and anionic (phosphate) backbones could be expected to have an improved cellular uptake while maintaining partial water solubility because of their anionic elements. Another possible advantage for the use of non-phosphate backbones is that they may be more economical to synthesize on a large scale than the phosphoramidites used for MP and PT synthesis.

[§] Dedicated to the memory of Professor Tohru Ueda for his contributions in nucleic acids research.

Nucleosides and Nucleotides as Antitumor and Antiviral Agents,
Edited by C.K. Chu and D.C. Baker, Plenum Press, New York, 1993

311

Natural phosphodiester linkage   Synthetic backbone modifications

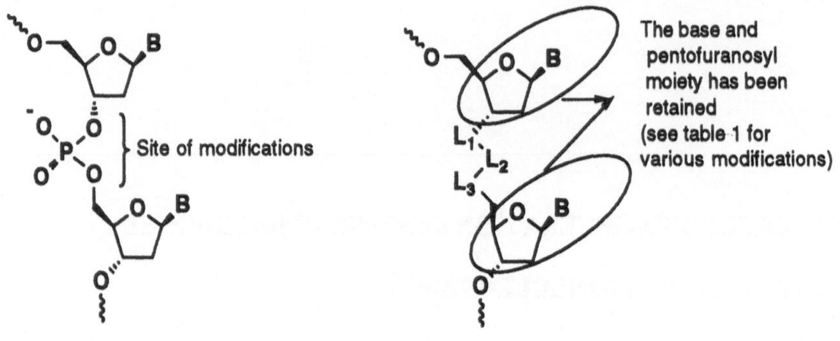

Site of modifications

The base and pentofuranosyl moiety has been retained (see table 1 for various modifications)

Figure 1

Table 1. Synthetic oligonucleosides and their properties

Name	$3' \rightarrow 5'$ Linker Atoms[1]			Tm^2	Nuclease[2] Resistance	Reference
	L_1	L_2	L_3			
Phosphate	O	PO_2	O	x	x	1b,k,l
Carbonate	O	C=O	O	–	–	1m, 5-7
Carbamate	O	C=O	NH	x	x	1m, 8-11
Silyl	O	SiR	O	x	x	1m,12,13
Sulfur	CH_2	S(O)n	CH_2	–	–	14-16
Sulfur	CH_2	CH_2	S	–	–	17
Sulfonate	O	SO_2	CH_2	x	x	18,19
Sulfonamide	O	SO_2	NH	x	x	20
Sulfonamide	NH	SO_2	CH_2	–	–	18c
Formacetal	O	CH_2	O	x	x	1m, 21-25
Thioformacetal	O	CH_2	S	x	x	24
Oxime	CH_2	=N	O	x	x	26
Methyleneimino	CH_2	NH	O	x	x	30
Methylene (methylimino) MMI	CH_2	NCH_3	O	x	x	26
Methylene (dimethyl-hydrazo) MDH	CH_2	NCH_3	NCH_3	x	x	31
Methyleneoxy (methylimino) MOMI	CH_2	O	NCH_3	x	x	32

[1] See figure 1 for positions of L_1, L_2, and L_3; [2] x indicates data available, – indicates results not available.

The techniques that have been reported for the synthesis of non-phosphate backbones are, in general, difficult and not versatile. Progress in developing techniques to replace the phosphate linkage with an isostere linkage has been slow over the years, as witnessed by only a handful of publications.[1e] The reported syntheses are not particularly suitable for automated synthesizers to the level at which all linkages in a sequence are modified. The most frequently used synthetic strategy is to prepare a nucleoside dimer containing backbone modification and then protect the 5'-hydroxyl with dimethoxytrityl and convert the 3'-hydroxyl to a phosphoramidite moiety. Subsequently, the blocked dimers are incorporated into a sequence at any site utilizing standard conditions in a DNA synthesizer. This approach provides modified oligonucleotides with a 50% reduction of the negative phosphodiester linkages as found in the wild-type sequences.

The potential therapeutic applications of antisense ONs have rekindled the interest of researchers, including those at ISIS, in the development of 'better backbones' for incorporation into ONs (Table 1). Comprehensive reviews of antisense ONs have been published,[1] and here we present major developments from our group and others. Furthermore, this account is limited to the synthesis of oligonucleosides with a four-atom linker connecting the two pentofuranosyl moieties of a nucleoside dimer.[1m] In each section we describe a different type of backbone linkage.

1. Carbonate backbones

Access to the backbone-modified dimer **6** (Figure 2) containing a 3', 5' carbonate linkage between nucleoside units was first gained by Mertes and Coats[5] in 1969. They obtained a 44% yield of dimer **6** by coupling 3'-chloroformate **1** with 3'-O-acetylthymidine **3** and then deblocking the protecting groups of **5**. In the same year, Jones and Tittensor[6] reported an alternative method. They reacted 5'-O-protected deoxyribonucleoside 3'-O-carbonate active ester **2** with 2',3'-O-protected ribonucleoside **4** to obtain dimer **7** in a higher yield, which on deblocking gave dimer **8**. The method was further extended to the preparation of a U*U*U trimer containing carbamate (represented by *) linkages.[7] The U*U*U trimer failed to stimulate the binding of yeast tRNA to ribosomes or to inhibit the binding of the yeast tRNA in the presence of polyU. The oligomers containing carbonate linkages were found to be unstable in aqueous solution at neutral pH and degraded to furnish hydrolyzed products. This instability limits the use of carbonate linkages in antisense ONs.

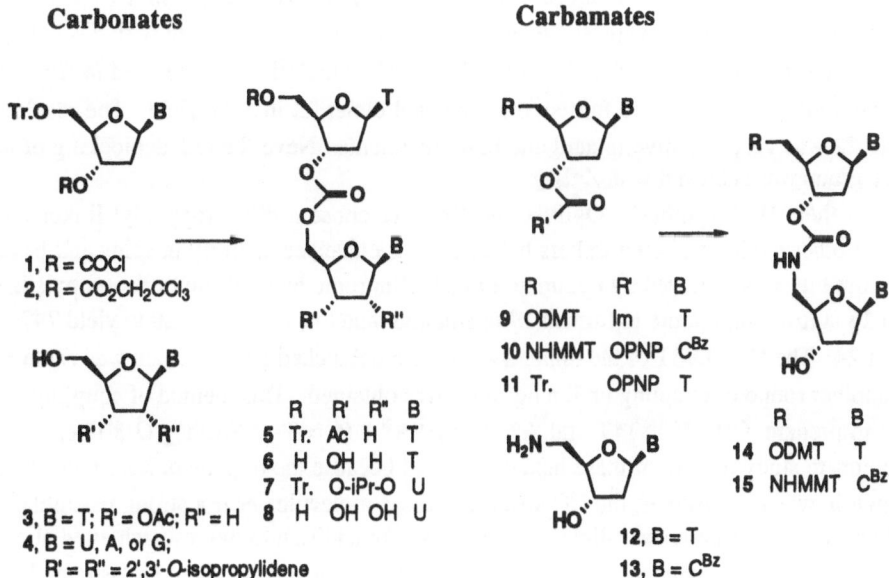

Carbonates

Carbamates

Figure 2

2. Carbamate backbones

Gait *et al.*[8] reported the first synthesis of a 3′ → 5′ carbamate-bridged dimer of thymidine in 1974. Dimer **14** (Figure 2) was obtained in 38% yield by coupling of 3′-*O*-(2,2,2-trichloroethyl)-carbonate of 5′-*O*-tritylthymidine with 5′-amino-5′-deoxythymidine and deblocking the trityl group. In 1977, Mungall and Kaiser[9] reported the synthesis of a trimer containing two carbamate moieties. They adopted an active ester method of polypeptide synthesis and condensed the 3′-*O*-*p*-nitrophenyl carbonate of 5′-*O*-tritylthymidine (**11**) with 5′-amino-5′-deoxythymidine (**12**) and isolated a crystalline dimer **14** in 71% yield. The utility of the procedure was further demonstrated by the conversion of dimer **14** into a trimer via another condensation.

An improved synthesis of the carbamate linkage was published in 1987 by two groups. Coull *et al.*[10] described the preparation of a thymidine hexamer utilizing a 3′-*O*-carbonyl- imidazolide moiety (as shown in **9**) to couple with nucleoside **12**. Stirchack *et al.*[11] described the preparation of a cytosine hexamer employing an activated 3′-*O*-*p*-nitrophenyl moiety (as shown in **10**) for coupling with amino nucleoside **13**. The hexamer of T prepared by Coull and co-workers failed to show any hypochromicity, which they thought may have been due to a possible failure in base stacking and restricted rotation about the trigonal carbonate linkage. However, the hexamer C*C*C*C*C*C (* represents a carbamate linkage) exhibited a T_m of 71 °C with p(dG$_6$), whereas the T_m of p(dC$_6$)·p(dG$_6$) was 30 °C. The reason for the unusually high binding was not given.[11] The nuclease sensitivity of the carbamate linkage was studied in the presence of snake venom phosphodiesterase or spleen phosphodiesterase, and no cleavage of the dimer was found.[9] In general, it appears that geometrical constraints imposed by the carbonate linkage on the backbone may restrict its use in antisense ONs.

3. Substituted-silyl backbones

In 1985, Ogilvie and Cormier[12] incorporated a diphenylsilyl linkage in a thymidine dimer for the first time. Three years later, in an extension of their earlier work, they reported the synthesis of a more stable diisopropylsilyl linkage in two different hexamers.[13a] In the first synthesis, dicholorodiphenylsilane was reacted with 5′-*O*-DMT-thymidine (**16**, Figure 3) to yield 3′-silyl chloride **17**, which then was reacted *in situ* with 3′-levulinylthymidine (**20**) to furnish the protected dimer **22** in 53% yield. The resulting dimer **22** was very sensitive to acid and base treatments. Nevertheless, deblocking of the DMT group was achieved with ZnBr$_2$.

In their 1988 synthesis, Ogilvie and Cormier chose a diisopropylsilyl linker over several other silyl-substituted linkers because of its enhanced stability in acids and bases. The coupling was achieved by treatment of bis(trifluoromethanesulfonyl)diisopropylsilane with **16** at low temperature to furnish **19** *in situ* and then the addition of **20** to yield 74% of dimer **24**. The 5′-*O*-DMT of the dimer **24** was then deblocked and the chain was extended via another round of coupling until a hexamer was obtained. This method of coupling was used to prepare T*T*T*T*T*T and A*A*A*A*A*A (* = 3′-*O*-Si(iPr)$_2$-*O*-5′) hexamers. Attempts to study the T_m of these hexamers failed because the oligomers were not soluble enough in water. However, the CD spectra of these molecules bear a strong resemblance to those of the phosphodiester-linker counterparts. In addition to the research cited above, a brief report on the synthesis of thymidine-3′-yl-(1,1-silacyclohexyl) or (1,1-silamethyl)-5′-thymidine via a bifunctional silylating reagent has been published.[13b]

In summary, due to their poor water solubility, the enhanced steric bulk of diisopropylsilyl group, and their acid/base lability, substituted-silyl backbones may not be useful for antisense ONs. However, a recent patent on a method of linking nucleosides with a silyl backbone indicates that such linkages may have some utility.[13c]

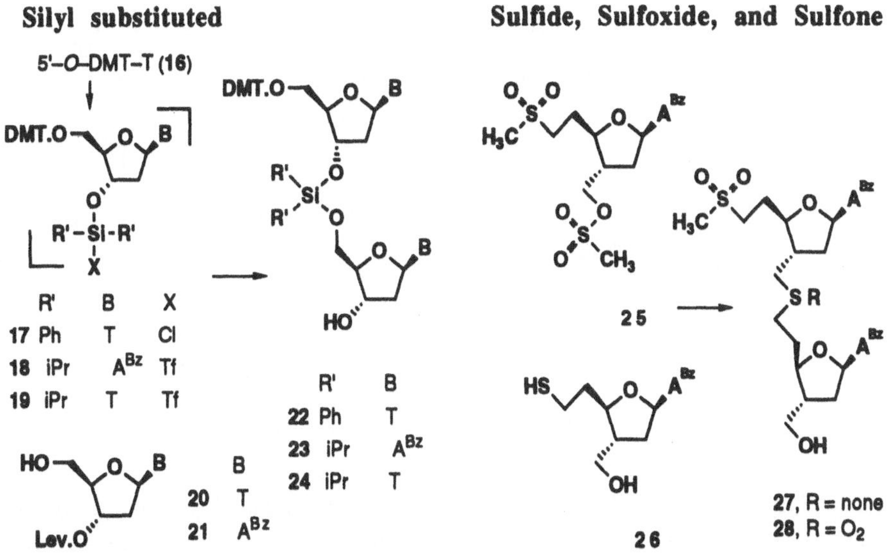

Silyl substituted   **Sulfide, Sulfoxide, and Sulfone**

R'	B	X
17 Ph	T	Cl
18 iPr	A^Bz	Tf
19 iPr	T	Tf

R'	B
22 Ph	T
23 iPr	A^Bz
24 iPr	T

B	
20	T
21	A^Bz

27, R = none
28, R = O₂

Figure 3

4. Sulfide, sulfoxide, and sulfone backbones

In 1990, Benner et al.[14] reported an elegant and elaborate incorporation of a sulfur moiety (CH$_2$SO$_2$CH$_2$) into a dimer in place of a phosphodiester moiety (OPO$_2$O). The rationale is to make nonionic, achiral, isosteric analogs of phosphate diesters that are stable to both chemical and biochemical hydrolysis. It is believed that multiple sulfone moieties in ONs could improve the cellular uptake of modified ONs, and this would make sulfone-modified ONs good candidates for antisense molecules. In a communication, Benner mentioned that tetramers[14] and octamers[15] containing a sulfur moiety have been made in satisfactory yields. Furthermore, sulfone analogs as short as tetramers form stable duplexes with natural oligonucleotides.[15] The enhanced binding is presumably due to the lack of interstrand charge repulsion.

In 1991, Benner and co-workers reported[16] a full account of the synthesis of deoxyadenosine dimer **28**. The coupling of bifunctional nucleoside **25** (Figure 3) with another bifunctional nucleoside **26**, in the presence of diazabicycloundecene in DMF, gave the thioether-linked dimer **27**, which on oxidation gave the dimer **28**.

A preliminary report on the synthesis of a positional isomer of thioether linkage of **27**, wherein the sulfur atom replaced the 5'-oxygen atom of the phosphodiester group was published by Just and co-workers[17a] in 1988. In a personal communication, Just acknowledged the synthesis of thymidine and cytosine oligomers **29** and **30** (Figure 4) containing multiple thioether linkages. Recently, a detailed synthesis of thymidine building blocks for the thioether linkage has been published.[17b,c] Since no one has yet reported complete biophysical data on these modifications, it is very difficult to judge how useful sulfide, sulfoxide, and sulfone backbone modifications are for antisense therapeutics.

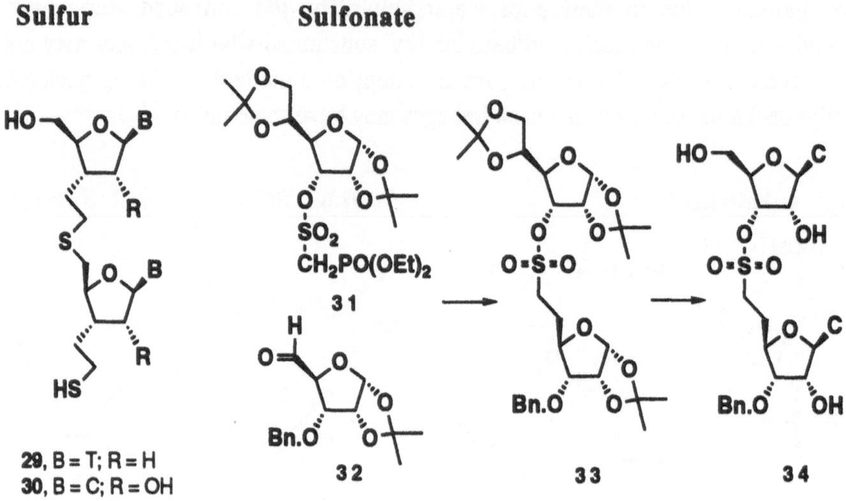

Sulfur Sulfonate

29, B = T; R = H
30, B = C; R = OH

31

32 **33** **34**

Figure 4

5. Sulfonate backbones

Musicki and Widlanski[18a] have synthesized a sulfonate-containing dimer of cytosine **34** (Figure 4). They developed a multistep synthesis in which the key reaction is coupling of phosphonate **31** with the aldehyde **32**, followed by reduction of the product to give the disaccharide **33** in 90% yield. Subsequent deblocking and glycosylation by standard methods gave a high yield of dimer **34**. The sulfonate linkage of dimer **34** was stable under basic conditions. They also synthesized an octamer of thymidine using a similar approach.[19] In recent communications an alternative synthesis of thymidine dimer containing the sulfonate linkage has been described.[18b] The results of hybridization studies of sulfonate-containing RNA are eagerly awaited.

6. Sulfonamide backbones

Another sulfur-based linkage was reported by Kirshenbaum et al.[20a] Oligomers in which one or more sulfamate linkages (OSO_2NH as shown by example **35a**, in Figure 5) replace the phosphodiester (OPO_2O) have been synthesized. Furthermore, the sulfamate linkage appears to be nuclease resistant and is capable of duplex formation, and thus holds a promise for antisense applications. A structurally similar sulfonamide linkage **35b** bridging a thymidine dimer has been reported.[18b]

7. Formacetal backbones

In 1990, Matteucci[21] and Van Boom and co-workers[22] independently reported a nonionic formacetal linkage that could serve as a simple, small, and achiral isostere for the phosphate linkage in ONs. Matteucci described the synthesis of a trimer of thymidine carrying two formacetal linkages and its incorporation into ONs. The coupling chemistry employed was similar to that used for disaccharide synthesis.[23] 3'-Methylthioacetal **36** (Figure 5) was activated with NBS/2,6-di-tert-butylpyridine and treated with a 3'-O-protected nucleoside **38** to furnish the dimer **40** with a 45% yield. Conventional deblocking of the 5'-OH from dimer **40** and another round of coupling with activated **36** yielded the protected trimer. The trimer was loaded on to CPG and, utilizing H-phosphonate chemistry, the oligomer I with a TTCCCTCTCTTT*T*T (* represents a formacetal linkage) sequence was synthesized. The oligomer **I** was stable to snake venom phosphodiesterase under conditions sufficient for the complete cleavage of the unmodified parent oligomer. The T_m of **I** with its RNA complement was 59 °C, whereas the T_m of the unmodified sequence with same complement was 59.5 °C. Matteucci also reported the hybridization properties of an oligomer **II**, TCTC'*TCTC'*TC'*TC'*TTTT (where C' is 5Me dC, and * is a formacetal linkage), which had a T_m of 59 °C when hybridized to a complementary RNA. The unmodified oligomer **II** had a T_m of 60 °C with its RNA complement. The formacetal modified sequence, compared with methylphosphonate (MP) and phosphoramidate (PA) modified sequences, hybridized better to complementary RNA than did the MP and PA analogs. The T_m of **II** with a complementary DNA was 39 °C, whereas the unmodified oligomer had a T_m of 49.5 °C with the same DNA. The reason for the drop in the hybridization affinity for DNA is rather surprising and no explanation was provided.

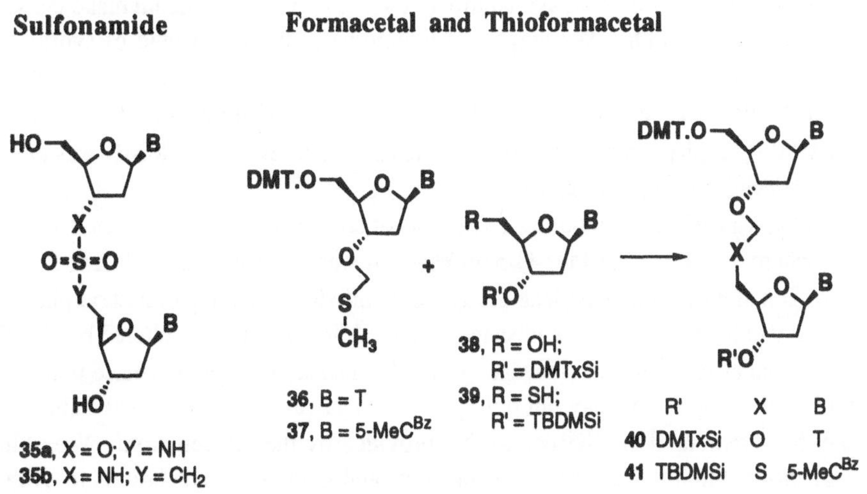

Sulfonamide   **Formacetal and Thioformacetal**

35a, X = O; Y = NH
35b, X = NH; Y = CH$_2$

36, B = T
37, B = 5-MeC^{Bz}

38, R = OH;
R' = DMTxSi
39, R = SH;
R' = TBDMSi

	R'	X	B
40	DMTxSi	O	T
41	TBDMSi	S	5-MeC^{Bz}

Figure 5

Matteucci also incorporated a thioformacetal linkage[24] (via a dimer **41**) in oligomer **II** in a similar manner to that of the formacetal linkage and at the same positions. A footprinting technique was used to determine the affinity and specificity of the thioformacetal-linked oligomer, which had a reduced binding relative to the parent formacetal linkage. It was reasoned that there may be a steric interaction between the 5'-sulfur of the thioformacetal linkage and the H-6 of the thymine moiety, which may destabilize the duplex.

Van Boom and co-workers[22,25a-d] described an efficient approach to the synthesis of thymidine dimers containing formacetal linkages. They have also improved the coupling conditions to furnish the dimer **40** (Figure 5) with a 85% yield. A systematic search for the activator required for the coupling lead to NIS and catalytic TfOH, which provided a rapid and efficient coupling. Preliminary results also indicated that the decamer [d(GCGTT*TTGCG)], containing one formacetal linkage, formed a stable duplex with a complementary DNA as determined by 600 MHz ^{1}H NMR spectroscopy.[25] Van Boom also prepared a mixed-base dimer[25d] and incorporated that into a decamer, [d(GCGC*TC*TGCG)] (where * is a formacetal linkage), using a standard phosphite triester protocol on a DNA synthesizer. In summary, the formacetal linkage appears to provide ONs with desirable qualities (duplex stability, nuclease resistance, and nonionic character for enhanced uptake) for use in antisense oligomers. However, the synthesis of antisense sequences uniformly modified with formacetal linkages has not been reported.

8. **Oxime backbones**
9. **Methyleneimino backbones**
10. **Methylene(methylimino) backbones (MMI)**
11. **Methylene(dimethylhydrazo) backbones (MDH)**
12. **Methyleneoxy(methylimino) backbones (MOMI)**

We have identified several potentially useful backbone linkages, and these were synthesized by a convergent approach in which the natural nucleosides are functionalized at the 3'- or 5'- terminus and coupled with another nucleoside to establish a new $3' \rightarrow 4'$ linkage between the pentofuranosyl moieties of the two nucleosides. The nucleoside approach for making dimers containing a modified linkage is an efficient and versatile method. In addition, commercial availability of natural nucleosides should make the scale-up of this process possible. We employed thymidine in the first round of synthesis of dimers because its base moiety does not require protection for automated synthesis of ONs. The synthesis of a novel 3'-de(oxyphosphinico)-3'-[methylene(methylimino)] thymidylyl($3' \rightarrow 5'$)thymidine (**58**, Figure 6, hereafter abbreviated as MMI) was the first dinucleoside synthesized in this series.[26]

In 1984, Mazur *et al.*[27] reported a synthesis of an isosteric phosphonic-acid analog of U*U*U, where * is a 3'-CH_2-PO_2-O-5' linkage. It was believed that replacement of the esteric oxygen atom with a methylene group would provide an analog of natural phosphate linkage that would retain the size and shape of natural linkage but with altered biochemical properties, such as resistance to cleavage by cellular nucleases. This suggested that a 3'-C-CH_2 moiety in a nucleotide may not only provide chemical and biochemical stability, but also may have structural similarities to that provided by the oxygen atom. We utilized computer model building, molecular mechanics, and dynamics calculations to evaluate whether the novel linkage of dimer **58** (Figure 6) can reside in ONs without causing major conformational distortions that could affect their efficiency to form duplexes.[28] Our studies suggested that replacement of a natural phosphate linkage by an MMI linker as in **58** resulted in a stable B-type duplex. The natural occurrence of an oxyimino bridge in antibiotic calicheamicin γ[29] indicated the acceptance of such an interglycosidic linkage in biological systems.

In the process of synthesizing the MMI linker, we have also incorporated the two precursors, namely, oxime and methyleneimino dimers, into various oligomers. In brief, the oxime dimer **50** (Figure 6) was obtained in 88% yield by coupling an aldehyde **42** and 5'-*O*-amino-thymidine derivative **47**.[26] Deprotection of the dimer **50** provided **51** as a mixture of *E/Z* isomers. Dimethoxytritylation of **51**, followed by phosphitylation of the product, gave a good yield of **52**. The amidite **52** was inserted into oligonucleotides by means of an automated DNA synthesizer. Hybridization studies indicated that a single incorporation of dimer **52** in the middle of a 16-mer sequence destabilized the duplex by about 3.2 °C. This is not surprising because of the geometrical restraints in the backbone due to *E/Z* isomers. The oxime dimer **50** was then reduced to furnish **53**, which was then protected with a phenoxyacetyl group to give **54**. Conventional deblocking of **54** gave **55**. The latter dimer was converted into amidite **56** with standard procedures. The amidite **56** was incorporated into oligonucleotides utilizing standard protocols. Our studies indicated that incorporation of the methyleneimino backbone into an oligomer lowers the T_m by about 1-2 °C per modification when hybridized with a complementary RNA.

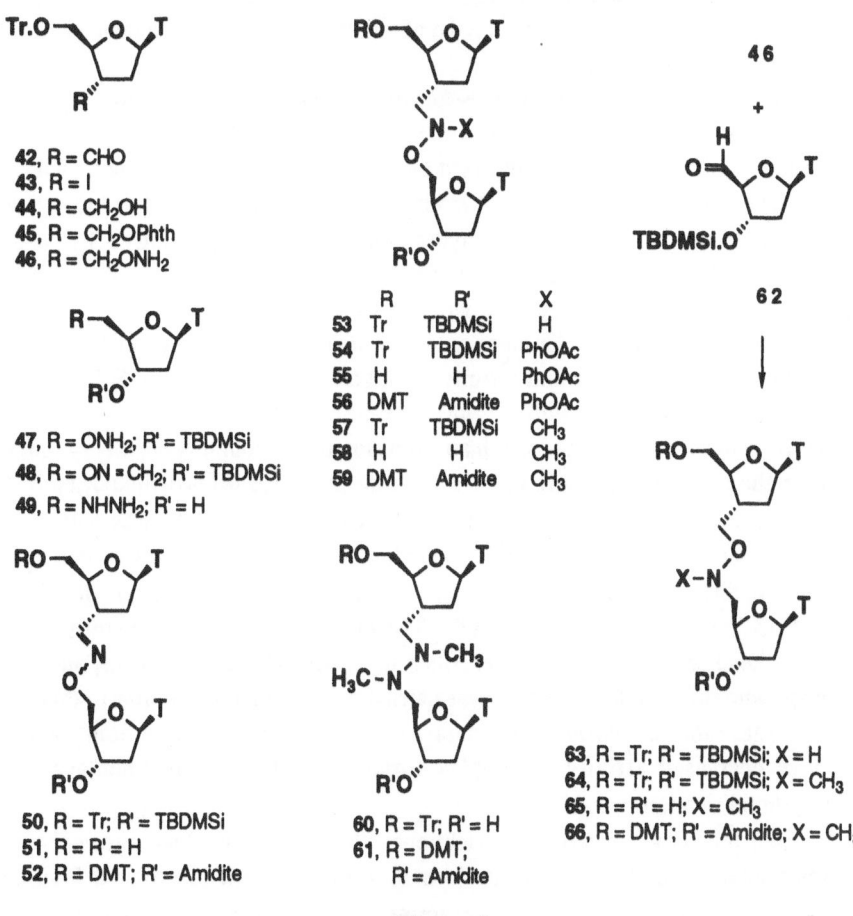

42, R = CHO
43, R = I
44, R = CH$_2$OH
45, R = CH$_2$OPhth
46, R = CH$_2$ONH$_2$

47, R = ONH$_2$; R' = TBDMSi
48, R = ON=CH$_2$; R' = TBDMSi
49, R = NHNH$_2$; R' = H

	R	R'	X
53	Tr	TBDMSi	H
54	Tr	TBDMSi	PhOAc
55	H	H	PhOAc
56	DMT	Amidite	PhOAc
57	Tr	TBDMSi	CH$_3$
58	H	H	CH$_3$
59	DMT	Amidite	CH$_3$

50, R = Tr; R' = TBDMSi
51, R = R' = H
52, R = DMT; R' = Amidite

60, R = Tr; R' = H
61, R = DMT; R' = Amidite

63, R = Tr; R' = TBDMSi; X = H
64, R = Tr; R' = TBDMSi; X = CH$_3$
65, R = R' = H; X = CH$_3$
66, R = DMT; R' = Amidite; X = CH$_3$

Figure 6

Using an intramolecular radical reaction, we have developed an alternative pathway for the stereoselective synthesis of the dimer **53**.[30] The dimer **53** was obtained by C–C bond formation between iodo nucleoside **43** and an oxime ether **48** in about 30% yield. We believe the oxime ether radical linkage methodology may have a general application in constructing other modified-backbone antisense ONs. The dimer **53** was methylated to furnish the desired MMI-linked dimer **57** in 87% yield. Two-step deprotection of **57** gave **58**, which was converted into amidite **59** with high overall yield (Figure 6). Using an automated synthesizer, amidite **59** was incorporated into a standard 16-mer oligonucleoside sequence from one to five times, with an average coupling efficiency of 98% for each dimer incorporation. An extensive hybridization study indicated that incorporation of MMI backbones into oligomers have a remarkably small effect on the stability of the duplexes formed between the oligomers bearing MMI backbones and their RNA complements. The average ΔT_m per modification is about -0.3 °C when compared to the unmodified DNA·RNA duplexes. In addition, the studies suggested that an alternating, uniform distribution of MMI linkage in an oligomer provided a more stable oligonucleoside·RNA duplex (ΔT_m/modification = +0.1 °C). The Watson–Crick base-pair specificity of oligomers containing MMI linkages was found to be as good as or better in some cases than the wild-type DNA. The uncompromised affinity and specificity of the

MMI linkage may be due to the lack of interstrand charge repulsion as demonstrated by Benner and co-workers with the sulfone analogs.[16] Furthermore, we expect that a neutral backbone such as MMI will enhance the uptake of antisense oligomers in a manner similar to that described for neutral MP analogs.

In order to explore the structure–activity relationship between backbone modifications and the altered biochemical and biophysical properties of oligomers containing such modifications, we synthesized two more novel linkages, namely, MDH[31] and MOMI[32] (see Table 1 and Figure 6). A straightforward synthesis of MDH linkage was achieved by coupling aldehyde **42** with hydrazinothymidine **49**, followed by *in situ* reduction and alkylation, to produce dimer **60**. Conventional deblocking of **60** and conversion of the product to amidite **61** worked very well. Incorporation of **61** into oligomers was carried out in a similar manner. Preliminary results of a hybridization study indicated that the duplex stability and affinity of oligomers containing MDH linkage are very similar to those of oligomers with the MMI linkage.

MOMI modifications were made with an extension of the coupling chemistry that we developed for the synthesis of MMI linkage. Synthesis of the MOMI-linked dimer **65** demonstrates the versatile use of the methodology. The 3′-precursor **46** was obtained in three convenient steps from aldehyde **42** in good yield. Coupling between **46** and **62** gave an oxime intermediate, which was reduced *in situ* to furnish **63** (Figure 6). Reductive alkylation of **63** gave the MOMI dimer **64**. The dimer **64** was then transformed into amidite **66** via the chemistry previously described for MMI linkage. The incorporation of MOMI linkage into oligonucleotides also was carried out in a manner similar to that used for MMI linkage. Initial hybridization data show that oligomers containing MOMI linkages have decreased affinity for duplex formation with a complementary RNA (average ΔT_m/modification = -1.44 °C).

At ISIS, all four types of modified backbones "**8–12**" were evaluated for enhanced nuclease resistance. The studies were performed with 10% heat inactivated fetal calf serum using a standard protocol. Our preliminary results indicated that a 3′-capped oligomer (5′-CGACTATGCAATT*TC-3′) containing MMI linkage has a half-life of 14 h, while the same oligomer with the MDH linkage has a half-life of 36 h. The half-life of the unmodified oligomer is 0.5 h.

In summary, the postulation that nuclease resistance can be conferred to ONs by the incorporation of 3′-C-CH$_2$ moieties such as MMI and MDH in the backbone linkage has been confirmed by experimental data. The hybridization data for ONs with the MMI linker clearly support the initial computer modeling that predicted the formation of a stable B-type duplex. ^{1}H NMR studies on T*T dimers containing MMI and MDH linkages have shown good stacking interaction of two thymine bases and a strong preference for an *S*-pucker (2′-endo), which is typical of a DNA structure.[28] We believe that the reactivity of the amino group in the methyleneimino linker could be utilized to attach various linkers, cleavers, intercalators, probes, etc., in order to improve the antisense properties of ONs. In general, the backbone linkers developed at ISIS appear to be promising candidates for tailoring antisense ONs to improve their pharmacokinetic and pharmacodynamic properties.

Future Directions

Synthetic backbone linkages have become an attractive area of antisense ONs research. We believe that these isosteres of natural backbone modifications as discussed in sections 1–12 should allow us to begin to understand the key structural requirements for a "perfect" backbone linkage. Incorporation of these linkers into ONs may open the way to a second generation of non-phosphate antisense molecules with potentially great therapeutic applications. Chemical methods for the synthesis of antisense ONs have improved to the point where ON sequences could become routinely available in the quantities required for therapeutic purposes. However, the efficacy and clinical prospect of antisense ONs depends largely on an efficient cellular uptake process. An optimum intracellular concentration of antisense ONs may be achieved by appropriate placement of nonionic and achiral linkages in a sequence. Furthermore, it has been demonstrated by us and several others that the natural phosphate-backbone can be changed from anionic to nonionic and from hydrophilic to lipophilic without seriously interfering with hybridization to the RNA target. An alteration of the total charge of an ON affects the efficiency of passive transport (as in methylphosphonates) and could affect active transport of the ON molecule into the cell.

Another approach that may enhance cellular penetration is to conjugate a cationic group, such as a polyamine, to an antisense ON. Several cationic phosphoramidites have been synthesized and converted to cationic oligonucleotides, but modifications of this type suffer from chirality and chemical hydrolysis problems. We believe that such problems may be eliminated by placing an achiral and non-phosphate cationic alkylamine linker in ONs. As a concluding comment, the encouraging results from the backbone modification work reviewed here have brought us closer to the rational design of useful antisense ONs for human therapeutics. It is very likely that once again medicinal chemistry will play an important role in the discovery and development of drugs for the treatment of viral infections, cancer, and other cellular diseases.

Acknowledgements

We would like to thank members of the Medicinal Chemistry and Cellular & Molecular Biology Departments at ISIS, especially the visiting scientists from CNRS (France), Drs. J.-J. Vasseur and Françoise Debart for their excellent efforts in the syntheses of backbone-modified dimers discussed in this article. Special thanks are due to Ms. Julie Walker for typing this manuscript.

REFERENCES

†   We refer to modified oligonucleotides that lack the phosphorus atom in the backbone linkage as oligonucleosides. Designation of the backbone linkage as the moiety that connects the 3'-carbon of one furanosyl ring with the 4'-carbon of another furanosyl ring is generally applicable in describing various backbone linkages.

1. Selected books and reviews published in 1991-1992 (refer to 1e, 1k and 1m for a complete list): (a) S.T. Crooke and B. Lebleu, *Antisense Research and Applications*, CRC Press, Inc.: Florida (1993); (b) F. Eckstein, Ed., *Oligonucleotides and Analogs: A Practical Approach*, IRL Press, New York (1991) (c) R.P. Erickson and J.G. Izant, Eds., *Gene Regulation: Biology of Antisense RNA and DNA*, Raven Press, Ltd., New York (1992); (d) E. Wickstrom, Ed., *Prospects for Antisense Nucleic Acid Therapy of Cancer and AIDS*, Wiley-Liss, New York (1991); (e) P.D. Cook, Medicinal Chemistry of Antisense Oligonucleotides-Future Opportunities, *Anti-Cancer Drug Des.* 6:585 (1991); (f) D.G. Knorre and V.V. Vlassov, Reactive oligonucleotide derivatives as gene-targeted biologically active compounds and affinity probes, *Genetica*, 85:53 (1991); (g) V. English and D.H. Gauss, Chemically modified oligonucleotides as probes and inhibitors, *Angew. Chem. Int. Ed., Engl.* 30:613 (1991); (h) M.J. Gait, DNA/RNA synthesis and labelling, *Current Opinion in*

Biotech., 2:61 (1991); (i) C. Helénè, The Anti-Gene Strategy: Control of Gene Expression by Triplex-Forming-Oligonucleotides, *Anti-Cancer Drug Des.* 6:569 (1991); (j) J.S. Cohen, Antisense oligonucleotides as antiviral agents, *Antiviral Res.*, 16:121 (1991); (k) S.L. Beaucage and R.P. Iyer, The synthesis of oligonucleotides via the phosphoramidite approach: advances and applications, *Tetrahedron*, 48:2223 (1992); (l) S.L. Beaucage and R.P. Iyer, Phosphoramidite derivatives and their synthetic applications, *Tetrahedron*, 49: in press (1993); (m) E. Uhlmann and A. Peyman, Oligonucleotide Analogs Containing Dephospho Internucleoside Linkages in *Methods in Molecular Biology*, Chapter 16: *Oligonucleotide Synthesis Protocols*, S. Agrawal, Ed., The Humana Press Inc., Totowa, New Jersey (1993) (The authors would like to thank Dr. Uhlmann for a preprint of foregoing article).

2.   G. Zon and T.G. Geister, Phosphorothioate oligonucleotides: chemistry, purification, analysis scale-up and future directions. *Anti-Cancer Drug Des.* 6:539 (1991).

3.   P. S. Miller, Oligonucleoside methylphosphonates as antisense reagents, *Biotechnology* 9:358 (1991).

4.   V. Glaser, Antisense faces first human clinical test, *Genetic Eng. News* 12:1 (1992); See reference 1a for details.

5.   M.P. Mertes and E.A. Coats, Synthesis of carbonate analogs of dinucleosides, *J. Med. Chem.* 12:154 (1969).

6.   D.S. Jones and J.R. Tittensor, The preparation of dinucleoside carbonates, *J. Chem. Soc., Chem. Commun.* 1240 (1969).

7.   J.R. Tittensor, The preparation of nucleoside carbonates, *J. Chem. Soc. C* 2656 (1971).

8.   M.J. Gait, A.S. Jones, and R.T. Walker, synthetic analogs of polynucleotides. XII. Synthesis of thymidine derivatives containing an oxyformamido linkage instead of a phosphodiester group, *J. Chem. Soc., Perkin Trans. 1*, 1684 (1974).

9.   W.S. Mungall and J.K. Kaiser, Carbamate analogues of oligonucleotides, *J. Org. Chem.* 42:703 (1977).

10.   J. M. Coull, D.V. Carlson, and H.L. Weith, Synthesis and characterization of a carbamate-linked oligonucleoside, *Tetrahedron Lett.* 28:745 (1987).

11.   J.E. Summerton, Stereoregular polynucleotide binding polymers, *PCT Int. Appl.* WO 86/05518, 25 September 1986; E.P. Stirchak, J.E. Summerton, and D.D. Weller, Uncharged stereoregular nucleic acid analogues, *J. Org. Chem.* 52:4202 (1987).

12.   K.K. Ogilvie and J.F. Cormier, Synthesis of a thymidine dinucleotide analogue containing an internucleotide silyl linkage, *Tetrahedron Lett.* 26:4159 (1985).

13.   (a) J.F. Cormier and K.K. Ogilvie, Synthesis of hexanucleotide analogues containing diisopropylsilyl internucleotide linkages, *Nucleic Acids Res.* 16:4583 (1988); (b) H. Seliger and G. Feger, Oligonucleotide analogs with dialkyl silyl internucleoside linkages, Nucleosides & Nucleotides 6:483 (1987); (c) A. L. Weis, A. Saha and F. H. Hausheer, A method of linking nucleosides with a siloxane bridge, PCT *Int. Appl.* WO 92/04364, 19 March, 1992.

14.   K.C. Schneider and S.A. Benner, Building blocks for oligonucleotide analogs with dimethylene-sulfide, -sulfoxide, and -sulfone groups replacing phosphodiester linkages, *Tetrahedron Lett.* 31:335 (1990).

15.   (a) Z. Huang and S.A. Benner, Non-ionic antisense oligonucleotides containing sulfide and sulfone linkages in place of phosphodiester groups in natural oligonucleotides, *J. Cell. Biochem.* (Abstracts, 20th Keystone Symposia 1991) CD 209, p 19; (b) S.A. Benner, Z. Huang, K.C. Schneider, and T. Arslan, Sulfone DNA:The antisense oligonucleotide analogue?, *203rd ACS meeting*, San Francisco, April 5-10, 1992, Abstract No. CARB 68.

16. S.A. Benner, Oligonucleotide analogs containing sulfur, *PCT Int. Appl.* WO 89/12060, 14 December 1989; Z. Huang, K.C. Schneider, and S.A. Benner, Building blocks for oligonucleotide analogues with dimethylene sulfide, sulfoxide, and sulfone groups replacing phosphodiester linkages, *J. Org. Chem.* 56:3869 (1991).

17. (a) S.H. Kawai, G. Just, and J. Chin, Single-stranded DNA & RNA binding: Backbone-modified polynucleotide analogues, *Third Chem. Cong. of North America*, Toronto, Canada, June 5-10, 1988, Abstract No. ORGN 318; (b) S. H. Kawai, D. Wang and G. Just, Synthesis of the thymidine building blocks for a nonhydrolyzable DNA analogue, *Can. J. Chem.* 70:1573 (1992); (c) G. Just, M. Damha, S. Kawai, B. Meng, and P. Giannaris, Synthesis, binding and nuclease stability of oligonucleotides containing a sulfide backbone, *Inserm/NIH Conference* on Antisense oligonucleotides and Ribonucleases H, Arcachon, France, Sepetember 27- October 2, 1992.

18. (a) B. Musicki and T.S. Widlanski, Synthesis of carbohydrate sulfonates and sulfonate esters, *J. Org. Chem.* 55:4231 (1990); Synthesis of nucleoside sulfonates and sulfones, *Tetrahedron Lett.* 32:1267 (1991); (b) R. C. Reynolds, P. A. Crooks, J. A. Maddry, M. S. Akhtar, J. A. Montgomery, and J. A. Secrist III, Synthesis of thymidine dimers containing internucleoside sulfonate and sulfonamide linkages, *J. Org. Chem.* 57:2983 (1992); P. A. Crooks, R. C. Reynolds, J. A. Maddry, M. S. Akhtar, J. A. Montgomery, and J. A. Secrist III, Nucleoside sultones: Synthesis for the preparation of novel nucleotide analogues. 1. Synthesis and ring-opening reactions, *J. Org. Chem.* 57:2830 (1992); J. A. Secrist III, P. A. Crooks, J. A. Maddry, R. C. Reynolds, R. C. Rathore, M. S. Akhtar, and J. A. Montgomery, Progress toward the synthesis of nonionic oligonucleotide analogues with sulfonate and sulfonamide internucleotide linkages, *Nucleic Acids Res. Symposium series* No. 24, 5 (1991).

19. Personal communication.

20. M.R. Kirshenbaum, E.M. Huie and G.L. Trainor, Novel oligonucleotide analogs with sulfur-based linkage, *The 5th San Diego Conference: Nucleic Acids: New Frontiers*, Poster abstract 28, November 14-16, 1990; Abstract CD 210 in reference 15a; E. M. Huie, M. R. Kirshenbaum, and G. L. Trainor, Oligonucleotides with a nuclease-resistant sulfur-based linkage, *J. Org. Chem.* 57:4569 (1992).

21. M. Matteucci, Deoxyoligonucleotide analogs based on formacetal linkages, *Tetrahedron Lett.* 31:2385 (1990).

22. G.H. Veeneman, G.A. van der Marel, Van der Elst and J.H. van Boom, Synthesis of oligodeoxynucleotides containing thymidines linked via an internucleosidic- (3'-5')-methylene bond, *Recl. Trav. Chim. Pays-Bas* 109:449 (1990).

23. K.C. Nicolaou, S.P. Seitz and D.P. Papahatjis, A mild and general method for the synthesis of O-glycosides, *J. Am. Chem. Soc.* 105:2430 (1983).

24. M. Matteucci, K-Y. Lin, S. Butcher, and C. Moulds, Deoxyoligonucleotides bearing neutral analogues of phosphodiester linkages recognize duplex DNA via triple-helix formation, *J. Am. Chem. Soc.* 113:7767 (1991); M. Matteucci, Hybridization properties of a deoxyoligonucleotide containing four formacetal linkages, *Nucleosides Nucleotides* 10:231 (1991); M. Matteucci, Oligonucleotide analogs with novel linkages, *PCT Int. Appl.* WO 91/06629, 16 May 1991.

25. (a) G.H. Veeneman, G.A. Van der Marel, H. Van der Elst, and J.H. Van Boom, An efficient approach to the synthesis of thymidine derivative containing phosphate-isosteric methylene acetal linkages, *Tetrahedron* 47:1547 (1991); (b) P.J.L.M. Quaedflieg, G.A. Van der Marcel, E. Kuyl-Yeheskiely, and J.H. Van Boom,

Synthesis of (3'-5') methylene acetal linked dinucleosides containing cytosine bases, *Recl. Trav. Chim. Pays-Bas* 110:435 (1991), (c) P.J.L.M. Quaedflieg, C. M. Timmers, V. E. Kal, G.A. Van der Marcel, E. Kuyl-Yeheskiely, and J.H. Van Boom, An alternative approach towards the synthesis of (3'→5') methylene acetal linked dinucleosides, *Tetrahedron Lett.* 33:3081 (1992); (d) In a personal communication with P.J.L.M. Quaedflieg, he acknowledged the preparation of all sixteen possible dimers d(B^1*B^2) (B^1 = B^2 = T, C, A, and or G) via trimethylsilyl trifluoromethanesulfonate activation of 3'-O-(di-n-butoxyphosphoryloxymethyl)-nucleoside donors. The work is submitted for publication in *Synthesis* entitled: An alternative route to the preparation of (3'→5') methylene acetal linked di- and trinucleosides. The same authors also presented a poster, which was entitled Synthesis and hybridization properties of (3'→5') methylene acetal analogs of DNA at the Tenth International Roundtable on Nucleosides, Nucleotides and their Biological Applications, Park City, Utah, USA (September 16-20, 1992). The foregoing work is submitted for publication as: Synthesis and physicochemical properties of decanucleotides containing (3'→5') methylene acetal linkages at predetermined positions, in *Recl. Trav. Chim. Pays-Bas.*

26. J-J. Vasseur, F. Debart, Y.S. Sanghvi, and P.D. Cook, Oligonucleotides: synthesis of a novel methylhydroxylamine linked nucleoside dimer and its incorporation into antisense sequences, *J. Am. Chem. Soc.* 114:4006 (1992).

27. A. Mazur, B.E. Tropp and R. Engel, Isosteres of natural phosphates. 11. Synthesis of a phosphonic acid analog of an oligonucleotide, *Tetrahedron* 40:3949 (1984).

28. Unpublished results.

29. J.M.J. Tonchet, Z-L. Guido, G-B. Griselda, and B-R. Francoise, Blocked disacchride analogs bearing an oxyimino interglycosidic bridge, *J. Carbohydr. Chem.* 10:723 (1991).

30. F. Debart, J-J. Vasseur, Y.S. Sanghvi, and P.D. Cook, Intermolecular radical C-C bond formation: synthesis of a novel dinucleoside linker for non-anionic antisense oligonculeosides, *Tetrahedron Lett.* 33:2645 (1992).

31. Y.S. Sanghvi, J-J. Vasseur, F. Debart, and P.D. Cook, Flexible strategy to nonionic antisense oligonucleotides: Synthesis and incorporation of novel dimers, PD29, Tenth International Roundtable on Nucleosides, Nucleotides and their Biological Applications, Park City, Utah, USA (September 16-20, 1992); J-J. Vasseur, F. Debart, Y.S. Sanghvi, and P.D. Cook, Synthesis and properties of antisense oligonucleotides containing (3'→5') methylene(dimethylhydrazo) linked thymidine dimers, manuscript in preparation (1992).

32. F. Debart, J-J. Vasseur, Y.S. Sanghvi, and P.D. Cook, Synthesis and incorporation of methyleneoxy(methylimino) (3'→5') linked thymidine dimer into antisense oligonucleosides, *Biomed. Chem. Lett.* 2: in press (1992); Y.S. Sanghvi, J-J. Vasseur, F. Debart, and P.D. Cook, Novel strategies towards construction of non-ionic and achiral backbone in antisense oligonucleotides, *Nucleic Acids Res. Symposium series* No.27, 133 (1992).

INDEX